New Approaches to Structural Mechanics, Shells and Biological Structures

SOLID MECHANICS AND ITS APPLICATIONS
Volume 104

Series Editor: G.M.L. GLADWELL
Department of Civil Engineering
University of Waterloo
Waterloo, Ontario, Canada N2L 3GI

Aims and Scope of the Series

The fundamental questions arising in mechanics are: *Why?*, *How?*, and *How much?* The aim of this series is to provide lucid accounts written by authoritative researchers giving vision and insight in answering these questions on the subject of mechanics as it relates to solids.

The scope of the series covers the entire spectrum of solid mechanics. Thus it includes the foundation of mechanics; variational formulations; computational mechanics; statics, kinematics and dynamics of rigid and elastic bodies: vibrations of solids and structures; dynamical systems and chaos; the theories of elasticity, plasticity and viscoelasticity; composite materials; rods, beams, shells and membranes; structural control and stability; soils, rocks and geomechanics; fracture; tribology; experimental mechanics; biomechanics and machine design.

The median level of presentation is the first year graduate student. Some texts are monographs defining the current state of the field; others are accessible to final year undergraduates; but essentially the emphasis is on readability and clarity.

For a list of related mechanics titles, see final pages.

New Approaches to Structural Mechanics, Shells and Biological Structures

Edited by

H.R. DREW

CSIRO Molecular Science,
North Ryde, NSW Australia

and

S. PELLEGRINO

University of Cambridge,
Department of Engineering,
Cambridge, United Kingdom

KLUWER ACADEMIC PUBLISHERS

DORDRECHT / BOSTON / LONDON

A C.I.P. Catalogue record for this book is available from the Library of Congress.

ISBN 1-4020-0862-7

Published by Kluwer Academic Publishers,
P.O. Box 17, 3300 AA Dordrecht, The Netherlands.

Sold and distributed in North, Central and South America
by Kluwer Academic Publishers,
101 Philip Drive, Norwell, MA 02061, U.S.A.

In all other countries, sold and distributed
by Kluwer Academic Publishers,
P.O. Box 322, 3300 AH Dordrecht, The Netherlands.

Printed on acid-free paper

Printed in the Netherlands.

CONTENTS

FOREWORD

This Festschrift marks the retirement of Professor Chris Calladine, FRS after 42 years on the teaching staff of the Department of Engineering, University of Cambridge. It contains a series of papers contributed by his former students, colleagues, and friends.

Chris Calladine's research has ranged very widely across the field of structural mechanics, with a particular focus on the plastic deformation of solids and structures, and the behaviour of thin-shell structures. His insightful books on Engineering Plasticity and Theory of Shell Structures have been appreciated by many generations of students at Cambridge and elsewhere. His scientific contribution outside engineering, in molecular structures, is at least as significant, and he is unique among engineers in having co-authored a book on DNA. Also, he has been keenly interested in the research of many students and colleagues, and on many occasions his quick grasp and physical insight have helped a student, and sometimes a colleague, find the nub of the problem without unnecessary effort. Many of the papers contained in this volume gratefully acknowledge this generous contribution.

We thank Professor G.M.L. Gladwell for reading through all of the contributions, Mrs R. Baxter and Mrs O. Constantinides for help in preparing this volume, Godfrey Argent Studio for permission to reproduce Calladine's portrait for the Royal Society, and Dr A. Schouwenburg —from Kluwer— for his assistance.

Horace R. Drew
Sergio Pellegrino

CHRIS CALLADINE

SOME THOUGHTS ON RESEARCH

C.R. Calladine
Department of Engineering, University of Cambridge
Trumpington Street, Cambridge, CB2 1PZ, UK
crc@eng.cam.ac.uk

1. INTRODUCTION

The Editors have invited me to write a few pages, on the basis of my own experience, on research methods "either in general or in relation to specific problems". Since most of my research has been done on a diverse sequence of specific problems in structural mechanics, I have usually had to tailor the research method to the problem in hand. Thus, when I first thought about what I might write on this occasion, it seemed unlikely that any general principles would emerge. However, having vegetated on these matters for a while, I can now see some implicit general points beginning to clarify themselves.

For example, one obvious requirement for fruitful research is to find a good problem to work on. One doesn't have to do much reading in the history of structural mechanics, or indeed any other subject in science or engineering, to realise that active research areas are rapidly-changing ones, in the sense that new directions and opportunities are constantly appearing at the frontier. So particular research problems, once solved, are overtaken by the emergence of other issues.

2. HOW DOES ONE FIND A GOOD PROBLEM?

In my experience there are three ways of finding a good research problem.

H.R. Drew and S. Pellegrino (eds.),
New Approaches to Structural Mechanics, Shells and Biological Structures, 1–10.

First, of course, *industry* is bristling with good research problems, if only one can locate them. My own experience of working in the nuclear power industry in the late 1950s furnished many exciting research problems in the field of structural mechanics: creep of pressure vessels, and structures in general, at high temperature; creep buckling of fuel rods; dynamic buckling of dropped control rods; energy absorbing devices to prevent dropped control rods from piercing the pressure-vessel; and unstable thermal bowing of fuel pins in advanced gas-cooled reactors. Similarly, the oil industry – particularly in its off-shore aspects – has provided (mostly here in Cambridge through the good offices of Andrew Palmer) a steady stream of challenging buckling problems, mainly concerned with different aspects of pipelines on the sea-bed.

A second way of finding problems is through personal contacts. For example, a chance meeting in our Departmental tea-room with a group from Ove Arup who had come to discuss with colleagues here the dynamic crushing of steel box structures led to some interesting work in dynamic plasticity. Others, I am sure, can testify to the fruitfulness of chance meetings in the tea-room. And the same goes for those unscheduled meetings which happen a lot in Cambridge colleges. They can often be of an interdisciplinary nature, of course: the groups of scholars working in different fields, that are to be found in colleges, provide excellent forums for the discussion of ideas of an interdisciplinary kind. My own work on the construction of bacterial flagella, which I shall mention again below, began in Peterhouse with an approach from Aaron Klug, who had recently returned from a conference at which he had spotted that the remarkable properties of these organelles posed a problem in structural engineering, as distinct from, say, crystallography or thermodynamics.

Going to conferences is also a good way of finding out about new problems, particularly when a conference extends into fields with which one is not particularly familiar. I remember clearly hearing a lecture by Professor Miura at a "space structures" conference in Osaka, about the design of extended spacecraft structures, such as solar sails, which are subjected to very low-level loading from such forces as solar pressure. This introduced me to a whole field of structures in which the design challenges were quite different from those with which we are familiar in terrestrial applications; and it has, of course, led to very fruitful collaboration between Professor Miura and our group. I also recall seeing some test specimens of failed double-layer grids at another "space structures" conference, in Guildford. That set me thinking about the smallest number of bars which can fail before such a structure collapses as a whole; which led to some very interesting work by my student Amer Affan.

3. GIVEN A GOOD PROBLEM, HOW DOES ONE SOLVE IT?

Unfortunately, there is no "universal method" of solving problems, even in our own rather circumscribed field. One has to take each problem as it comes, and devise a way of tackling it. Here are some examples that illustrate this point.

(i) In the work which David Miles and I did on the lateral buckling of heated pipelines on the sea-bed, we devised a method of doing small-scale experiments in the laboratory, using a long thin strip of silicone rubber lying on a block of polystyrene. The thermal expansion of the pipeline was simulated indirectly by a longitudinal compression of the polystyrene. This simple scheme successfully modelled the buckling of an elastic pipeline on a primarily frictional seabed. David was then able to set up a computer program which simulated closely our particular experiments, and which then enabled us to investigate a larger range of cases than were accessible with the experimental scheme alone. This in turn led to a parametric study which revealed the importance of certain dimensionless groups of variables which not only characterised the phenomena but also provided good guidance for thinking about design. (Incidentally, this work provides a counter-example to Hadamard's contention that if one finds after long labour a simple pattern of behaviour, one should be able to provide a different and simple explanation for it. So far we have been unable to understand why these lateral-buckling phenomena involve the particular dimensionless groups in the way that they evidently do.)

(ii) When we began to investigate "tensegrity" structures (after a meeting of colleagues which Aaron Klug arranged) the tool which we used was Maxwell's rule for the construction of stiff frames (which we found didn't work; but then Maxwell had anticipated such exceptional cases) together with physical models made from "Geo-D Stix" rods and joints, sometimes also with pieces of string. This led to a much more systematic treatment by Sergio Pellegrino, who showed us how classical linear algebra can be used directly to explain these subtle structural forms. An interesting feature of this work was that problems which had hitherto been regarded as essentially non-linear became linear when seen from a better viewpoint.

(iii) In my work on plasticity, I was greatly helped by the late Dan Drucker, who happened to spend a sabbatical term in the Department soon after I joined the teaching staff. I was trying to develop a method for performing simple calculations on the creep of structures by analogy with the well-established limit-analysis methods in plasticity. He spotted a theorem on "nesting surfaces of constant energy-dissipation in creep" which enabled us to interpolate between solutions for linear viscosity and perfect plasticity, respectively, in order to solve structural problems for a general "Norton" creeping material. A-level work that I had done at school, from one of Durell's books, on the "reciprocation" of a geometric figure with respect to a circle, was particularly useful in this investigation.

(iv) Our work on the construction of bacterial flagella required a quite different approach. What we needed was nothing less than a new structural concept, to do with the behaviour of assemblies of identical building blocks, each containing a bi-stable mechanical feature. We made a lot of physical models, beginning with a sort of tubular triangulated structure in which some of the bars could adopt two slightly different lengths thanks to a simple mechanical switching system which we devised. Unfortunately, that kind of model neither worked properly nor provided any illumination. After a while we realised that it would be necessary to think of the building blocks as being genuinely 3-dimensional objects, but with specific points of contact with each other. We devised an appropriate form of Maxwell's rule for such blocks and decided that a 4-cornered block, with attachments to neighbours at the corner points, might provide the way forward. Again, however, nothing useful came out of the physical models (which were provided by Mr A.A. Barker of the Engineering Department's workshop), even though at one stage we had an elaborate scheme of small permanent magnets which provided a particular switching feature. Eventually we realised that a 4-cornered block could build not only an arbitrary cylindrical structure, as distinct from the very specific one which we were trying to model, but also, if we so desired, a spherical structure with, say, icosahedral symmetry. Then, at last, we got on to the idea of a 6-cornered block, with one of the corners affording an ambiguous connection with its neighbour, and the way became open to develop a scheme which mimicked very closely the observed behaviour of the bacterial flagella, with their discrete family of helical wave forms. The design of the entire assembly was implicit in the geometric (and mechanical) design of the building-block.

(v) On my return to Cambridge from industry I became interested in the interpretation of the mechanical properties of saturated clay within the framework of classical plasticity theory. Roscoe, Schofield and Wroth had made some important advances in this area, and Andrew Schofield was keen to discuss these matters with a newcomer. An important problem was to understand the qualitative difference between a perfectly-plastic *metal*, which does not yield plastically under hydrostatic stress however large, and saturated clay which *does* yield under sufficient isotropic effective stress. The yield-surface in principal stress space is shaped like a bullet for clay but like an extended cylinder for metals. It so happened that in the late 1960s there was a shortage of mechanics lecturers in our Department; and I was dragooned into relinquishing my first-year structures lectures in order to teach mechanics to the first year instead. *Friction* was a part of the mechanics syllabus with which I was unfamiliar; and so I consulted Ken Johnson and began to read Bowden and Tabor. Then I got onto papers by A.P. Green, and there, in a paper about plasticity at a welded "junction" between two contacting asperities, a perfect bullet-shaped yield locus between normal and tangential forces leapt out the page. And this yield locus enlarged isotropically as the contact-area increased. I saw immediately that this "asperity junction" provided, at a stroke, a model for many of the observed characteristics of the mechanical properties of saturated clay; and I was able to develop these ideas soon afterwards on a sabbatical leave spent at Stanford – where I had the good fortune to be able to discuss these matters with John Christian, and also the opportunity to talk to David Tabor, who was on leave at Berkeley at the same time, and was most encouraging.

(vi) In relation to research on the performance of thin-shell structures, other types of approach were needed. I shall return to these in a later section.

(vii) Our work on the structure of DNA (mostly with Horace Drew) required us to develop yet other sets of ideas. Our main contributions mostly involved simple geometrical matters. We began by thinking about the well-known observed "switching" of DNA between the distinct "A" and "B" forms, as described in the first instance in the 1950s by Rosalind Franklin, who had noticed two distinct patterns of spots on her x-ray films, depending on the moisture-content of the DNA fibre specimens. By the early 1980s there were lots of papers in the literature on the distinctive geometry of these two forms, mostly in relation to the detailed torsion angles of the sugar-phosphate "backbones". The 1982 Dickerson/Drew crystal structure of DNA in the "B" form, found by x-ray diffraction of proper crystals made from a DNA 12-mer of specific sequence, showed structural irregularities which were quite unexpected on the basis of all previous studies from the diffraction of fibres. It was a discussion after lunch one day in Peterhouse with Michael Levitt which set Horace and myself on the track of thinking about the conformation of DNA viewed, as it were, from the "inside" in relation to the stacking of base-pairs in the distinct dinucleotide steps. We quickly found that all of the well-known "external" differences between the "A" and "B" forms could be explained rather simply in terms of the "Roll", "Slide" and "Twist" degrees of freedom between successive base pairs, thought of here as rigid blocks. This in turn led to an understanding of the geometry of bent DNA, with curvature being a manifestation of the Fourier component of "Roll" at a period equal to the helical repeat of the DNA. The addition of a constant roll angle at every step would not affect curvature; and this led to a simple explanation of the "paradox" that periodically placed runs of A-A dinucleotide steps would produce curvature, even though the A-tracts themselves were particularly straight, as revealed by direct x-ray crystallography. In this work the key step was to adopt a different point of view about the geometry of DNA: the molecule is best seen as a sequence of dinucleotide steps of particular geometry, connected rather loosely by sugar-phosphate chains, rather than as a set of helical "backbones" between which the base-pairs are slung in a more-or-less passive way.

4. THE ROLE OF MATHEMATICS IN STRUCTURAL MECHANICS

So far, I have not dwelt on the role of mathematics in research in our field, even though I have mentioned the "clean" mathematical concepts of geometry and linear algebra. The history of the development of structural mechanics as a field is, of course, packed with examples in which a coupling of mathematics to some particular phenomenon has provided deep and useful insights. And indeed, many workers in structural mechanics today, just as in the past, regard themselves primarily as applied mathematicians.

It must be clear from what I have said already that while mathematics may be a useful tool once one has decided on the nature of the problem in hand, it does not provide anything like a universal *method* for solving the kinds of problem which I have been discussing. But it is difficult to generalise here, of course, because different areas in structural mechanics relate to various branches of mathematics with different degrees of directness.

The field of shell structures is often regarded as a particularly "mathematical area" in the field of structural mechanics. And over the years, many workers have striven to set forth, in appropriate co-ordinate systems, etc., the fundamental governing equations for general thin-shell structures under arbitrary loading. But it seems to me that even in this case the mathematics should be brought in at a relatively late stage, after physical ideas and arguments have been deployed in relation, particularly, to experimental evidence. The well-known controversy between A.E.H. Love and Lord Rayleigh in 1888, about which I wrote in 1988, provides a fine example of what I mean. Thus, Love had set up, for the first time, some general equations for thin elastic shells. He had realised physically that the mechanical performance of shells depends on a coupling of both "bending" and "stretching" effects; and his equations embodied that simple yet profound idea. He then attempted to show that a rather straightforward solution which had been proposed by Lord Rayleigh to the problem of determining the fundamental natural frequency of vibration of a thin hemispherical bowl – it was based on the observational idea that the mode of deformation was essentially inextensional – was actually wrong, since Rayleigh's postulated mode of deformation did not satisfy Love's detailed equations at the free boundary. Here, Love fell into the trap of discussing the solution of his equations without actually having solved them. His own proposal for an *extensional* fundamental mode of vibration turned out to give a numerical prediction above the audible range for the fundamental frequency of vibration of a particular glass bowl which actually gave an audible tone! The paradox was soon resolved by Lamb and Basset, who found that if one takes a rectangular elastic sheet and rolls it into a cylindrical shape, there is a narrow "boundary layer" at the free edge: this is necessary, in effect, to satisfy Love's detailed equations, but it does not contribute any significant strain-energy to the inextensional fundamental mode.

The difficulties which Love experienced over this problem illustrate some of the subtle forms of behaviour that are shown by thin-shell structures, not only on account of the physical interaction between bending and stretching effects, but also through conditions of freedom or fixity at the boundaries. The striking qualitative differences in behaviour between a complete spherical shell and a hemishperical bowl are obvious to anyone who has handled these objects; but it is a very different matter to deduce such differences from a study of the relevant governing equations.

In this connection it is useful to visualise the thin-walled shell as being composed of two separate but coincident surfaces, one of which carries loads by membrane action and the other by bending moments and out-of-plane shear forces. In terms of statics there is a local interface pressure between these two surfaces, while the kinematic connection is best expressed in terms of a common local Gaussian curvature, which changes as the surfaces distort. I was fortunate to pick up clues about the importance of Gaussian curvature from Aaron Klug, who had used it in describing the assembly of doubly-curved virus coats; and he directed me to the seminal book of Hilbert and Cohn-Vossen. In simple thin-shell problems there is a close formal analogy between the equations of statical equilibrium and kinematic compatibility for the two surfaces; and indeed the interface pressure and the distortional change in Gaussian curvature of the two surfaces turn out to be dual variables in this scheme. But perhaps the most important aspect of the "two-surface" model is the way in which it enables one to discuss in qualitative terms the different mechanical phenomena that are seen in thin-walled shells of different shapes and with different boundary conditions.

An important lesson to be learned from the Love-Rayleigh controversy is the importance of actually *solving* the governing equations, rather than speculating about the general character of their solution. In elastic shells the governing differential equation is of eighth order, even for thin simple cylindrical shells under non-symmetric loading. However, when one comes to solve the equation for a particular case it almost always turns out that about half of the terms are practically insignificant. Consequently instead of four boundary conditions being required at each edge, as indicated strictly by the order of the equation, in practice only two are usually of any significance. So an important step in solving practical thin-shell problems is to have an understanding of which are the important boundary conditions in a given case, and which are the insignificant ones. In this way, physical arguments can lead to a big reduction in the difficulty of the equations which remain to be solved.

In problems of this kind, we find that the use of the appropriate dimensionless variables can be helpful. In general, the performance of cylindrical shells under asymmetric loading – whether in straightforward response to applied forces, or in buckling, or in vibration – depends on the values of two distinct dimensionless groups involving the leading dimensions of the structure. And one can make useful "maps" summarising the dominant mode of behaviour in different regions of the space of these two groups – somewhat like the maps which Mike Ashby has developed in the field of material properties. In some problems it turns out that the "two-surface" model leads directly to the appropriate dimensionless groups which characterise the behaviour.

One of the aims of some of those who specialise in setting down in great detail the governing equations for thin-shell structures is to provide a definitive statement from which, by a process of solving the equations that is straightforward in principle, the physical performance of the structures can be deduced. As I have explained above, usually only about half of the terms in such an equation are relevant to any particular physical problem. The general practice which I advocate of setting down equations only after a good deal of physical thinking and experimental work have been done, is in line with a long-practiced English tradition in applied mathematics. In this scheme one tries to eliminate unimportant variables *before* setting down the relevant equations and solving them. One can see signs of this same idea in examination questions set over the years at Cambridge University, in which the candidate is first asked to explain in words some general principle or result, before putting it into a mathematical form in order to solve a particular physical problem.

I would not like to suggest that experiences of this kind with shell theory are a model for *all* problems in structural mechanics. For example, I have already mentioned the way in which there is an extremely close correspondence between linear algebra on the one hand and the performance of tensegrity structures on the other. The details of states of self-stress, infinitesimal mechanisms, etc. are all revealed in a very remarkable and complete way by the standard scheme of linear algebra. It is all very neat and tidy in that particular field.

A good example of the opposite, i.e. a not-very-tidy piece of applied mathematics, may be found in Mustafa El Hassan's work on the geometry of curved DNA. I have already mentioned that the overall curvature of a piece of DNA can be expressed in terms of the Fourier component of the Roll angle at a period equal to the helical repeat. This idea may be shown most simply by starting with a piece of "classical" "B"-form DNA, in which all of the Roll angles are zero – i.e. the base pairs are all parallel to each other. If one then applies small perturbations in Roll angle the result quoted above emerges immediately, since *small* angles of rotation can be added vectorially. And it turns out the same idea is still valid in an approximate sense, even when the Roll angles are as large as 10 or 20 degrees – in which range one cannot, of course, invoke the convenient small-angle simplifications. El Hassan did many computations to try to derive more accurate empirical formulae for curvature in terms not only of Roll, but also of other parameters such as helical Twist. At the end of the day, however, all that could really be said was that the simple formula quoted above is a rather good first approximation; and it is hardly worth the effort of trying to produce more elaborate and complete formulae.

This brings me to my last point, in relation to geometrical problems. On many occasions I have found that the key to understanding some particular problem in 3D geometry is to make a simple physical model, either out of pieces of wood or cork cut at appropriate angles on a band saw; or from pieces of wire bent in a suitable way; or from thin card by cutting, folding and gluing. Although there was a good deal of geometry in the A-level curriculum when I went to school in the late 1940s and early 1950s, I think that my interest in 3-dimensional geometry was formed when, at the age of 10, I was confined to bed for several weeks with chicken-pox and discovered in Arthur Mee's *Children's Encyclopaedia* plans for constructing small-scale model buildings in order to create a model village. The plans for these buildings were set out to scale in the pages of the encyclopaedia; but somehow I obtained a supply of thick paper with squares printed on it (as I recall inches and eighths) on which I replicated the drawings at a suitable scale and then cut them up and glued them together as appropriate. I discovered that in many cases one could construct the entire building from a single piece of paper, rather than having separate pieces for, say, walls and roof. And indeed, I managed to make out of a single piece some quite elaborate buildings with rather complicated shapes and particularly sophisticated systems of gables and dormer windows. Anyone who can set out the drawings for such roofs and dormers will find that ordinary 3-dimensional geometry holds few terrors.

PUBLICATIONS BY C.R. CALLADINE

1. STRUCTURAL MECHANICS

Calladine, C.R. (1960) On the creep of a wrinkle, *Creep in Structures Colloquium*, Stanford, J. Springer, 245-2.

Calladine, C.R. (1960) The steady creep of shells: a method of analysis, *Proc. Symp. Nuclear Reactors and Pressure Vessels*, Butterworths, Glasgow, 411-431.

Calladine, C.R. and Drucker, D.C. (1962) A bound method for creep analysis of structures: direct use of solutions in elasticity and plasticity, *J. Mech. Eng. Sci.* **4**, 1-11.

Calladine, C.R. and Drucker, D.C. (1962) Nesting surfaces of constant rate of energy dissipation in creep, *Quart. Appl. Math.* **20**, 79-84.

Calladine, C.R. (1962) The effect of cross-section shape on the creep buckling behaviour of columns, *Int. J. Mech. Sci.* **4**, 387-407.

Calladine, C.R. (1963) A note on the bending of beams made of non-linear material, *J. Roy. Aero. Soc.* **67**, 124-125.

Calladine, C.R. (1963) Stress concentration in steady creep: Interpolation between solutions in elasticity and plasticity, *Proc. I. Mech. E.* **178**, part 3A, 2.59 - 2.68.

Calladine, C.R. (1963) The yielding of clay, *Gotechnique* **13**, 250-255.

Calladine, C.R. (1964) A rapid method for estimating the greatest stress in a structure subject to creep, *Proc. I. Mech. E.* **178**, part 3L, 198-206.

Calladine, C.R. (1968) Simple ideas in the large-deflection plastic theory of plates and slabs, in *Engineering Plasticity*, edited by Heyman, J. and Leckie, F.A., Cambridge University Press, 93-127.

Bunyaraksh, S. and Calladine, C.R. (1969) *A study of the fundamentals of wrinkling in fuel element cans*, U.K. Atomic Energy Authority, TRG Report 1499 (S/X).

H.R. Drew and S. Pellegrino (eds.),
New Approaches to Structural Mechanics, Shells and Biological Structures, 11–21.

Calladine, C.R. (1969) *Engineering Plasticity*, xiv + 318pp, Pergamon Press.

Calladine, C.R. (1969) Time-scales for redistribution of stress in creep of structures, *Proc. Roy. Soc. A* **309**, 363-375.

Calladine, C.R. (1971) A microstructural view of the mechanical properties of saturated clay, *Geotechnique* **21**, 391-415.

Calladine, C.R. (1973) A plastic theory for collapse of plate girders under combined shearing force and bending moment, *The Structural Engineer* **51**, 147-154.

Calladine, C.R. (1973) Inelastic buckling of columns: the effect of imperfections, *Int. J. Mech. Sci.* **15**, 593-604.

Calladine, C.R. (1973) Overconsolidated clay: a microstructural view, in *Plasticity in Soil Mechanics*, edited by Palmer, A.C., Cambridge University Engineering Department, 144-158.

Gilbert, R.B. and Calladine, C.R. (1974) Interaction between the effects of local and overall imperfections on the buckling of elastic columns, *J. Mech. Phys. Solids* **22**, 519-540.

Calladine, C.R. (1975) Collapse of old metal organ pipes : a classroom demonstration of the creep buckling phenomenon, *Int. J. Mech. Eng. Education* **3**, 183-187.

Woodhead, A.L. and Calladine, C.R. (1976) A novel finite-element method for use in classical elastic plate problems, *Int. J. Mech. Sci.* **18**, 357-363.

Calladine, C.R. (1978) Buckminster Fuller's "Tensegrity" structures and Clerk Maxwell's rules for the construction of stiff frames, *Int. J. Solids Structures* **14**, 161-172.

Calladine, C.R. and Greenwood, J.A. (1978) Line and point loads on a non-homogenous incompressible elastic half-space, *Quart. J. Mech. Appl. Maths* **31**, 507-529.

Calladine, C.R. (1982) Modal stiffnesses of a pre-tensioned cable net, *Int. J. Solids Structures* **18**, 829-846.

Calladine, C.R. (1983) An investigation of impact scaling theory, in *Structural Crashworthiness*, edited by Jones, N. and Wierzbicki, T., 169-174, Butterworths, London.

Calladine, C.R. and English, R.W. (1984) Strain-rate and inertia effects in the collapse of two types of energy-absorbing structure, *Int. J. Mech. Sci.* **26**, 689-701.

Pellegrino, S. and Calladine, C.R. (1984) Two-step matrix analysis of prestressed cable nets, in *Proc. Third International Conference on Space Structures*, edited by Nooshin, H., 744-749, Elsevier Applied Science Publishers, London.

Calladine, C.R. (1985) *Plasticity for Engineers*, Ellis Horwood, Chichester, (updated re-issue of Engineering Plasticity (1969)).

Calladine, C.R. (1985) The strength of thin plates in compression, in *Aspects of the analysis of plate structures*, edited by Dawe, D.J. et al., Oxford University Press, 271-293.

Affan, A. and Calladine, C.R. (1986) Structural Mechanics of double-layer grids, in *Shells, Membranes and Space Frames, I.A.S.S. Symposium, Osaka, 1986*, edited by Heki, K. Elsevier, Amsterdam, vol. 3 41-48.

Kandil, K.S. and Calladine, C.R. (1986) Classical local buckling of tubes having rectangular cross- sections, *Int. J. Mech. Sci.* **28**, 789-797.

Pellegrino, S. and Calladine, C.R. (1986) Matrix analysis of statically and kinematically indeterminate frameworks, *Int. J. Solids Structures* **22**, 409-428.

Kamalarasa, S. and Calladine, C.R. (1988) Buckle propagation in submarine pipelines, *Int. J. Mech. Sci.* **30**, 217-228.

Abbassian, F. and Calladine, C.R. (1989) On the deformation of the pipe wall during propagation of a ductile crack in a high-pressure gas pipeline, *ASME J. Pressure Vessel Technology* **3**, 47-57.

Affan, A. and Calladine, C.R. (1989) Initial bar tensions in pin-jointed assemblies, *Int. J. Space Structures* **4**, 1-16.

Lu, G. and Calladine, C.R. (1990) On the cutting of a plate by a wedge, *Int. J. Mech. Sci* **32**, 293-313.

Calladine, C.R. and Pellegrino, S. (1991) First-order infinitesimal mechanisms, *Int. J. Solids Structures* **27**, 505-515.

Pellegrino, S. and Calladine, C.R. (1991) Structural computation of an assembly of rigid links, frictionless joints, and elastic springs, *J. Appl. Mech.* **58**, 749-753.

Tam, L.L. and Calladine, C.R. (1991) Inertia and strain-rate effects in a simple plate-structure under impact loading, *Int. J. Impact Engineering* **11**, 349-377.

Calladine, C.R. (1992) Editorial for issue on Masonry construction: structural mechanics and other aspects, *Meccanica* **27** (3).

Calladine, C.R. (1992) Stability: background to codes (22pp) in *Storage Tank Design Philosophy and Practice*, University of Cambridge Programme for Industry intensive design course, edited by Palmer, S.C.

Calladine, C.R. and Pellegrino, S. (1992) Further remarks on first-order infinitesimal mechanisms, *Int. J. Solids Structures* **29**, 2119-2122.

Calladine, C.R. (1993) Some problems in propagating plasticity. In *Plasticity and Impact Mechanics*, Proceedings of Symposium held at IIT, Delhi, 11-14 December 1993, edited by Gupta, N.K., Wiley Eastern, New Delhi, 71-97.

Maltby, T.C. and Calladine, C.R. (1995) An investigation into upheaval buckling of buried pipelines: I. Experimental apparatus and some observations, *Int. J. Mech. Sci.* **37**, 943-963.

Maltby, T.C. and Calladine , C.R. (1995) An investigation into upheaval buckling of buried pipelines: II. Theory and analysis of experimental observations, *Int. J. Mech. Sci.* **37**, 965-983.

Calladine, C.R. (1996) *Mhendislikte Plastisite, Bilimsel ve Teknik Yayinlari eviri Vakfi* (Translation of Engineering Plasticity (1969) into Turkish by Inan, E. and Togrol, E.).

Calladine, C.R. (1996) Some paradoxical experiments on beams, *Int. J. Mech. Eng. Education* **24**, 37-48.

Calladine, C.R. (1996) Upheaval and lateral buckling of submarine pipelines, in *Advances in Steel Structures*, Proceedings of International Conference, Hong Kong, edited by Chan, S.L. and Teng, J.G., Pergamon Press 647-656.

Miles, D.J. and Calladine, C.R. (1999) Lateral thermal buckling of pipelines on the sea bed, *J. Appl. Mech.* **66**, 891-897.

Palmer, A.C., Tebboth, L., Miles, D.J. and Calladine, C.R. (1999) Instability of pipelines on slopes, *J. Appl. Mech.* **66**, 794-799.

Calladine, C.R. (2000) Deployable structures : what can we learn from biological structures? in *Deployable Structures: Theory and Applications*, Proc. IUTAM IASS Symposium, 6-9 Sept. 1998, edited by Pellegrino, S., and Guest, S.D., Kluwer Academic Publishers, Dordrecht, 63-76.

Guest, S.D. and Calladine, C.R. (2000) Matrices in the teaching of statically indeterminate structures, in *Civil and Structural Engineering Education in the 21st Century*, edited by Allen, H.G., University of Southampton, 285-296.

2. SHELL STRUCTURES

Calladine, C.R. (1964) Edge-load response of a thin cylindrical shell in creep, in *Non-classical Shell Problems*, North Holland Publishing Co., 384-406.

Calladine, C.R. (1964) Steady creep in a cylindrical shell: close upper- and lower-bound solutions for all values of the creep index, in *Proc. World Conference on Shell Structures*, N.A.S., 603-618.

Calladine, C.R. (1965) Design of nozzles in spherical pressure vessels, *The Engineer* **219**, 386-387.

Calladine, C.R. (1966) On the design of reinforcement for openings and nozzles in thin spherical pressure vessels, *J. Mech. Eng. Sci.* **8**, 1-14.

Calladine, C.R. (1966) Stress concentration in nonlinear creep of a simple shell, *J. Appl. Mech.* **33**, 322-326.

Calladine, C.R. (1967) Some calculations to assess the effect of the stress-strain rate relationship on creep in the neighbourhood of an opening in a pressurised thin spherical shell, *J. Mech. Eng. Sci.* **9**, 198-210.

Calladine, C.R. (1969) Lower-bound analysis of symmetrically loaded shells of revolution, *Proc. ASME Conf. on Pressure Vessel Technology*, Part 1, 335-343.

Calladine, C.R. and Goodall, I.W. (1969) Plastic behaviour of thin cylindrical pressure vessels with circular cutouts and radial branches, *J. Mech. Eng. Sci.* **11**, 351-363.

Morris, A.J. and Calladine, C.R. (1969) The local strength of a spherical shell loaded radially through a rigid boss, *Proc. ASME Conf. on Pressure Vessel Technology*, Part 1, 35-44.

Morris, A.J. and Calladine, C.R. (1971) Simple upper-bound calculations for the indentation of cylindrical shells, *Int. J. Mech. Sci.* **13**, 331-343.

Calladine, C.R. and Barber, J.N. (1970) Simple experiments on self-weight buckling of open cylindrical shells, *J. Appl. Mech.* **37**, 1150-1151.

Calladine, C.R. (1972) Creep in torispherical pressure-vessel heads, in *Creep in Structures*, edited by Hult, J., Springer-Verlag, 247-268.

Calladine, C.R. (1972) On the derivation of yield conditions for shells, *J. Appl. Mech.* **39**, 852- 853.

Calladine, C.R. (1972) Structural consequences of small imperfections in elastic thin shells of revolution, *Int. J. Solids Structures* **8**, 679-697.

Calladine, C.R. (1973) A new finite-element method for analysing symmetrically loaded thin shells of revolution, *Int. J. Numerical Methods in Engineering* **6**, 475-487.

Calladine, C.R. (1974) Flexibility of axially symmetric bellows under axial loading, *Int. J. Mech. Sci.* **16**, 843-853 (1974).

Calladine, C.R. and Paskaran, N. (1974) A re-appraisal of influence coefficients for the edges of thin elastic spherical shells subjected to symmetric loads, *Quart. J. Mech. and Appl. Maths* **27**, 1-15.

Calladine, C.R. (1974) Limit analysis of curved tubes, *J. Mech. Eng. Sci.* **16**, 85-87.

Calladine, C.R. (1977) The static-geometric analogy in the equations of thin shell structures, *Math. Proc. Camb. Phil. Soc.* **82**, 335-351.

Calladine, C.R. (1977) Thin-walled elastic shells analysed by a Rayleigh method, I*nt. J. Solids Structures* **13**, 515-530.

Reddy B.D. and Calladine, C.R. (1978) Classical buckling of a thin-walled tube subjected to bending moment and internal pressure, *Int. J. Mech. Sci.* **20**, 641-65.

Calladine, C.R. and Robinson, J.M. (1980) A simplified approach to the buckling of thin elastic shells, in *Theory of Shells*, Proceedings of IUTAM Symposium, edited by Koiter, W.T. and Mikhailov, G.K., North-Holland Publishing Company, 173-196.

Calladine, C.R. (1982) Natural frequencies of cooling tower shells, *J. Sound Vibration* **82**, 345-369.

Calladine, C.R. (1982) The theory of shell structures: aims and methods, *Int. J. Mech. Sci.* **24**, 219-230.

Calladine, C.R. (1983) Plastic buckling of tubes in pure bending, in *Collapse: the Buckling of Structures in Theory and Practice*, edited by Thompson, J.M.T. and Hunt, G.W., Cambridge University Press, 111-124.

Calladine, C.R. (1983) *Theory of Shell Structures*, xiv + 763pp, Cambridge University Press.

Calladine, C.R. (1985) Gaussian curvature and shell structures. In *The Mathematics of Surfaces*, edited by Gregory, J.A., Oxford University Press, 179-196.

Calladine, C.R. (1986) Analysis of large plastic deformations in shell structures. In *Inelastic Behaviour of Plates and Shells*, IUTAM Symposium, Rio de Janeiro, 1985, edited by Bevilacqua, L. et al., Springer, Berlin 69-101.

Calladine, C.R. (1988) The theory of thin shell structures 1888-1988, *Proc. I. Mech. E., Part A* **202**, 141-149.

Calladine, C.R. Stability of the "Endeavour" balloon. (1988) In *Buckling of Structures*, edited by Elishakoff, I. et al., Elsevier, Amsterdam, 133-149.

Kamalarasa, S. and Calladine, C.R. (1989) Geometry and strain in the transition region of a collapsing submarine pipe, *Int. J. Mech. Sci.* **31**, 207-218.

Phaal, R. and Calladine, C.R. (1992) A simple class of finite elements for plate and shell problems. I: Elements for beams and thin flat plates, *Int. J. Num. Meth. Eng* **35**, 955-977.

Phaal, R. and Calladine, C.R. (1992) A simple class of finite elements for plate and shell problems. II: An element for thin shells, with only translational degrees of freedom, *Int. J. Num. Meth. Eng* **35**, 979-996.

Pamplona, D.C. and Calladine, C.R. (1993) The mechanics of axially symmetric liposomes, *ASME J. Biomech. Eng.* **115**, 149-159.

Calladine, C.R. (1994) Causes of imperfection-sensitivity in the buckling of thin shells, Euromech Colloquium 317, Buckling Strength of Imperfection-Sensitive Shells, *Thin-Walled Structures* **3**, 215-235.

Holst, J.M.F.G. and Calladine, C.R. (1994) Inversion problems in thin elastic shells, *European J. Mechanics A/Solids* **13** No. 4, supplement 3-18.

Hambly, E.T. and Calladine, C.R. (1996) Buckling experiments on damaged cylindrical shells, *Int. J. Solids Structures* **33**, 3539-3548.

Pamplona, D.C. and Calladine, C.R. (1996) Aspects of the mechanics of lobed liposomes, *ASME J. Biomech. Eng.* **118**, 482-488.

Holst, J.M.F.G., Rotter, J.M. and Calladine, C.R. (1999) Imperfections in cylindrical shells resulting from fabrication misfits, *ASCE J. Eng. Mechanics* **125**, 410-418.

Calladine, C.R. (2000) A shell-buckling paradox resolved, in *Advances in the Mechanics of Plates and Shells*, edited by Durban, D., Givoli, G. and Simmonds, J.G., Kluwer Academic Publishers, Dordrecht, 119-134.

Holst, J.M.F.G., Rotter, J.M. and Calladine, C.R. (2000) Imperfections and buckling in cylindrical shells with consistent residual stresses, *J. Constructional Steel Research* **54**, 265-282.

Lancaster, E.R., Calladine, C.R. and Palmer, S.C. (2000) Paradoxical buckling behaviour of a thin cylindrical shell under axial compression, *Int. J. Mech. Sci.* **42**, 843-865.

Mandal, P. and Calladine, C.R. (2000) Buckling of thin cylindrical shells under axial compression, *Int. J. Solids Structures* **37**, 4509-4525.

Zhu, E., Mandal, P. and Calladine, C.R. Self-weight buckling of open-topped cylindrical shells, and its relevance to shell buckling more generally, *Int. J. Mech. Sci.* (in press).

3. BIOLOGICAL STRUCTURES

Calladine, C.R. (1975) Bacteria can swim without rotating flagellar filaments? *Nature* **249**, 385.

Calladine, C.R. (1975) Construction of bacterial flagella, *Nature* **255**, 121-124.

Calladine, C.R. (1976) Design requirements for the construction of bacterial flagella, *J. Theor. Biol.* **57**, 469-489.

Calladine, C.R. (1978) Change of waveform in bacterial flagella: the role of mechanics at the molecular level, *J. Mol. Biol.* **118**, 457-479.

Calladine, C.R. (1980) Toroidal elastic supercoiling of DNA, *Biopolymers* **19**, 1705-1713.

Calladine, C.R. (1982) Construction of bacterial flagellar filaments, and aspects of their conversion to different helical forms, in *Prokaryotic and Eukaryotic Flagella*, Proc. 35th Symp. Soc. Experimental Biology, edited by Amos, W.B. and Duckett, J.G., Cambridge University Press, 33-51.

Calladine, C.R. (1982) Mechanics of sequence-dependent stacking of bases in B-DNA, *J. Mol. Biol.* **161**, 343-352.

Calladine, C.R. (1983) Construction and operation of bacterial flagella, *Sci. Prog. Oxf.* **68**, 365- 385.

Calladine, C.R. and Drew, H.R. (1984) A base-centred explanation of the B-to-A transition in DNA, *J. Mol. Biol.* **178**, 773-782.

Calladine, C.R. and Drew, H.R. (1987) Principles of sequence-dependent flexure of DNA, *J. Mol. Biol.* **192**, 907-918.

Drew, H.R. and Calladine, C.R. (1987) Sequence-specific positioning of core histones on an 860 base-pair DNA : experiment and theory, *J. Mol. Biol.* **195**, 143-173.

Stewart, M., McLachlan, A.D. and Calladine, C.R. (1987) A model to account for the elastic element in muscle crossbridges in terms of a bending myosin rod, *Proc. Roy. Soc. B* **228**, 381- 413.

Calladine, C.R., Drew, H.R. and McCall, M.J. (1988) The intrinsic curvature of DNA in solution, *J. Mol. Biol.* **201**, 127-137.

Drew, H.J., McCall, M.J. and Calladine, C.R. (1988) Recent studies of DNA in the crystal, *Ann. Rev. Cell Biol.* **4**, 1-20.

Dickerson, R.E., Bansal, M., Calladine, C.R. and 14 others, (1989) Definitions and nomenclature of nucleic acid structure parameters, *EMBO J.* **8**, 1-4.
Also (1989) in *J. Biomolecular Structure Dynamics* **6**, 627-634
Also (1989) in *J. Mol. Biol.* **205**, 787-791
Also (1989) in *Nucleic Acids Research* **17**, 1797-1803.

Drew, H.R., McCall, M.J. and Calladine, C.R.(1990) New approaches to DNA in the crystal and in solution, in *DNA Topology and its Biological Effects*, edited by Cozzarelli, N.R. and Wang, J.C., Cold Spring Harbor Press, 1-56.

Calladine, C.R., Collis, C.R., Drew, H.R. and Mott, M.R. (1991) A study of electrophoretic mobility of DNA in agarose and polyacrylamide gels, *J. Mol. Biol.* **221**, 981-1005.

Calladine, C.R. and Drew, H.R. (1992) Curvature and Flexibility of DNA: sequence-directed effects seen from a structural mechanics viewpoint, in *Molecular Structures and Life*, edited by Kyogoku, Y. and Nishimura, Y. Japan Scientific Society Press, Tokyo/CRC Press, Boca Raton, 43-55.

Calladine, C.R. and Drew, H.R. (1992) *Understanding DNA: the Molecule and How It Works*, ix + 220pp, Academic Press.

Calladine, C.R. (1993) Application of structural mechanics to biological systems, in *Theoretical and Applied Mechanics 1992*, IUTAM Congress, Haifa, Israel, edited by Bodner, S.R., Singer, J., Solan, A. and Hashin, Z., Elsevier Science Publishers B.V., 205-220.

Mott, M.R., Cox, D., Drew, H.R., Collis, C.R. and Calladine, C.R. (1993) Review on visualization by electron microscopy of double stranded curved DNA in relation to migration through electrophoretic gels, *J. Computer-Assisted Microscopy* **5**, 131-134.

El Hassan, M.A. and Calladine, C.R. (1995) The assessment of the geometry of dinucleotide steps in double-helical DNA; a new local calculation scheme. *J. Mol. Biol.* **251**, 648-664.

Calladine, C.R. and H.R. Drew (1996) A useful role for "static" models in elucidating the behaviour of DNA in solution, *J. Mol. Biol.* **257**, 479-485.

Calladine, C.R. and Drew, H.R. (1996) DNA Structure, in *Encyclopedia of Molecular Biology and Molecular Medicine*, edited by Meyers, R.A. VCH, Weinheim, Vol. 2, 102-118.

Calladine, C.R. and Drew, H.R. (1996) *Understanding DNA*, (Japanese translation) vii + 200pp, Kyoritsu Shuppan, Tokyo.

El Hassan, M.A. and Calladine, C.R. (1996) Propeller-twisting of base-pairs and the flexibility of dinucleotide steps, *J. Mol. Biol.* **259**, 95-103.

El Hassan, M.A. and Calladine, C.R. (1996) Structural mechanics of bent DNA, *Endeavour* **20**, 61- 68.

Lutter, L.C., Halvorson, H.R. and Calladine, C.R. (1996) Topological measurement of protein- induced DNA bend angles, *J. Mol. Biol.* **261**, 620-633.

Calladine, C.R. and Drew, H.R. (1997) *Understanding DNA: the Molecule and How It Works, 2nd Edition*, xi + 283 pp, Academic Press.

El Hassan, M.A. and Calladine, C.R. (1997) Curvature and torsion of infinitely repeating DNA sequences, *Proc. Roy. Soc. A* **453**, 365-386.

El Hassan, M.A. and Calladine, C.R. (1997) Conformational characteristics of DNA; empirical classifications and a hypothesis for the conformational behaviour of dinucleotide steps, *Phil. Trans. Roy. Soc. A* **355**, 43-100.

El Hassan, M.A. and Calladine, C.R. (1998) Two distinct modes of protein-induced bending in DNA, *J. Mol. Biol.* **282**, 331-343.

Calladine, C.R., Sharff, A. and Luisi, B.F. (2001) How to untwist an α-helix: structural principles of an α-helical barrel, *J. Mol. Biol.* **305**, 603-618.

4. MISCELLANEOUS

Calladine C.R. and Heyman, J. (1962) The mechanics of the game of croquet, *Engineering* **193**, 861-863.

Calladine, C.R. (1969) On the teaching of inelastic column theory, *Bull. Mech. Eng. Educ.* **8**, 239-252.

Calladine, C.R. and Sherby, O.D. (1971) Conversion factors for stress and pressure, *Metal Progress* **99**, 86-87.

Calladine, C.R. (1974) A new thermodynamic diagram for representing steady one-dimensional compressible fluid flow, *J. Mech. Eng. Sci.* **16**, 192-195.

Calladine, C.R. (1990) The teaching of some aspects of the theory of inelastic collisions, *Int. J. Mech. Eng. Education* **18**, 301-310.

Calladine, C.R. (1992) How should structures be taught in universities? in *Education for Structural Engineering*, edited by McKay, D., Innovative Press, Sheffield, 37-44.

COLLAPSE BEHAVIOR OF EXPANDABLE SLOTTED TUBES

F. Abbassian and S. Willson
BP Exploration, Upstream Technology Group
501 Westlake Park Boulevard, Houston, Texas 77079, USA
abbassif1@bp.com

A.J.L. Crook
Rockfield Software Limited, Technium
Kings Road, Prince of Wales Dock, Swansea SA1 8PH, West Glamorgan, U.K.

Abstract The behavior of expandable slotted tubes under uniform external pressure is investigated in this paper. Various modes of tube deformation during the collapse process are addressed and correctly predicted by means of idealized mechanical models. The model predictions are compared with available experimental and finite-element based numerical data. Potential use of the models in design of slotted liners for oil field application is then described.

Keywords: Plastic hinge, Shell structures, Expandable slotted tubes, Tubular collapse

1. INTRODUCTION

This paper investigates the collapse behavior of slotted expandable tubes subjected to uniform external pressure. These are steel tubes with a number of longitudinally oriented, pre-cut, overlapping slots which can be mechanically expanded to diameters > *1.5* times the original unexpanded diameter: see Fig. 1. One means of mechanical expansion is by forcing a mandrel with a cone of a larger diameter through the bore of the tube. As the mandrel passes through and the tube swages open, like a fish net, the ligaments between the slots undergo considerable plastic deformation, Fig. 1.

A recent and novel application of expandable tubes in the oil industry is to control sand production. This is a process whereby, in rocks possessing little or no intrinsic strength, sand grains are entrained within the hydrocarbon fluid flow, so causing erosion, blockages in pipework, and other processing upsets in production facilities. Cutting an overlapping slot pattern along the length of a stainless steel tube makes an expandable slotted base

H.R. Drew and S. Pellegrino (eds.),
New Approaches to Structural Mechanics, Shells and Biological Structures, 23–41.

pipe. A combination of overlapping rectangular sheets of metal weave filters, attached to the expandable base pipe and encased within a protective metal shroud provides a barrier to sand ingress. During expansion, the base pipe and the protective shroud slots open up to expose a flow area through the metal weave filters which accommodate the expansion by sliding over each other while maintaining a tight overlap. On the other side of the dimensional spectrum, a bi-stable version of these tubes (*0.1* to *0.3* inches in diameter), referred to as stents, are used in cardiovascular surgery as support structures to prop open arteries (Al Suwaidi et al., 2000). Here, an angioplasty balloon is used to expand the stent against the blood vessel wall.

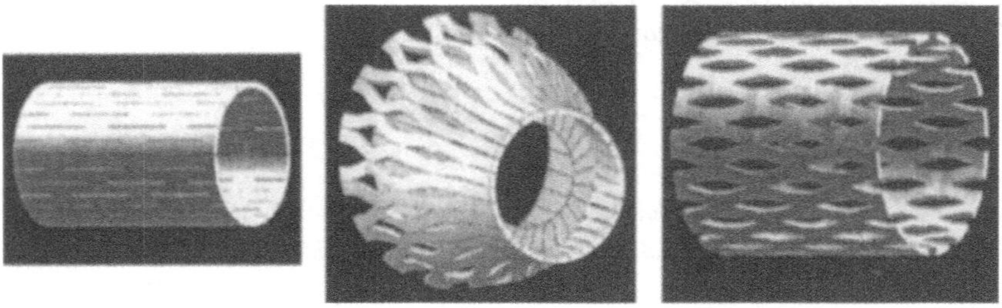

Figure 1. Expandable slotted tube configuration before, during and after expansion.

At the onset of production, when the reservoir begins to flow, there is little or no differential pressure loading across the expanded tube due to the large flow area provided by the expanded slots and a negligible pressure drop across the filtering shroud. However, during the production life of the reservoir, an expanded slotted tube may become subjected to loads induced by the deformation of the surrounding rock formation. There are two primary causes for the occurrence of this in-situ loading. First, plugging of the filter by fine particles released from the formation and transported in the flow. This causes an increase in the differential pressure load on the expanded tube. Second, and the more likely in practice, is the mechanical load transfer between the surrounding formation and the tube due to the compaction of the reservoir. Depending on its orientation, the tube becomes subjected to axial and/or "gripping" collapse loading. The two prevalent modes of tube failure arising from these loading scenarios are axial buckling and circumferential collapse. The integrity of the expanded tube over the production life of the field is of crucial importance, as any failure will interrupt production, causing an immediate impact on revenue.

Finite element analysis is currently the primary means of assessing the structural capacity of expandable slotted tubes. To supplement these necessarily complex analyses, this paper investigates the behavior of expandable slotted tubes under uniform external pressure loading by means of a series of idealized mechanical models. The primary aim is to develop an effective screening tool that enables efficient evaluation of the performance of different slotted tube designs prior to undertaking more time-consuming detailed evaluations using computational models. Consequently, the simplifying assumption of an elastic perfectly plastic material is adopted and the contribution of the outer shroud to the structural capacity is neglected, an assumption frequently made in the more complex computational models.

The various modes of tube deformation under uniformly applied external pressure are considered, and it is shown that the analytical model predicts axi-symetric closure and collapse loads that compare favorably with available experimental and finite-element based numerical data.

2. OBSERVATIONS

Full-scale collapse tests conducted on expanded and unexpanded slotted tubes indicate that the presence of the slots substantially reduces the tube's stiffness in circumferential bending, down to a few percent of the intact un-slotted tube stiffness. As a result, the residual collapse pressure of the tube is a very small percentage of that of the un-slotted tube. Figure 2 illustrates a typical collapse mode under applied pressure. Upon the application of pressure, the tube first ovalizes due to initial out-of-roundness. As the pressure increases further, the tube undergoes an elastic-plastic collapse ending up in a typical dog-bone geometry.

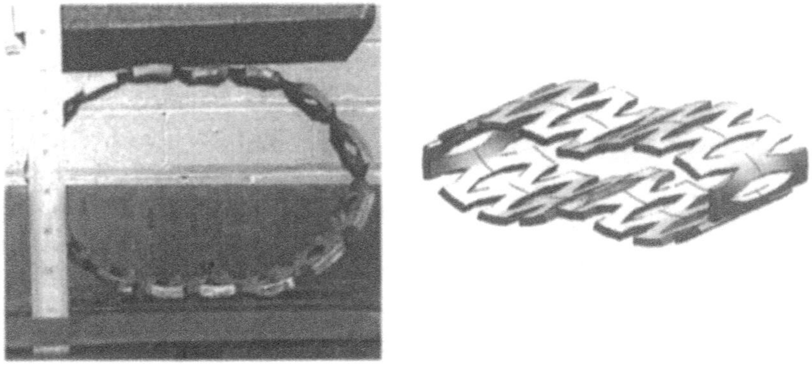

Figure 2. Typical collapse mode of an expandable slotted tube.

Detailed finite element analyses reported by Willson et al. (2002) and recent investigations carried out as part of the present study indicate that the plastic deformation during the expansion process is concentrated at the ends of the ligaments separating the adjacent overlapping slots. These concentrations in plastic strains are indicated in the equivalent von Mises plastic strain contours shown in Fig. 3(a). Furthermore, as the tube ovalizes upon application of the external pressure, the ligaments undergo considerable twist deformation as indicated from the plastic strain contours shown in Fig. 3(b), where the tube is in a post-collapse dog-bone configuration. As we will see later, this twisting action plays a major role in reducing both the circumferential bending stiffness of the tube during initial elastic ovalization and the eventual collapse pressure of the tube.

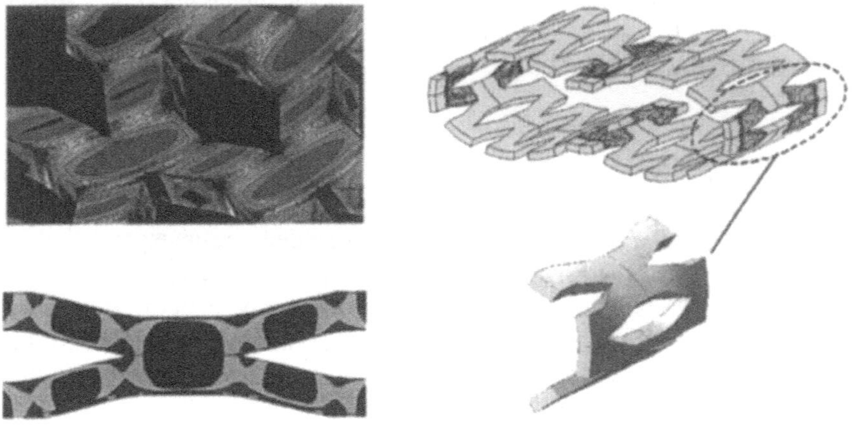

(a) (b)

Figure 3. Typical finite element results indicating (a) localized plastic bending deformation and (b) twisting of the ligaments.

3. IDEALIZED EXPANSION MODEL

The expansion process involves predominantly membrane deformation of the tube as the expansion cone is driven along the bore of the tube. In this section we shall introduce an idealized mode of deformation for the expansion process which we will later use to investigate axi-symetric behavior under uniform external pressure.

Based on the finite element results discussed in the previous section, an idealized model, schematically illustrated in Fig. 4, is proposed for the deformation of the tube during the expansion process. The expansion of the tube diameter is accommodated by the rotation at the plastic hinges formed at both ends of the ligament separating two adjacent overlapping slots. The tube material elsewhere remains rigid and undeformed.

The following geometrical relationship relates the expanded diameter D to the initial unexpanded diameter D_i:

$$D = D_i + \frac{2\Delta N \sin\alpha}{\pi} \quad (1)$$

N represents the number of slots around the circumference, Δ is the slot overlap and α the angular opening of the slots, as defined in Fig. 4.

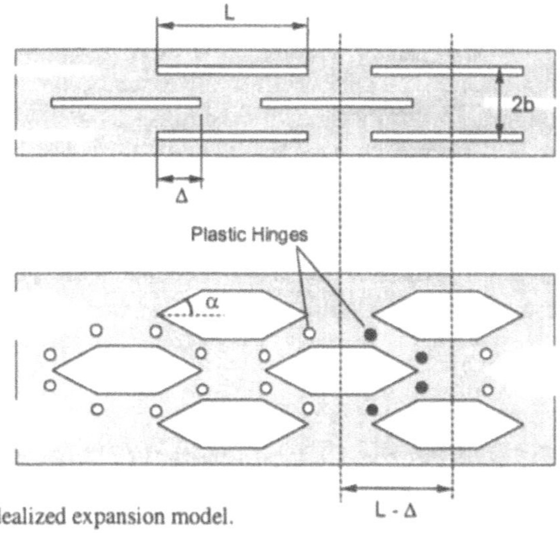

Figure. 4. Idealized expansion model.

4. AXI-SYMETRIC DEFORMATION

Aided by our simple expansion model, we will now investigate the behavior of the slotted tube, after it is fully expanded, under a uniform external pressure loading. Several modes of behavior could exist here, the simplest being uniform deformation, where the tube remains circular as its diameter reduces with pressure increase. This axi-symmetric mode of deformation excludes the possibility of tube ovalization due to out-of-roundness imperfections, which are normally present in the tube.

Within the context of the idealized expansion model described earlier, the change in tube diameter is accommodated through rotation at the plastic hinges and a reduction in the angle α. We will obtain the relationship between applied pressure, p, and displacement parameter, α, through the minimization of total potential energy for the system. The total potential energy concept, which is strictly valid for the analysis of elastic systems, can be applied here to investigate the loading portion of this single load system.

In deriving the total potential energy expression, we will consider a segment of the tube having length L-Δ.

The work done by the applied pressure, p, over the segment L-Δ can be expressed as:

$$W = \frac{\pi}{4} p(D_e^2 - D^2)(L - \Delta) \tag{2}$$

where D_e is the tube diameter immediately after full expansion and D is the reduced tube diameter under the applied pressure, p.

The above equation assumes that the pressure, p, is applied over the whole surface of the tube including the open area of the slots. This is a reasonable assumption, for example when the hydrostatic pressure is applied over a rubber membrane covering the slot openings.

The work done in the plastic hinges is given by:

$$W_p = 4N(\alpha_e - \alpha) M_{po} \tag{3}$$

where α_e is the angular opening of the slots immediately after expansion, and the full plastic moment, M_{po}, is given by

$$M_{po} = \frac{1}{4} b^2 t \sigma_y$$

Here, t is the tube wall thickness and b represents the width of ligament between two adjacent slots, as defined in Fig. 4.

The total potential energy, $PE=W_p$-W, per length L-Δ of the tube can then be expressed as:

$$PE = N b^2 t \sigma_y (\alpha_e - \alpha) - \frac{\pi}{4} p(D_e^2 - D^2)(L - \Delta) \tag{4}$$

where D and D_i are related through the geometrical relationship (1).

α is the sole displacement parameter in the total potential energy expression (4) and equilibrium condition is defined by $dPE/d\alpha$. This gives:

$$p = \frac{b^2 t \sigma_y}{D(L - \Delta)\Delta \cos\alpha} \tag{5}$$

In deriving the above equation, the pressure p has been assumed to act on the whole surface of the tube including the slot openings. If the pressure is only applied to the solid portion of the tube, then p as expressed by (5) should be scaled up by D/D_i representing the ratio between total and solid surface areas of the tube. Moreover, we have ignored the axial elongation of the tube as it circumferentially deforms under external pressure. It can however be shown that, for the deformation ranges of interest, this is a second order effect and its impact on the pressure, p, is negligible.

Equation (5) has been plotted in Fig. 5. Within the context of the idealized expansion model, the slots begin to close at a pressure p_{clo} that corresponds to the formation of plastic hinges at the ligaments separating the slots. From (5), this pressure is given by:

$$p_{clo} = \frac{b^2 t \sigma_y}{D_e (L-\Delta)\Delta \cos\alpha_e} \tag{6}$$

where α_e is obtained from (1) for $D=D_e$. Furthermore, the pressure at which the slots fully close up, p_{clf}, is obtained by substituting $D=D_i$ and $\alpha=0$ in (5).

As we shall explain in Section 5, the pressure at the onset of slot closure, p_{clo}, represents the limiting collapse pressure for the slotted tube.

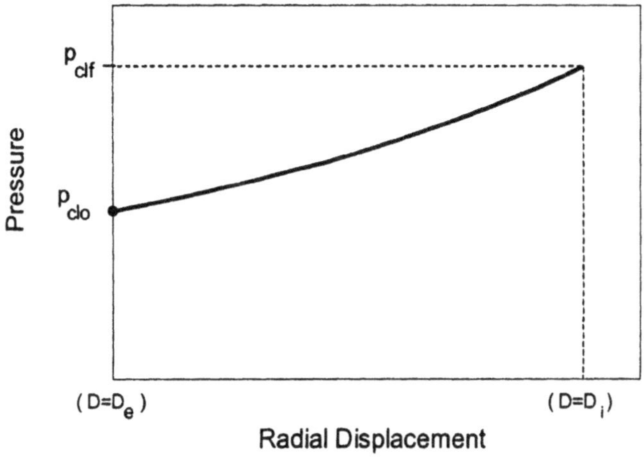

Figure 5. Prediction from equation (5) for axi-symmetric deformation under applied load.

4.1 Comparison with Numerical Solution

The closed-form solution (5) for pressure-deformation behavior obtained in the previous section has been compared in Fig. 6 against finite-element predictions for three different expanded slotted tube diameters. The input parameters used in these simulations are listed in Table 1. Solid finite elements were used to represent the tube and slot geometry, see Fig. 6 inset. With the exception of the initial rising part of the curve, excellent agreement is shown between the simple model predictions and detailed 3-D finite element simulations for all three cases.

The reason for the discrepancy at small radial displacements is the fact that finite-element solutions assume an elastic ideally-plastic material law whereas the model prediction is based on a rigid-plastic material.

It is noted that in both examples the slot closure begins at relatively low pressures, around *90-105* psi depending on tube diameter.

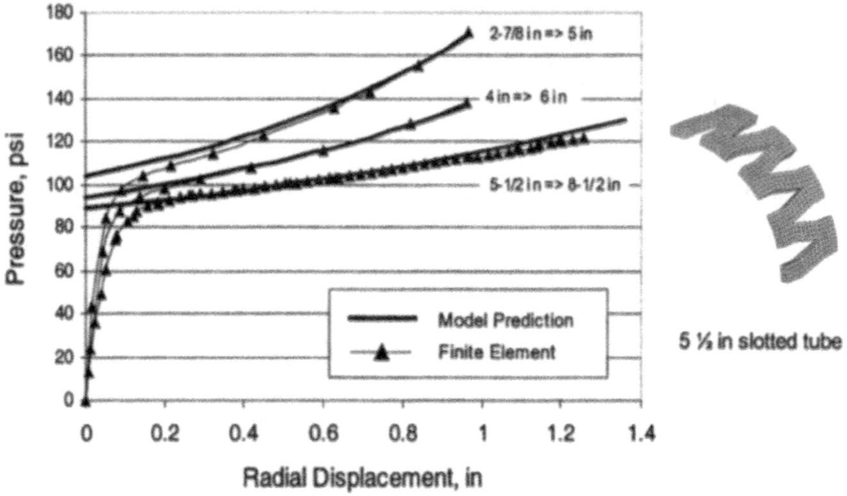

Figure 6. Axi-symmetric behavior: comparison between model predictions and finite- element results. Wall thickness ≅ 0.25 in: see Table 1 for details.

5. COLLAPSE UNDER UNIFORM PRESSURE

In the previous section an axi-symmetric mode of deformation was assumed, utilizing the idealized expansion model, to investigate the initial deformation of the expanded tube under uniform pressure. In general, however, a tube subjected to external pressure could undergo some ovalization due to inherent out-of-roundness imperfections in the tube. This ovalization, as it progresses, would eventually lead to tube collapse. For diameter/thickness ratios of interest, an expanded and unconfined slotted tube would, in general, find it easier to ovalize rather than deform in axi-symmetric fashion.

A new idealized model can be introduced to investigate the ovalization and collapse behavior of the slotted base pipe. The idealized model assumes that the collapse behavior under uniform external pressure consists of two distinct and independent behavioral regimes, namely, elastic ovalization and plastic collapse via a four-hinge mechanism. This approach has been previously applied by Abbassian and Parfitt (1998) to assess the collapse of solid tubes with considerable success.

Figures 7(a)-7(c) schematically illustrate the idealized models, the construction of the 4-hinge plastic collapse and elastic ovalization curves and the collapse pressure, p_c, defined by the intersection of the two curves.

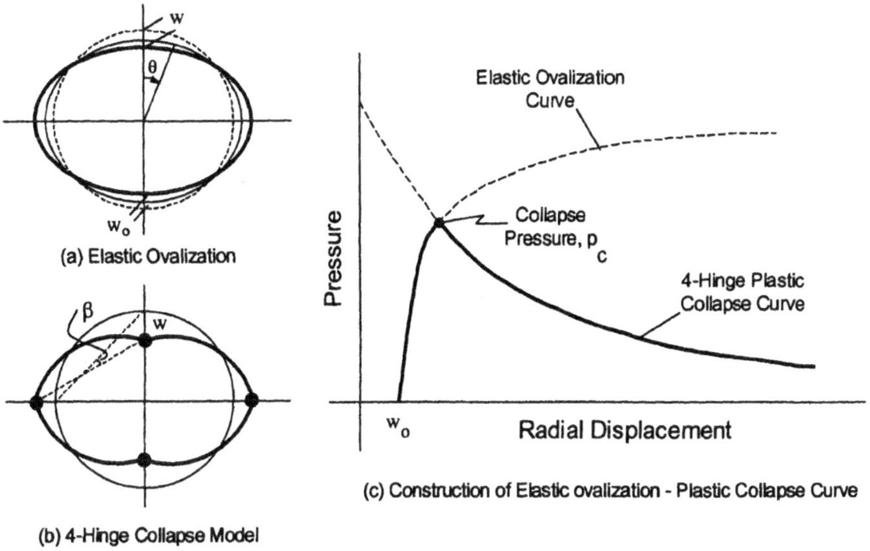

Figure 7. Idealized models for elasto-plastic collapse.

5.1 Elastic Ovalization

According to the small-displacement theory of thin shells, the load displacement relationship for a cylindrical tube with initial ovality with shape cos2θ and amplitude under externally applied uniform pressure, p, can be expressed as (Timoshenko and Gere, 1961):

$$P_{eo} = \frac{24B}{D^3}\left(1 - \frac{w_o}{w}\right) \tag{7}$$

where P_{eo} is the elastic ovalisation pressure, w is the radial displacement amplitude as defined in Fig. 7a, and B, the circumferential bending stiffness per unit length under a plane strain condition, defined as:

$$B = \frac{Et^3}{12(1-\nu^2)}$$

ν is the pipe's material Poisson's ratio and E, its Young's modulus.

In the next section we will investigate the effect of slots on the elastic ovalization via their impact on the circumferential stiffness B.

5.1.1 *Influence of the Slots on Elastic Ovalization*

The presence of the slots is expected to significantly reduce the flexural stiffness of the cylinder in the circumferential direction. Referring to Fig. 8(a), if every other longitudinal row of slots were shifted circumferentially by a distance b then, intuitively, the circumferential bending stiffness of the slotted tube would reduce to zero. It is therefore inferred that bending moment transfer in the circumferential direction is through the segment of the tube where the slots overlap.

Circumferential bending will produce zones of local bending and twisting as illustrated in Fig. 8(a). As the tube ovalizes the circumferential bending moment is transferred through the twisting action of the ligaments of overlapping slots, Fig. 8(b). The presence of this twisting action markedly reduces the circumferential flexural stiffness of the expanded slotted tube.

In our investigation of elastic ovalization, the slotted tube can be represented as an intact (un-slotted) tube with equivalent elastic bending stiffness B_E such that, when under load it behaves identically to the slotted tube. If the ligaments between overlapping slots in Fig. 8(b) were not allowed to twist when the tube undergoes circumferential bending, then the circumferential stiffness could be represented as λB where

$$\lambda = \left(\frac{a\,D_i}{2\,D_e}\right)\frac{1}{L-\Delta}$$

Here D_i/D_e accounts for the reduction in the second moment of area of the cross-section per unit length of the strip due to the slot openings and $a/2$ represents the proportion of the bending zone per length L-Δ of the expandable tube.

We can now obtain an estimate for the reduced circumferential bending stiffness due to ligament twist by considering the simple model shown in Fig. 8(c). In this model, a plate strip of width L-Δ and thickness t is considered, having a reduced flexural stiffness λB, representing the circumferential bending stiffness but ignoring the effect of ligament twist. The twisting action in the ligaments can be represented by rotational elastic springs of stiffness, k per unit length, as shown in Fig. 8(c). It can be shown that the combined bending stiffness of the equivalent model in Fig. 7(c) can be represented in terms of the stiffness parameters B and k as follows:

$$B_E = \frac{(D_e / D_i)\,b\,k}{1 + (D_e / D_i)\,b\,k / \lambda B} \tag{8}$$

where B_E is the equivalent bending stiffness per unit length of the slotted tube.

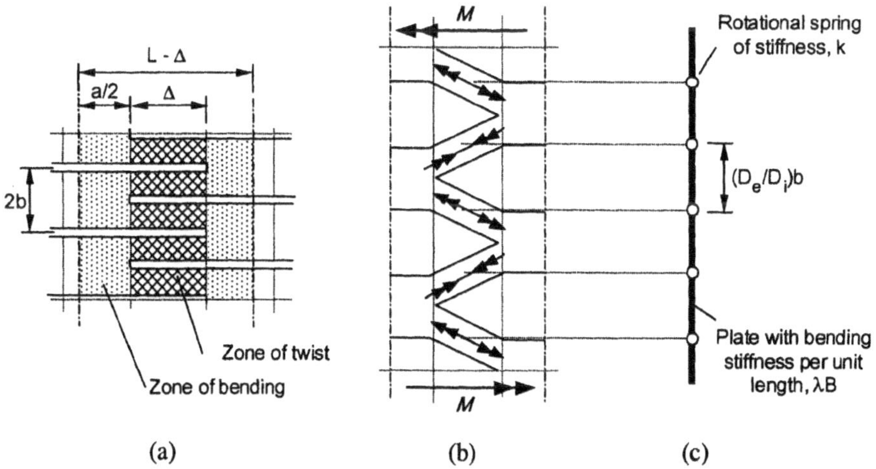

Figure 8. Model for calculating equivalent elastic parameters: (a) zones of bending and twisting, (b) transfer of circumferential bending through twist in overlapping slot ligaments, and (c) equivalent plate model.

We can now obtain an estimate for the elastic spring stiffness k by considering the stiffness of a beam of length Δ and cross-section $(b{\times}t)$ subjected to end twisting moments. Expressed in terms of a unit length of the strip L-Δ, we can then obtain:

$$k = \frac{KG}{\Delta(L-\Delta)} \qquad (9)$$

where $G = E/2(1+\nu)$ is the elastic shear modulus and the geometric factor K can be expressed (Timoshenko and Goodier, 1970) as:

$$K = \frac{1}{3}bt^3\left[1-0.63\frac{t}{b}\left(1-\frac{t^4}{12b^4}\right)\right]$$

For the parameter range of interest, $bk \ll \lambda B$. It therefore follows from (8) that,

$$B_E = \frac{D_e}{D_i}bk$$

which after substitution for k gives the following expression for B_E,

$$B_E = \frac{E b^2 t^3}{6(1+\nu)\Delta(L-\Delta)}\left(\frac{D_e}{D_i}\right)\left[1-0.63\frac{t}{b}\left(1-\frac{t^4}{12b^4}\right)\right] \tag{10}$$

The elastic ovalization curve is then given by (7) for $B=B_E$.

5.2 Plastic Collapse Curve

Within the context of the 4-hinge collapse mechanism, in Fig. 7(b), the formation of four plastic hinges allows collapse to occur while the four segments of the tube separated by the hinges remain rigid. Using an energy approach, taking into account the work done by the applied pressure and energy dissipated at the four plastic hinges, an expression can be obtained relating applied pressure, p, to rotation, β, at the plastic hinges (Abbassian and Parfitt, 1998):

$$P = \frac{8}{D^2 \sin 2\beta} M_{pl} \tag{11}$$

Here the parameter β is related to the radial displacement amplitude, w, as defined in Fig. 7(b), through:

$$\sin 2\beta = 4\frac{w}{D^2}(D - w)$$

and M_{pl} is the plastic moment capacity per unit length at hinge locations associated with the 4-hinge collapse model in Fig. 7(b). Note that M_{pl} is not otherwise related to the plastic moment M_{po}, discussed in Section 4, in association with the idealized expansion model shown in Fig. 4. Equation (11) can then be written in terms of w as:

$$p = \frac{2}{w(D-w)} M_{pl} \tag{12}$$

5.2.1 Influence of the Slots on Plastic Collapse

We saw in the previous section that the twist in the ligaments of the overlapping slots governs the circumferential bending during elastic ovalization. Detailed finite-element simulations, referred to in Section 2, also clearly indicate that the ligaments between the overlapping slots in the vicinity of the four plastic-hinge locations undergo plastic torsion as the post-collapse deformation of the tube takes place, Fig. 3(b).

Referring to Fig. 9, torsional plastic deformation in slot ligaments a, b, c and d would allow a rotational mechanism along the plastic hinge line. The plastic moment capacity per unit length along the longitudinal hinge line can then be expressed as:

$$M_{pl} = \frac{2}{L-\Delta} T_p \tag{13}$$

where T_p is the torque required to plastically deform the ligament.

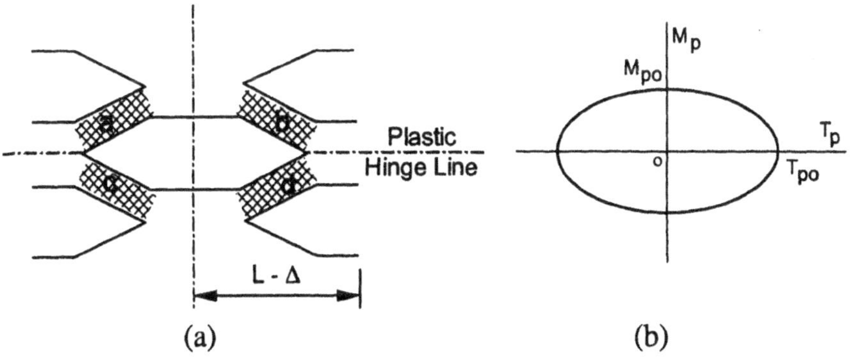

Figure 9. (a) 4-hinge collapse mechanism by plastic torsion in the ligaments in the vicinity of plastic hinge line, (b) interaction diagram representing safe combinations of torque and bending moment.

If the ligament between two overlapping slots is under pure shear, then $T_p = T_{po}$, the plastic moment in pure shear of a rectangular cross-section ($b \times t$). From the well-known "sand hill" analogy, a lower bound expression for T_{po} can be obtained from (Calladine, 1969):

$$T_{po} = \frac{1}{2}\tau_y \left(bt^2 - \frac{t^3}{3} \right) \tag{14}$$

where τ_y is the yield stress in pure shear. Assuming that the material follows a Tresca yield criterion, the pure-shear yield stress is given by $\tau_y = \sigma_y/2$.

The ligament, however, is under combined bending (due to axi-symmetric deformation) and torsion, which tends to promote the onset of yield and reduce the plastic shear moment. The following relationship accounts for this interaction between torsion and bending of the ligament (Calladine, 1969):

$$\left(\frac{T_p}{T_{po}} \right)^2 + \left(\frac{M}{M_{po}} \right)^2 = 1$$

where M is the bending moment induced at the ends of a given ligament due to axi-symmetric deformation and M_{po} is the ligament's full plastic moment used earlier in (3). Now, M/M_{po} can be taken as being proportional to p/p_{clo}, the ratio between the applied pressure, p, and the pressure at the onset of slot closure as defined by (6). We thus have:

$$T_p = T_{po}\sqrt{1-(p/p_{clo})^2} \tag{15}$$

From (13), (14) and (15) we obtain the following expression for plastic moment per unit length of tube at each plastic hinge location:

$$M_{pl} = \frac{2b}{L-\Delta}\sigma_y\left(t^2 - \frac{t^3}{3b}\right)\sqrt{1-(p/p_{clo})^2} \tag{16}$$

Substituting the above equation into (12) and solving the resulting quadratic equation for p, we obtain an expression for the plastic collapse curve.

5.3 Comparison with Numerical Solutions

The capability of the model to predict collapse loads was benchmarked against detailed 3-D finite element simulations for the three tube geometries set out in Table 1. Figure 10 shows a typical comparison for the expanded *5½* inch slotted tube with an initial out-of-roundness of amplitude w_o=*0.028* inches. The finite-element representation of the expanded tube and its displacement under applied pressure are shown in Fig. 11.

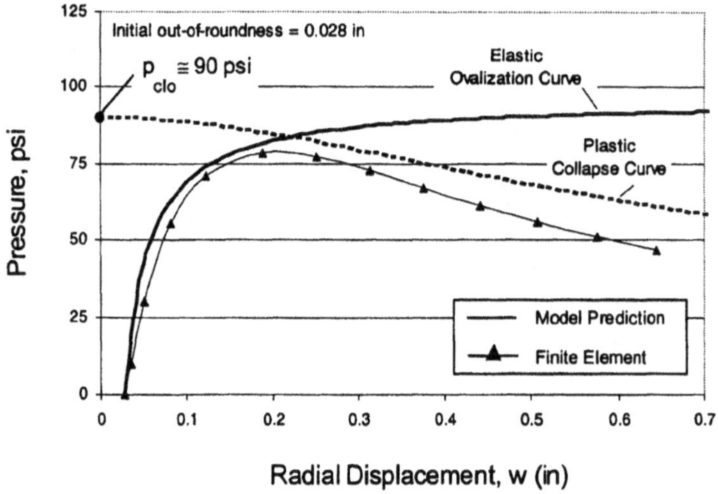

Figure 10. Model prediction compared with finite-element formulation.

In all the cases considered, model prediction for collapse load was within *10%* of the finite element solution. In particular, the elastic ovalization curve was accurately predicted by the model. In this example, the presence of the slots reduces the elastic ovalization stiffness by over *95%* of the intact un-slotted tube.

There is, however, a discrepancy in the post-collapse regime between the model prediction and results from finite-element simulation, which worsens at higher displacements. One reason for this stiffer response could be attributed to the deformation model assumed by the 4-hinge collapse mechanism namely, formation of localized plastic hinge along four longitudinal hinge lines in an otherwise rigid tube.

It is also noted that within the context of 4-hinge plastic collapse mechanism, the post-collapse behavior is independent of initial out-of-roundness. Therefore, it is expected that the model will progressively overestimate collapse at higher values of initial out-of-roundness.

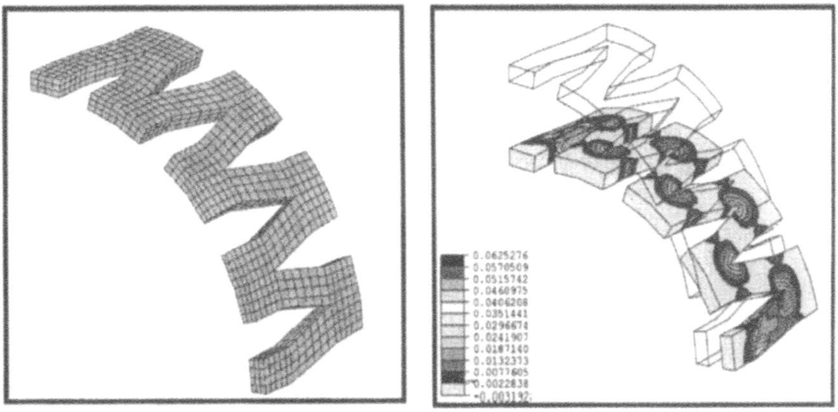

Figure 11. 3-D Finite-element discretisation and typical displacement mode.

As a result of the torque-bending moment interaction (15), the plastic-collapse curve in Fig. 10 intersects the pressure axis at p_{clo} ($\cong$ *90* psi) defined by (6). This in turn implies that the model prediction for collapse pressure p_c (defined as the interaction between the elastic-ovalization and plastic-collapse curves) will be always less than p_{clo}, and approaching p_{clo} as initial out-of-roundness reduces. Indeed, numerous finite-element simulations performed on the three slotted tube configurations in Table 1 with various magnitudes of initial out-of-roundness, support this assertion.

5.4 Influence of Tube Out-of-Roundness

Figure 12 highlights the sensitivity of *2⅞* , *4* and *5½* inch expanded slotted tubes to initial out-of-roundness w_o = cos2θ. Note that at w_o=*0*, the collapse pressure p_c = p_{clo}. It is seen that larger-size slotted tubes are more sensitive to out-of-roundness imperfections which is expected due to their relatively larger diameter/thickness ratio. Typical out-of-roundness imperfections left in the tube after the expansion process are expected to be in the range of $w_o/D_i=0.01$ to 0.04. From Fig. 12, this would result in a degradation in collapse pressure of *15-35%* for *5½* inch, *5-20%* for *4* inch, and less than *5%* for *2⅞* inch tube.

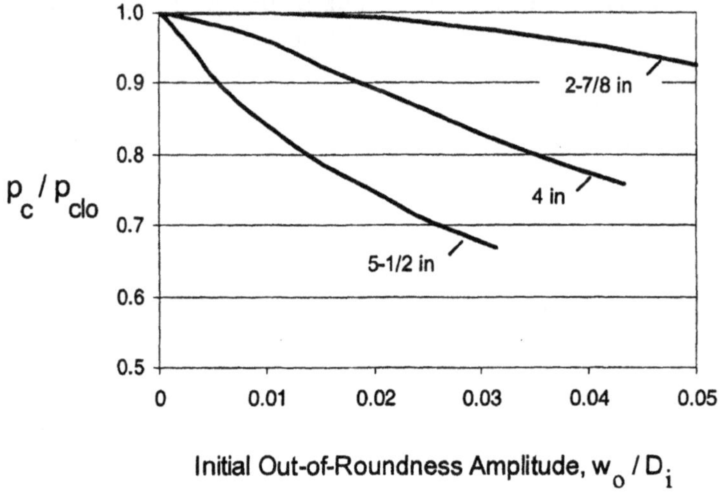

Figure 12. Imperfection sensitivity.

6. CLOSURE BEFORE COLLAPSE

Using the simple idealized mechanical models described in this paper, the full behavior of an expanded slotted liner under uniform external pressure can be investigated. In general, three behavioral regimes exist, namely: axi-symmetric deformation leading to slot closure; elastic ovalization; and post-collapse plastic deformation idealized by a 4-hinge collapse mechanism.

In the case of little or no initial out-of-roundness, the tube is expected to undergo axi-symmetric deformation with little ovalization. Slot closure would take place as the applied pressure exceeds p_{clo}, given by (6). After the slots fully close up (i.e. when load p_{clf} is reached), the tube stiffness against uniform radial deformation increases markedly. In practice, as pressure increases further, the inherent out-of-roundness imperfections in the tube will promote tube ovalization leading to its eventual collapse.

With the presence of any out-of-roundness in the tube, ovalization will be a preferred mode of deformation, during which little or no slot closure occurs. This ovalization leads to the eventual tube failure by collapse through excessive plastic deformation, triggered by twist in the ligaments separating the slots. In this situation, collapse occurs with little or no slot closure.

In practice, slot closure prior to tube collapse would be a precursor to in-situ failure through a drop in production rate allowing time to re-enter the well for any remedial operation; e.g. strengthening the section by some kind of cladding. Collapse before slot closure, on the other hand, would result in permanent interruption in production terminating the life of the well. In stiff formations, where the tube is confined by the surrounding formation, slot closure prior to collapse is likely as the tube is expected to be insensitive to any inherent out-of-roundness. In weak uncemented formations, where the tube is laterally less constrained and hence "free" to ovalize, tube collapse would be accompanied with little or no slot closure.

From a design standpoint, in these situations we would want to install tubes with higher collapse pressure to compensate for the degradation in collapse resistance due to any out-of-roundness. As the pressure at the onset of axi-symmetric slot closure, p_{clo}, provides the limiting collapse pressure, (6) indicates that fewer slots around the circumference (i.e. smaller b), heavier wall thickness, higher yield material, shorter slot length and greater slot overlap, should be amongst the design considerations.

7. CONCLUDING REMARKS

The analysis methodology presented in this paper has shown that expandable slotted tubes are inherently weak when subjected to external pressure, as shown in Figs 6 and 10. One might question then, how advisable is it to install such a weak component in oil wells several kilometers deep, where in-situ stresses are obviously higher than the low collapse pressures predicted. The reason these components are robust enough for oil-field use arises from the nature of the loading mechanism. The non-linear interaction of the rock with the slotted tubular, results in the *near wellbore* loading on the tube being significantly lower than the far-field in-situ stresses.

The two sources of loading are the hydraulic pressure imposed during production and the formation loading gradually exerted on the tube by the surrounding formation over the production life of the well. Under normal operating conditions, the expanded tube with its outer sand retention screen (shroud) is not subjected to significant hydraulic pressure, as sufficient in-flow area exists to make the hydraulic pressure-drop across the tube negligible. If the screen were to become fully plugged for any reason – such as by the build-up of agglomerated fine particles unable to pass through the mesh filter – then the screen would most likely collapse as the applied fluid pressure is normally much greater than the relatively low collapse pressures predicted. However, plugging is unlikely as measures are normally taken during screen design to prevent such an event.

Therefore, the only source of loading on tube is that imposed by the surrounding formation. By implication, these formations are typically weak and possess a low stiffness (otherwise sand retention would not be required). Excavation of the borehole during the drilling process creates a yielded zone in the near-wellbore region whose properties are, if anything, weaker and more compliant than the undisturbed far-field material. Within this yielded ('failed') zone, the material has insufficient strength to sustain large differential stresses (Bradford and Durban, 1998). The stress relaxation resulting from the presence of the yielded zone surrounding the screen results in low contact pressures being developed, which permit the screen to retain its expanded diameter once the well is put onto production.

As production takes place and reservoir pore fluid pressure drops, the effective stress in the formation increases, causing an increase in pressure loading on the tube and consequently a closure of the borehole. The tube responds to this increase by deforming, thereby reducing the magnitude of the formation-induced loading, which is subsequently built up as production continues. This non-linear coupling between the formation loading and tube displacement results in a rather slow progressive collapse over a period of time. For this reason, these expandable systems have proven successful in field application. The operating limits of these systems are still being proven, as these sand control screens become deployed in geomechanically more hostile hydrocarbon reservoirs.

Table 1. Parameters for the expandable slotted liners.

	2⅞" => 5"	*4" => 6"*	*5½" => 8½"*
D_o = Tube diameter before expansion (in)	*2.88*	*4*	*5.56*
D_e = Tube diameter after expansion (in)	*4.88*	*6*	*8.23*
t = Wall thickness (in)	*0.276*	*0.226*	*0.258*
L = Slot Length (in)	*4.33*	*4.33*	*4.33*
N = Slots around the circumference	*10*	*12*	*16*
Δ = Slot Overlap (in)	*1.38*	*1.38*	*1.38*
σ_y = Material yield stress (psi)	*40,600*	*40,600*	*40,600*
E = Young's Modulus (psi)	30×10^6	30×10^6	30×10^6
ν = Poisson's ratio	*0.3*	*0.3*	*0.3*

SI Metric Conversion Factor

1 in. = *2.54* cm
1 lbf = *4.45* N
1 psi = 6.89×10^{-3} N/mm²

Acknowledgements

The authors would like to thank BP for permission to publish this paper, and Steve Parfitt of BP for his valuable comments.

References

Abbassian, F., and Parfitt, S.H.L. (1998) A simple model for collapse and post collapse behavior of tubulars with application to perforated and slotted liners, *SPE Drilling & Completions*, 190-196.

Al Suwaidi, J., Berger, P., and Holmes, D.R. (2000) Coronary artery stents, *Journal of American Medical Association* **284**, No. 14, 1828-1836.

Bradford, I.D.R., and Durban, D. (1998) Stress and deformation patterns around a cylindrical cavity embedded in a pressure sensitive elasto-plastic medium, *Journal of Applied Mechanics* **65**, 374-379.

Calladine, C.R. (1969) *Engineering Plasticity*, First Edition, Pergamon Press, Oxford.

Timoshenko, S.P, and Goodier, J.N. (1970) *Theory of Elasticity*, Third Edition, McGraw-Hill.

Timoshenko, S.P, and Gere, J.M. (1961) *Theory of Elastic Stability*, Second Edition, McGraw-Hill.

Willson, S. Crook, T., Yu, J-G. Stenebråten, J., Gilchrist, J., and Tiffin, D. (2002) Assuring the mechanical integrity of expandable sand screens, paper OTC 14314, presented at the 2002 Offshore Technology Conference, Houston, TX, 6–9 May.

LAWS OF EVOLUTION OF CRACKS IN PLATES IN TERMS OF STRESS RESULTANTS

G. Augusti, P.M. Mariano and F.L. Stazi
Dipartimento di Ingegneria Strutturale e Geotecnica
Università di Roma "La Sapienza", via Eudossiana 18, I-00184 Roma Italy
gaugusti@uniroma1.it

Abstract The laws of evolution of a crack that starts at the edge of a plate and progresses toward its interior are derived in terms of stress and moment resultants written in the reference configuration of the plate. A modified form of the J-integral in terms of stress and moment resultants is also derived.

Keywords: Crack evolution, Plate structures

1. INTRODUCTION

In the present paper, the equations that govern the evolution of cracks in plates are presented in a concise way in terms of stress and moment resultants. We treat the plate as a Cosserat surface, following Ericksen and Truesdell (1958) and Naghdi (1972), and consider the framework of configurational forces discussed in Gurtin and Podio-Guidugli (1996) and also in Maugin (1992; 1994), and make use of results for cracks evolving in continua with substructure, obtained in earlier papers (see Mariano, 2001).

For the sake of simplicity of presentation, this paper is limited to flat plates of constant thickness: however, the treatment could be easily extended to more general cases of shells.

In Section 2 the standard kinematics of plates and cracks is summarized: cracks are assumed to originate at the edge of the plate and progress into it, without cutting the plate completely. Sections 3 and 4 present the local and global balance equations of both the standard stress and moment resultants and of the configurational forces generated by the evolution of the crack. Finally, in Section 5 these configurational

H.R. Drew and S. Pellegrino (eds.),
New Approaches to Structural Mechanics, Shells and Biological Structures, 43–56.

forces are expressed in terms of stress and moment resultants by the use of a dissipation inequality; the explicit expression of the J-integral in terms of stress and moment resultants is derived, and the conditions for its path-independence in the quasi static case are stated.

2. KINEMATICS

2.1 Standard Kinematics of Plates

We shall deal with finite plates defined as follows. Let π be a plane in the 3-D Euclidean point space $\mathcal{E}^3$ and Ω a regular subset of π bounded by a piecewise smooth curve $\partial\Omega$. Points of Ω are indicated with $\mathbf{X}$. A unit vector $\tilde{\mathbf{e}}$ indicates the direction orthogonal to π. We identify the *reference configuration* of the plate with the set, Fig. 1a,

$$\mathcal{B} = \left\{ \bar{\mathbf{X}} = \mathbf{X} + \xi\tilde{\mathbf{e}} \mid \mathbf{X} \in \Omega, \xi \in \left[-\frac{h}{2}, \frac{h}{2} \right] \right\} \tag{1}$$

where the constant h is the thickness of the plate, $h << diam\Omega$, where $diam\Omega$ is the 'diameter' of Ω. The *edge* of the plate is the set

$$\left\{ \bar{\mathbf{X}} = \mathbf{X} + \xi\mathbf{e} \mid \mathbf{X} \in \partial\Omega, \xi \in \left[-\frac{h}{2}, \frac{h}{2} \right] \right\}$$

defined in Fig.1b. At each point $\mathbf{X}$ of Ω, the vector $\tilde{\mathbf{e}}$ describes the material cross-fiber of the plate.

We consider standard finite deformations that bend the plate and rotate each cross-section, without forming self-contact points. We then consider two mappings defining the current configuration of the deformed plate:

- a continuous and continuously differentiable one-to-one mapping $\hat{\mathbf{x}}(\cdot)$ from Ω into $\mathcal{E}^3$ $\left(\hat{\mathbf{x}}(\cdot) : \Omega \to \mathcal{E}^3\right)$ such that $\hat{\mathbf{x}}(\Omega)$ is a smooth surface;

- a continuous and continuously differentiable mapping $\hat{\mathbf{t}}(\cdot) : \Omega \to S^2$, where S^2 is the unit sphere in $\mathbb{R}^3$ defined by $S^2 = \{\mathbf{t} \in \mathbb{R}^3 | \ \ \|\mathbf{t}\| = 1$

The *current configuration* of the deformed plate is given by

$$\left\{ \bar{\mathbf{x}}(\mathbf{X}) = \mathbf{x}(\mathbf{X}) + \xi\mathbf{t}(\mathbf{X}) \mid \mathbf{X} \in \Omega, \xi \in \left[-\frac{h}{2}, \frac{h}{2} \right] \right\} \tag{2}$$

as indicated in Fig. 2. The value $\mathbf{x}(\mathbf{X})$ provides the current placement of $\mathbf{X}$. The actual representation of $\mathbf{t}(\mathbf{X})$, namely a vector $\bar{\mathbf{t}}(\mathbf{x})$, describes the material cross-fiber in the current configuration. The vector

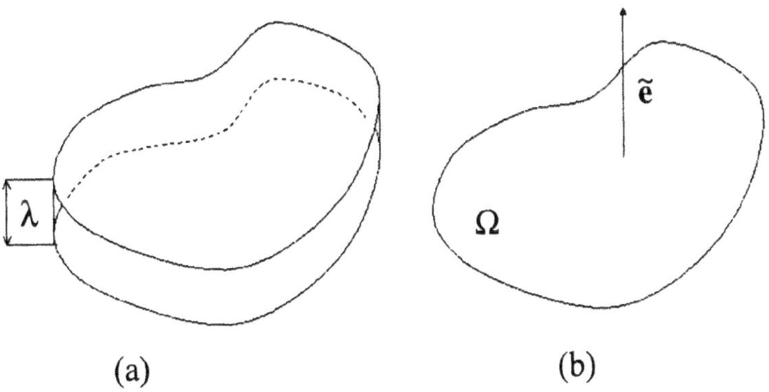

Figure 1. (a) Plate in 3D space; (b) plate as a Cosserat surface.

$\bar{\mathbf{t}}(\mathbf{x})$ is not necessarily normal to $\hat{\mathbf{x}}(\Omega)$ at $\mathbf{x}(\mathbf{X})$; namely, when $\bar{\mathbf{t}}(\mathbf{x})$ is not normal to $\hat{\mathbf{x}}(\Omega)$ at $\mathbf{x}(\mathbf{X})$, shear deformation has taken place. For this reason, in the reference configuration, at any given point $\mathbf{X} \in \Omega$ the vector $\mathbf{t}(\mathbf{X})$ may or may not coincide with $\tilde{\mathbf{e}}$ because it is the material representation of the descriptor $\bar{\mathbf{t}}(\mathbf{x})$ of the material fiber in the current configuration. We call $\mathbf{t}$ a *director* and take $\mathbf{x}$ and $\mathbf{t}$ as kinematical descriptors of the plate, following in this way the standard *director approach* (Ericksen and Truesdell, 1958). Basically, we consider the plate as a Cosserat surface, Fig. 2.

For a given coordinate system in Ω, the point $\mathbf{X}$ has coordinates X_1 and X_2. We assume that

$$(\partial_{X_1}\mathbf{x}\times\partial_{X_2}\mathbf{x})\cdot\mathbf{t} > 0 \tag{3}$$

In Equation (3) and in the following, $\partial_y c$ indicates the derivative of any function c with respect to the argument y. With the relation (3) we impose that $\mathbf{t}$ cannot be tangent to $\mathbf{x}(\mathcal{B})$ to avoid the unreasonable physical situation of infinite shearing deformations (Simo and Fox, 1989).

We also indicate with ∇_Ω the gradient operator calculated with respect to $\mathbf{X}$ and with Div_Ω its trace, i.e. the divergence.

Motions are time parametrized mappings $\mathbf{x}_t$ and $\mathbf{t}_t$, where t denotes time, $t \in \left[0, \bar{t}\right]$. The current placement at the instant t of a material element positioned at $\mathbf{X}$ in the reference configuration and the corresponding director are given by $\mathbf{x}(\mathbf{X}, t)$ and $\mathbf{t}(\mathbf{X}, t)$, respectively. Motions

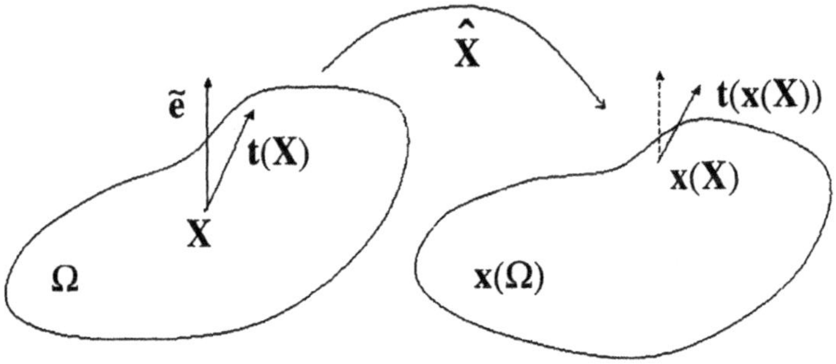

Figure 2. Deformation of the plate.

are assumed to be twice differentiable with respect to time. Material descriptions of the velocity fields are indicated with $\dot{\mathbf{x}}(\mathbf{X}, t)$ and $\dot{\mathbf{t}}(\mathbf{X}, t)$.

2.2 Kinematics of Cracks

We consider in our treatment only cracks that cut completely the thickness of the plate in the current configuration but do not cross Ω, i.e. do not separate the plate into two different pieces.

When a crack occurs in the current configuration, the map $\hat{\mathbf{x}}(\cdot)$ is no more one-to-one in all points $\mathbf{X} \in \Omega$: rather it is one-to-one in $\Omega \backslash C$, i.e. in all points $\mathbf{X} \in \Omega$ except the points of a curve $C \subseteq \Omega$. This curve is non-material and is the image of the crack in the reference configuration. C can be represented by a smooth function $\mathbf{r}(s)$ parametrized by the arc-length $s \in [0, \bar{s}]$; $\bar{s}$ is the length of C. The intersection of C with the boundary $\partial\Omega$ is assumed as the point $\mathbf{r}(0)$, while $\mathbf{r}(\bar{s}) = \mathbf{Z}$ is the *tip* of the crack. At each $s \in (0, \bar{s})$, the tangent to the curve is indicated with $\mathrm{t} = \partial_s \mathbf{r}$; the normal to C in Ω is indicated with $\mathbf{m}$, Fig. 3.

We assume the existence of the limit $\mathrm{t}(s) \to \mathrm{t}_{\mathbf{Z}}$ as $s \to \bar{s}$.

Let $\lambda(\mathbf{X})$ be any continuous field on $\Omega \backslash C$ and ε a real parameter. The limits $\lambda^{\pm}(\mathbf{X}) = \lim_{\varepsilon \to 0;\ \mathbf{X} \in C} \lambda(\mathbf{X} \pm \varepsilon \mathbf{m})$, if they exist, define the *jump* and the *mean value* of λ across C respectively as

$$[\lambda] = \lambda^{+} - \lambda^{-} \quad ; \quad \langle \lambda \rangle = \frac{1}{2}\left(\lambda^{+} + \lambda^{-}\right) \tag{4}$$

For any pair of fields λ_1 and λ_2 with the same properties of λ, the jump of the product is given by $[\lambda_1\lambda_2] = \langle\lambda_1\rangle[\lambda_2] + [\lambda_1]\langle\lambda_2\rangle$, when the product makes sense. If the crack is closed in the current configuration, the physical requirement of impenetrability of the margins of the crack implies

$$[\mathbf{x}] \cdot \mathbf{m} = 0 \quad ; \quad [\mathfrak{t}] = 0 \tag{5}$$

The second condition in Equation (5) derives from the fact that the plate has a non-vanishing thickness.

For the sake of simplicity we may reduce to the case of closed cracks without a loss of generality in the kind of results obtained. In fact, we are basically interested in the evolution of the tip of the crack, not in the behavior of the edges of the crack.

In general, in the current configuration the crack grows; hence its non-material picture in the reference configuration C depends on time: $C(t)$, with $C(t_1) \subseteq C(t_2)$ for any pair of instants $t_1 \leq t_2$. We study the evolution of the crack in a time interval in which the crack itself does not cut the plate in two different pieces. The tip varies in time; we write $\mathbf{Z}(t)$ and define the *intrinsic tip velocity* $\mathbf{v}_{tip}$ as

$$\mathbf{v}_{tip} = \frac{d\mathbf{Z}(t)}{dt} \tag{6}$$

We assume that $\mathbf{v}_{tip} = V\mathbf{t}_{\mathbf{Z}}$ and call $\mathbf{t}_{\mathbf{Z}}$ *direction of propagation of the crack.*

For any continuous field $\lambda(\mathbf{X}, t)$ in $\Omega \backslash C(t) \times [0, \bar{t}]$, if there exists any function of time $\hat{\lambda}(\mathbf{Z}, t)$ such that $\lambda(\mathbf{X}, t) \to \hat{\lambda}(\mathbf{Z}, t)$ as $\mathbf{X} \to \mathbf{Z}(t)$ uniformly in time, we say that λ has uniform limit at the tip and write $\lambda(\mathbf{Z}, t)$ instead of $\hat{\lambda}(\mathbf{Z}, t)$. So, we may define a *rate of the director at the tip* $\mathbf{w}_{tip} = \dot{\mathfrak{t}}(\mathbf{Z}, t)$. Rates *following the crack tip* are given by

$$\mathbf{x}^{\diamond} = \dot{\mathbf{x}} + (\nabla_{\Omega}\mathbf{x})\,\mathbf{v}_{tip} \quad ; \quad \mathfrak{t}^{\diamond} = \dot{\mathfrak{t}} + (\nabla_{\Omega}\mathfrak{t})\,\mathbf{v}_{tip} \tag{7}$$

The fields defined in (7) are velocities as seen by an observer placed at the tip of the crack. These velocities also have uniform limits at the tip, indicated with $\tilde{\mathbf{v}}_{tip}$ and $\tilde{\mathbf{w}}_{tip}$ respectively.

We call *part* any regular subset $\mathfrak{b}$ of Ω (with geometrical properties analogous to Ω), Fig. 3. When we consider any part $\mathfrak{b}(t)$ varying in time, the points of its boundary $\partial\mathfrak{b}(t)$, a closed curve in Ω, are identified with $\mathbf{X}(\mathfrak{s}, t)$, $\mathfrak{s}$ being the arc-length parametrization of $\partial\mathfrak{b}(t)$. The velocity of any point $\mathbf{X} \in \partial\mathfrak{b}(t)$ is given by $\mathbf{u} = \frac{d\mathbf{X}(\mathfrak{s},t)}{dt}$; indicating by $\mathbf{n}$ the outward normal to $\partial\mathfrak{b}(t)$, only the normal component $U = \mathbf{u} \cdot \mathbf{n}$ is independent of the parametrization $\mathfrak{s}$. We may then define velocity fields as seen by observers sitting on $\partial\mathfrak{b}(t)$ as

$$\mathbf{x}^{\circ} = \dot{\mathbf{x}} + (\nabla_{\Omega}\mathbf{x})\,\mathbf{u} \quad ; \quad \mathfrak{t}^{\circ} = \dot{\mathfrak{t}} + (\nabla_{\Omega}\mathfrak{t})\,\mathbf{u} \tag{8}$$

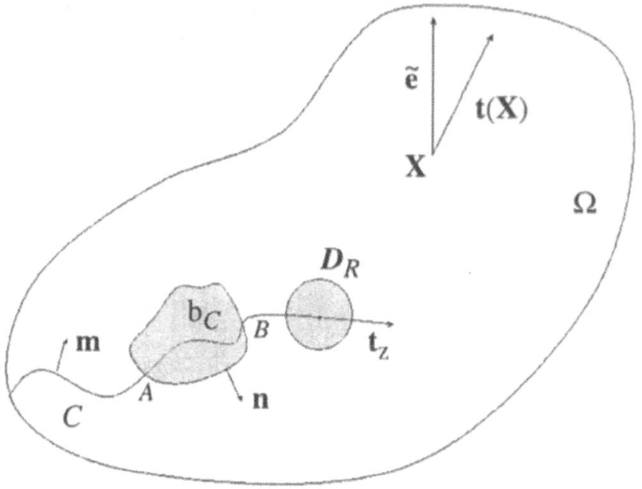

Figure 3. The crack in the plate.

Finally we need to recall two integral formulas which will be used in the following. For any scalar smooth field $a(\mathbf{X}, t)$ on $\Omega \backslash C(t) \times [0, \bar{t}]$ and any disc $D_R \subseteq \Omega$ centered at the tip $\mathbf{Z}$ and of radius R, we have

$$\frac{d}{dt}\int_{D_R(t)} a = \int_{D_R(t)} \dot{a} + \int_{\partial D_R(t)} aU + \int_{tip} aV \tag{9}$$

$$\int_{D_R(t)} \nabla a = \int_{\partial D_R(t)} a\mathbf{n} - \int_{C\cap D_R(t)} [a]\,\mathbf{m} - \int_{tip} a\mathbf{n} \tag{10}$$

To obtain these formulas, we take another disc D_r in Ω, centered at the tip $\mathbf{Z}$ and of radius $r < R$, write standard transport theorems on $D_R \backslash D_r$ and shrink D_r to $\mathbf{Z}$ by letting $r \to 0$. We put $\int_{tip}(\cdot) = \lim_{r\to 0}\int_{D_r}(\cdot)$. The other integrals in (10) and (11) are calculated on different parts of $D_R \backslash D_r$ letting $r \to 0$ and assuming that all the integrands are integrable uniformly in time and also the limits exist uniformly in time.

3. BALANCE OF STRESS AND MOMENT RESULTANTS

Consider any part of the plate in the current configuration, the *contact interactions* of that part with the rest of the plate and the environment are represented through standard stress and moment resultants;

while *bulk interactions* are measured through standard bulk forces and moments. We prefer to write the balance equations in the reference configuration.

Given any part $\mathfrak{b} \subseteq \Omega$, we indicate with $\mathbf{N}$ and $\mathbf{M}$ the *stress resultant* and the *moment resultant* in the reference configuration along the boundary $\partial\mathfrak{b}$, for the explicit definition of $\mathbf{N}$ and $\mathbf{M}$ in terms of averages along the thickness of the plate involving the Piola-Kirchhoff stress, see e.g. Naghdi (1972), Simo and Fox (1989), Simo et al. (1988). Basically, $\mathbf{N}$ and $\mathbf{M}$ are the counterparts in the reference configuration of the stress and moment resultants $\mathbf{N}_c$ and $\mathbf{M}_c$ in the current configuration. We know that for each $\mathbf{X} \in \partial\mathfrak{b}$ we have $\mathbf{N} = \hat{\mathbf{N}}(\mathbf{X}, t; \mathbf{n})$ and $\mathbf{M} = \hat{\mathbf{M}}(\mathbf{X}, t; \mathbf{n})$ and also that stress and moment resultants are linear in $\mathbf{n}$, $\mathbf{n}$ being the outward normal to $\partial\mathfrak{b}$; so that linear operators (second order tensors) $\mathbb{N}$ and $\mathbb{M}$ such that $\mathbf{N} = \mathbb{N}\mathbf{n}$ and $\mathbf{M} = \mathbb{M}\mathbf{n}$ exist. We indicate with $\bar{\rho}$, $\mathbf{f}$ and *sll* the density of mass, the body forces and the body couples per unit area, averaged along the thickness of the plate.

The above interactions satisfy the following integral balance equations:

$$\int_{\mathfrak{b}} \bar{\rho}\mathbf{f} + \int_{\partial\mathfrak{b}} \mathbb{N}\mathbf{n} = \mathbf{0} \tag{11}$$

$$\int_{\mathfrak{b}} \left((\mathbf{x} - \mathbf{x}_0) \times \bar{\rho}\mathbf{f} + \mathbf{t} \times l\right) + \int_{\partial\mathfrak{b}} \left((\mathbf{x} - \mathbf{x}_0) \times \mathbb{N}\mathbf{n} + \mathbf{t} \times \mathbb{M}\mathbf{n}\right) = \mathbf{0} \tag{12}$$

which can be derived by requiring the invariance of the external power under classical rules of changes of observers. The arbitrariness of $\mathfrak{b}$ and the application of Gauss theorem imply from (11) the local balance

$$\bar{\rho}\mathbf{f} + Div_{\Omega}\mathbb{N} = \mathbf{0} \tag{13}$$

The same procedure applied to (12) and the validity of (13) lead to

$$\mathsf{e}\mathbb{N}\nabla_{\Omega}\mathbf{x}^T = \mathbf{t} \times l + \mathbf{t} \times Div_{\Omega}\mathbb{M} + \mathsf{e}\nabla_{\Omega}\mathbf{t}^T\mathbb{M} \tag{14}$$

where e is the alternating indicator (Ricci's tensor). The validity of the pointwise balance of moments requires the existence of a vector field *slm* with the meaning of a moment per unit area (named *director moment*) such that

$$\mathbf{t} \times m = \mathsf{e}\left(\mathbb{N}\nabla_{\Omega}\mathbf{x}^T - \nabla_{\Omega}\mathbf{t}^T\mathbb{M}\right) \tag{15}$$

By substituting (15) in (14) we obtain

$$l + Div_{\Omega}\mathbb{M} - m = \mathbf{0} \tag{16}$$

Equations (13), (15) and (16) coincide with equations (9.77)-(9.79) in Naghdi (1972).

We assume that the bulk terms $\mathbf{f}$ and l are the sum of inertial (in) and non-inertial (ni) components, namely $\mathbf{f} = \mathbf{f}^{ni}+\mathbf{f}^{in}$ and $l = l^{ni} + l^{in}$, and identify $\mathbf{f}^{in} = -\ddot{\mathbf{x}}$ and $l^{in} = -\mathcal{I}\ddot{\mathbf{t}}$, with $\mathcal{I}$ the inertia coefficient of the cross section per unit length. Consequently, equations (13) and (16) reduce respectively to

$$\bar{\rho}\mathbf{f}^{ni}+Div_{\Omega}\mathbb{N}=\bar{\rho}\ddot{\mathbf{x}} \quad ; \quad l^{ni} + Div_{\Omega}\mathbb{M} - m = \mathcal{I}\ddot{\mathbf{t}} \tag{17}$$

To obtain balance equations at the sides of the crack we consider a part $\mathfrak{b}_C$ such that $\partial\mathfrak{b}_C$ cuts C away from the tip in two points A and B (Fig. 3). Then we write the integral balances (11) and (12) on $\mathfrak{b}_C$ and shrink it to the crack. As $\mathfrak{b}_C \to \mathfrak{b}_C \cap C$, the integrals on $\mathfrak{b}_C$ tend to 0, while the integrals on $\partial\mathfrak{b}_C$ tend to the integral on $\mathfrak{b}_C \cap C$ of the jump of the integrands on $\mathfrak{b}_C \cap C$. We obtain then the integral balances

$$\int_{\mathfrak{b}_C \cap C} [\mathbb{N}]\,\mathbf{m} = \mathbf{0} \quad ; \quad \int_{\mathfrak{b}_C \cap C} \mathbf{t}\times [\mathbb{M}]\,\mathbf{m} = \mathbf{0} \tag{18}$$

The arbitrariness of $\mathfrak{b}_C$ implies the pointwise balances

$$[\mathbb{N}]\,\mathbf{m} = \mathbf{0} \quad ; \quad [\mathbb{M}]\,\mathbf{m} = m_C \tag{19}$$

with m_C a vector such that $\mathbf{t}\times m_C = 0$.

To derive balance equations at the tip $\mathbf{Z}$ we write the balance equations on a disc D_R centered in $\mathbf{Z}$ (and such that C intersects the boundary $\partial D_R(t)$ in one point for any t; Fig. 3), adding two terms of inertial nature, $\mathbf{f}_{tip}$ and $\mathbf{t}\times l_{tip}$, to (11) and (12) respectively. Then we shrink the disc to the tip by letting $R \to 0$, uniformly in time and obtain

$$\mathbf{f}_{tip}+\int_{tip} \mathbb{N}\mathbf{n} = \mathbf{0} \quad ; \quad l_{tip} + \int_{tip} \mathbb{M}\mathbf{n} = m_{tip} \tag{20}$$

with m_{tip} a vector such that $\mathbf{t}\times m_{tip} = 0$.

To identify explicitly $\mathbf{f}_{tip}$ and l_{tip}, we write two inertial balances on the previously considered tip disc $D_R(t)$. These balances prescribe that the sum of the inertial forces is equal to the rate of change of momentum in $D_R(t)$, namely

$$\mathbf{f}_{tip} + \int_{D_R(t)} \bar{\rho}\mathbf{f} = -\frac{d}{dt}\int_{D_R(t)} \bar{\rho}\dot{\mathbf{x}} + \int_{\partial D_R(t)} \bar{\rho}\dot{\mathbf{x}}U \tag{21}$$

$$l_{tip} + \int_{D_R(t)} \bar{\rho}\mathbf{f} = -\frac{d}{dt}\int_{D_R(t)} \mathcal{I}\dot{\mathbf{t}} + \int_{\partial D_R(t)} \mathcal{I}\dot{\mathbf{t}}U \tag{22}$$

As $R \to 0$ uniformly in time, we obtain $\mathbf{f}_{tip} = \int_{tip} \bar{\rho}\dot{\mathbf{x}}\,(\mathbf{v}_{tip}\cdot\mathbf{n})$ and $l_{tip} = \int_{tip} \mathcal{I}\dot{\mathbf{t}}\,(\mathbf{v}_{tip}\cdot\mathbf{n})$. Consequently, equation (20) reduces to

$$\int_{tip} \mathbb{N}\mathbf{n} = \mathbf{v}_{tip} \int_{tip} \mathbf{n}\otimes\bar{\rho}\dot{\mathbf{x}} \quad ; \quad \int_{tip} \mathbb{M}\mathbf{n} - m_{tip} = \mathbf{v}_{tip} \int_{tip} \mathbf{n}\otimes\mathcal{I}\dot{\mathbf{t}} \tag{23}$$

The observation of the real behavior of cracks in plates suggests that stresses and moments are bounded up to the tip, i.e. $\int_{\partial D_R(t)} |\mathbb{N}\mathbf{n}|$ and $\int_{\partial D_R(t)} |\mathbb{M}\mathbf{n}|$ are bounded as $R \to 0$ uniformly in time. This implies that

$$\int_{tip} \mathbb{N}\mathbf{n} = \mathbf{0} \quad ; \quad \int_{tip} \mathbb{M}\mathbf{n} = \mathbf{0} \tag{24}$$

Moreover, in the real word the director - representing the cross section of the plate - does not rotate with infinite speed at the tip: hence we may also conjecture that $\int_{\partial D_R(t)} \mathbf{n}\otimes\mathcal{I}\dot{\mathbf{t}}$ is also bounded as $R \to 0$, which implies

$$\int_{tip} \mathbf{n}\otimes\mathcal{I}\dot{\mathbf{t}} = \mathbf{0} \tag{25}$$

As a consequence

$$\int_{tip} \mathbf{n}\otimes\bar{\rho}\dot{\mathbf{x}} = \mathbf{0} \quad ; \quad m_{tip} = 0 \tag{26}$$

4. FORCES DURING THE CRACK GROWTH

In this treatment, the physical growth of the crack is represented by the evolution of C in the reference configuration. This defines a kinematic path in the reference configuration, which is usually fixed in space and time, in absence of crack propagation. New interactions must be associated to the kinematics generated in the reference configuration by the crack growth, as dual objects of the peculiar velocities. These interactions (usually called *configurational*) develop power in the additional kinematics, but they are apparent because the kinematics of C is non material. Hence they must be expressed in terms of standard stress and moment resultants because $\mathbf{N}_c$ and $\mathbf{M}_c$ are the sources of the actual crack growth in the current configuration. Configurational interactions can be measured through a bulk stress $\mathbb{P}$ (a second order tensor associating vectors to covectors in Ω), internal and external bulk forces $\mathbf{g}$ and $\mathbf{e}^*$ (vectors in Ω), a line stress $\sigma\mathbf{t}$ along C (with σ a constant because C grows but any piece of it far from the tip does not increase its area), an internal force $\mathbf{g}_C$ along C, internal and external tip forces $\mathbf{g}_{tip}$ and $\mathbf{e}^*_{tip}$ respectively (vectors). In particular, we assume that $\mathbf{e}^*_{tip}$ is only of inertial nature. These interactions satisfy the following balance equations:

$$Div_{\Omega}\mathbb{P}+\mathbf{g}+\mathbf{e}^{*}=\mathbf{0} \quad \text{in } \Omega \tag{27}$$

$$[\mathbb{P}]\,\mathbf{m}+\mathbf{g}_{C}+\sigma\mathfrak{h}=\mathbf{0} \quad \text{along } C \tag{28}$$

$$\mathbf{g}_{tip}+\mathbf{e}_{tip}^{*}-\sigma_{tip}\mathbf{t}\mathbf{z}+\int_{tip}\mathbb{P}\mathbf{n}=\mathbf{0} \tag{29}$$

These balance equations can be derived in different ways, see e.g. Gurtin and Podio-Guidugli (1996) or Gurtin (2000).

In the following section, with the aid of a mechanical dissipation inequality, we shall identify the expressions of the forces due to crack growth in terms of stress and moment resultants and of the free energy of the plate. Before this, let us derive an explicit expression for $\mathbf{e}_{tip}^{*}$. Consider a time dependent tip disc $D_R(t)$; equate the rate of the kinetic energy in it with the power of all inertial forces, including $\mathbf{e}_{tip}^{*}$, and let $R \to 0$. Taking into account the results at the end of previous section, namely that $\int_{tip}\mathbf{n}\otimes\mathcal{I}\dot{\mathbf{t}}=\mathbf{0}$ and $\int_{tip}\mathbf{n}\otimes\bar{\rho}\dot{\mathbf{x}}=\mathbf{0}$, we thus obtain

$$\mathbf{e}_{tip}^{*}=\int_{tip}\frac{1}{2}\left(\bar{\rho}\dot{\mathbf{x}}\cdot\dot{\mathbf{x}}+\mathcal{I}\dot{\mathbf{t}}\cdot\dot{\mathbf{t}}\right)\mathbf{n} \tag{30}$$

5. DISSIPATION INEQUALITY AND J-INTEGRAL

A mechanical dissipation inequality states that for any choice of the velocity fields the rate of the free energy of a given evolving part of Ω less the power of all interactions is smaller than or equal to zero. Here we consider first a part $\mathfrak{b}(t)$ away from the crack and write

$$\frac{d}{dt}\int_{\mathfrak{b}(t)}\psi-\int_{\mathfrak{b}(t)}\left(\bar{\rho}\mathbf{f}\cdot\dot{\mathbf{x}}+l\cdot\dot{\mathbf{t}}\right)-\int_{\partial\mathfrak{b}(t)}\left(\mathbb{N}\mathbf{n}\cdot\mathbf{x}^{\circ}+\mathbb{M}\mathbf{n}\cdot\mathbf{t}^{\circ}+\mathbb{P}\mathbf{n}\cdot\mathbf{u}\right)\le 0 \tag{31}$$

where ψ is the free energy density per unit area, averaged along the thickness. Note that the forces $\mathbf{g}$ and $\mathbf{e}$ do not develop power because they are internal forces. By substituting (8) into (31) and developing the time derivative of the integral, we obtain

$$\int_{\mathfrak{b}(t)}\dot{\psi}+\int_{\partial\mathfrak{b}(t)}\psi U-\int_{\mathfrak{b}(t)}\left(\mathbf{b}\cdot\dot{\mathbf{x}}+l\cdot\dot{\mathbf{t}}\right)-$$

$$-\int_{\partial\mathfrak{b}(t)}\left(\mathbb{N}\mathbf{n}\cdot\dot{\mathbf{x}}+\mathbb{M}\mathbf{n}\cdot\dot{\mathbf{t}}+\left(\mathbb{P}+\nabla_{\Omega}\mathbf{x}^{T}\mathbb{N}+\nabla_{\Omega}\mathbf{t}^{T}\mathbb{M}\right)\mathbf{n}\cdot\mathbf{u}\right)\le 0 \tag{32}$$

Since only the normal component of $\mathbf{u}$ is independent of the parametrization of $\partial\mathfrak{b}(t)$, which is not physical, the vector $\left(\mathbb{P} - \nabla_\Omega \mathbf{x}^T \mathbb{N} - \nabla_\Omega \mathbf{t}^T \mathbb{M}\right)\mathbf{n}$ must be normal to $\partial\mathfrak{b}$. This reasoning, firstly proposed in Gurtin (1995), allows us to write $\mathbb{P} - \nabla_\Omega \mathbf{x}^T \mathbb{N} - \nabla_\Omega \mathbf{t}^T \mathbb{M} = \omega \mathbf{I}$, with ω an undetermined scalar. When we introduce ω in (27), we obtain a term of the type $(\psi - \omega) U$. Moreover, since (32) must hold for any choice of the velocity fields, and $\mathfrak{b}(t)$ is arbitrary, we must have $\psi = \omega$, a result that allows us to express explicitly the stress $\mathbb{P}$ in terms of the other quantities as

$$\mathbb{P} = \psi \mathbf{I} - \nabla_\Omega \mathbf{x}^T \mathbb{N} - \nabla_\Omega \mathbf{t}^T \mathbb{M} \tag{33}$$

Equation (33) is a special form of the extended Eshelby tensor derived in Mariano (2000; 2001) within the setting of multifield theories, whose general framework includes the director description of structural two-dimensional elements and can be found in (Capriz, 1985; 1989).

At this point, if we consider an expression of the free energy density of the type $\psi = \hat{\psi}(\mathbf{X}, \nabla_\Omega \mathbf{x}, \mathbf{t}, \nabla_\Omega \mathbf{t})$, substitute it into (32) and take into account (33) and the balances (13) and (16), we obtain an inequality which is linear in the rates $\nabla_\Omega \dot{\mathbf{x}}$, $\dot{\mathbf{t}}$, $\nabla_\Omega \dot{\mathbf{t}}$, and implies, due to the arbitrariness of $\mathfrak{b}(t)$ and of the rates, the standard relations

$$\mathbb{N} = \partial_{\nabla_\Omega \mathbf{x}} \psi \quad ; \quad \mathbb{M} = \partial_{\nabla_\Omega \mathbf{t}} \psi \quad ; \quad m = \partial_{\mathbf{t}} \psi \tag{34}$$

By inserting (33) and (34) in (27) and considering the balances (13) and (16), we are able to identify the forces $\mathbf{g}$ and $\mathbf{e}$ as

$$\mathbf{g} = -\partial_{\mathbf{X}} \psi \quad ; \quad \mathbf{e}^* = -\bar{\rho} \nabla_\Omega \mathbf{x}^T \mathbf{f} - \nabla_\Omega \mathbf{t}^T l \tag{35}$$

Notice that for an homogeneous plate $\mathbf{g}$ vanishes identically.

To identify the meaning of the line stress $\sigma \mathbf{t}$, we take a part $\mathfrak{b}_C(t)$ crossing the crack away from the tip, Fig. 3, and write for it a relation analogous to (31) in which we include the contribution of $\sigma \mathbf{t}$ and substitute $\int_{\mathfrak{b}(t)} \psi$ with $\int_{\mathfrak{b}_C(t)} \psi + \int_{\mathfrak{b}_C(t) \cap C} \phi$.

ϕ is a constant line energy density along the margins of C, continuous up to the tip. As a final step, we shrink $\mathfrak{b}_C(t)$ to $\mathfrak{b}_C(t) \cap C$ uniformly in time and obtain

$$(\phi - \sigma)(\mathbf{t} \cdot \mathbf{u}) \Big|_{A(t)}^{B(t)} - \int_{\mathfrak{b}_C(t) \cap C} \left([\mathbb{N}\mathbf{n} \cdot \dot{\mathbf{x}}] + [\mathbb{M}\mathbf{n} \cdot \dot{\mathbf{t}}] \right) \leq 0 \tag{36}$$

where $A(t)$ and $B(t)$ are the intersections of $\partial\mathfrak{b}_C(t)$ with C. Since the mechanical dissipation inequality must hold for any choice of the velocity fields, we must have

$$\phi = \sigma \tag{37}$$

Finally, to obtain information on the tip force $\mathbf{g}_{tip}$ we write the mechanical dissipation inequality with reference to a tip disc $D_R(t)$, taking into account $\mathbf{g}_{tip}$. As $R \to 0$, we obtain the reduced dissipation inequality $\mathbf{g}_{tip} \cdot \mathbf{v}_{tip} \leq 0$. As a consequence, recalling that $\mathbf{t_Z}$ has been assumed as the direction of propagation in Ω of the crack, $\mathbf{v}_{tip} = V\mathbf{t_Z}$, whence

$$(\mathbf{g}_{tip} \cdot \mathbf{t_Z})\, V \leq 0$$

Putting $g_{tip} = \mathbf{g}_{tip} \cdot \mathbf{t_Z}$, a solution of the previous inequality is

$$g_{tip} = \tilde{g}_{tip} V \tag{38}$$

with $\tilde{g}_{tip}$ some negative function of the state variables

$$\tilde{g}_{tip} = \tilde{g}^*_{tip}\left(\nabla_\Omega \mathbf{x}, \mathbf{t}, \nabla_\Omega \mathbf{t}; V\right) \tag{39}$$

By taking into account (30), (33) and (37), we may write the tip balance (29) in the following form:

$$-\phi_{tip}\mathbf{t_Z} + \int_{tip} \left(\psi\mathbf{I} + \frac{1}{2}\left(\bar{\rho}\dot{\mathbf{x}} \cdot \dot{\mathbf{x}} + \mathcal{I}\dot{\mathbf{t}} \cdot \dot{\mathbf{t}}\right) - \nabla_\Omega \mathbf{x}^T \mathbb{N} - \nabla_\Omega \mathbf{t}^T \mathbb{M}\right)\mathbf{n} = -\mathbf{g}_{tip} \tag{40}$$

The term

$$\mathsf{j} = \int_{tip} \left(\psi\mathbf{I} + \frac{1}{2}\left(\bar{\rho}\dot{\mathbf{x}} \cdot \dot{\mathbf{x}} + \mathcal{I}\dot{\mathbf{t}} \cdot \dot{\mathbf{t}}\right) - \nabla_\Omega \mathbf{x}^T \mathbb{N} - \nabla_\Omega \mathbf{t}^T \mathbb{M}\right)\mathbf{n} \tag{41}$$

is the *tip traction* in Ω averaged on the thickness of the shell and exerted on the tip by the surrounding material. The component of (41) along the direction of propagation of the crack is given by

$$-\phi_{tip} + \mathbf{t_Z} \cdot \mathsf{j} = -\mathbf{t_Z} \cdot \mathbf{g}_{tip} \tag{42}$$

If we indicate with $\mathsf{J} = \mathbf{t_Z} \cdot \mathsf{j}$ the *dynamic* J*-integral*, we interpret the difference

$$f = \mathsf{J} - \phi_{tip} \tag{43}$$

as the *force driving the tip of the crack.* By comparison with (39), we obtain that

$$fV \geq 0 \tag{44}$$

i.e. when the crack grows ($V > 0$), the inequality (45) implies that the driving force f must be non negative.

Previous results allow us to write the *balance of energy at the tip* in the form

$$\int_{tip} \left(\left(\psi + \frac{1}{2}\left(\bar{\rho}\dot{\mathbf{x}} \cdot \dot{\mathbf{x}} + \mathcal{I}\dot{\mathbf{t}} \cdot \dot{\mathbf{t}}\right)\right) V + \mathbb{N}\mathbf{n} \cdot \dot{\mathbf{x}} + \mathbb{M}\mathbf{n} \cdot \dot{\mathbf{t}}\right) - \phi_{tip} V = fV \tag{45}$$

When the inertia effects are negligible, the evolution of the crack is quasi static and the J-integral reduces to

$$\mathsf{J}_{q-static} = \mathbf{t_Z} \cdot \int_{tip} \left(\psi \mathbf{I} - \nabla_\Omega \mathbf{x}^T \mathbb{N} - \nabla_\Omega \mathbf{t}^T \mathbb{M}\right) \mathbf{n} = \mathbf{t_Z} \cdot \int_{tip} \mathbb{P}\mathbf{n} \tag{46}$$

The quasi static J-integral for plates, written in terms of stress and moment resultants, is *path-independent* when the material is homogeneous, the sides of the crack are straight and free of stresses and moments (i.e. $[\mathbb{N}]\,\mathbf{m} = \mathbf{0}$ and $[\mathbb{M}]\,\mathbf{m} = \mathbf{0}$) and $\left(\nabla_\Omega \mathbf{t}^T\right) m_C = \gamma \mathbf{m}$. The sketch of the proof of this statement is as follows. First we notice that in absence of inertia effects and for homogeneous materials, the balance (27) reduces to $Div\mathbb{P} = 0$. Then, taken any tip disc we must evaluate the difference

$$\mathbf{t_Z} \cdot \left(\int_{\partial D_R} \mathbb{P}\mathbf{n} - \int_{tip} \mathbb{P}\mathbf{n} \right) \tag{47}$$

The use of the divergence theorem (10) and the hypotheses $[\mathbb{N}]\,\mathbf{m} = \mathbf{0}$, $[\mathbb{M}]\,\mathbf{m} = \mathbf{0}$ and $\left(\nabla_\Omega \mathbf{t}^T\right) m_C = \gamma \mathbf{m}$ allows us to obtain

$$\mathbf{t_Z} \cdot \left(\int_{\partial D_R} \mathbb{P}\mathbf{n} - \int_{tip} \mathbb{P}\mathbf{n} \right) = 0 \tag{48}$$

which implies obviously the path-independence of $\mathsf{J}_{q-static}$.

6. CONCLUSIONS

Basic laws of crack propagation in plates undergoing large deformations have been deduced in terms of stress and moment resultants along the thickness of the plate. The results are computationally useful because they allow us to write directly algorithms in terms of stress and moment resultants, obtaining thus a reduction of the computational cost.

Moreover, the procedure used here opens the way to analyze more complicated situations occurring in plates when, for example, one considers deformable thickness and/or (standard or non-local) plastic deformations.

In the present paper, the whole treatment has been developed with respect to the reference configuration of the plate. Details on the derivation of formulae and examples of application will be presented in a follow-up paper (Mariano et al., 2002).

Acknowledgments

Financial support in the form of a research grant awarded by the Università "La Sapienza" (Ricerca di Ateneo, 2000) is gratefully acknowledged. P.M.M. acknowledges also the support of the Italian National Group of Mathematical Physics (GNFM).

References

Capriz, G. (1985) Continua with latent microstructure, *Arch. Rational Mech. Anal.* **90**, 43-56.

Capriz, G. (1989) *Continua with Microstructure*, Springer-Verlag, Berlin.

Ericksen, J. L. and Truesdell, C. A. (1958) Exact theory of stress and strain in rods and shells, *Arch. Rational Mech. Anal.* **1**, 295-323.

Gurtin, M. E. (1995) The nature of configurational forces, *Arch. Rational Mech. Anal.* **131**, 67-100.

Gurtin, M. E. (2000) *Configurational Forces as Basic Concepts in Continuum Physics*, Springer-Verlag, Berlin.

Gurtin, M. E. and Podio-Guidugli, P. (1996) Configurational forces and the basic laws for crack propagation, *J. Mech. Phys. Solids* **44**, 133-154.

Mariano, P. M. (2000) Configurational forces in continua with microstructure, *Z. angew. Math. Phys.* ZAMP **51**, 752-791.

Mariano, P. M. (2001) Multifield theories in mechanics of solids, *Adv. Appl. Mech.*, **38**, 1-93.

Mariano, P. M., Augusti, G. and Stazi, F. L. (2002) Cracks in plates: the driving force in terms of stress and moment resultants, *in preparation.*

Maugin, G. A. (1992) *The Thermomechanics of Plasticity and Fracture*, Cambridge University Press, Cambridge.

Maugin, G. A. (1994) On the J-integral and energy-release rates in dynamical fracture. *Acta Mech.* **105**, 33–47.

Naghdi, P. M. (1972) The theory of shells. In *Handbuk der Physics*, VIa/2 (C. Truesdell ed.), 425-640, Springer-Verlag, Berlin.

Simo, J. C. and Fox, D. D. (1989) On a stress resultant geometrically exact shell model. Part I: Formulation and optimal parametrization, *Comp. Meth. Appl. Mech. Engng* **72**, 267-304.

Simo, J. C., Marsden, J. E. and Krishnaprasad, P. S. (1988) The Hamiltonian structure of nonlinear elasticity: The material and convective representation of solids, rods and plates, *Arch. Rational Mech. Anal.* **104**, 125-183.

NESTING SURFACES AND CONSTITUTIVE LAWS FOR THE INELASTIC BEHAVIOUR OF COMPOSITE MATERIALS

Alan C.F. Cocks
Department of Engineering, University of Leicester, Leicester LE1 7RH, UK
acfc1@le.ac.uk

Abstract This paper examines the structure of constitutive laws for composite materials making use of the concepts of nesting surfaces and reference stress methods originally developed for the study of creeping structures. The relationship between different bounding approaches is examined. It is demonstrated how a suitable reference stress method can be employed to interpolate between perfectly plastic and linear viscous material responses.

Keywords: Constitutive Laws, Bounding Theorems

1. INTRODUCTION

The general field of applied mechanics is driven by the problems of the day. In the 1960's one of the major driving forces was the development of design strategies for power-generating plant operating at high temperatures. Today, mechanics is being used at a smaller scale to aid the development of constitutive laws for a wide range of material systems. In this paper we examine how concepts developed in the 1960's can be used to provide new insights into the structure of constitutive laws for rate-dependent composite systems. In particular we examine how the concept of nesting surfaces developed by Calladine and Drucker (1962a, b) and the reference stress approach developed for creeping structures can be used to guide the development of constitutive laws. We compare the resulting structure and analysis to those presented by Ponte Castaneda (1991) and Suquet (1994).

H.R. Drew and S. Pellegrino (eds.),
New Approaches to Structural Mechanics, Shells and Biological Structures, 57–66.

In the following section we present the basic equations and inequalities which underpin the analyses presented in subsequent sections. In Section 3 we derive expressions for the rate of energy dissipation in creeping structures and review the "nesting surface" analysis of Calladine and Drucker (1962a, b) and the reference stress method of Leckie (1971). In Section 4 we treat a composite material as a structure and extend these theorems to provide bounds and approximations of stress and strain-rate potentials which describe the deformation response of the composite. This approach is used to obtain constitutive laws for porous materials in Section 5.

2. FUNDAMENTAL EQUATIONS

In this paper we consider a power-law creeping material, whose constitutive response can be expressed in the form:

$$\dot{\varepsilon}_{ij} = \frac{\partial \phi}{\partial \sigma_{ij}}; \ \sigma_{ij} = \frac{\partial \psi}{\partial \dot{\varepsilon}_{ij}} \tag{1}$$

where ϕ and ψ are homogeneous convex strain-rate and stress potentials, which we express in the form

$$\phi = \frac{1}{n+1}\dot{\varepsilon}_o \sigma_o \left(\frac{\sigma_e}{\sigma_o}\right)^{n+1} \quad \text{and} \quad \psi = \frac{n}{n+1}\dot{\varepsilon}_o \sigma_o \left(\frac{\dot{\varepsilon}_e}{\dot{\varepsilon}_o}\right)^{\frac{n+1}{n}} \tag{2}$$

where σ_e, is an effective measure of stress, such as the von Mises effective stress and $\dot{\varepsilon}_e$ is the conjugate effective strain-rate; $\dot{\varepsilon}_o$ is the strain-rate in a uniaxial test at a stress σ_o and n is the creep exponent. If $\dot{d}$ is the energy dissipation rate per unit volume, then

$$\dot{d} = \phi + \psi \tag{3}$$

The results presented in this paper are based on two fundamental relationships. If f is a convex function of α_i, then (see Cocks, 1994, for example)

$$\left(\alpha_i^1 - \alpha_i^2\right)\frac{\partial f^1}{\partial \alpha_i} - \left(f^1 - f^2\right) \geq 0 \tag{4}$$

where the superscripts 1 and 2 refer to quantities determined at states 1 and 2 respectively. The second fundamental relationship is Hardy, Littlewood and Polya's (1934) general properties of means, which is a special case of Holder's inequality. Consider the situation where a quantity g is defined over a domain Ω and

$$\int_{\Omega} q d\Omega = 1$$

then, if $0<r<s$,

$$\int_{\Omega} q g^{r} d\Omega \leq \left(\int_{\Omega} q g^{s} d\Omega \right)^{\frac{r}{s}} \tag{5}$$

3. NESTING SURFACES AND REFERENCE STRESS METHODS

In this section we use the equations and inequalities presented in the previous section to develop the nesting surface results of Calladine and Drucker (1962a, b). We then use these results to explore the reference stress method of Leckie (1971). Consider the situation where two identical structures of volume V are manufactured from power-law creeping materials with creep exponents, n_1 and n_2, where $n_2>n_1$. The two structures experience loads which produce the same normalised energy dissipation rate $\dot{D}/\dot{\varepsilon}_o\sigma_o V = 1$. For simplicity we represent these as a series of loads F_i and let $\dot{u}_i$ represent the displacement rates in the directions of the loads. Applying the convexity condition of (4) to the strain-rate potential for material 1, identifying stress state 1 with the exact field and stress state 2 with the exact field obtained for structure 2 and integrating over the volume gives

$$\int_{V} \left(\sigma_{ij}^{1} - \sigma_{ij}^{2}\right)\dot{\varepsilon}_{ij} dV \geq \frac{1}{n+1}\dot{\varepsilon}_o\sigma_o V\left[\int_{V} V^{-1}\left(\frac{\sigma_e^1}{\sigma_o}\right)^{n_1+1} dV - \int_{V} V^{-1}\left(\frac{\sigma_e^2}{\sigma_o}\right)^{n_1+1} dV\right]$$

Using the principle of virtual work for the terms to the left of the inequality and inequality (5) for the last term gives

$$\left(F_i^1 - F_i^2\right)\dot{u}_i^1 \geq \frac{1}{n+1}\dot{\varepsilon}_o\sigma_o V\left[\int_V V^{-1}\left(\frac{\sigma_e^1}{\sigma_o}\right)^{n_1+1} dV - \left(\int_V V^{-1}\left(\frac{\sigma_e^2}{\sigma_o}\right)^{n_2+1} dV\right)^{\frac{n_1+1}{n_2+1}}\right]$$

$$= \frac{1}{n+1}\dot{\varepsilon}_o\sigma_o V\left[\frac{\dot{D}^1}{\dot{\varepsilon}_o\sigma_o V} - \left(\frac{\dot{D}^2}{\dot{\varepsilon}_o\sigma_o V}\right)^{\frac{n_1+1}{n_2+1}}\right] = 0 \qquad (6)$$

This inequality simply states that the surfaces of constant energy dissipation rate in force space (or F-space) are convex, with the strain-rate normal to the surface. Any set of forces which lie on the surface for n_2 lies inside the surface for n_1, i.e. the surface of constant energy dissipation rate for n_2 nests inside that for n_1. Since the values of n_1 and n_2 are arbitrary, this implies that surfaces of constant energy surfaces nest inside each other in order of decreasing n, with the surface for $n \to \infty$ forming the innermost surface and that for $n=1$ forming the outermost sur face. A similar analysis can be conducted starting with the stress potential, ψ, of (2). We now find

$$\left(\dot{u}_i^2 - \dot{u}_i^1\right)F_i^2 \geq 0 \qquad (7)$$

i.e., surfaces of constant energy dissipation rate in displacement rate space (or U-space) are convex. The force vector is normal to the surface and the surfaces again nest inside of each other, with the surface for $n \to \infty$ now forming the outer surface.

For many practical situations F-surfaces for different values of n lie close to each other. If they are co-incident with each other then

$$\dot{D} = \dot{\varepsilon}_o\sigma_o V = \dot{\varepsilon}_R\sigma_R V$$

Since $\dot{\varepsilon}_o$ and σ_o are arbitrary, for a given applied loading these quantities can be chosen such that the above relationship is satisfied. The resulting reference stress is denoted as σ_R and the reference strain-rate $\dot{\varepsilon}_R$ can be determined from a reference uniaxial test at this stress. The reference stress is independent of n and can be determined from the solution for $n \to \infty$ (Sim, 1968; Leckie, 1971; Leckie and Martin, 1967). It is simply the value of the yield stress which is required to make the limit load surface pass through the loading point in stress space.

If the surfaces of constant energy dissipation rate do not exactly coincide with each other, dimensional arguments require that the energy dissipation rate should be written in the form (Leckie, 1971)

$$\dot{D} = \dot{\varepsilon}_o \sigma_o V_R = \dot{\varepsilon}_R \sigma_R V_R \tag{8}$$

where V_R is a reference volume. If the reference stress is determined as before, the reference volume can be determined by matching (8) to the solution for n=1. The energy dissipation rate predicted by (8) then provides the exact solution for n=1 and $n \rightarrow \infty$.

4. CONSTITUTIVE LAWS FOR COMPOSITES

In this section we apply the concepts described in the previous two sections to composite systems. We consider the situation where the composite consists of two phases, a power-law creeping phase and either a vacuous or rigid phase, which occupies a volume fraction f of the body. We describe the loading on a macroscopic element of volume V in terms of macroscopic stresses Σ_{ij} and the resulting strain-rate as $\dot{E}_{ij}$. For a power-law creeping material represented by (1) and (2) these are related to each other through an analogous set of equations

$$\dot{E}_{ij} = \frac{\partial \Phi}{\partial \Sigma_{ij}} \; ; \Sigma_{ij} = \frac{\partial \Psi}{\partial \dot{E}_{ij}} \tag{9}$$

where now

$$\Phi = (1-f) \int_{1-f} \phi dv \; ; \; \Psi = (1-f) \int_{1-f} \psi dv \tag{10}$$

$$\dot{D} = \Phi + \Psi \tag{11}$$

and $dv = dV / V_m$. We can now follow the procedures of the previous section to determine the relationship between the solutions for different values of n. Rather than express the response in terms of surfaces of constant energy dissipation rate it is more instructive to express the response in terms of surfaces of constant Φ in Σ_{ij} space (equivalent to F-surfaces) or surfaces of constant Ψ in $\dot{E}_{ij}$ space (equivalent to U-surfaces). Alternatively, following Cocks (1989), Ponte Castaneda (1991) and Suquet (1993) we can express these potentials in the form

$$\Phi = \frac{1}{n+1}\dot{\varepsilon}_o\sigma_o(1-f)\left(\frac{\overline{\Sigma}}{\sigma_o}\right)^{n+1} ; \Psi = \frac{n}{n+1}\dot{\varepsilon}_o\sigma_o(1-f)\left(\frac{\overline{\dot{E}}}{\sigma_o}\right)^{\frac{n+1}{n}} \quad (12)$$

where $\overline{\Sigma}$ and $\overline{\dot{E}}$ are scalar measures of stress and strain. Now, we find that surfaces of constant $\overline{\Sigma}$ in stress space nest inside each other with the surface for n=1 forming the outer surface, i.e. for a prescribed macroscopic stress state

$$\overline{\Sigma}^1 \le \overline{\Sigma}^{n_1} \le \overline{\Sigma}^n \le \overline{\Sigma}^\infty \quad (13)$$

where the superscript identifies the value of n at which $\overline{\Sigma}$ is determined. Similarly, surfaces of constant $\overline{\dot{E}}$ nest inside each other in strain-rate space, with the surface for $n \to \infty$ forming the outer surface, i.e. for a given macroscopic strain-rate

$$\overline{\dot{E}}^1 \ge \overline{\dot{E}}^{n_1} \ge \overline{\dot{E}}^{n_2} \ge \overline{\dot{E}}^\infty \quad (14)$$

(13) contains the bound of Ponte Castaneda (1991) who showed that bounds to the macroscopic strain-rate potential can be obtained from the solution of an equivalent elastic material, i.e.

$$\Phi \ge \frac{1}{n+1}(1-f)\dot{\varepsilon}_o\sigma_o\left(\frac{\overline{\Sigma}^1}{\sigma_o}\right)^{n+1}$$

The nesting character of the surfaces of constant $\overline{\Sigma}$ has been demonstrated independently by Michel and Suquet (1992) and Cocks (1994).

Cocks (1989) has demonstrated how bounds to the macroscopic potentials can be obtained by assuming a limited class of displacement rate fields. For any assumed compatible field which produces local strain-rates $\dot{\varepsilon}_e$

$$\Psi \le A\int \dot{\varepsilon}_e^{\frac{n+1}{n}} dv \le A\int \dot{\varepsilon}_e^{1\,\frac{n+1}{n}} dv \le A\left[\int \dot{\varepsilon}_e^{1\,2} dv\right]^{\frac{n+1}{2n}}$$

where $A = \frac{n}{n+1}(1-f)\dot{\varepsilon}_o\sigma_o$ and $\dot{\varepsilon}_e^{\,1}$ is determined from the optimum field within the limited set for n=1. Thus the kinematic bounds satisfy the relationship

$$\bar{\dot{E}}^{n} \leq \bar{\dot{E}}^{n1} \leq \bar{\dot{E}}^{1}$$

where the superscripts n, $n1$ and 1 relate to the optimum bound for a given value of n, the bound obtained using the field from a linear solution and the bound for the linear material respectively. Equivalent results for the macroscopic effective stresses determined from these fields (Cocks, 1989 and 1994) satisfy the relationship

$$\bar{\Sigma}^{n} \geq \bar{\Sigma}^{n1} \geq \bar{\Sigma}^{1}$$

These results can be extended in a number of ways. It is evident, however, that surfaces of constant $\bar{\Sigma}$ and $\dot{E}$ determined from a given class of displacement-rate fields nest inside each other in the same way as the exact surfaces. In practice, the constitutive law for a composite system is determined in this way.

It is evident that if the nesting surfaces lie close to each other, then the linear solution provides a reasonable bound. For situations where the surfaces are not particularly close, a better estimate of the macroscopic potential can be obtained by adopting a procedure equivalent to the reference stress method described in Section 3. Following the analysis presented there we can write the macroscopic strain-rate potential in the form

$$\Phi = \frac{1}{n+1}(1-f)\dot{\varepsilon}_R \sigma_R \nu_R = \frac{1}{n+1}(1-f)\dot{\varepsilon}_o \sigma_o \left(\frac{\sigma_R}{\sigma_o}\right)^{n+1} \nu_R$$

where $(1-f)\nu_R$ is an effective volume fraction of matrix material. We determine the reference stress from the solution for $n \to \infty$, ie $\sigma_R = \bar{\Sigma}^{\infty}$, and ν_R by matching the solution for n=1. Then

$$\Phi \approx \frac{1}{n+1}(1-f)\dot{\varepsilon}_o \sigma_o \left(\frac{\bar{\Sigma}^{\infty}}{\sigma_o}\right)^{n-1} \left(\frac{\bar{\Sigma}^{1}}{\sigma_o}\right)^{2}$$

i.e.

$$\bar{\Sigma} \approx \bar{\Sigma}^{\infty \frac{n-1}{n+1}} \bar{\Sigma}^{1 \frac{2}{n+1}} \qquad (15)$$

This result reduces to those for $n \to \infty$ and n=1 in these two limits.

5. POROUS MATERIALS

In this section we consider the response of a power-law creeping material which contains a volume fraction f of spherical pores. This situation has been examined extensively in the literature. Bounds and approximate results have been obtained assuming a restricted class of velocity fields for the deforming material; see for example Cocks (1989) and Michel and Suquet (1992). It is not possible to obtain closed-form expressions for the full range of values of n, but numerical results are well approximated by Sofronis and McMeeking (1992) and Suquet (1993)

$$\overline{\Sigma} = \left(\alpha_n \Sigma_m{}^2 + \beta_n \Sigma_e{}^2\right)^{\frac{1}{2}} \tag{16}$$

where $\alpha_n = \dfrac{9}{4}\dfrac{1}{(1-f)^{\frac{2}{n+1}}\left(n\left(f^{-\frac{1}{n}}-1\right)\right)^{\frac{2n}{n+1}}}$ and $\beta_n = \dfrac{1+\frac{2}{3}f}{(1-f)^2}$.

In the limit of $n \to \infty$ $\alpha_\infty = \dfrac{9}{4}\dfrac{1}{(\ln f)^2}$. The stresses Σ_e and Σ_m are the von Mises effective and mean stresses respectively.

Surfaces of $\overline{\Sigma} = 1$ predicted by (16) are plotted in Figure 1 for three values of n and two values of the void volume fraction f. The nesting character of these surfaces is clearly demonstrated. The surfaces for low values of f are well spaced in the vicinity of the mean stress axis, but they become closer together as f is increased. The surface for n=3 lies closer to the perfect plasticity limit than the linear limit. The dashed curves on Figure 1 represent the prediction of the reference stress result of (15). This provides a reasonably accurate result and is always much better than the prediction of the linear solution. For this material the exact result lies between the perfectly plastic limit and the approximate result of (15). Clearly, for this material it is most appropriate to use the analytical expressions of (16), but in situations where only results for the linear and perfectly plastic limits are available (15) provides a good approximation, even when the surfaces of constant $\overline{\Sigma}$ are well spaced.

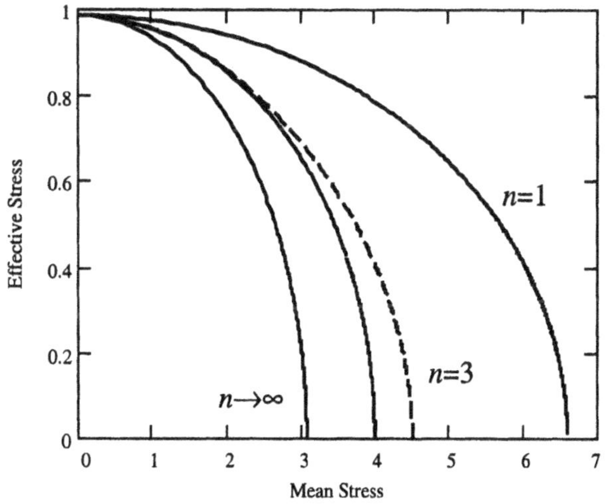

(a)

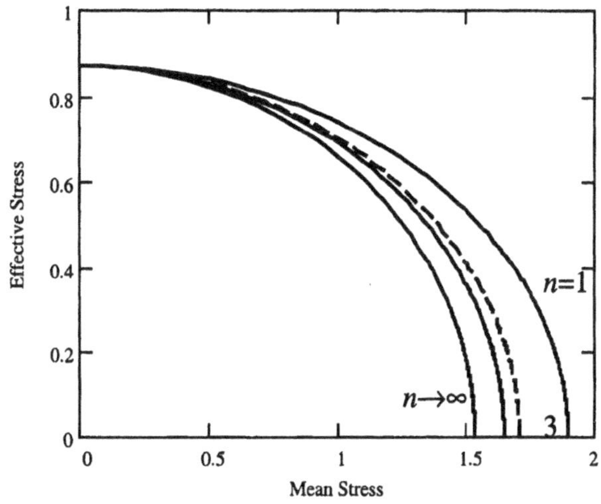

(b)

Figure 1. Surfaces of constant effective stress for (a) f=0.01 and (b) f=0.1. The nesting character of these surfaces is demonstrated. The dashed curve is the prediction of the reference stress approach for n=3.

6. CONCLUDING COMMENTS

In this paper we have demonstrated how the concepts of nesting surfaces and reference stress, which were originally developed to model the creep of structural components, can be used to guide the development of constitutive laws for the time-dependent behaviour of composite systems. The nesting surfaces analogy results in models which are consistent with more recent theorems (see Ponte Castaneda (1991) and Suquet (1993) for example). In general it is much easier to develop models in the limits of linear viscous and perfectly plastic behaviour. The reference stress method provides a consistent approach for interpolating between these two limits.

References

Calladine, C.R. and Drucker, D.C. (1962a) Nesting surfaces of constant rate of energy dissipation in creep, *Quarterly of Applied Mathematics* **20**, 79-84.

Calladine, C.R. and Drucker, D.C. (1962b) A bound method for the creep analysis of structures: direct use of solutions in elasticity and plasticity, *Jnl. Mech. Eng. Sci.* **4**, 1-11.

Cocks, A.C.F. (1989) Inelastic deformation of porous materials, *J. Mech. Phys. Solids* **37**, 693-715.

Cocks, A.C.F. (1994) The structure of constitutive laws for the sintering of fine grained materials, *Acta metall. mater* **42**, 2191-2210.

Hardy, G., Littlewood, J.E. and Polya, G. (1934) Inequalities, Cambridge University Press, Cambridge, 143.

Leckie, F.A. (1971) Some structural theorems of creep and their implications, in *Advances in Creep Design*, Applied Science Publishers, London.

Leckie, F.A. and Martin, J.B. (1967) Deformation bounds for bodies in a state of creep, *Jnl. Appl. Mech.* **34**, 411-417.

Michel, J.C. and Suquet, P. (1992) The constitutive law of nonlinear viscous and porous materials, *J. Mech. Phys. Solids* **40**, 783-812.

Ponte Castaneda, P. (1991) The effective mechanical properties of nonlinear isotropic composites, , *J. Mech. Phys. Solids* **39**, 45-71.

Sim, R.G. (1968) PhD thesis University of Cambridge.

Sofronos, P. and McMeeking, R.M. (1992) *Jnl. Appl. Mech.* **59**, 88.

Suquet, P.M. (1993) Overall potentials and extremal surfaces of power law or ideally plastic composites, *J. Mech. Phys. Solids* **41**, 981-1002.

Williams, J.J. and Cocks, A.C.F. (1979) Reference stress and temperature for non-isothermal creep of structures, *Jnl. Appl. Mech.* **46**, 795-799.

ON THE ROTATION CAPACITY OF REINFORCED CONCRETE STRUCTURAL ELEMENTS

Mario Como
Department of Civil Engineering, University of Rome "Tor Vergata"
Via del Politecnico 1, 00133 – Rome, Italy
como@ing.uniroma2.it

Abstract A simple expression of the rotation capacity is derived for structural reinforced concrete elements. The model, that assumes simple schemes of the constitutive behaviour of the materials, is able to describe the dependence of the average curvature between the cracks on the extension of the yield penetration in the tension steel. The different roles played by all the geometrical and mechanical quantities involved are pointed out and some examples are developed.

Keywords: Ductility, Rotation capacity, Reinforced concrete structures

1. INTRODUCTION

Many studies have been devoted to the evaluation of the plastic rotation capacity of reinforced concrete structural elements starting from the fundamental paper by Bachmann[1]; a wide survey of references on the subject can be found in Ref.[2]. The problem is particularly important and current also because of the property changes of construction materials due to recent technological developments. Improved bond properties, increased strength and weldability of steel reinforcement are positive aspects of these changes; the strong reduction of the strain hardening of the steel is, on the contrary, a less desiderable consequence. This reduction, in fact, results in a localization of strains at the cracks at the bending failure of the members[4].

Many geometrical and mechanical quantities affect the rotation capacity of reinforced concrete elements and it is not yet quite clear which role is played by each quantity. The aim of this paper is to propose a simple model to attempt to point out the main aspects of the problem. By means of simple analytical formulations, the direct dependence of the rotation capacity of

H.R. Drew and S. Pellegrino (eds.),
New Approaches to Structural Mechanics, Shells and Biological Structures, 67–79.

plastic hinges on the extent of the yield penetration into the steel bars will be shown to be the key point that controls the ductility of reinforced concrete elements. The rotation capacity of the member is in fact due to a sequence of small plastic bends distributed along small plastic lengths across the cracks. The analysis assumes simple schemes for the constitutive behaviour of the materials: an elasto-plastic stress strain relation for the steel with constant hardening, a stress block with a uniform strength σ_{co} in the compression concrete zone, a linear elastic distribution of tensile stresses in the concrete between the cracks, and a constant plastic bond shear stress τ_o between the tension steel and the surrounding concrete.

2. ULTIMATE PLASTIC CURVATURES IN SIMPLE FLEXURE

2.1 Stresses and Strains in Neighborhood of Cracked Sections

We focus our attention on the behaviour of a reinforced concrete beam element under the increasing action of a bending moment, constant along its length, Fig. 1(a). Vertical cracks will be present in the member with a more or less regular pattern and a characteristic length λ. Under the increasing bending, the steel tension bars will reach the yielding stress σ_y. Yielding in the steel is first attained at the cracked sections of the beam element where tensile stresses reach their highest values. Between the cracks, because of tension stiffening effect, tractions are gradually transferred from the reinforcement to the concrete by bond shear stresses.

As the loading progresses after the yielding of the tension steel, at the cracked sections the neutral axis will reduce its depth and the concrete will begin to suffer high compressive stresses. Within the concrete surrounding the tension steel bars the limit bond strength τ_0 will be also reached and further slips will develop at constant bond shear stress τ_0. Both steel and compressed concrete will be thus in a plastic state. This condition can effectively be reached under increasing bending if the steel content μ is not larger than the balanced steel content

$$\mu_b = \frac{4}{5}\frac{\sigma_{co}}{\sigma_{sr}}\frac{\varepsilon_{cr}}{\varepsilon_{cr}+\varepsilon_{sr}} + \mu'\frac{\sigma_{sy}}{\sigma_{sr}} \tag{1}$$

This corresponds to the simultaneous tension and compression failure of the critical section. In Equation (1) σ_{sr}, σ_{sy} and ε_{sr} are respectively the tensile strength, the yield stress, the failure strain of the steel and ε_{cr} is the crushing strain of the concrete. At the plastic state of the element the stresses in the tension steel and in the extreme compressed fiber of the concrete, will be located at points P* on the stress strain diagrams of Fig.

1(b) and 1(c).

At the same time in the neighborhood of the cracks the bond shear stresses will be given by point P* of the $\tau - s$ diagram of Fig. 1(d).

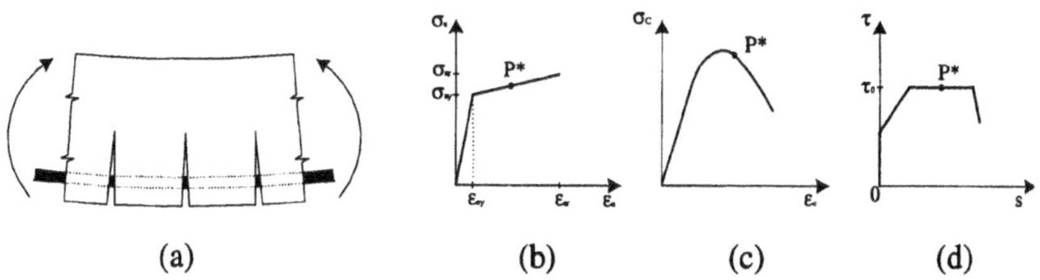

Figure 1. Plastic state in the steel, in the concrete and, near the cracks, between bars and concrete

To describe the state of stress at the critical section we can use, for the sake of simplicity, the well known stress block diagram with the conventional compression strength σ_{co}. The depth x_c of the neutral axis is thus related to the depth a of the compressed zone according to the well known simplifying relation

$$a(0) = \frac{4}{5} x_c(0) \tag{2}$$

where the depth $a(0)$ is given by the equilibrium equation

$$\frac{a(0)}{h} = \mu \frac{\sigma_s(0)}{\sigma_{co}} - \mu' \frac{\sigma_{sy}}{\sigma_{co}} \tag{2a}$$

In Equation (2a) μ and μ' indicate the tension and compression steel contents; it has been also assumed, with good approximation, that the steel in compression has yielded. The critical section is defined by $z = 0$ and all the various quantities with the index *(0)* are referred to the cracked section. The corresponding bending moment, with the ordinary meaning of symbols, is

$$M = \sigma_s(0) A_s h [1 - \frac{a(0)}{2h}] + \sigma_{sy} A_s' [\frac{a(0)}{2} - \delta] \tag{3}$$

To evaluate the stress state between the cracks, the effect of the bond shear stresses acting between the tension bars and the surrounding concrete has to be taken into account. Thus tension steel stress at distance z from the cracked section is

$$\sigma_s(z) = \sigma_s(0) - k \frac{z}{h} \tag{4}$$

where

$$k = \frac{4\tau_o h}{d_s} \tag{5}$$

The plastic depth d_{op}, with $d_{op} \leq \lambda/2$, is defined as the distance $z = d_{op}$ from the cracked section where the stress in the steel reduces to σ_{sy}. Thus

$$d_{op} = \frac{\sigma_s(0) - \sigma_{sy}}{4\tau_o} d_s \tag{6}$$

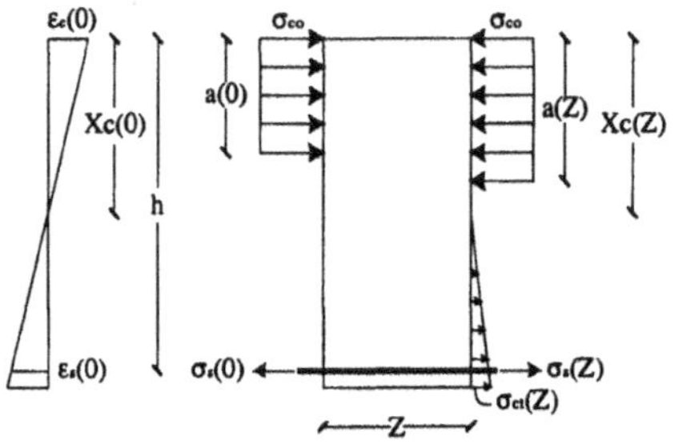

Figure 2. The stress state in the neighborhood of the cracked section

In the neighborhood of the cracked section the concrete is still in a plastic state. Thus at an internal section z the compression stress in the concrete, acting over a zone of unknown depth $a(z)$, is σ_{co}. Tensile stresses in the concrete fibers, on the other hand, will be distributed along the lower zone of the section z because of the tension stiffening effect (Fig. 2). A linear tensile stress distribution is assumed to approximate this distribution of tensile stresses and $\sigma_{ct}(z)$ is the traction occurring on the lowermost concrete fiber. Plastic slips at constant bond shear τ_o will occur between the tension steel bars and the surrounding concrete to correct the different elongations between bars and concrete fibers. The following equilibrium equations relate the unknown quantities $a(z)$ and $\sigma_{ct}(z)$ at section z

$$a(z)b\sigma_{co} + \sigma_{sy}A_s' = [\sigma_s(0) - kz]A_s + b\sigma_{ct}(z)[h - x_c(z)]/2 \tag{7a}$$

$$\begin{gathered}[\sigma_s(0) - kz]A_s[h - a(z)/2] + \sigma_{sy}A_s'[a(z)/2 - \delta] + \\ +b\sigma_{ct}(z)[h - x_c(z)]\{h - [h - x_c(z)/3]/3 - a(z)/2\}/2 = M\end{gathered} \tag{7b}$$

At $z = 0$ these equations reduce to (2a) and (3). Eqs 7(a) and 7(b) represent a non-linear algebraic system whose solution can be obtained in the neighborhood of the cracked section, $z = 0$, by means of the linear approximation of $a(z)$ and $\sigma_{ct}(z)$,

$$\frac{a(z)}{h} \cong \frac{a(0)}{h} + ha'(0)\frac{z}{h} \qquad \sigma_{ct}(z) \cong \sigma_{ct}'(0)z \qquad ()' = d()/dz \tag{8}$$

This approximation is justified by the closeness of z/h to 0. The depth $a(z)$ of the compression zone and the maximum tensile concrete stress $\sigma_{ct}(z)$ are thus given by

$$a(z) = a(0) + \eta(0)\Phi_1 z \qquad \sigma_{ct}(z) = \sigma_s(0)\eta(0)\Phi_2 \frac{z}{h} \tag{9}$$

where

$$\eta(0) = \frac{kh}{\sigma_s(0)} \qquad q_s(0) = \frac{\sigma_s(0)\mu}{\sigma_{co}} - \frac{\sigma_{sy}\mu'}{\sigma_{co}} \tag{10}$$

and $\Phi_1[q_s(0),\mu',\beta_{sc}]$, $\Phi_2[q_s(0),\mu,\mu',\beta_{sc}]$, defined as the first and second parameter functions of the plastic stiffening effect, are

$$\Phi_1 = \frac{q_s(1/3 - 5q_s/12)}{(2/3 - 7q_s/12) + \mu'\beta_{sc}/2} \qquad \Phi_2 = \frac{2\mu(1-q_s) + \mu\mu'\beta_{sc}}{(1-5q_s/4)(2/3 - 7q_s/12 + \mu'\beta_{sc}/2)} \tag{11}$$

In the expressions (11) the index (0) has been omitted and

$$\beta_{sc} = \frac{\sigma_{sy}}{\sigma_{co}} \tag{12}$$

Both functions Φ_1 and Φ_2 increase with μ.

2.2 Distance Between Cracks Near Failure

Additional cracks can form during the plastic bending process of the beam element. The distance λ_o from the initial cracked section to the section where the tensile strength σ_{crt} is once again reached is obtained by the condition

$$\sigma_{ct}(\lambda_o) = \sigma_s(0)\eta(0)\Phi_2 \frac{\lambda_o}{h} = \sigma_{crt} \tag{13}$$

The distance λ_o is thus given by

$$\lambda_o = \frac{ds}{4}\frac{\sigma_{crt}}{\tau_o}\frac{1}{\Phi_2[q_s(0),\mu,\mu',\beta_{sc}]} \tag{14}$$

The function Φ_2 increases with $q_s(0)$, i.e. also with the steel stress $\sigma_s(0)$ at the cracked section. We can evaluate now the limit distance λ_{os}^{r} from $z = 0$ where the tensile concrete strength σ_{crt} is reached *at collapse* of the cracked section. This collapse can occur for the failure of either the steel *($\mu < \mu_b$)* or of the concrete *($\mu > \mu_b$)*. At failure of the cracked section the tension steel stress is thus

$$\sigma_s^r = \begin{matrix} \sigma_{sr}(\mu < \mu_b) \\ \sigma_{src}(\mu > \mu_b) \end{matrix} \tag{15}$$

and the corresponding mechanical steel content becomes

$$q_s^r = \mu \frac{\sigma_s^r}{\sigma_{co}} - \mu' \frac{\sigma_{sy}}{\sigma_{co}} \tag{10'}$$

From (14) thus we get

$$\lambda_{os}^r = \frac{d_s}{4} \frac{\sigma_{crt}}{\tau_o} \frac{1}{\Phi_2[q_s^r(0), \mu, \mu', \beta_{sc}]} \tag{16}$$

where $\Phi_2[q_s^r(0), \mu, \mu', \beta_{sc}]$ is the value taken by the function Φ_2 at q_s^r, i.e. at failure of the element, when the tension steel stress is σ_s^r. Hence, if $\lambda_s^r < \lambda / 2$ additional cracks can develop before collapse.

2.3 Yield Penetration Across Cracks

According to the assumed linear hardening of the steel, sketched in Fig.1b, the steel strain in the plastic zone at distance z from the cracked section is

$$\varepsilon_s(z) = \varepsilon_{sy} + \frac{\sigma_s(z) - \sigma_{sy}}{E_s^*} \tag{17}$$

for $0 \le z \le d_{op}$. In the remaining portion of the bar the strain is still elastic. We can evaluate the propagation depth d_{opr} of plastic strains in the steel bars at failure. We have in fact that *at the steel failure ($\mu < \mu_b$)*, since $\sigma_s(0) = \sigma_s^r = \sigma_{sr}$

$$d_{opr} = d_{oprs} = (\psi_{sr} - 1) \frac{\sigma_{sy}}{4\tau_o} d_s \tag{18}$$

where

$$\psi_{sr} = \frac{\sigma_{sr}}{\sigma_{sy}} \tag{19}$$

is the ratio between the failure and yielding steel stresses of the steel.

At concrete failure ($\mu > \mu_b$), $\sigma_s^r = \sigma_{src}$ and σ_{src} is connected to the strain ε_{src} by means of the constitutive equation

$$\sigma_{src} = \sigma_{sy} + E_s^*(\varepsilon_{src} - \varepsilon_{sy}) \tag{20}$$

corresponding to the assumption of constant hardening. Defining

$$e_{sr} = \frac{\varepsilon_{sr}}{\varepsilon_{sy}} \qquad e_{src} = \frac{\varepsilon_{src}}{\varepsilon_{sy}} \qquad e_{cr} = \frac{\varepsilon_{cr}}{\varepsilon_{sy}} \qquad \overline{q}_{sr} = \mu\frac{\sigma_{sr}}{\sigma_{sco}} \tag{21}$$

the strain ε_{src} is obtained by solving the following algebraic equation

$$e_{src}^2(\psi_{sr} - 1) + e_{src}\{(e_{sr} - \psi_{sr}) + (\psi_{sr} - 1)[e_{cr} - \frac{\mu'}{\mu}(e_{sr} - 1)]\}$$

$$+ \{e_{cr}[(e_{sr} - \psi_{sr}) - \frac{\mu'}{\mu}(e_{sr} - 1)] - 4/5\frac{\psi_{sr}}{\overline{q}_{sr}}(e_{sr} - 1)\} = 0 \tag{22}$$

Thus we have

$$d_{opr} = d_{oprc} = (\psi_{sr}\alpha - 1)\frac{\sigma_{sy}}{4\tau_o}d_s \tag{23}$$

where

$$\alpha = \frac{\sigma_{src}}{\sigma_{sr}} \tag{24}$$

Since $\psi_{sr}\alpha > 1, \alpha < 1, d_{oprc} < d_{opsr}$, at concrete failure, i.e. for $\mu > \mu_b$, the penetration depth of plastic strains is thus lower than at steel failure. In any case, this length is strongly dependent on the hardening ratio ψ_{sr}. The non dimensional plastic penetration depth at failure is

$$\delta_{opr} = \frac{d_{opr}}{h} = \frac{\psi_s^r - 1}{\beta_s^r \psi_s^r} \tag{25}$$

where, according to (18) and (23),

$$d_{opr} = \begin{matrix} d_{oprs}(\mu < \mu_b) \\ d_{oprc}(\mu > \mu_b) \end{matrix} \tag{26}$$

2.4 Local and Average Plastic Curvatures Between Cracks

The strains in the concrete tension fibers, that slip with respect to the steel bars and are cut by cracks, cannot represent the curvature of the reinforced concrete beam.

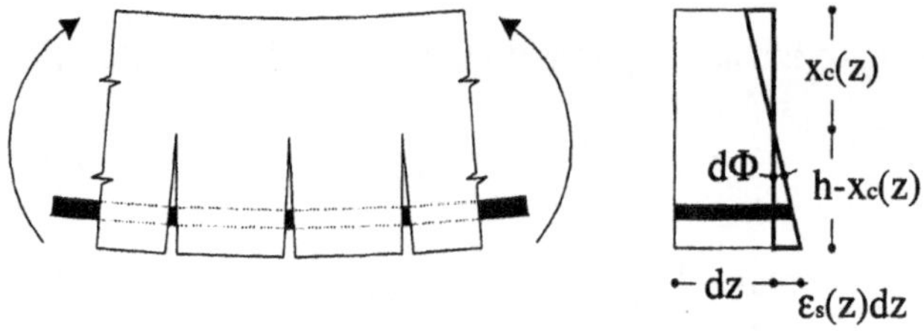

Figure 3. Local curvature between the cracks.

Steel bars, on the contrary, run with continuity along the beam and are suitably anchored at the beam ends. Their strains can thus be used to define the curvature of the beam and we can write, (Fig. 3)

$$\frac{d\phi}{dz}(z) = -\frac{\varepsilon_s(z)}{h - x_c(z)} \tag{27}$$

From Equation (27) we can obtain a corresponding expression of the plastic curvature at the section z. In fact, if $\varepsilon_s^p(z)$ indicates the plastic share of the total steel strain $\varepsilon_s(z)$, the *local plastic curvature* of the beam element is

$$\frac{d\phi_p}{dz}(z) = -\frac{\varepsilon_s^p(z)}{h - x_c} \tag{28}$$

The plastic strain $\varepsilon_s^p(z)$ can be obtained subtracting from the total strain $\varepsilon_s(z)$ the elastic strain $\varepsilon_s^e(z)$ Thus, with the previous notations we get

$$\varepsilon_s^p(z) = \frac{1}{E_s}\frac{e_{sr} - \psi_{sr}}{\psi_{sr} - 1}\{[\sigma_s(0) - \sigma_{sy}] - kz\} \tag{29}$$

The plastic curvature at section z can thus be obtained by substitution of (27) into (26). Considering only the linear terms in z, the plastic curvature becomes

$$\frac{d\phi_p}{dz}(z) = -\frac{1}{E_s}\frac{e_{sr} - \psi_{sr}}{\psi_{sr} - 1}\frac{[\sigma_s(0) - \sigma_{sy}] - kz}{[h - 5/4a(0)]^2}\{[h - \frac{5}{4}a(0)] + \frac{5}{4}a'(0)z\} \tag{30}$$

The *average* plastic curvature between the cracks at a general stage of plastic deformation can be evaluated as

$$(\frac{d\phi_p}{dz})_m = \frac{2}{\lambda_o}\int_0^{d_{op}} \frac{d\phi_p}{dz}(z)dz \tag{31}$$

because, for $d_{op} \le z \le \lambda_o/2$, i.e. along the elastic portion of the bar, plastic deformations vanish. Thus after integration, we have

$$(\frac{d\phi_p}{dz})_m \frac{\lambda_o}{2} = -\delta_{op}\varepsilon_{sy} \frac{\psi}{(1-5/4q_s)^2} \frac{e_{sr}-\psi_{sr}}{\psi_{sr}-1}\{(1-\frac{5}{4}q_s)(1-\frac{1}{\psi})$$
$$+\frac{\beta}{2}[\frac{5}{4}(1-\frac{1}{\psi})\Phi_{1s} - (1-\frac{5}{4}q_s)]\delta_p - \frac{5}{12}\beta^2\Phi_{1s}\delta_p^2\} \tag{32}$$

where

$$\psi = \psi(0) = \frac{\sigma_s(0)}{\sigma_{sy}} \qquad \beta = \beta(0) = \frac{kh}{\sigma_s(0)} \tag{33}$$

are mechanical parameters related to the steel stress $\sigma_s = \sigma_s(0)$ in the cracked section, Φ_{1s} is the value of the function (11) at $q_s = q_s(0)$, given by (10) and

$$\delta_{op} = \frac{d_{op}}{h} \tag{34}$$

indicates the non-dimensional yield penetration depth.

2.5 Ultimate Average Plastic Curvature

A simpler expression of the average plastic curvature at the failure can be obtained by using the following parameters

$$\psi_s^r = \frac{\sigma_s^r}{\sigma_{sy}} \qquad \beta_s^r = \frac{k}{\sigma_s^r} \tag{35}$$

We have in fact *at failure*

$$(\frac{d\phi_p}{dz})_{mr} = \frac{2}{\lambda_{os}^r}\chi_{mr} \tag{36}$$

where

$$\chi_{mr} = -\frac{\varepsilon_{sy}}{2\beta_s^r}\frac{e_{sr}-\psi_{sr}}{\psi_{sr}-1}\frac{(\psi_s^r-1)^2}{\psi_s^r(1-5/4q_s^r)}[1+\frac{5}{12}\frac{\Phi_{1s}^r}{(1-5/4q_s^r)}\frac{\psi_s^r-1}{\psi_s^r}] \tag{36'}$$

and $\Phi_{1s}^r = \Phi_{1s}(q_s^r)$ is the value of the function Φ_{1s} at q_s^r.

The average plastic curvature quickly decreases by decreasing the hardening ratio ψ_{sr}, in consequence of the corresponding reduction of the plastic penetration depth d_{opr}. In particular,

$$\lim_{\psi_{sr}\to 1} \chi_{mr} = 0 \tag{37}$$

because when $\psi_{sr} \to 1$ the steel plastic strains localize in a thinner and thinner neighborhood of the cracks and, in the limit, the length d_{opr} becomes zero.

3. ROTATION CAPACITY OF PLASTIC HINGES

3.1 Slenderness Ratio and Different Types of Plastic Hinges

Reinforced concrete structures, such as slabs, beams or frames, are subjected to a more or less moderate shear, in addition to a bending moment. The bending moment is thus, as a rule, variable along the beam. The plastic curvature, spread in the neighborhood of the cracks over the length of the beam at which the bending moment exceeds the yield moment, gives rise to a plastic rotation. Depending on the magnitude of the shear forces in the critical region, two significantly different types of plastic flexural hinges can develop:

1) - flexural hinges with vertical cracks under moderate shear;

2) - flexural hinges with diagonal cracks under strong shear.

For flexural hinges (1) cracking geometry is not influenced by the moderate action of the shear and cracks remain vertical as in the case of pure bending. In case (2), on the contrary, because of the strong action of the shear, cracks take a diagonal orientation. The slenderness ratio

$$l^*/h^* \tag{38}$$

where h^* is the internal lever arm of the cross section and l^* the distance between critical section and the point of contraflexure, can be used to describe the intensity of the bending moment gradient, i.e. the intensity of shear. In presence of constant shear the slenderness ratio takes the form

$$\frac{l^*}{h^*} = \frac{M_u}{T_u h^*} \tag{39}$$

where T_u is the shear at failure. The smaller is $T_u h^*$, the larger is the slenderness ratio l^*/h^*. High values of l^*/h^* thus correspond to small shear and the collapse of the reinforcement member is a typical bending failure with vertical cracking. By gradually reducing the ratio l^*/h^* the shear influence becomes more relevant. Shear cracks begin to take a diagonal orientation when, at the bending failure and at points along the neutral axis, the principal stress acting in a diagonal direction reaches the tensile concrete strength $\bar{\sigma}_{crt}$ in pure shear. Thus we have

$$\frac{T_u}{bh^*} = \bar{\sigma}_{crt} \tag{40}$$

The slenderness ratio attains the limit value

$$\zeta_L = \mu \frac{\sigma_s^r}{\bar{\sigma}_{crt}(1 - q_s^r)} \tag{41}$$

Considering the scope of this work, we will analyse the ductility properties of reinforced concrete members where the bending effects dominate, i.e. flexural members with slenderness ratio larger than ζ_L.

3.2 Rotation Capacity in Presence of Moderate Shear

The previous expressions for the average plastic curvature of a reinforced concrete member in simple flexure can be used to describe the rotation capacity of plastic hinges in the presence of moderate shear, i.e., for values of the slenderness ratio $l^*/h^* > \zeta_L$. We assume, for the sake of simplicity, that the shear is constant across the member. Along the linear bending moment diagram we can consider the critical section F-F where the collapse bending moment M_u is reached and the section Y-Y where the yielding of the tension steel bars has been just reached, Fig. 4. Let l_p be the distance between these sections: it measures the *length of the plastic hinge* and is given by

$$l_p = l^*(1 - \frac{M_y}{M_u}) = l^*(1 - \frac{1}{\psi_s^r}) \tag{42}$$

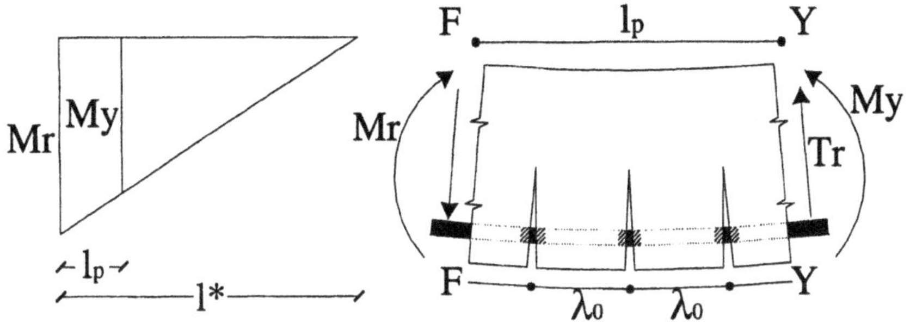

Figure 4. Plastic hinge

Let us proceed along the plastic zone, of length l_p, from the section Y-Y to the critical section F-F. We encounter, across the cracks, a sequence of segments of increasing penetration depth of the plastic strains. Only at the critical section F-F the plastic penetration depth reaches the ultimate value d_{opr} given by (18) or (23). We encounter a sequence of small plastic bends, distributed along the cracks, whose average curvatures can be evaluated by means of equation (32). We can distribute these average curvatures along the plastic length l_p by assuming a linear variation across the cracks; thus we have

$$\left(\frac{d\phi_p}{dz}\right)_m(\zeta)=\frac{\zeta}{l_p}\left(\frac{d\phi_p}{dz}\right)_{mr} \qquad 0<\zeta<l_p \tag{43}$$

In Equation (43) the abscissa ζ increases from $\zeta = 0$, corresponding to section Y-Y, to $\zeta = l_p$, corresponding to section F-F. Thus we have

$$\int_0^{l_p}\left(\frac{d\phi_p}{dz}\right)_m(\zeta)d\zeta=\frac{l_p}{2}\left(\frac{d\phi_p}{dz}\right)_{mr}=\phi_p(l_p)-\phi_p(0)=-\theta_{pr} \tag{44}$$

because $\phi_p(l_p) = 0$. Taking into account Equation (36) finally we obtain an expression for the rotation capacity θ_{pr}

$$\theta_{pr}=-\frac{l_p}{2\lambda^r_{os}}\chi_{mr} \tag{45}$$

where λ^r_{os} is the crack distance at failure, obtained by (14) taking $q_s(0)=q^r_s$. More extensively, with (36') we have

$$\theta_{pr}=\frac{l^*}{2h}\varepsilon_{sy}\frac{\sigma^r_s}{\sigma_{crt}}\Phi^r_{2s}\frac{e_{sr}-\psi_{sr}}{\psi_{sr}-1}\frac{(\psi^r_s-1)^3}{\psi^{r2}_s}\frac{1}{1-5/4q^r_s}\left(1+\frac{5}{12}\frac{\Phi^r_{1s}}{1-5/4q^r_s}\frac{\psi^r_s-1}{\psi^r_s}\right) \tag{46}$$

that can be specified for the case of weak ($\mu < \mu_b$) or strong steel content ($\mu > \mu_b$), as follows.

Steel failure ($\mu < \mu_b$). In this case $\sigma^r_s = \sigma_{sr}$. Starting from $\mu = \mu_b$, we gradually reduce μ. The parameter q_{sr}, given by (10') with $\sigma_s(0)=\sigma_{sr}$, decreases and with it also the function Φ^r_{1s}. The same happens for the quantity $(1-\frac{5}{4}q_{sr})^{-1}$. Hence, the rotation capacity decreases by reducing μ from μ_b.

Concrete failure $(\mu > \mu_b)$. In this case $\sigma^r_s = \sigma_{src}$ where σ_{src} is determined by solving Equation (22). By increasing the steel content μ from the balanced value μ_b, the steel strain ε_{src} at the concrete failure decreases while the neutral axis depth x_c increases. The plastic penetration depth d_{oprc} at the concrete failure thus decreases and the rotation capacity decreases. Furthermore, the average plastic curvature vanishes when $\psi_{sr} \to 1$.

To conclude, the rotation capacity reaches a maximum at the balanced geometrical percentage μ_b, and decreases when μ is both reduced or increased. Increasing the compression steel content increases the ductility of the element.

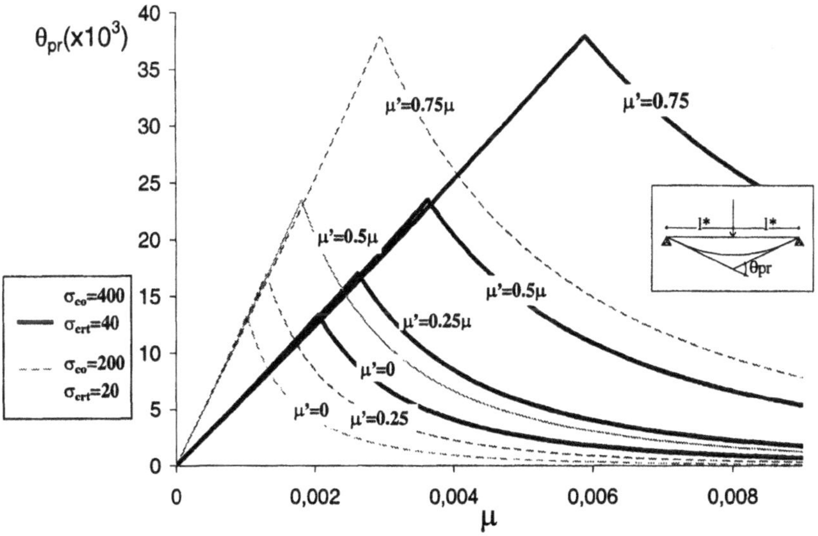

Figure 5. Rotation capacity for varying steel contents μ, μ' and concrete strengths $\sigma_{co}, \sigma_{crt}$.

At the same time an increase of compression concrete strength produces moderate increments of the ductility. On the other hand, to lower values of the tensile concrete strength σ_{crt} correspond to closer cracks and more rotation capacity. For flexural hinges subjected to moderate shear, i.e. for $l^*/h^* > \zeta_L$, the rotation capacity is proportional to the slenderness ratio l^*/h^*; thus if the length l* remains equal, the rotation capacity is inversely proportional to the beam height. Figure 5 gives an example of the evaluation of the rotation capacity of reinforced concrete elements by application of expression (46) to the case of a simply supported beam with one load applied in the mid span. The assumed material constants are: Steel: σ_{sr}=575 N/mm^2; ψ_{sr}=1.15; ε_{sr}=9%; ε_{sy}=2,43×10^{-3}; Concrete: σ_{co} = 40 N/mm^2 (20 N/mm^2), σ_{crt} = 4 N/mm^2 (2 N/mm^2), ε_{cr}=0.35%. The assumed slenderness ratio is l^*/h^* = 9.5. The example simulates one of the cases studied in References [3,5].

References

[1] Bachmann H., Influence of Shear and Bond on Rotational Capacity of Reinforcment Concrete Beams, *IABSE Publications,* 30, II, 1970

[2] CEB, Bulletin d'Information n°242, *Ductility of Reinforced Concrete Structures*, Part 1 and Part 2, 1998

[3] Bigaj A. J., *Structural Dependence of Rotation Capacity of Plastic Hinges in RC Beams and Slabs*, Delft University Press, 1999

[4] Rinaldi Z., Duttilità e resistenza di elementi in c.a.: influenza della localizzazione delle deformazioni nell'acciaio, *Tesi di Dottorato*, Università di Roma "Tor Vergata", 1998

[5] Riva P., Cohn M.Z., Rotation capacity of structural concrete members, *Magazine of Concrete Research*, 46 (1994) 223-234

THE TOPOLOGY OF CELLULAR STRUCTURES

N. A. Fleck, M. F. Ashby and V. S. Deshpande
Department of Engineering, University of Cambridge,
Trumpington Street, Cambridge, CB2 1PZ, U.K.

Abstract: Cellular solids can deform by either the bending or stretching of the cell walls. While most cellular solids are bending-dominated, those that are stretching-dominated are more weight-efficient for structural applications. In this study we have investigated the topological criteria that dictate the deformation mechanism of a cellular solid by analysing the rigidity (or otherwise) of pin-jointed frameworks comprising inextensional struts. We show that the minimum node connectivity for a special class of lattice structured materials to be stretching-dominated is 6 for 2D foams and 12 for 3D foams.

Keywords: stretching-dominated structures, node connectivity, static determinacy, Maxwell's criterion.

1. INTRODUCTION

Low-density cellular structures appear widely in nature and are manufactured on a large scale by man. The cytoskeleton of living cells (Stamenovic, 2000; Kamm 2002), the structures of cork, of wood, of trabecular bone, of coral (Gibson and Ashby, 2000) and of many other growing organisms (Vincent, 1990) are described as "cellular" – a word coined by Hooke (1664) to describe the structure of cork. Synthetic polymer foams, now a mature industry, are widely used in cushioning, packaging and energy-absorption applications (Hilyard, 1982). More recently, a variety of polymeric, metallic and ceramic foams have been developed with the aim of manufacturing lightweight structures – notably sandwich structures – that are both stiff and strong (Gibson and Ashby, 1997; Ashby et al, 2000;

H.R. Drew and S. Pellegrino (eds.),
New Approaches to Structural Mechanics, Shells and Biological Structures, 81–89.

Banhart et al 1999, 2001). In all these examples the mechanical properties are of central importance.

Cellular materials can be thought of as 'composites' made up of solid and space. The solid forms a set of connected struts or plates connected at their ends or edges, to give an open structure through which fluids can flow, or one in which the cells have faces and form closed spaces. Cellular materials are characterised by the solid of which they are made, by their relative density $\bar{\rho}$ (the fraction of space occupied by the solid), by the cell size and wall thickness, and by the connectivity and regularity of the cell walls and edges. We refer to the last two as 'cell-topology', meaning the aspects of the structure that depend on shape but not on scale. This paper concerns cell topology and its influence on mechanical properties

Numerous studies on open-cell foams have shown that both their stiffness and strength are governed by cell-wall bending for all loading conditions; the stiffness scales as $\bar{\rho}^2$ and the strength as $\bar{\rho}^{3/2}$ (Gibson and Ashby, 1997). Most closed-cell foams also follow these scaling laws, at first sight an unexpected result because the cell faces must carry membrane stresses, and this should lead to a linear dependence of both stiffness and strength on $\bar{\rho}$. The explanation lies in the fact that the cell faces are very thin; they buckle or rupture at stresses so low that their contribution to stiffness and strength is small, leaving the cell edges to carry most of the load. Bending is a 'soft' mode of deformation; if the cell edges carried significant tensile loads both the stiffness and strength would be larger. So it is interesting to examine why bending dominates the deformation of almost all common foam-like structures, and whether other topologies exist in which bending is replaced by stretching.

An open-cell foam can be treated as a connected set of pin-jointed struts by the following argument. Consider the pin-jointed frames shown in Figure 1. The frame in Figure 1(a) is a mechanism. When loaded, the struts rotate about the joints and the frame collapses; it has neither stiffness nor strength. The triangulated frame shown in Figure 1(b) is a structure: when loaded the struts support axial loads, tensile in some, compressive in others. Thus, the deformation is stretching-dominated and the frame collapses by stretching of the struts. Imagine now that the joints of both frames are frozen to prevent free rotation of the struts. On loading the first frame, the struts can no longer rotate. The applied load induces bending moments at the frozen joints, and these cause the struts to bend. This is the situation in most foam structures. However, freezing the joints of the triangulated structure has little effect on the macroscopic stiffness or strength; although the struts bend, the frame is still stretching-dominated and the collapse load is dictated mainly by the axial strength or buckling load of the struts.

Foams that are stretching-dominated are more efficient from a weight standpoint; for example, a stretching-dominated foam is expected to be about ten times as stiff and about three times as strong as a bending-dominated foam for a relative density $\bar{\rho} = 0.1$. This can obviously be achieved by aligning one set of edges along a single axis, as they are along the prism-axis of honeycombs or along the grains of wood. But these structures are exceedingly anisotropic, and when loaded normal to the axis, the cell walls bend. In this paper we discuss the topological criteria that dictate whether a foam will be bending or stretching dominated, with the overall aim of developing an understanding of the microstructure that maximises the stiffness and strength-to-weight ratio of a nearly isotropic cellular solid.

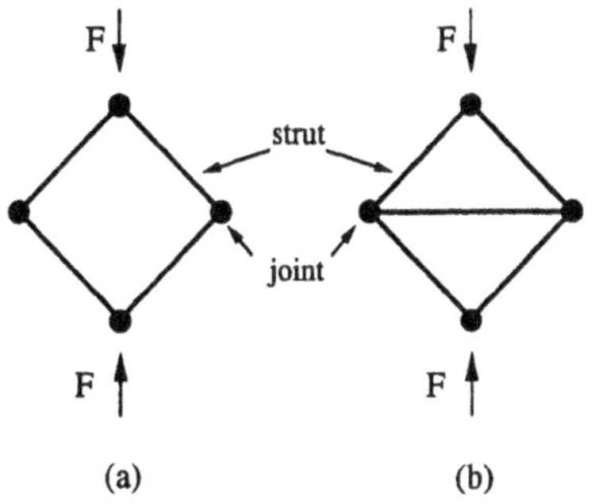

Figure 1. (a) A mechanism; (b) a structure.

2. RIGIDITY OF FRAMES

To establish the topological criteria for stretch-dominated behaviour, we proceed to analyse the rigidity (or otherwise) of an assembly of inextensional pin-jointed struts.

2.1 Maxwell's criterion

Maxwell (1864) suggested an algebraic rule setting out the condition for a pin-jointed frame of b struts and j frictionless joints to be both statically and kinematically determinate i.e. to just be rigid. In 2 and 3 dimensions the criteria are:

$$b = 2j - 3 \tag{1a}$$

and

$$b = 3j - 6, \tag{1b}$$

respectively. These criteria are necessary conditions for rigidity, but are not in general sufficient conditions as they do not account for the possibility of states of self stress (a self-equilibrated state of strut tensions in the absence of external load) and of mechanisms. A generalisation of the Maxwell rule in 3D is given by Calladine (1983):

$$b - 3j + 6 = s - m \quad (2)$$

where s and m count the states of self stress and mechanisms, respectively, and each can be determined by finding the rank of the equilibrium matrix that describes the frame in a full structural analysis (Pellegrino and Calladine, 1986). A just-rigid framework (i.e. a framework that is both statically and kinematically determinate) has $s = m = 0$. The nature of Maxwell's rule as a necessary rather than sufficient condition is made clear by examination of (2): vanishing of the LHS in (2) only implies that the number of mechanisms and states of self-stress are equal, not that each equals zero.

The infinite 2D square framework sketched in Figure 2(a) has a connectivity $Z = 4$ and the unit cell of the framework admits one mechanism and one state of self stress. Thus, while the framework satisfies the Maxwell criterion it is not rigid. On the other hand, an infinite 2D square framework with edge cells braced as shown in Figure 2(b) also has an average connectivity $Z = 4$ and represents a rigid framework that satisfies the Maxwell criterion. Similarly, a 3D rigid framework that satisfies the Maxwell criterion is a cubic framework, provided it is supported by 3 orthogonal planes of braced cells.

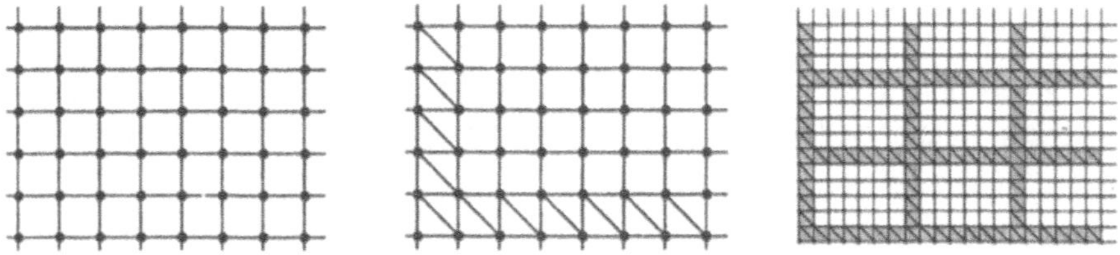

(a) (b) (c)

Figure 2. Three types of frameworks, (a) a mechanism that satisfies Maxwell's criterion, Eqn. (1b); (b) a statically determinate structure that satisfies Maxwell's criterion, Eqn. (1b) and (c) a statically indeterminate structure that is "minimal" stretch-dominated.

2.2 Periodic frameworks

Open-cell foam microstructures can be treated as periodic frameworks with rigid joints. Here we identify classes of unit cells from which rigid periodic structures can be synthesised. Evidently, periodic frameworks constructed from rigid unit cells will necessarily be rigid. The converse may not hold: frameworks constructed from non-rigid unit cells may or may not be rigid. For example, the unit cell of the framework in Figure 3 is not rigid while the framework synthesised from this cell is rigid. This is because the rigid braced square sub-cells link up to form a rigid skeleton as depicted in Figure 3.

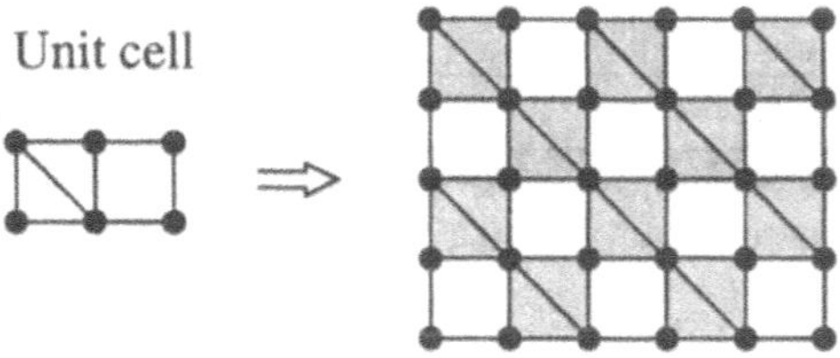

Figure 3. Example of a non-rigid unit cell that can be stacked to form a rigid framework.

Examples of some idealised cell shapes are shown in Figure 4. Isolated cells that satisfy Maxwell's criterion and are rigid are labelled "YES" while "NO" means the Maxwell condition is not satisfied and that the cell is a mechanism. It is worth mentioning here that any convex simply-closed polyhedron with triangular faces satisfies the Maxwell criterion and is rigid (see Appendix 9 in Calladine, 1983). It is generally assumed that the best model for a cell in a foam approximates a space filling shape. However, none of the space filling shapes (indicated by numbers 2, 3, 4, 6, 7 and 8) are rigid. In fact, we could not identify any rigid space filling polyhedral cell in the 3D case and only succeeded in synthesising rigid periodic 3D frameworks from combinations of rigid cells (e.g. the tetrahedron and octahedron in combination fill space to form a rigid framework).

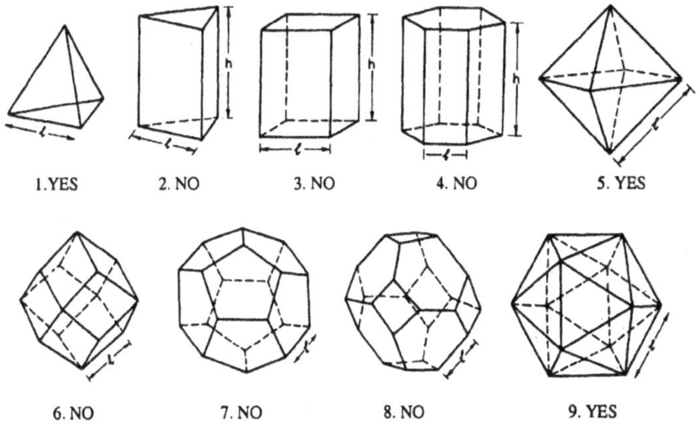

Figure 4. Three dimensional polyhedral cells that do, or do not, satisfy the Maxwell criterion.

The point has been made that stretch-dominated structures offer greater stiffness and strength per unit weight than those in which the dominant mode of deformation is by bending. It should be realised that minimal stretch-domination offers only marginal gain; for the full gain to be realised, the structure must be predominantly stretch-dominated. The point is brought out by Figure 2(c) which shows a two-dimensional "material" made by repeating a unit cell with a geometry like that of Figure 2(b). On a length-scale large compared with the size of the unit cell, the structure is homogeneous and it cannot be deformed without stretch-deformation of the struts that form the shaded square grid of sub-cells. However, the full gain possible by stretch-domination is not realised in this case because the interior of each square still deforms by bending of cell edges. The structure resembles that of a composite made up of a grid of stiff material containing regularly dispersed soft "inclusions". The upper bound for the modulus is then given by a rule of mixtures (or, more accurately, by the Hashin-Shtrikman (1963) upper bound, which is lower); and as the area fraction of this stiffer phase is reduced, the stiffness falls to that of the less stiff material. The same considerations hold in three dimensions. If a space-filling unit cell is made by assembling stretch and bending dominated sub-units, and these are stacked (as in Figures 2 and 3) to retain some stretch domination, the overall stiffness may still lie well below the "ideal", meaning the best achievable. It is therefore relevant to examine the optimisation of 3D lattice structures to maximise stretch-dominated behaviour. This we explore in the next section.

2.3 A Special Class of Periodic Frameworks

In this section we analyse a special class of frameworks with nodes which are all similarly situated; nodes are said to be similarly situated if the rest of the framework appears the same and in the same orientation if viewed from any of the nodes (Nye, 1985). We now proceed to determine the minimum connectivity Z for frameworks of this class to be rigid. Consider an infinite framework of inextensional struts with similarly situated nodes and a connectivity Z. Without any loss of generality all the struts are assumed to be of unit length. We define a Cartesian co-ordinate system x_i with its origin at a node 0. For an applied macroscopic strain field E_{ij}, the displacement $u_i^{(k)}$ of the ends of each strut k connected to 0 is given by

$$u_i^{(k)} = E_{ij}\, n_j^{(k)} \qquad k = 1, 2 ... Z \qquad (3)$$

where $n_j^{(k)}$ is a unit vector in the direction of strut k. However, as the struts are inextensional the displacements (3) must satisfy each of the constraints

$$u_i^{(k)}\, n_i^{(k)} = 0 \qquad k = 1, 2 ... Z \qquad (4)$$

The framework under consideration has similarly situated nodes. Thus, each strut shares a node with a collinear partner, which implies that (4) represents only Z/2 independent equations. Combining the Z/2 independent equations in (4) with the corresponding equations in (3) we get

$$E_{ij}\, n_i^{(k)} n_j^{(k)} = 0 \qquad k = 1, 2 \ldots \frac{Z}{2}. \tag{5}$$

where $n_j^{(k)}$ are now all non-collinear. We proceed by writing (5) in the matrix form:

$$\mathbf{N}\mathbf{E} = 0 \tag{6}$$

where

$$\mathbf{N} = \begin{pmatrix} n_1^{(1)^2} & n_1^{(1)} n_2^{(1)} & n_2^{(1)^2} \\ n_1^{(2)^2} & n_1^{(2)} n_2^{(2)} & n_2^{(1)^2} \\ \ldots & \ldots & \ldots \\ \ldots & \ldots & \ldots \\ n_1^{(Z/2)^2} & n_1^{(Z/2)} n_2^{(Z/2)} & n_2^{(Z/2)^2} \end{pmatrix} \tag{7a}$$

and

$$\mathbf{E} = \begin{pmatrix} E_{11} \\ 2E_{12} \\ E_{22} \end{pmatrix} \tag{7b}$$

in the 2D case. Similarly, in the 3D case $\mathbf{N}$ is a $Z/2 \times 6$ matrix formed from the unit vectors $n_j^{(k)}$ while **E** is a 6 x 1 matrix of the 3D strain components. The necessary and sufficient condition for rigidity is that the only solution to the homogeneous system of linear equations (6) is $E_{ij} = 0$ i.e. $\text{rank}(\mathbf{N}) = 3$ and 6 in the 2D and 3D cases, respectively. Using the observation that $\mathbf{N}$ is formed from non-collinear vectors it can be easily shown that

$$\text{rank}(\mathbf{N}) = \frac{Z}{2} \le 3 \text{ for the 2D case, and} \tag{8a}$$

$$\text{rank}(\mathbf{N}) = \frac{Z}{2} \le 6 \text{ for the 3D case.} \tag{8b}$$

Thus, the necessary and sufficient condition for rigidity of 2D and 3D frameworks is that the connectivity $Z = 6$ and $Z = 12$, respectively. Note that if Z exceeds these values the framework is redundant from a rigidity standpoint. On the other hand, if for example a 3D framework has $Z = c < 12$ then the solutions to (6) form a vector space of dimension $(6 - c/2)$ corresponding to $(6 - c/2)$ independent mechanisms. Thus, a 3D cubic framework with $Z = 6$ has 3 independent mechanisms.

Examples of frameworks that satisfy the above rigidity conditions are a 2D fully triangulated framework and a 3D "face-centred cubic" lattice structure (see Deshpande et al., 2001 for a detailed analysis of the "Octet-truss" lattice material with a FCC microstructure). It is worth mentioning here that in 2D, a triangle is the only rigid polygon that can be constructed from inextensional pin-jointed struts. Thus, a rigid 2D framework with equal connectivity at all nodes is necessarily fully triangulated.

2.3.1 Comparison with optimal microstructures

Isotropic cellular materials that attain the Hashin-Shtrikman (1963) upper bounds for the bulk and shear moduli of a voided solid maximise the stiffness to weight ratio. Francfort and Murat (1986) suggested the so called "rank" laminates which attain both the bulk and shear Hashin-Shtrikman bounds with a finite number of layering directions. Rank laminates are obtained by a sequential process where at each stage the previous laminate is laminated again with a single phase (always the same) in a new direction. Thus, a rank-n laminate is produced by n such successive laminations. Francfort and Murat showed that while in the 2D case, isotropic rank-3 laminates have the extremal bulk and shear moduli, in the 3D case rank-6 laminates are the optimal microstructures. However, no single length-scale microstructure that attains both the bulk and shear Hashin-Shtrikman bounds in 3D has been proposed to date. It is conjectured that the FCC microstructure designed using the rigidity prescription detailed above is close to the optimal single length-scale cellular microstructure.

3. DESIGN IMPLICATIONS

The modulus and initial yield strength of a stretching-dominated cellular solid are much greater than those of a bending-dominated cellular material of the same relative density. This makes stretching-dominated cellular solids attractive alternatives to bending-dominated foamed materials for lightweight structural applications. However, unlike bending dominated foams, in compression the stretching-dominated materials have a softening post-yield response due to the buckling of the struts (Deshpande et al, 2001). Thus these materials may be less attractive as energy-absorbers since this application requires a stress-strain response with a long, flat plateau.

References

Ashby, M.F., Evans, A.G., Fleck, N.A. Gibson, L.J., Hutchinson, J.W. and Wadley, H.N.G. (2000), *Metal Foams: A Design Guide*. Butterworth Heinemann, Oxford UK.

Banhart, J., Ashby, M.F. and Fleck, N.A. (eds) (1999), Proc. Int. Conf. *Metfoam '99,* MIT Verlag, Bremen, Germany.

Banhart, J., Ashby, M.F. and Fleck, N.A. (eds) (2001), Proc. Int. Conf. *Metfoam '01,* MIT Verlag, Bremen, Germany.

Calladine, C.R. (1978), Buckminster Fuller's "Tensegrity" structures and Clerk Maxwell's rule for the construction of stiff frames. *International Journal of Solids and Structures*, **14**, 161.

Calladine, C.R. (1983), *Theory of Shell Structures.* Cambridge University Press, Cambridge, U.K.

Deshpande, V.S. Fleck, N.A. and Ashby, M.F. (2001), Effective properties of the octet-truss lattice material. *Journal of the Mechanics and Physics of Solids*, **49,** 1747.

Francfort, G. and Murat F. (1986), Homogenization and optimal bounds in linear elasticity. *Archive for Rational Mechanics and Analysis*, **94**, 307.

Gibson, L.J. and Ashby, M.F. (1997), *Cellular solids: Structure and Properties,* 2nd edition, Cambridge University Press, Cambridge UK.

Hashin, Z. and Shtrikman, S. (1963), A variational approach to the theory of the elastic behaviour of multi-phase materials. *Journal of the Mechanics and Physics of Solids,* **11,** 127.

Hilyard, N.C. (ed.) (1982), *Mechanics of Cellular Plastics*, Applied Science Publishers, Barking, UK.

Hooke, R. (1664), *Micrographia*, The Royal Society, London, p. 112.

Kamm, R.D. (2002), *Mechanics of the Cytoskeleton*, to be published.

Maxwell, J.C. (1864), On the calculation of the equilibrium and stiffness of frames. *Phiosophical. Magazine,* **27**, 294.

Nye, J.F. (1985), *Physical Properties of Crystals.* Clarendon Press, Oxford.

Pellegrino, S. and Calladine, C.R. (1986), Matrix analysis of statically and kinematically indeterminate frameworks. *International Journal of Solids and Structures,* **22**, 409.

Stamenovic, D. and Wang, N. (2000), Engineering Approaches to Cytoskeletal Mechanics, *Journal of Applied Physics*, **89**, 2085.

Vincent, J.F.V. (1990), *Structural Biomechanics*, 2nd edition, Princeton University Press, Princeton NJ, USA.

Wicks, N. and Hutchinson, J.W. (2001), Optimal truss plates. *International Journal of Solids and Structures*, **38**, 5165.

SYMMETRY ANALYSIS OF THE DOUBLE BANANA AND RELATED INDETERMINATE STRUCTURES

P.W. Fowler
School of Chemistry, University of Exeter
Stocker Road, Exeter EX4 4QD, UK
P.W.Fowler@exeter.ac.uk

S.D. Guest
Department of Engineering, University of Cambridge
Trumpington Street, Cambridge, CB2 1PZ, U.K.
sdg@eng.cam.ac.uk

Abstract Symmetry arguments characterise the mechanisms and states of self-stress of the 'double banana' and its n-fold generalisations (bar-and-joint assemblies of n distorted bipyramids linked at shared apical vertices). The scalar form of Maxwell's rule shows only that any states of self-stress and mechanisms of these systems are equal in number, whereas the symmetry extension of the rule shows further that they are either equisymmetric (for odd n) or differ in two specific one-dimensional representations (for even n). General symmetry criteria are stated for the indeterminacy produced by similar condensations of rigid frameworks with suppression of two joints.

Keywords: Symmetry, rigidity, mechanism, self-stress

1. INTRODUCTION

The 'double banana' (Figure 1) is a classic example of a structure that satisfies Maxwell's rule for the rigidity of frames and yet is clearly indeterminate, since the two banana units are free to rotate about the line joining their common points (the so-called 'implied edge' (Graver, Servatius and Servatius, 1993)). It has been shown (Fowler and Guest, 2000) that a symmetry extension of Maxwell's rule can provide extra information that detects and characterises such indeterminancies in favourable

H.R. Drew and S. Pellegrino (eds.),
New Approaches to Structural Mechanics, Shells and Biological Structures, 91–100.

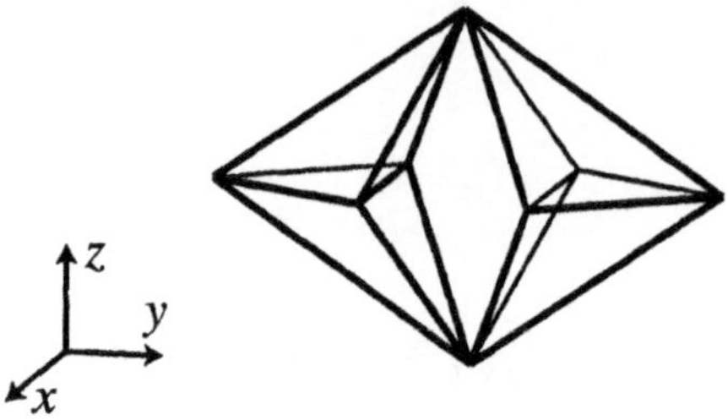

Figure 1. The 'double banana'.

cases. Here we show that the application of symmetry analysis of this type detects the mechanism and its associated state of self-stress in the double banana. A full symmetry description of mechanisms and states of self stress is also presented for the generalised n-fold 'multiple banana', and for some related frameworks produced by augmentation of polyhedra with banana subunits on all edges.

2. THE DOUBLE BANANA

Maxwell's rule (Maxwell, 1864), that a statically and kinematically determinate framework with b bars and j joints obeys

$$b = 3j - 6, \tag{1}$$

is clearly satisfied by the single 'banana' formed by face-fusion of two tetrahedra, which has $b = 9$, $j = 5$. When the double banana is created by joining two such units, with net elimination of two vertices, Figure 1, the Maxwell rule is still satisfied ($b = 18$, $j = 8$) but the composite structure now has an obvious mechanism. The two bananas can rotate freely about the line joining their common vertices. Calladine's statement of an extended Maxwell rule (Calladine, 1978),

$$s - m = b - 3j + 6, \tag{2}$$

where s and m count states of self-stress and mechanisms, respectively, implies the existence of a corresponding state of self-stress. This is easily seen to be the state in which one banana is in tension and the other is in compression. Counting structural components alone cannot deduce $s = m = 1$ from $s - m = 0$, but in this case, consideration of component symmetries can.

The symmetry extension of Maxwell's rule (Fowler and Guest, 2000) is

$$\Gamma(s) - \Gamma(m) = \Gamma(b) - [\Gamma(j) \times \Gamma_T - \Gamma_T - \Gamma_R], \tag{3}$$

where $\Gamma(s)$, $\Gamma(m)$ etc. are the representations of s states of self-stress, m mechanisms, b bars and j joints, and Γ_T, Γ_R are the translational and rotational representations, all within the point group of the undistorted frame. Several examples of the use of this formula are given in the original reference.

In the present case, the point group of the double banana is D_{2h} and the evaluation of the term in square brackets on the RHS of (3) proceeds as

D_{2h}	E	$C_2(x)$	$C_2(y)$	$C_2(z)$	i	$\sigma(xy)$	$\sigma(xz)$	$\sigma(yz)$
$\Gamma(j)$	8	2	2	0	0	6	2	4
$\times\Gamma_T$	3	−1	−1	−1	−3	1	1	1
=	24	−2	−2	0	0	6	2	4
$-\Gamma_T - \Gamma_R$	−6	2	2	2	0	0	0	0
=	18	0	0	2	0	6	2	4

which reduces to

$$4A_g + 2B_{1g} + B_{2g} + 2B_{3g} + A_u + 2B_{1u} + 3B_{2u} + 3B_{3u}.$$

Similarly, the first term on the RHS of (3) gives

D_{2h}	E	$C_2(x)$	$C_2(y)$	$C_2(z)$	i	$\sigma(xy)$	$\sigma(xz)$	$\sigma(yz)$
$\Gamma(b)$	18	0	2	0	0	6	0	6

which reduces to

$$4A_g + 2B_{1g} + B_{2g} + 2B_{3g} + A_u + 2B_{1u} + 4B_{2u} + 2B_{3u}.$$

Hence, by subtraction, $\Gamma(s) - \Gamma(m) = B_{2u} - B_{3u}$, and the separate symmetries of the state of self-stress and the mechanism are $\Gamma(s) = B_{2u}$ and $\Gamma(m) = B_{3u}$, respectively.

B_{2u} is the symmetry of a vector pointing along the y axis (Atkins, Child and Phillips, 1970), and matches the dipolar state of self-stress in which the two bananas have equal and opposite bar forces. Likewise, B_{3u} is the symmetry of a vector pointing along the x axis, and matches the mechanism, in which the two bananas have equal and opposite rotational displacements about the common axis. Figure 2 represents these symmetries in a symbolic notation which will be useful for larger systems.

This linearised analysis shows the existence of a local, possibly infinitesimal, possibly finite, mechanism. In the special case of the double banana it is possible to deduce that the mechanism is in fact finite. This

Figure 2. Representation of (a) the single state of self-stress, and (b) the single mechanism, of the double banana. Each vertex represents a banana: in (a), the + or − represents an overall state of tension, or compression; in (b), the arrow represents rotation of the banana around the shared implied edge.

follows from the fact that the symmetries of the mechanism and the state of self-stress remain distinct along the path followed by the mechanism (Guest and Kangwai, 1999), which preserves C_{2v} symmetry. In this lower group, the mechanism is totally symmetric, whereas the state of self-stress is antisymmetric under the C_2 rotation and one reflection operation.

Although the banana unit considered here has a triangular cross-section, the analysis is easily seen to generalise to bipyramidal units based on any cross-section, including the pentagon of the biological banana, as long as the two units remain symmetry-equivalent under a reflection or C_2 rotation. The following section shows how far these considerations may be extended for systems with more than two banana subunits.

3. THE MULTIPLE BANANA

An obvious generalisation of the double banana is the *multiple banana*: a set of n bananas in which exactly two joints are common to all n subunits (Figure 3). The multiple banana satisfies Maxwell's rule (with $b = 9n$, $j = 3n + 2$) and obeys Calladine's extended version with $s = m$, but it is also intuitively clear that $m = n-1$ (each banana is free to rotate about the axis defined by the two common joints, but concerted rotation of all n constitutes a rigid-body motion rather than a mechanism), and hence $s = n - 1$ (tensions can be freely chosen for $n - 1$ subunits, that in the final subunit being constrained by the requirement of preserving equilibrium).

An analysis similar to that in the previous section can be performed for any particular value of n, but a more general formulation is also straightforwardly constructed. For this purpose, the full three-dimensional n-fold multiple banana framework can be replaced by an n-vertex polygon, where each vertex stands for a banana subunit. A local tension in a subunit is then shown as a scalar at the relevant vertex, and the rotational

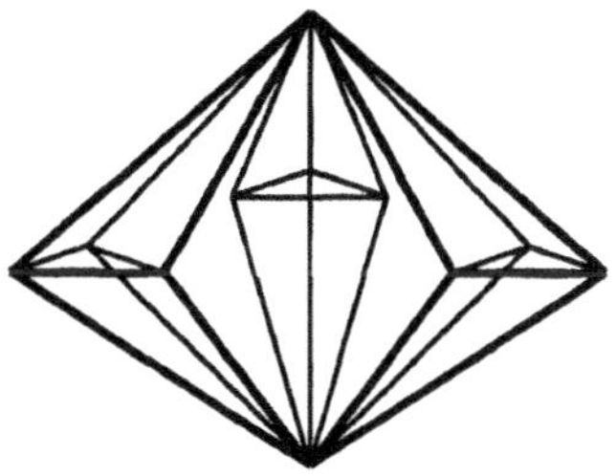

Figure 3. An example of the multiple banana, here a triple banana with three sub-units.

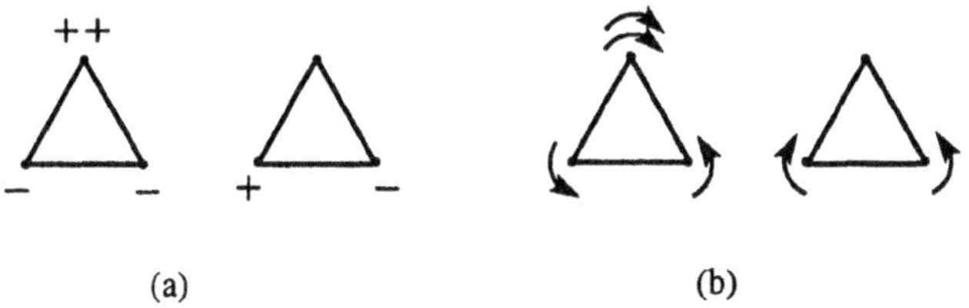

Figure 4. (a) The two states of self-stress, and (b) the two mechanisms, of the 3-banana, described using the notation introduced in Figure 2. A double + or arrow represents a quantity with double magnitude.

motion of a single subunit as a tangential vector at that vertex, Figure 4.

The representation of the $n - 1$ states of self-stress in the symmetry group that describes the undistorted structure (in this case D_{nh} for an n-fold banana) is then seen to be

$$\Gamma(s) = \Gamma_\sigma(v) - \Gamma_0 \tag{4}$$

where $\Gamma_\sigma(v)$ is the permutation representation of the vertices of the polygon, and Γ_0 is the totally symmetric representation, which has character $+1$ under every operation of the symmetry group. The interpretation of this group-theoretical formula is that the states of self-stress comprise all the independent combinations of local tensions (vertex scalars) that sum overall to zero, or in other words all those that are orthogonal to the totally symmetric $(1, 1, 1, \ldots)$ combination.

Likewise, the representation of the mechanisms of the n-fold multiple banana is

$$\Gamma(m) = \Gamma_\rightarrow(v) - \Gamma_{R_z} \tag{5}$$

where $\Gamma_\rightarrow(v)$ is the representation of a set of in-plane tangential vectors, one for each vertex, and Γ_{R_z} is the representation of a rotation about the axis of the structure. The relation (5) expresses the fact that any combination of such vectors, with the exception of the concerted rigid-rotation, stands for an internal mechanism.

$\Gamma_\sigma(v)$ is a familar object, tabulated in the chemical literature, but $\Gamma_\rightarrow(v)$ is more unusual. However, its evaluation can be side-stepped by use of the identity (Fowler and Quinn, 1986)

$$\Gamma_\sigma(v) \times \Gamma_T = \Gamma_\sigma(v) + \Gamma_\pi(v) \tag{6}$$

where $\Gamma_\pi(v)$ is the representation of the set of pairs of orthogonal tangential vectors at vertex sites. In systems with a horizontal mirror plane, the π space splits into disjoint representations of the in-plane and out-of-plane tangential vectors:

$$\Gamma_\pi(v) = \Gamma_\uparrow(v) + \Gamma_\rightarrow(v). \tag{7}$$

The part of $\Gamma_\pi(v)$ that is antisymmetric with respect to the horizontal plane is

$$\Gamma_\uparrow(v) = \Gamma_\sigma(v) \times \Gamma_{T_z} \tag{8}$$

where T_z is a translation along the principal axis, so that the desired symmetry $\Gamma_\rightarrow(v)$ can be found by subtraction as

$$\Gamma_\rightarrow(v) = \Gamma_\pi - \Gamma_\sigma(v) \times \Gamma_{T_z} = \Gamma_\sigma(v) \times \Gamma_{T_x,T_y} - \Gamma_\sigma(v), \tag{9}$$

where Γ_{T_x,T_y} is the representation of the in-plane translations T_x and T_y. It is therefore possible to express the symmetries of the mechanisms in terms of $\Gamma_\sigma(v)$ only:

$$\Gamma(m) = \Gamma_\sigma(v) \times \{\Gamma_{T_x,T_y} - \Gamma_0\} - \Gamma_{R_z}. \tag{10}$$

Equation (10) is completely general for all n, and, taken with (4), satisfies the target of giving an explicit expression for both $\Gamma(s)$ and $\Gamma(m)$ in terms of a property of the polygon (namely $\Gamma_\sigma(v)$), but it can be given a more transparent form if the two cases of odd and even n are treated separately. It turns out then that

$$(n \text{ odd}) \qquad \Gamma_\rightarrow(v) = \Gamma_\sigma(v) - \Gamma_0 + \Gamma_{R_z} \tag{11a}$$

and

$$(n \text{ even}) \qquad \Gamma_\rightarrow(v) = \Gamma_\sigma(v) \times \Gamma_{R_z}. \tag{11b}$$

Both statements are easily proved by considering characters under the possible symmetry operations of D_{nh} for odd and even n. It follows that

$$(n \text{ odd}) \qquad \Gamma(m) = \Gamma(s) \tag{12a}$$

and

$$(n \text{ even}) \qquad \Gamma(m) = \Gamma(s) \times \Gamma_{R_z}. \qquad (12b)$$

Equations (4) and (12) give a complete symmetry specification of the indeterminacy of the multiple banana. From (12), the representation $\Gamma(s) - \Gamma(m)$ vanishes identically for odd n and for even n is given by

$$(n \text{ even}) \qquad \Gamma(s) - \Gamma(m) = \{\Gamma_\sigma(v) - \Gamma_0\} \times \{\Gamma_0 - \Gamma_{R_z}\}. \qquad (13)$$

A further simplication is possible. Equation (13) apparently has a linear dependence on n, through $\Gamma_\sigma(v)$. In fact, $\Gamma_\sigma(v)$ can be eliminated altogether if we define a representation $\Gamma_\star$, which is the symmetry of a decoration of the vertices of the polygon with consistently alternating + and − signs all the way around the ring. In a chemical context $\Gamma_\star$ is a powerful tool in the study of unsaturated systems; its relevance follows from the bi-partite nature of the n-polygon, as the sign decoration can be achieved only for even values of n. In terms of the $\Gamma_\star$ representation,

$$(n \text{ even}) \qquad \Gamma(s) - \Gamma(m) = \Gamma_\star \times \{\Gamma_0 - \Gamma_{R_z}\}. \qquad (14)$$

Equivalence of (13) and (14) follows from the angular momentum properties of $\Gamma_\sigma(v)$, which consists, apart from Γ_0 and $\Gamma_\star$ of pairs $\Gamma_{\pm\Lambda}$ that are self-conjugate under R_z and so make no net contribution to the RHS of (13). The equivalence can also be proved by direct comparison of characters.

In point groups C_{nv}, D_n, D_{nh} and D_{nd}, Γ_{R_z} is non-degenerate and distinct from Γ_0, and the direct inference from the negative sign on the RHS of (14) is that, for even n, $\Gamma(m)$ always contains the one-dimensional representation $\Gamma_\star \times \Gamma_{R_z}$, whereas $\Gamma(s)$ contains $\Gamma_\star$ itself.

The implications for the application of the Maxwell rule and its various extended forms to these systems can be summarised as four subcases.

$n = 1$. Both counting and symmetry versions of the rule agree in correctly predicting that $s = m = 0$.

$n = 2$. Simple counting fails, but the symmetry version correctly predicts $s = m = 1$ with $\Gamma(s) = B_{2u}$ and $\Gamma(m) = B_{3u}$.

odd $n > 1$. Counting correctly predicts $s = m$, and the symmetry version extends this to the equisymmetry $\Gamma(s) = \Gamma(m)$. Further considerations give $s = m = n - 1$ and $\Gamma(s) = \Gamma(m) = \Gamma_\sigma(v) - \Gamma_0$.

even $n > 2$. Counting correctly predicts $s = m$, and the symmetry version extends this to $\Gamma(s) - \Gamma(m) = \Gamma_\star - \Gamma_\star \times \Gamma_{R_z}$, with $\Gamma(s)$ containing $\Gamma_\star$ and $\Gamma(m)$ containing $\Gamma_\star \times \Gamma_{R_z}$. Further considerations fix the full sets of symmetries as $\Gamma(s) = \Gamma_\sigma(v) - \Gamma_0 = \Gamma(m) \times \Gamma_{R_z}$.

This completes the characterisation of the indeterminacy of the multiple banana. Similar reasoning can be extended to frameworks based on attachment of banana units to polyhedral frameworks, as will be shown in the next section.

4. OTHER AUGMENTED FRAMEWORKS

The reason for the celebrity in rigidity theory of the double banana is that condensation of two rigid frameworks introduces freedom in this simple system, without disturbing the Maxwell bar count. Any two determinate frameworks can be condensed in this way, across any pair of joints, whether linked by a bar or not, with consequent introduction of a degree of freedom. Condensation across pairs of joints directly linked by a bar can also be used to introduce additional freedoms in cases where the original frameworks were not themselves rigid. Multiple bananas are the result of repeated formal condensation of this last type, all based on the single implied edge of the double banana.

Another family of statically indeterminate bar-and-joint assemblies can be constructed when augmentation with banana subunits is carried out globally, once per bar, on a framework. Symmetry arguments can again give a helpful improvement over the simple Maxwell counting procedure for these augmented systems. Specifically, let the initial framework be a polyhedron P. Each additional subunit spanning an edge of P contributes an extra state of self-stress associated with a compression or tension that can be denoted by a scalar associated with that edge, and the total set of such states therefore has symmetry

$$\Gamma(s_{\text{extra}}) = \Gamma_\sigma(e), \tag{15}$$

i.e. the permutation representation of the edges (bars) of P. Each subunit also introduces a rotational freedom about the augmented edge, and the total set of such mechanisms spans the symmetry

$$\Gamma(m_{\text{extra}}) = \Gamma_\perp(e), \tag{16}$$

the representation of a set of tangential vectors across the edges of P. Thus the Maxwell indeterminacy symmetry for all such systems is the difference of a scalar and a vector representation defined on the polyhedron edges:

$$\Gamma(s_{\text{extra}}) - \Gamma(m_{\text{extra}}) = \Gamma_\sigma(e) - \Gamma_\perp(e). \tag{17}$$

Both $\Gamma_\sigma(e)$ and $\Gamma_\perp(e)$ can be expressed in terms of other structural representations (Ceulemans and Fowler, 1991), and in specialised forms when P belongs to the class of deltahedra, or alternatively of their duals, the trivalent polyhedra.

Three examples of the augmentation of initially determinate structures (so that $\Gamma(s) = \Gamma(s_{\text{extra}})$, $\Gamma(m) = \Gamma(m_{\text{extra}})$) are presented. Augmentation of a simple triangle leads to a 30-bar, 12-joint system that is completely specified by the symmetry-extended Maxwell rule. In the D_{3h} group, with z taken along the three-fold axis, evaluation of (17) gives

$$\Gamma(s) - \Gamma(m) = A_1' + E' - A_1'' - E'', \tag{18}$$

and hence

$$\begin{aligned} \Gamma(s) &= A_1' + E', \\ \Gamma(m) &= A_1'' + E'', \end{aligned} \tag{19}$$

as $\Gamma_\perp(e) = \Gamma_\sigma(e) \times \Gamma_{T_z} = \Gamma_\sigma(e) \times A_1''$ for the underlying triangle.

Augmentation of a tetrahedron leads to

$$\begin{aligned} \Gamma(s) &= A_1 + E + T_2, \\ \Gamma(m) &= T_1 + T_2, \end{aligned} \tag{20}$$

and as the s and m have a representation in common, their difference alone, in the symmetry-extended Maxwell rule, would give only partial information on each.

Similarly, augmentation of an octahedron yields a framework with

$$\begin{aligned} \Gamma(s) &= A_{1g} + E_g + T_{2g} + T_{1u} + T_{2u}, \\ \Gamma(m) &= A_{2u} + E_u + T_{1g} + T_{2g} + T_{1u}, \end{aligned} \tag{21}$$

and again the pair of representations carry more information than the Maxwell difference alone.

5. CONCLUSION

The aim of the present paper has been to confirm the utility of symmetry arguments as an adjunct to static and kinetic analysis of frameworks. Symmetry extension of the Maxwell rule is sufficient to provide a full description of the double banana. The outcome of the detailed analysis in Section 3 is an essential symmetry distinction between odd-n and even-n multiple-banana systems. When n is odd, the mechanisms and states of self-stress are equisymmetric; when n is even, their symmetries differ only by two (one-dimensional) representations. This distinction further defines the amount of information that can be gained from the symmetry-extended Maxwell equation in such cases.

Finally, it was noted that the multiple banana is only one, highly symmetric example of a more general class of problem. Any case where linkage of two fully determinate frameworks is carried out by identifying a pair of vertices introduces equal numbers of states of self-stress and balancing mechanisms, as does each repetition of the operation.

Symmetry considerations of the structural representations $\Gamma_\sigma(v)$, $\Gamma_\sigma(e)$, and their vector derivatives will identify completely $\Gamma(s)$ and $\Gamma(m)$ for augmented frameworks, and the symmetry-extended Maxwell rule will improve on the counting version in many cases. The degree of advantage will of course depend on the amount of symmetry in the structure, but where such symmetries exist, their incorporation in the analysis will always be fruitful.

References

Atkins, P.W., Child, M.S. and Phillips, C.S.G. (1970) *Tables for Group Theory*, OUP, Oxford.

Calladine, C.R. (1978) Buckminster Fuller's 'Tensegrity' structures and Clerk Maxwell's rules for the construction of stiff frames, *Int. J. Solids Structures*, **14**, 161–172.

Ceulemans, A. and Fowler, P.W. (1991) Extension of Euler's theorem to the symmetry properties of polyhedra, *Nature* **353**, 52–4.

Ceulemans, A. and Fowler, P.W. (1995) Symmetry extensions of Euler's theorem for polyhedral, toroidal and benzenoid molecules, *J. Chem. Soc. Faraday Trans.* **91**, 3089–93.

Fowler, P.W. and Guest, S.D. (2000) A symmetry extension of Maxwell's rule for rigidity of frames, *Int. J. Solids Structures* **37**, 1793–1804.

Fowler, P.W. and Quinn, C.M. (1986) σ, π and δ representations of the molecular point groups, *Theor. Chim. Acta* **70**, 333–350.

Graver, J.E, Servatius, B., and Servatius, H., (1993), *Combinatorial Rigidity*, Graduate Studies in Mathematics, AMS, Providence.

Maxwell, J.C. (1864) On the calculation of the equilibrium and stiffness of frames, *Phil. Mag.* **27**, 294–299.

Kangwai, R.D. and Guest, S.D. (1999) Detection of finite mechanisms in symmetric structures, *Int. J. Solids Structures* **36**, 5507–5527.

LARGE DEFLECTIONS OF A CANTILEVER

J.A. Greenwood
Peterhouse
Trumpington Street, Cambridge, CB2 1RD, U.K.
jag@eng.cam.ac.uk

Abstract Attempts to find a convenient trial shape for use in estimating the (large) deflections of the non-uniform cantilevers used as springs in electronic connectors are described. It becomes clear that the lower bound property of the total potential energy method is lost when the load-deflection curve is non-linear.

Keywords: Cantilever, Large deflections, Total potential energy.

1. INTRODUCTION

Real structural engineers tend to design their beams so that they are not in the 'large deflection' regime; it takes an Euler to get involved in such matters. Or a Calladine, when wearing his second hat as a molecular biologist and trying to explain the crossbridge mechanism in vertebrate skeletal muscle [1]. In references [1,3], the result for a cantilever of length with an end-load P is given: that

$$\ell\sqrt{\frac{P}{EI}} = K(k) - F(k,\alpha);$$

$$\delta / \ell = 1 - 2[E(k) - E(k,\alpha)]/[K(k) - F(k,\alpha)]$$

where E and F are, respectively, incomplete elliptic integrals of the first and second kinds and their arguments k and α are related to the end-rotation φ_E of the cantilever, [$k = \sqrt{(1+\sin\varphi_E)/2}$, $\sin\alpha = 1/\sqrt{(1+\sin\varphi_E)}$]. Or more precisely, incomplete *Legendre* elliptic integrals: for not merely are Legendre forms not the only forms, but, as the Calladine paper hints, *Carlson* elliptic integrals are more convenient. We shall return to this later.

My own interest in large-deflection theory comes from the days when I worked for an electrical contact manufacturer. Separable connectors make

H.R. Drew and S. Pellegrino (eds.),
New Approaches to Structural Mechanics, Shells and Biological Structures, 101–114.

electrical contact by pushing a pin against a spring: and the design of the spring to provide an adequate contact load while not demanding an excessive insertion force, and indeed, while not providing a force which damages the electroplating during insertion or removal, is a fine art. The springs are convoluted; the cross-section is as non-uniform as can conveniently be produced by stamping out from a flat strip – typically a 'top-hat' cross-section – and the deflections are large. Even 1 mm is not unknown, which is more impressive when we recall that the spring is perhaps 1 cm long, and has been bent to fit into a 'box' perhaps 3 mm long. Accordingly, we developed a spring deflection program to calculate the deflections of highly convoluted variable cross-section cantilevers. A versatile program: it was even used in some of the preliminary work on the crossbridges in skeletal muscle!

Our method of determining the shape was to integrate along the beam from its root, having estimated the bending moment there: of course, if that bending moment is correct, the solution is exact. An obvious test case is a uniform straight cantilever of bending stiffness EI and length ℓ taken to lie along the x-axis with its root at $x = 0$. A load P is applied *at the tip*, wherever that may lie, acting in the y-direction: Figure 1 shows a result of assuming the root bending moment to be that found by applying an end-load at the initial position of the cantilever tip. When the load is large, in this case equal to the Euler buckling load $P_E \equiv \pi^2 EI/\ell^2$, the curvature at

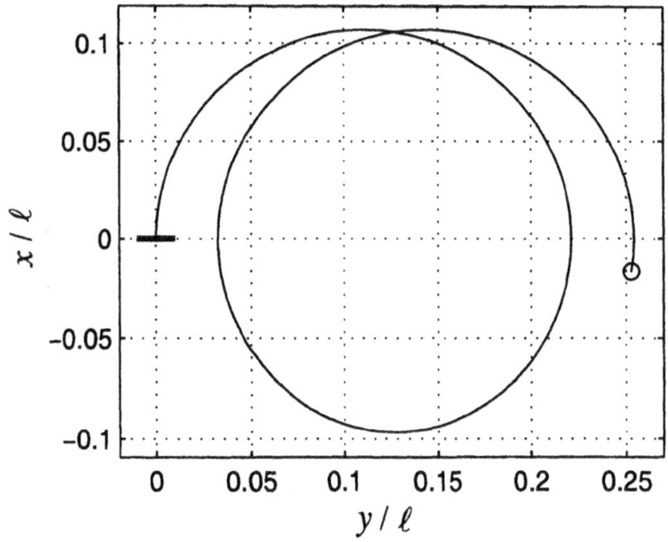

Figure 1. Shape calculated assuming root bending moment is $P_E \times \ell$

the root is so large that the beam bends itself up into a coil spring: not a useful first step in the iterative process. We need a method of estimating the root bending moment.

My forays into elasticity have all been attempts at exact solutions of idealised problems: I tended to dismiss approximate methods until Prof. Calladine brought the 'four energy theorems' firmly to my notice - over after-lunch coffee. But of course all we need to do is to guess a suitable form for the beam shape, and minimising the total potential energy will give us the deflection, and, hence, the moment arm [2].

2. EXACT SOLUTION [3]

If the tip of the straight cantilever in the test case is deflected to (d,δ) – both unknown – then the bending moment at (x,y) will be $M = P(d-x)$ and the curvature will be given by $M = EI\kappa = EI\dfrac{d\varphi}{ds}$.

Differentiating,

$$EI\frac{d^2\varphi}{ds^2} = -P\frac{dx}{ds} = -P\cos\varphi$$

so that integrating, and noting that the curvature will be zero at the tip:

$$s\sqrt{\frac{P}{EI}} = \int_0^{\varphi}\frac{d\varphi}{\sqrt{2[\sin\varphi_E - \sin\varphi]}}$$

and here is our elliptic integral. The end–angle φ_E is given by

$$\ell\sqrt{\frac{P}{EI}} = \int_0^{\varphi_E}\frac{d\varphi}{\sqrt{2[\sin\varphi_E - \sin\varphi]}}$$

which will turn out to be a complete elliptic integral. Finding x is easy:

$x = \int_0^{\varphi}\cos\varphi\, ds$ so that

$$x\sqrt{\frac{P}{EI}} = \int_0^{\varphi}\frac{\cos\varphi\, d\varphi}{\sqrt{2[\sin\varphi_E - \sin\varphi]}} = \sqrt{2[\sin\varphi_E - \sin\varphi]}\Big|_0^{\varphi}$$

Thus

$$(d-x)\sqrt{\frac{P}{EI}} = \sqrt{2[\sin\varphi_E - \sin\varphi]};$$

and

$$\boxed{d\sqrt{\frac{P}{EI}} = \sqrt{2\sin\varphi_E}}.$$

This equation, well-known in the fracture mechanics literature because of its importance in the 'peel test', is worth noting: once we know – or can guess – the end rotation of the cantilever, we immediately know the root moment: $M(0) = Pd = \sqrt{2PEI\sin\varphi_E}$.

The y-integration is harder. Since $y = \int_0^\varphi \sin\varphi\, ds$ we have

$$y\sqrt{\frac{P}{EI}} = \int_0^\varphi \frac{\sin\varphi d\varphi}{\sqrt{2[\sin\varphi_E - \sin\varphi]}}$$

and the deflection is

$$\delta\sqrt{\frac{P}{EI}} = \int_0^{\varphi_E} \frac{\sin\varphi d\varphi}{\sqrt{2[\sin\varphi_E - \sin\varphi]}}.$$

How do express these as elliptic integrals? Carlson elliptic integrals [4] are defined as

$$R_F(a,b,c) = \frac{1}{2}\int_0^\infty \frac{du}{\sqrt{[u+a][u+b][u+c]}} \qquad \text{and}$$

$$R_D^*(a,b\,;c) = \frac{1}{2}\int_0^\infty \frac{du}{\sqrt{[u+a][u+b][u+c]^3}}.$$

These look undesirable when it is recalled that the usual Legendre integrals are a function of a single parameter, compared with the three arguments here. However, the Carlson integrals replace the incomplete Legendre integrals, which do already have two arguments: the complete Legendre integrals correspond to one of the three arguments of the Carlson form becoming zero. This still leaves one additional argument, and we recall the table-maker's maxim "To tabulate a function of one variable takes a page, a function of two takes a book; a function of three a library". But a physical problem requiring the complete elliptic integral often actually involves two arguments b and a on an equal basis; for example, the Hertz elastic contact problem involves the integral $\int_0^\infty \frac{d\psi}{\sqrt{\psi(a^2+\psi)(b^2+\psi)}}$. While Legendre cleverly reduces this to a single argument (b/a) *with a requirement that* $b/a < 1$, transforming the answer for $b/a > 1$ can be remarkably difficult, and there are distinct advantages in a form which retains the equal basis – particularly when the Carlson forms have the delightfully simple scaling rules

$$R_F(\lambda a, \lambda b, \lambda c) = \lambda^{-1/2} R_F(a,b,c);$$
$$R_D(\lambda a, \lambda b; \lambda c) = \lambda^{-3/2} R_D(a,b;c)$$

so that one argument could readily be taken as unity if a table were required. But it isn't: the real beauty of the Carlson forms is the existence of straightforward, highly convergent, iterative processes by which they may be evaluated on a programmable calculator. Note that the arguments of R_K are completely interchangeable, but only the first two of R_D are: the semi-colon is to emphasise this. Note also that I cannot bring myself to follow Carlson in his definition of R_D as three times that given here and identified as R_D^*: beware when using published routines!

[The iterative process is essentially that of the arithmetic-geometric mean described in the NBS tables for evaluating the complete Legendre integrals K and E. But the corresponding process for the incomplete integrals is messy, while the Carlson process remains straightforward.]

To write the expressions for ℓ and δ as Carlson elliptic integrals is easy: simply put $\sin\varphi = \dfrac{u}{1+u}\sin\varphi_E$ which removes the trigonometric functions and converts the range to $0-\infty$. Then

$$\ell\sqrt{\frac{P}{EI}} = \frac{\sqrt{\sin\varphi_E}}{\cos\varphi_E}\frac{1}{\sqrt{2}}\int_0^\infty \frac{du}{\sqrt{[u+(1-\sin\varphi_E)^{-1}][u+(1+\sin\varphi_E)^{-1}][u+1]}}$$
$$= \frac{\sqrt{2\sin\varphi_E}}{\cos\varphi_E} R_F\left((1-\sin\varphi_E)^{-1}, (1+\sin\varphi_E)^{-1}, 1\right)$$

while

$$\delta\sqrt{\frac{P}{EI}} = \sqrt{2\sin\varphi_E}\,\tan\varphi_E \times$$
$$\left[R_F\left((1-\sin\varphi_E)^{-1}, (1+\sin\varphi_E)^{-1}, 1\right) - R_D^*\left((1-\sin\varphi_E)^{-1}, (1+\sin\varphi_E)^{-1}; 1\right)\right]$$

Note that these are immediately functions of the end-angle φ_E: there is no need for the auxiliary arguments k, α of the Legendre form. Note also that the scaling laws give a somewhat more convenient form for the first:

$$\ell\sqrt{\frac{P}{EI}} = \sqrt{2\sin\varphi_E}\,R_F\left((1+\sin\varphi_E), (1-\sin\varphi_E), \cos^2\varphi_E\right).$$

Figure 2 (below) shows the results of the exact theory. We may note, on the Calladine principle of forming meaningful non-dimensional groups, that the Euler buckling load is $P_E = \pi^2 \dfrac{EI}{\ell^2}$, so the result may be written $\delta/\ell = F(\varphi_E) = G(P/P_E)$; and that when P/P_E is well below 1, the usual strength of materials formula for the deflection holds, but for $P/P_E \sim 1$ the results are very different. The Euler load is not only of significance in compression!

3. ESTIMATING THE ROOT MOMENT BY TOTAL POTENTIAL ENERGY

If we can guess the shape adopted by the cantilever, then differentiation gives the curvatures, and we can readily find the strain energy $U = \int_0^\ell \frac{1}{2} EI\kappa^2 ds$. The potential energy lost by the load in adopting this shape is $P\delta$, (and knowing the shape implies a knowledge of δ.) The correct shape is the one which maximises the 'total potential energy'[1] $C \equiv P\delta - U$; so if the shape contains an adjustable parameter, differentiation of C with respect to the parameter will give the best estimate of the parameter and so of the deflection – and the root moment. The art lies in choosing a suitable shape!

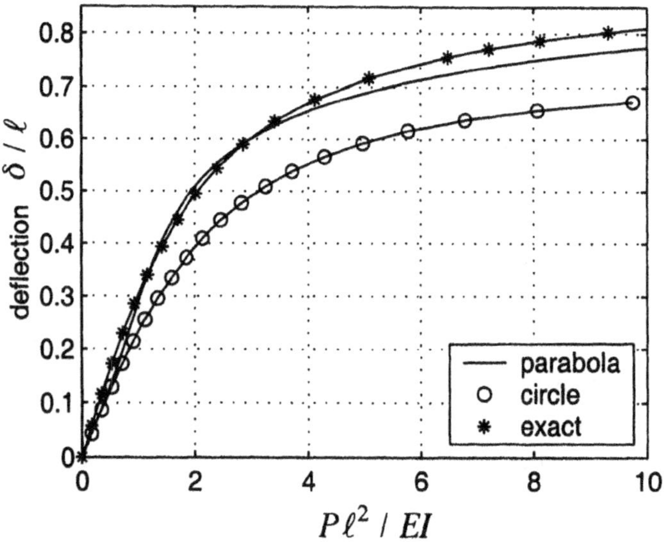

Figure 2. Exact deflections and results using parabola or circular arc.

[1] Strictly, the potential energy is $V \equiv U - P\delta$, and should be *minimised*. But then V would be a negative quantity....

An obvious candidate, at least to students attempting the corresponding problem of estimating the lowest natural frequency of a cantilever, is the parabola $y = ax^2$. Certainly this satisfies the most obvious conditions of being built-in at $x = 0$, and is easy to differentiate: $y'' = 2a$.

Unfortunately, for large deflections the second derivative is not the curvature, nor is the deflection δ equal to $a\ell^2$. But let's do the small deflection case first: we readily find $U = \frac{1}{2}EI(2a)^2\ell$: $C = P \cdot a\ell - 2EIa^2\ell$.

This is a maximum when $a = P\ell/4EI$ so that $\delta = \dfrac{P\ell^3}{4EI}$ - or better, $\dfrac{\delta}{\ell} = \dfrac{1}{4}\dfrac{P\ell^2}{EI} = \dfrac{\pi^2}{4}\dfrac{P}{P_E}$.

As innumerable attempts to estimate the fundamental vibration frequency have found, this is wrong: too small by a factor 3 / 4: the curvature should not be constant, but should decrease to zero at the end.

At least for the large deflection case the curvature does reduce along the cantilever: $\kappa = \dfrac{2a}{\left(1+(2ax)^2\right)^{3/2}}$, so things may get better. We find, easily, that $U = EIa\int_0^{\varphi_E} \cos^3\varphi \, d\varphi = EIa\left[\sin\varphi_E - \frac{1}{3}\sin^3\varphi_E\right]$ and the moment arm is $d = \int \cos\varphi ds = \dfrac{1}{2a}\tan\varphi_E$, so that the end deflection is

$$\delta = ad^2 = \frac{1}{4a}\tan^2\varphi_E \quad \text{and}$$

$$C \equiv P\delta - U = \frac{P}{4a}\tan^2\varphi_E - EIa\,[\sin\varphi_E - \tfrac{1}{3}\sin^3\varphi_E].$$

Unfortunately we need also the relation between the two parameters a and φ_E. This comes from the length;

$$\ell = \int ds = \int_0^{\varphi_E} \frac{1}{2a}\sec^3\varphi \, d\varphi = \frac{1}{4a}\left[\sec\varphi_E \tan\varphi_E + \ln\left(\frac{1+\sin\varphi_E}{\cos\varphi_E}\right)\right]$$

which we can use to eliminate a before differentiating C with respect to φ_E. Realistically, the maximum is located numerically; back substitution leads to surprisingly good answers when the deflection is large, (Fig. 2) particularly when we recall that the aim is to estimate the moment arm, *not* to calculate the actual deflection.

It is time to recall that the object of this work is to develop an approximation for a cantilever of variable section; and of course the section variation is specified as a function of the distance s along the beam, not as a function of x. The parabola is not really a convenient shape.

3.1 Circular Arc

The easiest possible shape for the strain energy calculation with variable EI is to have a constant curvature; so consider a shape $s = R\varphi$ where $\ell = R\varphi_E$ – no problem here in relating the length to the rotation! Then $C = P\ell(1-\cos\varphi_E)/\varphi_E - EI\,\varphi_E^2/2\ell$, which is a maximum when $\dfrac{P\ell^2}{EI} = \dfrac{\varphi_E^3}{\varphi_E \sin\varphi_E - 1 + \cos\varphi_E}$. This must be solved to find φ_E, and hence δ and $d = R\sin\varphi_E = \ell\sin\varphi_E/\varphi_E$.

The deflections are shown on Fig. 2. Of course they are again too low by a factor $3/4$ for small deflections: we have in fact reproduced the parabola analysis. But now the results for the deflection, not surprisingly, are consistently poor – though the results for the moment arm are consistently rather good!

3.2 Catenary

It is becoming clearer what we need. (a) The curvature must reduce as we go along the beam, even if not to the theoretical zero. (b) The shape must be defined in terms of distance along the beam, so that stiffness variations can be readily included – we need a shape defined by its intrinsic equation. Which leads directly to the catenary $s = a\tan\varphi$. Then

$U = \frac{1}{2}EI\int \kappa^2 ds = \frac{1}{2}EI\int \kappa\, d\varphi = \dfrac{EI}{4a}[\varphi_E + \sin\varphi_E\cos\varphi_E]$. Also $\ell = a\tan\varphi_E$

and $\delta = \int \sin\varphi\, ds = a[\sec\varphi_E - 1] = \ell\tan(\varphi_E/2)$, so that

$$C = P\delta - U = P\ell\tan(\varphi_E/2) - \frac{EI}{4\ell}\left[\varphi_E\tan\varphi_E + \sin^2\varphi_E\right]$$

$$= P\ell\tan(\varphi_E/2) - \frac{EI}{4\ell}\left[\varphi_E\tan\varphi_E + \sin^2\varphi_E\right].$$

This can be maximised analytically, leading to $P\ell^2/EI = \frac{1}{2}\left[\tan\varphi_E + \varphi_E\sec^2\varphi_E + \sin(2\varphi_E)\right]\cos^2(\varphi_E/2)$: although for given values of $P\ell^2/EI$ a numerical solution is needed. The results are shown on Fig. 3. Once again, the deflection is low by a factor $3/4$ for small loads, but is very close for large loads. The moment arm is consistently good.

3.3 Linear curvature variation

Have we already, mistakenly, excluded the best method? One direct generalisation of $\kappa = a(\ell - x)$, is $\kappa = a(\ell - s)$. Then $\varphi = \beta\left[1 - (1 - s/\ell)^2\right]$ where β is the end-rotation $\left(\beta = \frac{1}{2}al^2\right)$. Then

$$U = \tfrac{1}{2} EI \int_0^\ell \beta^2 (1 - s/\ell)^2 \, ds = \tfrac{2}{3} EI\beta^2/\ell \ .$$

This is easily modified to allow for variable stiffness: it becomes $U = 2(\beta/\ell)^2 \int_0^\ell EI(s)(\ell - s)^2 \, ds = \frac{2}{3}\langle EI\rangle \beta^2/\ell$ where $\langle EI\rangle$ is a weighted average of $EI(s)$, and is readily found: $\langle EI\rangle = \dfrac{3}{\ell^3}\int_0^\ell EI(s)(\ell - s)^2 \, ds$.

The drawback is the calculation of the deflection: $\delta = \int_0^\ell \sin\varphi \, ds = \int_0^\ell \sin\left[\beta[1 - (1 - s/\ell)^2]\right] ds$. If we set $t = 1 - s/\ell$ this becomes $\delta/\ell = \int_0^1 \sin\left[\beta\left(1 - t^2\right)\right] dt$, which is a Fresnel integral but may readily be evaluated numerically (or from tables). Importantly, it is not concerned with whether or not the stiffness varies. Thus, we arrive at

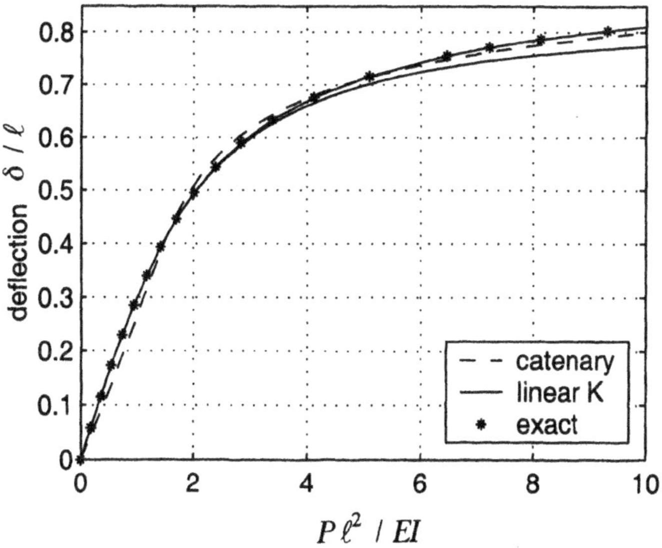

Figure 3. Exact deflections compared with results using a catenary and 'linear curvature'.

$C \equiv P\delta - U = P\ell f(\beta) - \frac{2}{3}\langle EI \rangle \beta^2/\ell$, and optimising gives us β as a function of $P\ell^2/\langle EI \rangle$ and hence the deflection as a function of this variable.

Figure 3 shows that the estimate of the deflection is reasonable: for the first time we get the correct behaviour for small loads: at higher loads the behaviour is close to that for the parabolic shape, but around the 'Euler load' the values are slightly worse than for the parabola (which was worse than the catenary). The moment arm approximations are much the same, and quite satisfactory.

3.4 An Elementary Approximation

If the deflection is appreciable, then it is obvious that the moment arm d must be less than the length ℓ: indeed, since the beam is smoothly curved, it must be less than $\sqrt{\ell^2 - \delta^2}$. What happens if we apply the small deflection equations to the unknown length d, and then use this condition to find d ? We have $\delta = Pd^3/3EI$; put $\delta/\ell = \sin\alpha$; $d/\ell = \cos\alpha$. Then $\sin\alpha = (P\ell^2/3EI)\cos^3\alpha$ which can be written $\tan^3\alpha + \tan\alpha = (P\ell^2/3EI)$ and is easily solved by Newton's method: from this the deflection may be found. The results have been added to Fig. 3, and are clearly a very satisfactory approximation for such a simple scheme.

3.5 Variable Cross-Section and Effective Stiffness

The real aim is to devise a scheme which can be applied to a cantilever whose stiffness changes along its length. As a test case, consider a beam whose stiffness reduces to one-fifth at the mid-point. (This would occur in practice by indenting the strip to a top-hat shape at the root end.). The model above suggests that the effective stiffness will be

$$\langle EI \rangle = \frac{3}{\ell^3}\int_0^\ell EI(s)(\ell - s)^2\, ds = [7EI_{root} + EI_{tip}]/8 = 0.9EI_{root}.$$

But surely the correct answer is some weighted average of $(EI)^{-1}$? Small deflection theory readily gives a more plausible estimate of the effective stiffness: $\langle EI \rangle^{-1} = [(1 - \ell_{root}^3/\ell^3)(EI_{root})^{-1} + (\ell_{root}^3/\ell^3)(EI_{tip})^{-1}]$ or $\langle EI \rangle = (2/3)EI_{root}$: very considerably lower!

Figure 4 shows the deflection curve for the beam mentioned above, where the tip half has one-fifth the stiffness of the root half. Also shown are the results for a uniform beam with the root stiffness. In the small deflections regime it is clear that using the 'equivalent' stiffness

$\langle EI \rangle = (2/3)EI_{root}$ would indeed reproduce the results found: but in the large deflection region the deflection for the beam whose stiffness reduces is **less** than that for the uniform beam: a better estimate of the deflection is found by ignoring the reduced stiffness!

Can this be correct? How can reducing the stiffness reduce the deflection? A hint is given by considering the extreme case where the tip stiffness is zero, which is equivalent to applying the load to a rope (of length $\ell/2$) attached to a beam of length $\ell/2$. The end of the rope will then be deflected to the point $\ell/2$ below the tip of this shorter beam: and as shown in Fig. 4, this deflection is indeed somewhat less than the deflection of the beam of uniform stiffness. So perhaps the calculations are correct!

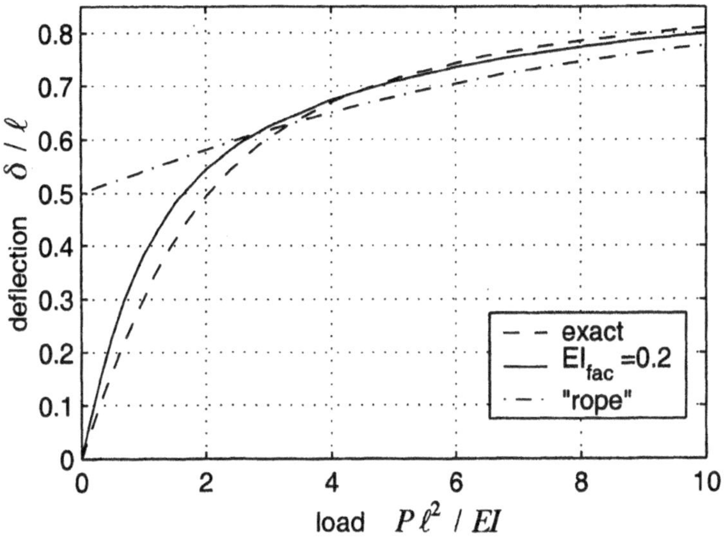

Figure 4. Deflections of a beam whose stiffness halves for the tip half.

4. THE LOWER BOUND THEOREM

The expert will have deduced that all the calculations above are wrong! For of course the total potential energy theorem not only says that the exact solution maximises the total potential energy but that any trial solution (if found correctly!) *provides a lower bound on the exact deflection.* Well, yes: but not in the form of the principle we have used. The correct form is that the maximum total potential energy must be less than or equal to the true value. What we have done is to reject the rather accurate estimate of the maximum potential energy in favour of the relatively inaccurate estimate of the deflection at which this occurs. We should have used $\int_0^{P_1} \delta \, dP \geq C(P_1)\big|_{max}$ – it will be clear why we did not! In fact, with some reluctance, this method was programmed for the catenary, only to find that the answers were virtually the same. But was this necessary? Is $\delta_2 \equiv \left.\frac{dC_{max}}{dP}\right|_{P_1}$ *in principle* a better estimate than the value δ_1 at which the maximum occurred? Did we really need to go through the process of numerical differentiation of the maximum total potential energy with respect to the load in order to find the deflection?

For a linear system there is certainly no gain; the calculation is always as follows:

$$C = P\delta - k\delta^2;$$

so that C is a maximum when

$$\delta = \delta_1 = P/2k.$$

Substituting this value in C gives

$$C_{max} = P(P/2k) - k(P/2k)^2 = (P^2/4k),$$

and now the same, unchanged, answer: $\delta_2 = P/2k = \delta_1$ is found, whether $\delta = \delta_2 \equiv 2C_{max}/P$ or $\delta = \delta_2 \equiv \frac{dC_{max}}{dP}$ is used.

Is this a general result?

In fact, it is readily seen to be true. Write $C = P\delta - f(\delta)$, so that the maximum is when $P = f'(\delta)$. Solving this equation gives $\delta_1 = g(P)$, and substituting gives $C_{max} = Pg(P) - f(g(P))$.

Hence

$$\delta_2 \equiv \frac{dC_{max}}{dP} = g(P) + Pg'(P) - f'(g(P))g'(P)$$

and since $P = f'(\delta_1)$, the second and third terms cancel to leave $\delta_2 = g(P) = \delta_1$:

back-substitution and differentiating is a waste of effort.

Or not entirely: for it is only by virtue of this process that we can establish the result that *for a linear system* δ_2, *and* δ_2 *therefore also,* is a lower bound.

But is it true that δ_2 is a lower bound for a non-linear system? Are the results in the figures above wrong or not? Careful analysis of the steps of the proof of the total potential energy theorem (see [2], or the appendix) shows that it is not affected by non-linear deflections (but linear elasticity): it still follows that the true TPE $C \equiv P\delta - U$ exceeds that for any fictitious, admissible shape, $C^* \equiv P\delta - U^*$. Only the final stage of interpreting the

LHS as $\int_0^{P_1} \delta\, dP$ rather than $\frac{1}{2}P_1\delta_2$ differs. But the fact that $\int_0^{P_1} \delta\, dP \geq C^*{}_{\max}$ does ***not*** prove that $\delta(P_1) \geq \delta_2$: if there is a range of loads with deflections substantially below the exact values, the integral has a balance in hand and can then, over a limited range, tolerate values above the exact values. Which is just what we found using a parabola or a catenary. On the other hand, the linear curvature shape is asymptotically correct at low loads: the integral has no balance in hand, and the results are indeed always below the exact values. So the results are correct, and

there is no lower bound principle for a non-linear system.

APPENDIX: The Theorem of Total Potential Energy

Consider a kinematically admissible set of strains and displacements (κ^*, δ^*), the true set being (κ, δ). Let (M, P) be the true equilibrium set of loads and moments. Applying the virtual work theorem twice and subtracting:

$$P(\delta^* - \delta) = \int M(\kappa^* - \kappa)ds \ .$$

Non-linearity between force and deflection does not affect this.

For each element it is easily shown that $dU(e^*) - dU(e) \geq M(\kappa^* - \kappa)ds$, (where $dU(e^*) = \frac{1}{2}EI\kappa^{*2}ds$) so that

$$U^* - U \geq \int M(\kappa^* - \kappa)ds \, .$$

The moment/curvature relation is still linear.

Hence $U^* - U \geq P(\delta^* - \delta)$ or $P\delta - U \geq P\delta^* - U^*$

Just as for a linear system.

For a non-linear system we need to be clear that each side of the equation represents the area *above* the load–deflection curve: the 'co-energy', ie the complementary energy but written in terms of displacements not loads. It therefore equals $\int \delta\, dP$, not $\int P d\delta$ – though each would equal $\frac{1}{2}P\delta$ for a linear system. As discussed in the text, this is irrelevant to the application of the theorem: clearly the RHS should be maximised, and the best answer we can obtain from our trial shape is the value which gives this maximum.

References

1. Stewart, M., McLachlan, A.D. & Calladine, C. R (1987) A model to account for the elastic element in muscle crossbridges. *Proc. Roy. Soc.* **B229** pp 381-413.
2. Calladine, C. R (1999) Lecture notes, University of Cambridge Engineering Tripos, Part II Module R4: Structural Theorems and Concepts.
3. Bisshopp, K.E. & Drucker D.C. (1945) Large Deflections of Cantilever Beams. *Q. Appl. Math.* **3** pp 272-275. An outline is given in Timoshenko, S. *Theory of Elastic Stability,* McGraw-Hill 1936
4. Carlson, B. C. (1964) Normal Elliptic Integrals of the First and Second Kinds. *Duke Math. J.* **31** pp 415-419. See also notes on NAG Library Routines.

ROSE WINDOWS

Jacques Heyman
Department of Engineering, University of Cambridge
Trumpington Street, Cambridge, CB2 1PZ, U.K.
jh10002@eng.cam.ac.uk

Abstract The significant load on a large glazed window arises from the pressure or suction of wind. The masonry in a rose window resists this loading by developing flat-arch action, and thrusts with considerable force against the surrounding walls. Thus connexions at the periphery must be firm, and the masonry of the window itself must be of a satisfactory geometrical form to accommodate any irregularities which may develop at the boundary.

Keywords : Masonry structures, Windows

1. INTRODUCTION

It seems clear that a hole may be punched through a masonry wall with relative impunity. If the hole is rectangular, then the head will need a monolithic lintel to 'relieve' the weight of the masonry in the wall above; more sophisticatedly, the monolith may be replaced by a flat arch of several 'voussoirs'. As the rectangular window becomes larger, then the flat arch develops, in Gothic, into a pointed form, and the eye is given a good indication of the 'flow' of forces round the window.

If the hole is circular, then the opening has been called, for a century and a half, an oculus, or eye. The circular hole again needs reinforcement round the perimeter; the oculus is bounded by a circular arch in the upper half, and a reflected arch in the lower. Unglazed oculi admitted both light and air to large public rooms, whether in churches or private houses; they were not used in small chambers such as bedrooms where light was not needed and air could be admitted through an open doorway. Oculi can be found in the gables of Romanesque churches up to the late twelfth century, but larger round windows were already developing in early Gothic. An open oculus, of say 2 m diameter, admits some light, but not enough – it also admits rain.

As oculi became larger during late Romanesque, so the need for primitive glazing arose, and this glazing required a frame. The simplest possible frame is formed by a single vertical member (a mullion), or by a

H.R. Drew and S. Pellegrino (eds.),
New Approaches to Structural Mechanics, Shells and Biological Structures, 115–125.

mullion and a horizontal member (a transom), placed diametrally to form a cross. Further 'spokes' were soon added to form the wheel window, with 6, 8, 12, 16 or even more radial members running from a central 'hub' (which could be a glazed ring of stone) to the circumference (Cowen, 1979).

The static vertical loading on a spoke is negligible, whether the member runs horizontally, vertically, or is inclined. The critical loading on the members of a masonry window arises from transverse wind forces, and this loading is many times greater than the self-weight of the material. Thus gravity forces may be neglected for a primary analysis, and only the response to horizontal wind investigated.

2. THE FLAT ARCH

A large window in a cathedral may be approximated by a rectangle, say 12 m x 9 m. There may perhaps be two main mullions, together with subsidiary mullions and horizontal transoms, but essentially the two main mullions receive the wind forces from the glazing housed into them, and transmit those forces to their ends at top and bottom of the window, Fig. 1. The mullions may be thought to act as beams standing vertically and subject to a uniform horizontal load, just as architraves in Greek trabeated architecture may be thought to act as horizontal beams subject to their own vertical weight.

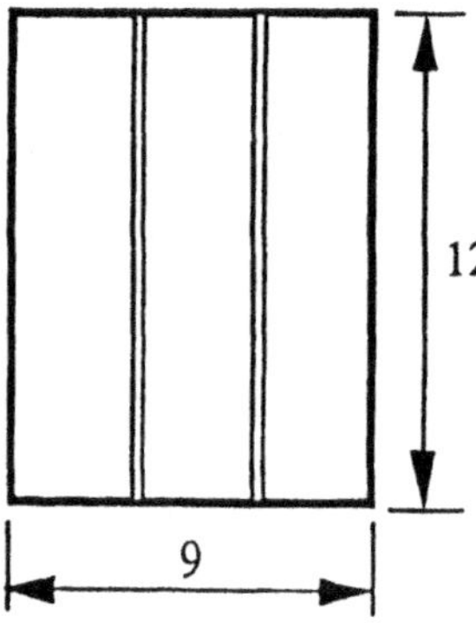

Figure 1. A rectangular window.

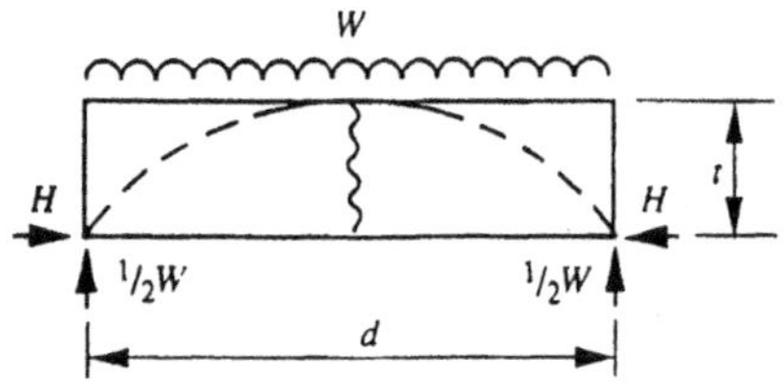

Figure 2. A Greek architrave.

A simple architrave is represented in the sketch of Fig. 2, and the (self-weight) load W induces vertical end reactions of $\frac{1}{2}W$. If the architrave is really acting as a beam in bending, then tensile strains will be present on the under surface, and corresponding compressive strains on the top. Such actions are in fact inadmissible for masonry (Heyman, 1995); individual stones may be strong in tension but, once cracked, they can transmit only compressive forces between those portions which remain in contact. Many massive Greek architraves are cracked through near their centres, as indicated in Fig. 2, and there is no possibility of regarding this cracked structure as a beam in bending – it is a flat arch formed of two 'voussoirs' (Heyman, 1972). Such a flat arch will remain stable if it is prevented from spreading; the horizontal thrusts H in Fig. 2 are supposed to be provided (in a Greek temple) by adjacent architraves. A very small increase in span will induce a small sag in the architrave, and the central crack will develop into a 'hinge'; forces within the masonry, the thrust line, are indicated by the broken line in Fig. 2.

Thus an uncracked architrave subjected to a uniform load W may be supported at its end by purely vertical forces $\frac{1}{2}W$. Once cracked, however, the architrave can only remain in equilibrium if a horizontal force H is developed; by simple statics,

$$H = \frac{Wd}{8t}. \tag{1}$$

Each of the two mullions in Fig. 1 will be made from several pieces of stone, pinned together at their centres by metal dowels (of iron, phosphor bronze, stainless steel). The dowels can contribute little to bending strength, and the mortar (if any) in the joints can develop at best feeble tensions. The mullion is acting as a (vertical) flat arch composed of several 'voussoirs', the voussoirs being connected to each other and to the masonry frames at top and bottom by dowels, so that the stones cannot slip; the mullion is acted upon by (horizontal) wind forces, and behaves exactly as the Greek architrave of Fig. 2. If the transverse wind pressure is taken as 1 kN/m^2 (20 lb/ft^2, a slightly low figure), then each mullion in Fig. 1 will be subjected to a uniform loading of 36 kN (i.e. 3.6 tonnes). Thus if d/t is 24 (say), then $H = 3W$ from equation (1); a transverse load of 36 kN will engender end thrusts of 108 kN. Each mullion is sustained in equilibrium by horizontal reactions of 18 kN (the end reactions of $\frac{1}{2}W$ of Fig. 2), and the end dowels serve to pass these reactions to the main fabric; and the wind also induces vertical end reactions (H of Fig. 2) of value 108 kN (if the mullion has proportions d/t = 24), as indicated in Fig. 3. Each mullion will contain within its thickness a curved line of thrust (the broken line of Fig. 2); this thrust line is quasi two-dimensional, lying in a vertical plane parallel to the direction of wind.

An obvious requirement is therefore that a mullion, and in general any member or sub-member of a framed masonry window, be essentially straight; that is, only gentle curves are permitted. The requirement is familiar; for example, the scissor braces connecting the crossing piers at Wells Cathedral are certainly not straight, but they are effective because straight 'props' may be contained within their profiles.

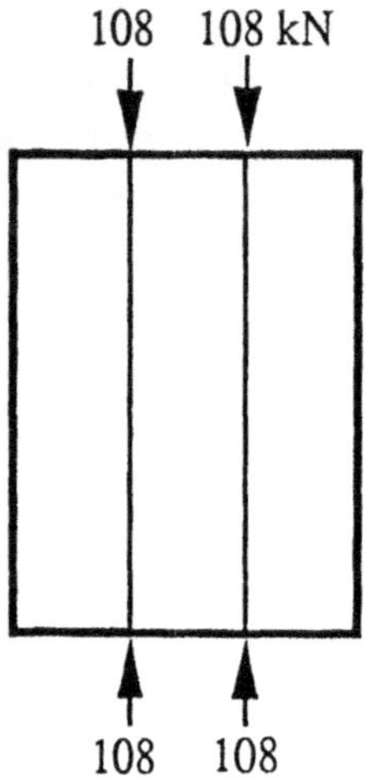

Figure 3. End loads on mullions.

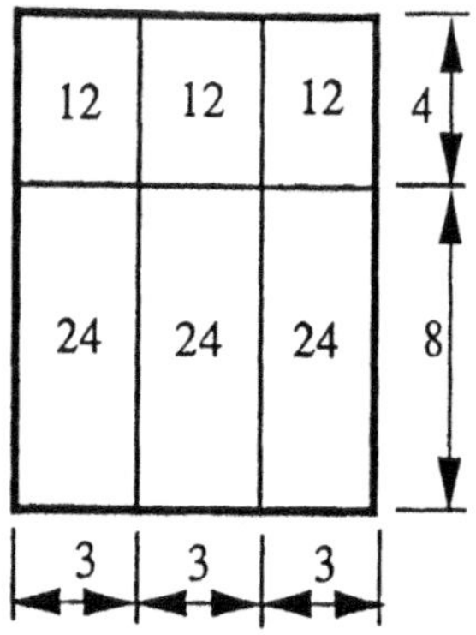

Figure 4. Mullions and transom.

3. EQUILIBRIUM ANALYSIS

The above numerical example used, without comment, an assumption of a kind usually made by structural analysts. The wind acts upon the glazing of the window, and the glazing bars transmit the wind forces to the mullions, and so to the frame of the window; it was assumed that the loading on a mullion was distributed uniformly. Such an assumption is made in the absence of any specific information about the disposition of the internal framing of the window – in fact the problem is hyperstatic in the extreme, and the best the structural analyst can do in such cases is to make a 'reasonable' guess about some of the parameters entering into the calculations. Such guesses do not invalidate the process, and it is a matter of fact that structural engineers have always worked in this way, and have relied, unknowingly for the most part, on the comfort of the plastic theorems.

The lower bound, or 'safe' theorem, is simple – if the analyst can find a state of the structure in which the structure is 'comfortable', then there cannot exist any state which might lead to collapse. The power of this theorem is that the analyst is not required to determine the 'actual' state of a structure under the given loading – if any one equilibrium state can be found, then this is sufficient proof that the structure itself will exist in a (certainly different) state of safety.

Thus for the rectangular window of Fig. 3 acted upon by a transverse wind load of 1 kN/m^2, each of the two mullions *could* be subjected to *uniform* loading of 36 kN, and this loading will engender the end thrusts shown of 108 kN. The forces shown in Fig. 3 represent one possible state of equilibrium for the window – if the end thrusts can be sustained by the mullions (and this is a question of stress analysis rather than overall structural behaviour), then the calculations have led to the assurance that the window can carry the required wind load. (For a mullion of suitable dimensions in Fig. 3, the average compressive stress arising from the end thrust is about 1 N/mm^2, which is an acceptable value.)

The point can be developed if a transom is added to the windows at two-thirds of the height of the mullions, as sketched in Fig. 4. The three upper panels of glazing will then be subjected to wind forces of 12 kN, and the lower to 24 kN. Once again an assumption has to be made as to how these forces are carried by the transom and mullions. One possibility, from the point of view of static equilibrium, is that all the wind forces are transferred to the transom (and to the top and bottom frames of the window), so that each of the three portions of the transom is subjected to a uniform (!) load of 18 kN, Fig. 5. The transom could be considered as a long member spanning 9 m from side to side, and receiving no support from the mullions; since the transom will be made of short lengths of stone, dowelled together, this is an unreasonable assumption. A much more reasonable view of the action of the transom is that it acts as three flat arches of the type sketched in Fig. 2, in which case the equilibrium forces of Fig. 5 may be determined.

From this figure it will be seen that each mullion is acted upon by a transverse point load of 18 kN (and by no other wind load) as sketched in Fig. 6. The thrust lines in this (cracked) mullion are straight and, for $d/t = 24$

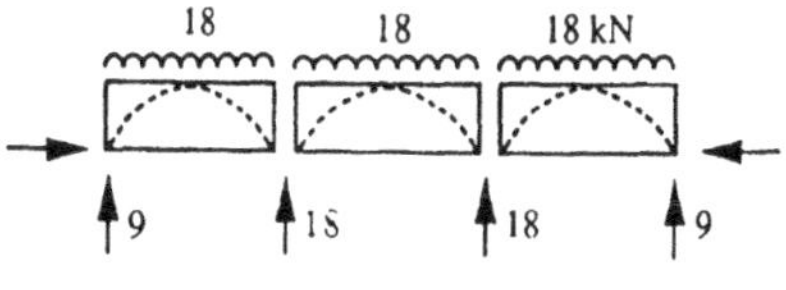

Figure 5. Equilibrium of transom.

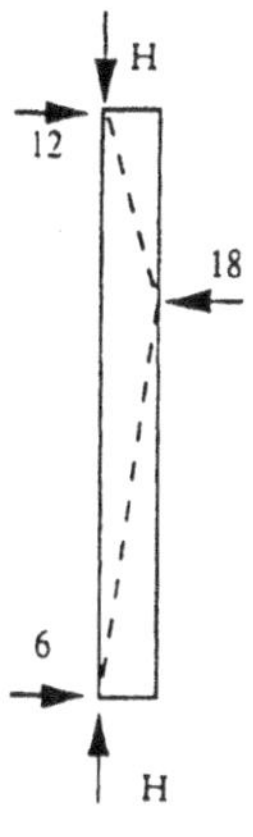

Figure 6. Equilibrium of mullion.

again, the value of H is determined as 96 kN (compared with 108 kN without a transom). In actual fact the transom and mullions will act partially together in some unknown way, but it must be emphasized that the 'reasonable' calculations leading to Figs 5 and 6 are safe. Consideration of further and much more complex sub-framing of the window (additional smaller mullions and further transoms) leads to the same conclusion that the mullions are subject to end forces of about 100 kN.

As an example of interactive response of the members of a window, the artificial framing of Fig. 7 may be considered. The wheel window shown is supposed to have only 3 'spokes', each of length 2 m, supporting the glazing. As before, the wind load will be taken as 1 kN/m^2; thus the window as a whole is subject to a load of 4π or 12.6 kN. Some of this load may pass directly to the circumference of the window, and some may be carried by the three spokes – a worst case would arise if the ferramenta distributed the load only to the spokes, so that each carried 4.2 kN. This load will not be distributed uniformly, but in the absence of information about the disposition of the glazing bars, the equilibrium of a spoke will be considered as sketched in Fig. 8. If the ratio d/t has value 10 (i.e. a stone spoke of length 2 m and depth 200 mm) the value of H is determined as 21 kN.

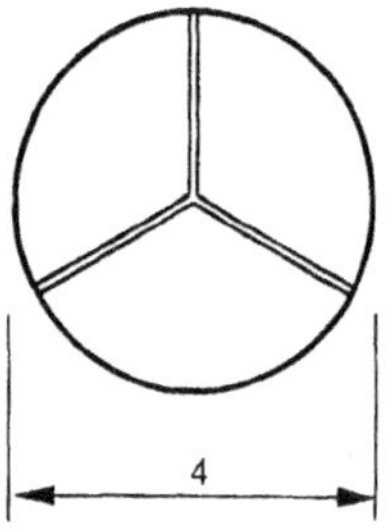

Figure 7. Basic wheel window.

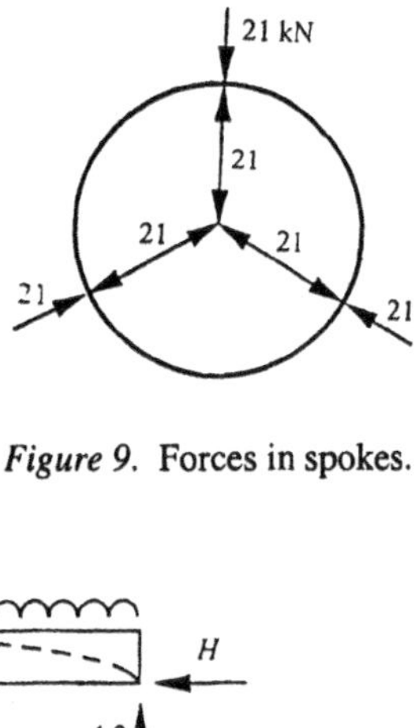

Figure 9. Forces in spokes.

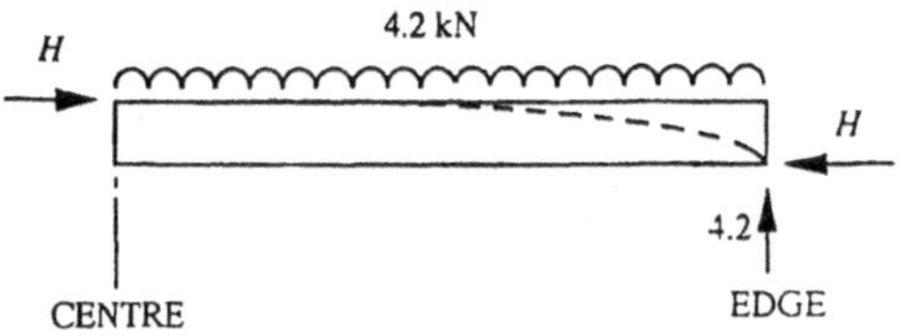

Figure 8. Equilibrium of a spoke.

Each of the three spokes thrusts at the centre of the window with a force H (= 21 kN); the three forces lie in a plane and are inclined to each other at 120°, and are therefore self-equilibrating. Figure 9 displays the equilibrium state of the window under the transverse loading of 12.6 kN, and is a simple and artificial but archetypal example of the kind of analysis that is needed for the assessment of the stability of rose windows. The 'spokes' of such a window thrust against the surrounding masonry, and the values of those thrusts (e.g. $H = 21$ kN) must be determined. Further, there must exist flat planes of thrust within the members of the window, even if these members show some moderate curvature. The internal thrusts may of course change their directions at junctions between members. The paths will appear straight in an elevation such as Fig. 9; within individual members, the thrusts will follow paths such as that shown by the broken line in Fig. 8.

4. EQUILIBRIUM OF ROSE WINDOWS

A simple wheel window is sketched in Fig. 10, and is supposed to be subjected to transverse wind load. It is clear that, once the reactions at the ends of the spokes have been found, a satisfactory pattern of equilibrium forces can be found for the whole window. The end reactions will be transmitted directly down the spokes; at their inner ends, the spokes are shown in Fig. 10 as terminating against an inner ring of masonry. The forces from the spokes will put the ring into (almost) uniform circumferential compression; in fact, a regular twelve-sided figure, a dodecagon, may be imagined as contained within a circular ring as recording the traces of the internal compression. Thus the geometry of an arrangement such as that shown in Fig. 10 is satisfactory.

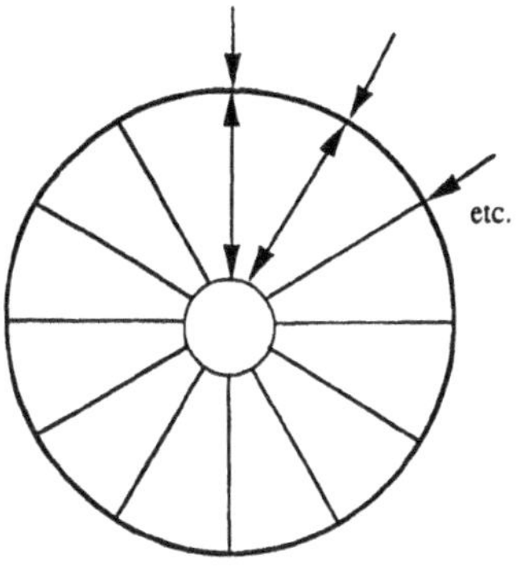

Figure 10. Simple wheel window.

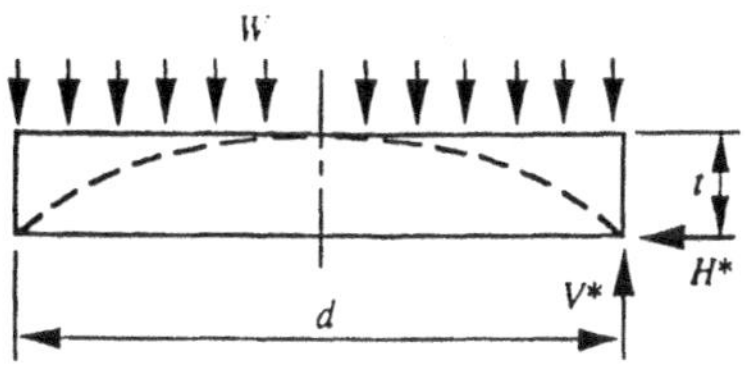

Figure 11. Wheel window with infinitely many spokes.

In order to check stresses, the end reactions at the spokes must be determined, and a very simple approximate formula may be found. Figure 11 imagines the window of Fig. 10 to be 'smeared' into a continuous circular slab, where the external wind load W is resisted by the domical thrusts indicated by the broken line. The sum of the horizontal reactions at the perimeter of the window of diameter d must of course equal W; the reaction V^* is supposed to be distributed round the window. Similarly the thrust H^* of the window against its circular frame is also distributed round the circumference. For small values of t/d it is easy to show that

$$H^* = \frac{Wd}{4t}. \tag{2}$$

Thus for the three-spoke example of Fig. 7, for which $d = 4$ m and $t = 200$ mm, the value of H^* is $5W$, and for $W = 12.6$ kN (as in the numerical example), $H^* = 63$ kN. This *total* reaction is provided by 3 spokes, so that the thrust contributed by each spoke is determined as 21 kN, exactly as shown in Fig. 9.

4.1 A Numerical Example

The rose window of Fig. 12 has diameter 6 m and has 12 'petals'. The thickness of the masonry is supposed uniform at 300 mm. The wind loading is 2 kN/m^2; thus the total load on the window is $W = 56.5$ kN. The ratio $d/4t$ has value 5, so that H^* from equation (2) totals 283 kN. Since the window

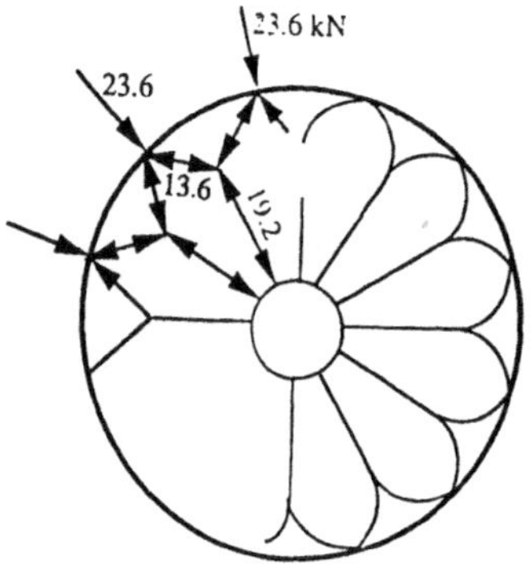

Figure 12. Forces in simple rose window.

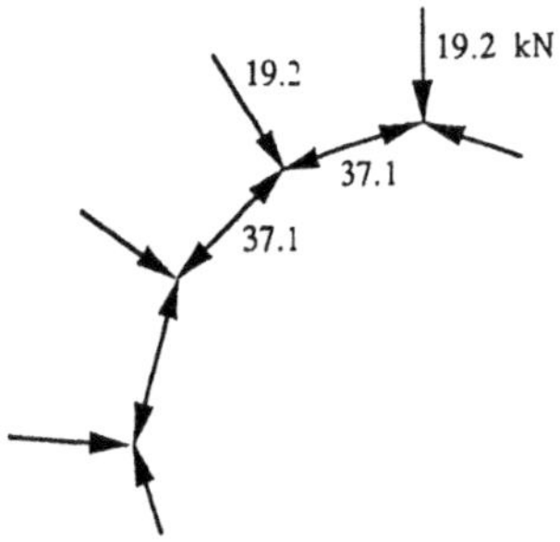

Figure 13. Forces in inner ring.

has 12 compartments, the reaction at each of the petal tips is 23.6 kN, as shown in Fig. 12. Each of the petal tips can be approximated by an equilateral triangle lying within the boundaries of the curved masonry, so that the analysis can be extended to the inner members of the rose, giving the compressive forces marked in Fig. 12. Finally, the conditions in the inner dodecagon are as sketched in Fig. 13. The force transmitted round the circle has value $\frac{1}{2}(19.2)$ cosec 15°, or 37.1 kN, and a masonry section 150 x 300 would keep average compressive stresses below 1 N/mm^2.

Different notional triangulations are possible for the petal tips, but different assumed angles lead to the same conclusion that loads in the 'spokes' and petal 'triangles' are about 20 kN, and perhaps double this in the central ring (cf. the values marked in Fig. 12).

4.2 Notre Dame de Mantes

The window at Mantes (c. 1180) sketched in Fig. 14 has a diameter of 8 m; all masonry will be taken to have dimensions 250 x 250. At a wind loading of 2 kN/m^2, the total transverse load W is 32π or 100.5 kN. Since $d/4t$ = 8, equation (2) gives H^* = 804 kN, and each of the 12 outer spokes will thrust against the containing masonry with a force of 67.0 kN. On the left-hand side of Fig. 14 is shown a 'reasonable' skeletal structure consisting of straight members containable within the actual curved masonry, and simple triangulation of forces leads to the values marked in the figure. The force round the inner ring (the dodecagon) is calculated as 91.6 kN; average compressive stresses in the members are of the order of 1 N/mm^2.

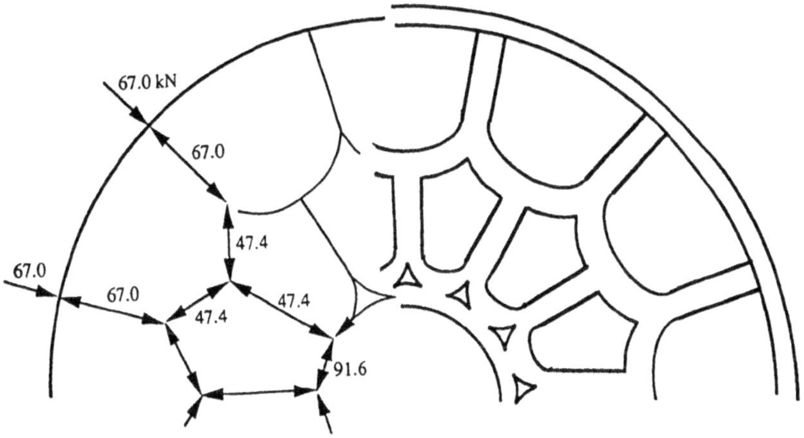

Figure 14. The rose window of Notre Dame de Mantes.

5. THE GABLE END

Figure 15 shows a rose window inserted in a gable end – perhaps to a transept without side aisles, so that the supporting buttresses may be placed against the masonry without the need for flyers. With or without flyers, the buttressing system will, inevitably, give way, and an initially circular window will become oval; the ovality may not be appreciated by the eye, but may in practice amount to a few centimetres out of circular. The framing of the window is represented schematically in Fig. 16 by a simple cross; the distortion of the window, greatly exaggerated, will tend to crush the vertical 'spokes', while the horizontal spokes are left poorly supported at their ends.

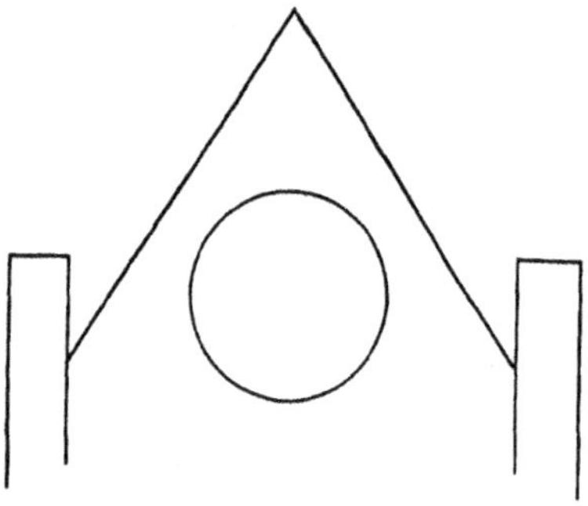

Figure 15. Rose window in a gable.

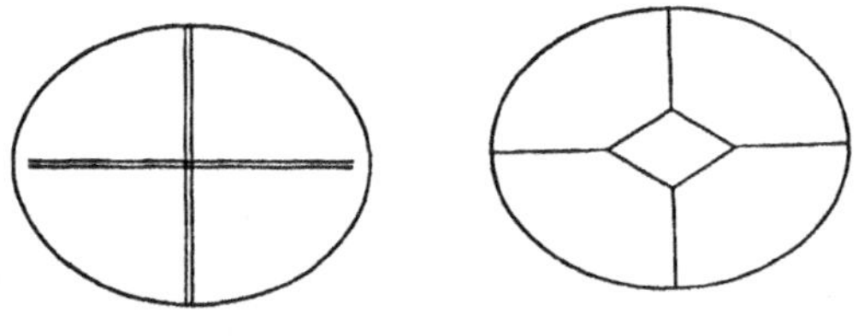

Figures 16 and 17. Ovalization of a window.

By contrast if the framing had incorporated a central ring of masonry, the scheme of Fig. 17, then it is evident that ovalization of the window could be accommodated flexibly, with the central ring also distorting. In Fig. 17 the central square has become a lozenge; in Figs 12 and 13 the regular dodecagon contained within the circle would flatten, but remain satisfactory as a load-bearing rib.

As a matter of interest, had the cross of Fig. 16 been rotated through 45° (at least one example of such a window exists) then the now diagonal spokes would, to the first order, be unaffected by ovalization of the frame.

6. CONCLUSION

The significant loading on a rose window arises from the action of wind. The members of the window are made from relatively small pieces of stone, and cohesion of the assembly must be assured by using connecting dowels, so that transverse forces arising from the wind may be transmitted from stone to stone. Similarly, the peripheral stones of the window must be firmly dowelled to the surrounding masonry.

The transverse wind loading is resisted by flat-arch action of the window as a whole, and a pattern of compressive forces is generated within the components of the window. These components may be curved visually, but they must have geometries such that a compressive 'truss' of straight members may be imagined to be contained within their widths. The members of the window thrust against the surrounding masonry, and they must remain 'tight' in order to transfer the wind loading. In practice the geometry of the opening surrounding the window will distort as a result of shrinkage of mortar, foundation settlements and other unpredictable imperfections, and the rose window itself must have a flexible geometry to accommodate such movements. Such flexibility is inherent in a window in which the spokes are supported by a small inner ring of masonry.

References

Cowen, Painton (1979) *Rose Windows*, Thames and Hudson, London.
Heyman, J (1972) 'Gothic' construction in Ancient Greece, *J.Soc. Architect. Hist.*, **31**, 3-9.
Heyman, J (1995) *The Stone Skeleton*, Cambridge University Press, Cambridge.

JOINING OF METAL MATRIX COMPOSITES

F.A. Leckie
College of Engineering
University of California
Santa Barbara, CA 93106 USA
Leckie@engineering.ucsb.edu

Abstract: The strength and failure modes of metal matrix joints are investigated. The metal matrix composite (MMC) consists of alumina fibers in an aluminum alloy matrix. Butt joints and joints with attachments are manufactured and subsequently tested in tension to failure. One set of joints is manufactured from aluminum matrix material and another with aluminum alloy reinforced by silicon carbide particles. The initiation of failure in the alloy matrix joints occurs at the transverse interface of the MMC with the joint. Large insertion lengths are needed for the aluminum attachment to reach the composite strength. The addition of particles suppresses failure at the interface and only a small insertion length is needed to match the full strength of the composite.

Key words: Joints, Metal Matrix Composites.

1. INTRODUCTION

Tensile elements in steel structures can be joined by butt welds without loss of strength or increase in weight. They set a standard of effectiveness. However the interface strength of many joints is small. In such circumstances joints are designed to avoid tensile stress at the interface and load is transferred by a combination of compression and shear.

Historic examples are the medieval bridge at Kappellbrücke in Switzerland and the numerous covered bridges in the United States built during the 19th Century. The diagonal members of the truss illustrated in Fig.1a are in compression and the joint simply positions the brace (Fig. 1b). The long horizontal members are made by joining timbers end-to-end by scarf joints, Fig. 1c, in which the interface is subject to a large shear force

H.R. Drew and S. Pellegrino (eds.),
New Approaches to Structural Mechanics, Shells and Biological Structures, 127–138.

and a small normal tensile force. Tongued and grooved joins support the shear and metal bands, wrapped around the perimeter, resist the tensile interface force.

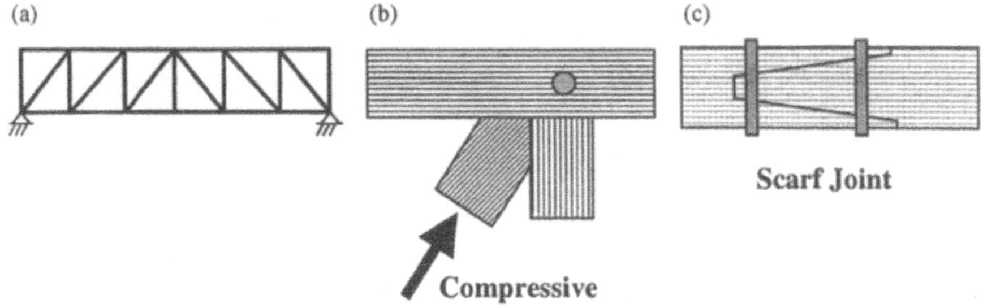

Figure 1. Joints in wooden bridges.

Metal trusses are longer than those made of wood and failure by buckling of the diagonals becomes more likely. Buckling in the diagonals is avoided by arranging them to carry tensile forces as illustrated in Fig. 2a. However an attachment must be added and load is transferred to the main structure by a bolted plate connection, Fig. 2b, by compressive bearing and shear stresses in the bolts.

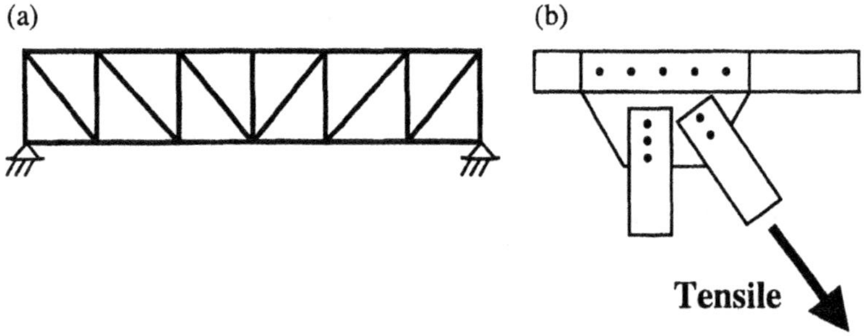

Figure 2. Joints in metal bridges.

The specific elastic and strength properties of Metal Matrix Composites (MMC) make them attractive candidates when structural weight is very important. The advantage can only be realized when the direction of the fiber and load coincide. For biaxial loading, when fibers are needed in two directions the specific properties are effectively halved since the load carried by the fibers is much greater than that carried by the matrix. MMCs are most effective for use in elements in which the stress is essentially tensile. The standard tensile test of unidirectional MMCs is difficult and special testing fixtures are needed to avoid premature failure at locations other than within the composite. The difficulties evident when performing tensile tests are likely to be repeated in joints connecting MMC elements.

A representative design configuration is shown in Fig. 3a. The load is transferred from the uni-axially reinforced MMC tie to the main structure by means of a monolithic sub-element formed from the matrix metal. The arrangement takes advantage of the properties of both materials. The MMC strut has good specific stiffness and strength properties aligned with the load. By contrast, the specific properties of the metal matrix are less favorable, but the isotropic matrix performs better under the multi-axial stress states occurring in the sub-element. Because the interface properties are not well understood, the interface strength normal to the fiber ends is usually ignored. Force transfer is by shear along the insertion length. Such joints have been demonstrated to be reliable in practice but their additional weight diminishes the advantage of the specific properties of the MMC.

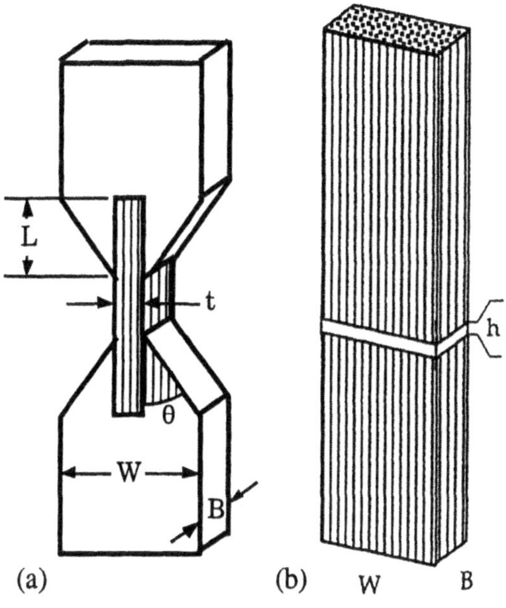

Figure 3. Butt and Attachment Joints.

Model MMC joints were manufactured and subsequently tested so that the strength and failure modes could be investigated. The Metal Matrix Composite (MMC) consists of alumina (Al_2O_3) fibers in a matrix of an aluminum alloy (Al-4.5Mg). Adding silicon carbide (SiC) particles to the aluminum alloy increases the stiffness and strength properties of the matrix. When particles are added to the aluminum alloy the matrix is said to be reinforced. Because the joints were produced within a laboratory setting good control of the process eliminated evidence of manufacturing defects. It is then possible to manufacture a series of model joints for which both geometry and matrix properties could be controlled. Two types of joints are investigated.

The butt joint, shown in Fig. 3b, transfers the applied load by tension normal to the transverse interface. The results of the tests are expected to provide information about the strength of the transverse interface. The attachment joints shown in Fig. 3a transfer load by a combination of tensile stresses at the transverse interface and shear along the longitudinal interface. The results of the tests are expected to provide information on how load is distributed between the transverse and longitudinal interfaces.

2. MATERIALS AND MANUFACTURE

The fiber of the MMC is a continuous polycrystalline α alumina Al_2O_3 of nominal diameter 10μm. The volume ratio is 0.6. Two different matrix materials are used. In one set of experiments the matrix is an aluminum alloy Al-4.5 wt % Mg. In the second set 500-gritSiC particles are added to the aluminum alloy. The volume ratio of the particles within the reinforced matrix is 53%. The stress/strain relations for the fiber composite, the aluminum alloy matrix and the particle-reinforced matrix are shown in Fig. 5.

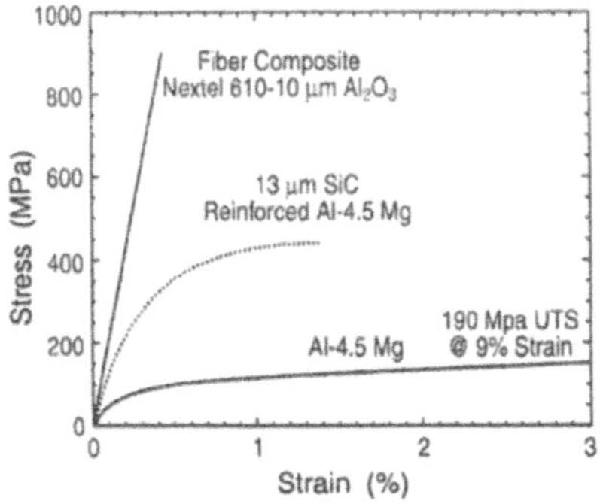

Figure 4. Stress/Strain of Joint Materials.

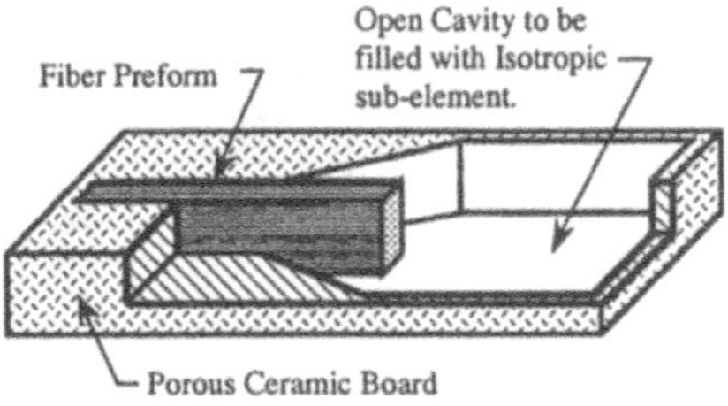

Figure 5. Porous Ceramic Mold.

The elastic modulus of the MMC is 200 GPa and the response is elastic until failure occurs at an average stress of 950 MPa. The aluminum matrix is ductile. The elastic modulus is 70 GPa, the yield stress 100 MPa, the ultimate stress 192 MPa and the failure strain 8.8%. The particle-reinforced matrix is semi-ductile. The elastic modulus is 166 GPa, the yield stress 300 MPa, the ultimate stress 440 MPa and the failure strain 1.4%.

The manufacturing technique involves the preparation of a suitable fiber preform followed by pressurized melt infiltration. Bars with a packing density of vol-60% are prepared by winding continuous Al_2O_3 with a polymer solution binder around a racetrack-shaped mandrel. After hardening the straight sections are cut and the resulting bars are encapsulated in a porous mold as shown Fig 5. The whole assembly is subsequently heated to 680°C to burn off the binder. The heated mold and preform are then placed in a cylindrical die which is filled with the molten matrix material and pressurized to 180 MPa. After solidification the castings are cut and machined to the required dimensions.

3. BUTT JOINTS

Tensile tests on the butt joints shown in Fig 3(b) have been reported by Brink [1]. Typically the specimens are 85 mm long, 11 mm wide and 6.5 mm broad (B). The thickness (h) of the joints is 150 μm and the aspect ratio B/h greater than 36. Typical graphs of the applied tensile stress and strains in the alloy and reinforced joints are shown in Fig. 6 (solid curve). Considerable scatter of +/- 15% characterizes the tests, the consequence-presumably-of unintended variation in manufacture. The elastic behavior is consistent with the elastic modulus of the corresponding matrix. The alloy matrix joints fail at 300 MPa or 3 times the yield stress of the matrix. Failure occurs at the interface and micrographic examination reveals the primary failure mechanism is debonding at the fiber ends, followed by void growth in the matrix adjacent to the fiber ends. In the case of the reinforced matrix (broken curve) the failure stress is 450 MPa and approximates the strength of the reinforced matrix. The failure path is through the middle of the reinforced matrix and the interface remains intact.

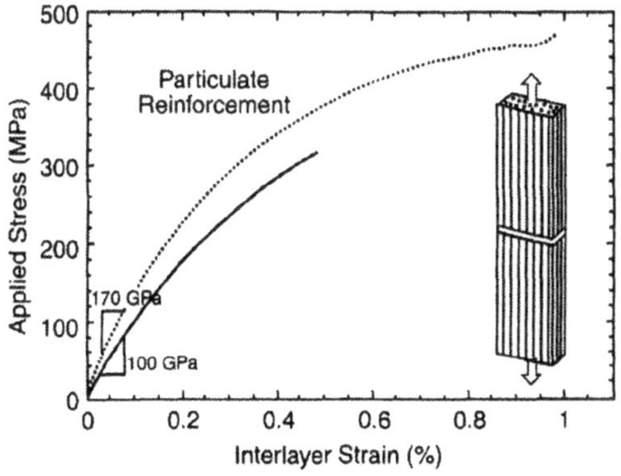

Figure 6. Stress/Strain of Butt Joints.

Finite element estimates of the normal stress and the plastic strains at the transverse interface are shown in Fig 7. The analysis includes the hardening properties of the joint matrix material measured in tensile tests, Fig. 4. The distributions of the normal stress and the effective strain at the joint interface shown in Fig. 7 correspond to the failure load of 300 MPa for the alloy joints and 450MPa for the reinforced. In the alloy joints there is an edge region whose length is approximately 1.6h in which there is a concentration of plastic strain with a maximum value of 1.6%. Within the core, which forms the greater portion of the joint, the stress and strain distributions are almost constant. By contrast the distribution of stress and strain in the reinforced joint is almost uniform and edge effects are small. The plastic strain at failure in the reinforced joint is an order of magnitude less than of the alloy joint confirming that the reinforcing particles constrain the growth of plastic deformation.

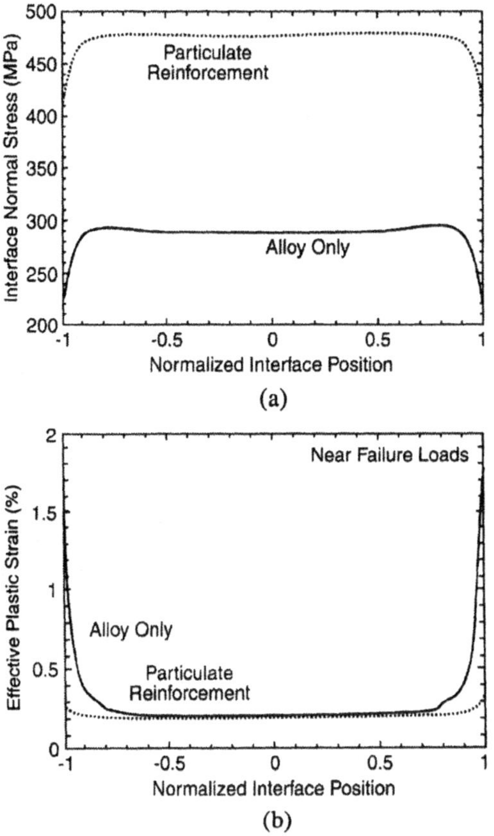

Figure 7. Stress and Stress Distributions at Interface in Butt Joints.

The butt joints do not match the strength of the composite material. To achieve the composite strength attachments are needed to transfer a portion of the load by shear along the longitudinal interface.

4. JOINTS WITH ATTACHMENTS

4.1 Preliminaries

In the tapered attachments shown in Fig. 3a part of the load transfer is by tension across the transverse interface and part by shear along the two longitudinal interfaces. In order to attain a joint strength equal to the strength of the MMC a simple estimate of the insertion length L is,

$$\frac{L}{t} = \frac{(\sigma_{UTS}/K_1) - \sigma_T}{\left(2/\sqrt{3}\right)\sigma_o} \tag{1}$$

In this relation σ_{UTS} is the strength of the composite, σ_T the strength of the transverse interface and σ_o the yield stress of the matrix. The coefficient K_l is the stress concentration factor at the junction of the taper with the composite. From finite element calculation the value of the stress concentration factor of the alloy attachment is estimated to be between K_l=1.2 and K_l=1.35 for the reinforced attachment. The stress concentration means the maximum load the attachment can support is σ_{UTS}/K_l and indicates the importance of detailed deign to keep the stress concentration small. Decreasing the taper and rounding out the sharp intersection angle can reduce the stress concentration. Both measures introduce complexity to the manufacturing process and are not attempted at this time.

If it is assumed the interface strength of the alloy attachment is limited by debonding at the fiber ends as observed in the butt joints, an interface strength of σ_T =300 MPa might be postulated. With this value of σ_T the estimate of the insertion length is L/t=4.07. If it is assumed the interface has no strength the insertion length becomes L/t=7.0. Reinforced matrix butt joints do not fail at the interface suggesting the interface strength σ_T is greater than 450 MPa. Using this value for σ_T in Eq. 1, the estimate of the insertion length is L/t=0.73. Neglecting the interface strength the estimate of the insertion length is L/t=2.03. It is evident that the insertion length when using the reinforced matrix is substantially less than that of the alloy matrix.

Tensile tests on MMC elements with tapered attachments have been reported in Ref. [2]. The joints shown in Fig 3a are of average thickness t=3.0mm and breadth B=9.0.mm. and the taper angle is 30°. The overall joint width W is sufficiently large, in all cases, to prevent failure of the attachment at positions removed from the transition.

4.2 Experiments on Alloy Attachments

For the alloy attachments tests are performed on joints with insertion ratios L/t of 1.0, 2.0 and 3.0. The preliminary estimates suggest values of the insertion length ratio L/t of between 4.0 and 7.0 are needed to reach the strength of the MMC. The tooling used in the manufacture is such that it is not possible to manufacture attachments for which L/t is greater than 3.0. Consequently the realization of full composite strength is unlikely in the alloy attachments. In situ optical observation revealed a consistent failure mode. Substantial plasticity was observed in the attachment at the intersection of the longitudinal and transverse interfaces. Failure initiates at the corners and continues by progressive separation of the interface between the attachment and the fiber ends. When the transverse interface is fully separated final failure occurs by shear-out along the longitudinal interface. The relation between the measured failure load and the insertion ratio L/t is shown in Fig. 8(a). The failure stress for L/t=3.0 is slightly less than the failure stress when L/t=2.0. This surprising result prompted closer examination of the fracture surface to reveal this specimen had a ribbon-shaped defect along the longitudinal interface where infiltration had not occurred. This lack of infiltration reduced the effective shear transfer area by approximately 30%. If the experimental failure stress for L/t=3.0 is modified to account for the reduction of the area, the failure stress is consistent with the other experimental results. Since the experiments are few the modification must be considered with caution.

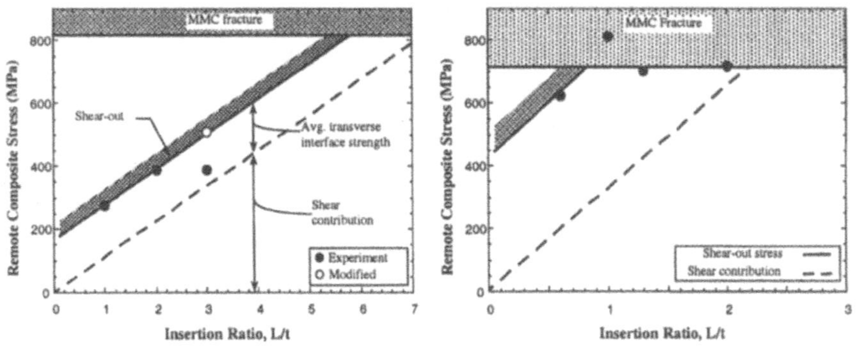

(a) (b)

Figure 8. Failure Load/Insertion Length for (a) Alloy Attachments and (b) Reinforced Attachments.

4.3 Experiments on Reinforced Attachments

Tests on reinforced attachments are performed with insertion ratios L/t of 0.6,1.0, 1.3 and 2.0. The preliminary calculations suggest that insertion ratios L/t >2.0 do not add to the strength. Unlike the alloy attachments it may be possible for the reinforced attachments to reach full composite strength.

The relation between the failure stress of the reinforced specimens and the insertion ratio L/t is given in Fig. 8b. When the insertion ratio L/t=0.6 the failure load is 625 MPa. The fracture path is everywhere within the reinforced matrix (Fig. 9a). The major portion of the path is parallel and close to, but not on, the transverse interface. The path then passes through the shoulders of the attachment at an angle of approximately 45°. There is no evidence of the shear-out failure observed in the alloy attachments. When the insertion length ratios L/t are greater than 1.0 failure occurs in the MMC at the intersection of the attachment with the composite (Fig. 9b) No damage is observed at the transverse interface between fiber ends and the reinforced attachment.

The failure stresses are 70-80% of the tensile strength of the MMC and are consistent with the knock down associated with the stress concentration.

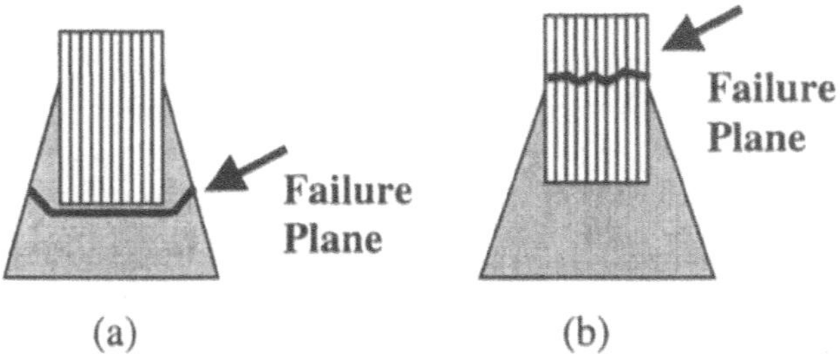

Figure 9. Failure Modes in Reinforced Attachments for Insertion Ratio (a) 0.6 and (b) 1.0.

4.4 Finite Element Calculations

The results of finite element calculations for the insertion ratio L/t=1 are shown in Fig. 10. The results are typical for all insertion ratios. The distributions correspond to the failure load of 288 MPa for the alloy attachment and 625 MPa for the reinforced attachment. In the case of the alloy attachments the normal stress and plastic strain vary considerably, with maximum values occurring at the intersection of the transverse and longitudinal interfaces. By contrast, the distributions in the reinforced attachments are largely uniform, with none of the concentrations evident in the alloy attachments. The maximum plastic strain in the reinforced attachment is less than one half of that in the alloy, illustrating the ability of the particle reinforcement to inhibit plastic deformation.

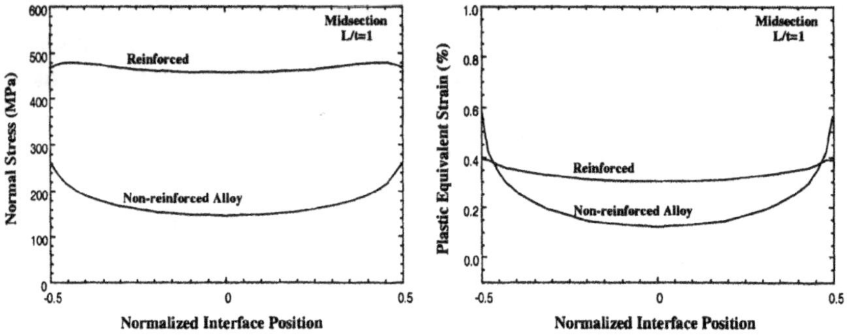

Figure 10. Stress/Strain Distributions in Attachment Joints.

4.5 Conclusions

A convenient description of the strength of the alloy attachment is presented in Fig. 8(a). Neglecting the transverse interface strength and assuming that load transfer occurs only by shear at the longitudinal interface, the estimated failure load is shown by the dotted straight line. The experimental results (including one modified) fall on a straight parallel to the shear only line with an offset of approximately 180 MPa. The result suggests a transverse interface strength of 180 MPa, which is less than the value of 290 MPa of the butt joint. That this should be is, perhaps, the consequence of the stress concentration observed in Fig 10 at the sharp corner of the MMC insert. This suggestion is consistent with experimental observation that the sharp corner acts as an initiation site for failure. Another location of stress concentration is the intersection of the attachment with the composite. The stress concentration is about 1.20 and when applied to the failure strength of the composite defines the maximum attainable joint strength. If the transverse interface strength is neglected the insertion length to reach full strength is L/t=7.0. Including the transverse interface strength of 180 MPa reduces the insertion ratio to 5.5, which is not a significant improvement. The strength of the attachment is dictated mostly by shear at the longitudinal interface and the traditional design assumption of no transverse interface strength is appropriate.

The strength of the reinforced attachments in terms of the insertion length is shown in Fig. 8(b). The stress concentration at the intersection of the attachment and the composite is 1.35. When applied to the failure strength of the composite the maximum attainable joint strength is 704 MPa. With the exception of L/t=0.6 the joints fail within the composite at the junction of the attachment and the MMC, with no evidence of failure at the transverse interface. When L/t=0.6 failure occurs within the attachment at a slightly reduced stress. A portion of the fracture path is parallel to, but not on, the transverse interface. The remaining portion is in the shoulder of the attachment. The experimental evidence suggests that the addition of reinforcing particles suppresses failure of the transverse interface, and that attachment strength is dictated by the stress concentration at the junction of the attachment with the composite.

5. SUMMARY

Failure of all the alloy joints is initiated by debonding and subsequent void growth at the fiber ends. The transverse strength of the butt joint interface is 285 MPa, and that of the attachment is 180 MPa. The likely cause for the strength reduction is the stress concentration at the corner edge of the transverse interface of the attachment. It is not possible therefore to assign a fixed value of the interface strength, which is dictated by local values of stress and strain. For the attachment joints to achieve full composite strength a large insertion ratio of between 5.5 and 7.0 is necessary, depending on the assumption of the strength of the transverse interface. The interface strength does not affect greatly the insertion length since the major portion of load transfer is by shear along the longitudinal interface. The attachments are long and add bulk.

The addition of particle reinforcement increases substantially the strength of both butt and attachment joints. A notable result is the suppression of failure at the transverse interface. Failure of the butt joints occurs within the reinforced layer at a stress approximately equal to the failure strength of the reinforced matrix. Provided the insertion ratio L/t is greater than 0.6, failure of the attachment joints occurs within the MMC at the location of stress concentration of the attachment with the MMC. The stress concentration can be reduced by the introduction of a rounding radius.

ACKNOWLEDGEMENTS

The author and Professor Carlos C. Levi cooperated on this study, based on the work of my students D.R. Brink and S.A. Waltner. The investigation was sponsored by the National Science Foundation under grant CMS-9634927.

References

[1] Brink, D. (2000) *The Interplay of Microstructure, Geometry, and Stress State on the Mechanical Performance of MMC Joints*, Ph.D., University of California Santa Barbara.

[2] Waltner, S.A., Brink, D., Levi, C.G. and Leckie, F.A. (2001) Directional-to-isotropic Transitions in metal/matrix composite joints, *Composites Science and Technology* **61**, 531-543.

UTILISING TENSION IN LIGHTWEIGHT STRUCTURES

Ian Liddell
Buro Happold, Camden Mill, Lower Bristol Road, Bath and Department of Engineering, University of Cambridge.
Ian.liddell@burohappold.com

Abstract: The paper discusses the uses of tension elements in a range of lightweight and deployable structures and particularly the impact on this technology of the introduction of high-performance fibres that are increasingly available. It is concluded that the materials are generally used well below their breaking strengths and are selected for their other valuable properties.

Key words: Tension structures, High performance fibres

1. INTRODUCTION

Tension elements are the simplest form of load-carrying members and the most easily understood. Everyone is familiar with the principle of a crane carrying a load by a rope, or a washing line supporting clothes hanging out to dry. Although tension elements can be rods they are generally composed of bundles of fibres or linked chains that are flexible. In fact the lack of bending stiffness is an important characteristic. Woven fabric is also a tension element, but two-dimensional. In this case the lack of bias or shear stiffness in the cloth is also an important feature that allows the Gaussian curvature of the surface to change. By contrast, a sheet material with shear stiffness can be a three-dimensional synclastic surface, that is, a rigid shell that can carry compression.

Since buckling instabilities are avoided, tension elements can potentially utilise the full tensile capacity of high-performance fibrous materials. However there are limits to the stress levels to which these materials can

H.R. Drew and S. Pellegrino (eds.),
New Approaches to Structural Mechanics, Shells and Biological Structures, 139–153.

safely be used. Very often stiffness and low weight are the important factors in the selection. There can also be problems with inspection and maintenance that restrict their use in long-life structures.

Tension can be used in structures as flexible ropes for hauling or lifting, as taut ropes for rigging or supporting masts or spars, or as tensioned cables loaded normally to their span where the load is carried by the tension times the curvature. In this last category there have been several recent advances in applications.

The earliest tensile elements were ropes made from vegetable fibres. There is evidence of these being used in 3000 B.C. for sailing boats in the Mediterranean. The earliest representation of rope-making is on a tomb of the 5th dynasty in Thebes. This shows 3 men making rope using papyrus fibres or possibly flax. One is twisting the yarns using a whirling device. In the centre is a man who is controlling the twisting together of the yarns while a third is counter-twisting the finished rope. There was a continuous history of rope making using similar techniques up to the start of the 17th century when spinning machines were first developed by Joseph Huddart, a captain in the East India Company. Huddart's aim as described in Gourock (1950) was to get more uniform tension and twist in the fibres so that the full strength of the fibre bundle could be achieved in the larger diameter ropes. In the 1940's man-made fibres, initially Nylon, were introduced. Up to that time the preferred fibres came from hemp, and lower-grade fibres came from Sisal. Since then other manmade fibres with higher stiffness and strength have been introduced.

1.1 Lifting Tackles

The main uses of ropes from the earliest times would have been for lifting and pulling things and for sailing boats. There can be no doubt that the Egyptians would have used rope tackles with blocks and sheaves for lifting with mechanical advantage. The story of the Vatican Obelisk Parsons (1939) is quite interesting. It was cut by the King of Egypt in the 10th century B.C. In the year 41 A.D. it was transported from the City of Heliopolis to Rome by the order of the Emperor Caligula and set up in the subsequently named Circus of Nero. In 1585 it was moved to the Vatican Square by Domenico Fontana. He got the job after what amounted to an engineering competition that attracted 500 entries. The obelisk was 78.9 ft high and 9.1 ft square at the base and weighed 308 t. It was moved a distance of 260 ft and finally lifted into place by hemp rope tackles from timber scaffold towers on each side. The total weight of the lift was estimated to be 362 t. We are fortunate to know about this as the event occurred after the introduction of printing and Fontana chose to publish his work with detailed drawings and a comprehensive description of the process which included some basic calculations and testing.

1150 years earlier around A.D. 430 an obelisk was placed in the Hippodrome in Constantinople. On the side of the base is a carving indicating how it was moved and apparently raised by means of ropes and windlasses. There also exists a plaster plaque from the time of Sennacherib (B.C. 705 – 681) that shows men hauling on ropes to raise a bull colossus. Lifting and lowering obelisks by tension is very much safer than the methods alleged to have been used by the Egyptians. Since there had been little change in the available technology over the preceding 2,500 years there is every reason to believe that the Egyptians could well have used similar methods. Unfortunately the only information on the technologies of those times that has survived are carvings in stone, and the artists were much more interested in telling the stories of the genealogy of the Pharaohs than explaining the craft skills and engineering feats of the people of the time.

2. MATERIAL SELECTION

Tension tests were probably the first experiments on the strength of materials. They would have been done to establish the safety of lifting very large loads as described above. Since the earliest uses of plant and animal materials as tension elements, the range of materials has expanded hugely. Iron rods were introduce in the 17th century or possibly earlier and iron and then steel wire was introduced in the 19th. Since the introduction of Nylon in the 1930's, the range of synthetic materials has increased rapidly and the developments are still continuing. The following materials are commonly used for structural purposes.

Plant: Flax, Cotton, Hemp, Sisal, Wood (Cellulose),

Animal: Spider web, Silk, Wool, Gut, Hide, Tendon

Metalic: Steel, Stainless steel, Bronze.

Mineral: Glass

Synthetic: Acrylic, Polyamide, Polyester, Polyethylene, Polypropylene Carbon, Aramid, HMPE, Vectran, Zylon

Table 1. Typical properties of fibres used for flexible tensile elements.

	density	modulus	tenacity	strength	elongation
	g/cm3	Gpa	N/tex	Gpa	%
Natural fibres					
cotton	1.4	5	0.21	0.3	6
flax	1.6	24	0.39	0.6	2.5
Metal fibres					
drawn carbon steel wire	7.8	200	0.26	2	2
stainless steel	7.8	200	0.19	1.45	
Mineral fibres					
E glass 20 microns	2.5	73	1.15	2.8	3.8
S glass 9 microns	2.5	85	1.84	4.6	5.4
Synthetic fibres					
polyethylene tape PE	0.95		0.39-0.53	0.38-0.50	12
polypropylene PP	0.91		0.75	0.68	
polyester PET	1.38	7.6	0.66	0.91	12
polyamide PA	1.14	5.5	0.75	0.85	18
aramid AR	1.44	62	1.93	2.8	4
high-mod aramid HMAR	1.44	131	1.93	2.8	2.5
Vectran	1.41	65	2	2.8	3.3
high-mod polyethylene HMPE	0.97	110	2.64	3.5	3.5+
carbon	1.8	230	1.94	3.5	1.5
Zylon PBO	1.54	280	3.7	5.8	3.5

2.1 Properties of Materials

Yarn weight or count is usually quoted in Tex (g/km) or in old-fashioned units Denier (g/9000m). The strength of a yarn can be calculated by multiplying the Tenacity (N/tex) by the yarn count with an additional factor for twist. Twist is essential for yarns made from short staple fibres, but the new synthetic materials come in continuous filaments and need only sufficient twist to hold them together, Liddell (1988).

The properties of thermoplastic materials are improved by stretching them during the spinning process so that the molecules are oriented. This process reduces creep and raises the breaking stress but is reversible if the material is heated whereupon the fibres shrink. High-modulus polyethylene is a new development where ultra-high molecular weight PE is dissolved in a solvent before spinning. This results in a fibre with very high tenacity and modulus. It is very flexible and resistant to abrasion and ultra-violet (UV) radiation but suffers from creep and creep failure, and loses strength and stiffness at temperatures above 40°C. It is widely used in the marine industry for towropes and sailboat sheets, and production is being increased.

Aramid and PBO fibres are based on benzene molecules. They have high strength and stiffness properties but are poor in compression and are damaged by flexing and UV. Along with the similar materials Nomex and PBI, they are fire resistant and are often used for safety clothing. Carbon fibres are also of high stiffness but easily damaged by flexing and are not often used for flexible elements.

Vectran is described as a liquid crystal polymer based on polyester-polyarylate. It has very good abrasion and folding resistance, medium temperature performance and very low creep. The UV resistance is similar to aramids.

Of the high-tension capacity materials drawn, steel wire remains the preferred choice largely because of its low cost in relation to tension capacity and stiffness. Its one big drawback is corrosion. Apart from that, it has a sufficient ductility to eliminate local overstressing; and there is over 100 years' experience of its use.

Stress levels in tension elements are normally related to the minimum breaking load (MBL). For over 100 years the minimum factors of safety applied to steel strand are 2 against MBL and 1.5 against yield. I n the AISC guide lines the recommended factor of safety (FOS) is 2.25 against MBL. Fatigue tests indicate that at this stress level the life is $>10^7$ cycles. For cable structures we use a FOS of 2 for short term structures and 2.5 for long term structures or for cables with clamps on them.

None of the materials listed is completely problem-free. Steel corrodes in air. Stainless steel corrodes if immersed in salt water or if air is excluded. Synthetic fibres are damaged to varying degrees by UV light and all have some other disadvantage such as creep failure at medium loads or damage from flexing or cyclic loading. For structures where safety is important we have to learn from experience what the safety factors should be and how to ensure safety.

3. SAIL BOAT RIGGING

The classic examples of lightweight deployable structures are the masts and rigging of sailing ships. These need to be lightweight firstly to prevent the ships from capsizing, and secondly to make the sails easier to handle. The rope rigging had two functions; standing rigging to hold the masts up and running rigging to control the sails. Until the development of reliably corrosion protected metallic wire ropes, the standing rigging consisted of multiple shrouds of hemp rope protected by dipping in tar. This multiple arrangement provided the necessary stiffness and accommodated the variable strength of the hemp ropes. It also allowed for damage to ropes by rot or gunfire, and the sailors of the 18th century were able to maintain and replace them while at sea. The cross yards supporting the sails enabled sail to be increased or reduced on a piece-by-piece basis and also allowed for the limited strength of the canvas of that time.

Big changes in rigs were enabled by the introduction of metallic wire-rope shrouds and improvements in the quality of canvas from machine spinning and weaving. The square-rigged ships gave way to fore-and-aft rigs by the 1860's. These reached their pinnacle of perfection in the giant J Class yachts built between 1900 and 1930 recently seen racing in the Solent. The introduction of stainless steel rigging and polyester cloth enabled further advances to be made in the 1950's. 50 years later we are now building boat hulls of glass and sometimes carbon fibre composites. Masts and spars are usually made in aluminium but for high performance racing yachts carbon fibre is frequently used. Rigging is generally stainless steel strand but for ultra high performance, cobalt or nitronic steel rods are occasionally used, and for super-light dinghies aramid fibre pultrusions or braided rope. Sails and ropes are frequently made of aramid, fibre cloth but this tends to be damaged by sharp flexing. Other high performance fibres used for ropes include Dyneema (high modulus polyethylene), Vectran and Zylon (PBO). HMPE ropes suffer from creep and also creep failure but otherwise they have good properties. They are frequently used for marine towropes.

The sizing of rope for rigging was traditionally done by rules of thumb. The 19th century guidance on this subject is given in the book on shipbuilding published in 1866 and edited by W.J.Macquorn Rankine who recommended a working load of a quarter of the proof load. The breaking load for iron was 2 – 2.5 and for hemp was 2 – 3 times the nominal proof load. The high factor of safety was to allow for "shocks and irregular strains". This practice continues today for most sailing boats which are built to class rules where the rigging sizes would be specified, or for cruising boats the builders would follow dimensions used on similar boats. The exception to this is on very high performance yachts, such as those for the America's cup, where no expense is spared to maximise the performance and that usually means minimising the weight by using very high performance materials. On these boats the rigs are analysed using non-linear finite element modelling programmes that include the masts sails and rigging and also model the wind flow and pressures under various positions of sailing. Such a program, "Relax", was written by Peter Heppel who was a post-graduate student in the Cambridge University Engineering Department at the time.

Aramid and PBO fibre ropes are being introduced for standing rigging on ocean going multi-hull yachts but at present the sailors who are using these materials are using twin lines for safety instead of the single rod rigging favoured by the large monohulls.

3.1 Sails

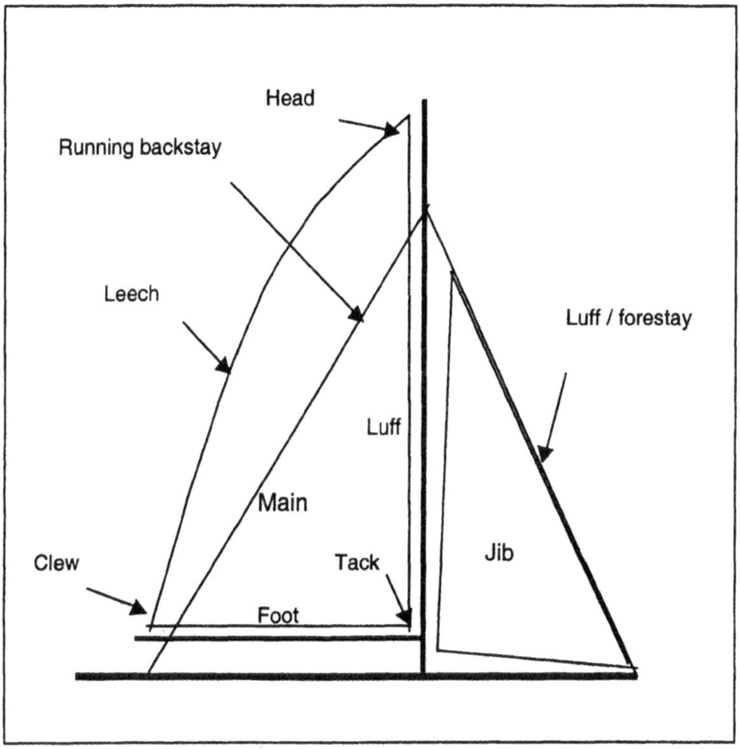

Figure 1. Sail terminology.

Sails are structures that are deployable from a furled or rolled bundle and can be adjusted by large changes in angle relative to the hull and with fine adjustments of surface shape. They are tensile elements, sheet surfaces, that are loaded normally by wind pressure. When sailing upwind, the aim is to achieve the curvature that will optimise the lift and hence boat speed in relation to the angle to the wind. The curvature of the sails will increase as increasing wind pressure stretches the cloth. However, for better performance the sail should be flatter. To achieve this the tension in the sail must be increased. So sails are adjustable fabric structures with the shape being controlled by the applied tension rather than just changes in dimension.

Modern high-performance sailing boats use high-performance materials for their sails. For smaller boats, the sails are generally made from high shrinkage polyester fibres, while for larger ones aramid fibres are used for critical areas. The choice of materials is not based on strength but on stiffness. The sailor's aim is to get the optimum shape of sails for all conditions of apparent wind direction and speed. Mason (1975) The shape of a main sail is defined by the angle of the chord to the centreline of the boat and the depth and position of the maximum camber from the chord. This latter is often called the 'draft'. The chord will usually have some twist, i.e. the chord angle will change from boom level to the head. The mainsail is continuously attached to the mast and is attached to the boom at the end by the clew. It can also be attached to the boom throughout its length but not usually in a way that affects the shape of the sail. Sails are cut to provide an optimum shape with a straight mast but the shape can be adjusted by the following

> The luff tension. This varies the length of the sail up the mast and causes the position of the maximum draft to move forward.
>
> The out-haul tension. The out-haul moves the clew of the sail towards the end of the boom, thus reducing the depth of the camber.
>
> Increasing the vertical load on the boom. This can be done by the main sheet when it is pulled in tight or by the vang or kicking strap, a short tension element close to the mast, when it is not. This action tightens the leech of the sail reducing the twist at the top.
>
> Bending the mast backwards at the top by increasing the tension on the back stay. This has the effect of flattening the sail in the central area and allowing the leech or trailing edge to open at the top (more twist). There are other ways of adjusting mast bend with rams at deck level or check-stays associated with running back-stays.

Similar controls exist for the jib or genoa but in this case it is not possible bend the fore-stay to which the jib is attached in the same way. What can be done is to increase the tension in the fore-stay thus reducing the sag in the cable line. This has to be done by increasing the back-stay tension or in the case of rigs where the fore-stay does not extend to the top of the mast by tensioning the running back-stays that are attached to the mast at the top of the fore-stay.

As the wind speed increases, the sail is required to be flatter, so adjustments can be made using the controls as above. However the increase in pressure on the cloth causes it to distort in a way that moves the draft aft and this can have a negative effect on the sail shape, so additional compensations are required. Another problem with sailcloth is the low shear or bias stiffness of the weave. Attempts are made to increase this by heat sealing the cloth and this helps the sail to develop a good shape. This makes the cloth smooth and shiny but handling from the necessary business of taking down and raising the sails breaks down the shear stiffness and causes local creases that reduce the performance of the sails. In the days of cotton sails this problem was overcome by getting the crew men to wash and iron the sails between races. Modern sail cloths are selected for stiffness but have a short life because the cloth is easily distorted in the bias direction and this effect is not elastically recoverable.

To improve the shape control over the life of the sails, special sails are made from polyester foil with reinforcing tapes laid to follow the trajectories of principle stresses, the tapes being made from high stiffness fibres such as aramid and recently PBO or carbon . In some sail lofts these sails are made on a curved mould with the reinforcing added in the form of tapes glued on to the base foil. Needless to say sails made in this way are extremely costly with very short lives and also require expert sailors to get the most out of them.

In modern racing yachts the shrouds do not prevent the mast from moving forward. The shrouds are in a plane with the mast and forward resistance is provided by the backstay or twin running backstays. The backstays enable the tension in the forestay to be adjusted and allows for bend control of the mast but it is necessary to take care when gybing. If the backstays are both released together it is likely that the mast will fall down.

4. SUSPENSION BRIDGES

4.1 Early Applications

There were parallel early developments in rope technology in other parts of the world besides Egypt. In South America there was a tradition of making rope bridges to span gorges in the Andes. These were made from selected grasses that were dried and cleaned and twisted into ropes. There was a similar tradition in China and northern Eurasia. The ropes had a very limited life so the bridges had to be repaired or rebuilt by the local people on a regular basis.

These bridges were supported by draped cables that would have considerable sag in comparison to their span, say around 1:8. Such a cable has the optimum stiffness for uniform loading but for non-uniform loading it can move freely. The problem of deflections under non-uniform loading is dealt with by having a stiffening girder within the deck. The early bridges did not have such a girder. They probably had a smaller dip and either no deck, or the deck would have been made from linked timber planks, and under pedestrian loading they would have moved significantly. There was a more recent example of such a bridge built in 1826 across the Tweed at Melrose. This had eyebar chains and a linked timber deck. Significantly it was known to the locals as the Swing Bridge, and it had a notice at each end limiting the number of users to 8. Sadly it has recently been renovated and heavily stiffened and its charm has been lost.

4.2 Wire Rope Bridges

After hand-made rope suspension bridges the earliest suspension bridges were built using chains of linked plates or eye-bars. Wire or iron strip bundles appear to have been first used between 1810 and 1820. Wire cables were used on several bridges in France in the 1820's, and the problem of corrosion was noted by Vicat in 1831. Sutherland (1988). The corrosion protection of those times was to soak the wires in hot linseed oil to which some carbon and litharge, white lead oxide, was added. Vicat introduced the practice of protecting the underground lengths of the cables by grouting them with hydraulic lime. In 1850 the Basse-Chaine bridge collapsed. The cause was corrosion of the wires near the anchorages. This led to the inspection and discovery of similar corrosion in several other wire bridges and the decline in suspension bridge building in France. (Sutherland 1988).

Corrosion problems reappeared on the Manhattan bridge opened in 1909. In 1988 it was discovered that the eye-bars that anchored the strands of one of the main cables had suffered more than27% loss by corrosion. Mayrbaurl (1988) The eye-bars are connected to the strands, and are anchored into the concrete anchorage block. They had corroded just above the surface of the concrete. In a remarkable piece of engineering, the strands were re-connected to the anchor block in a process involving changing the splay casting, re terminating the outer strands with white metal sockets, and connecting them to new anchors. Some of the wires from the remaining strands were removed from their sockets and reconnected to the vacated eye-bars. Part of this process involved repairing or reconnecting damaged wires. The bridge was kept open with reduced traffic during the operation.

Because of the otherwise reliable properties of galvanised steel wire, it has continued to be the preferred material for long-span suspension bridges up to the present time. Corrosion and fatigue have also continued to be the major problem with the use of the material, and a range of preventive measures has been tried. The latest technique used in Japan is to wrap and seal the cables and pump in dried air. The use of synthetic fibres for suspension bridge ropes has been proposed as a way of avoiding corrosion problems and reducing weight for very long spans; but has yet to be tried at a serious scale.

5. HYPER-TENSIONED CABLE STRUCTURES

A cable that is tensioned between two fixed points can carry loads applied at right angles to it, but only if it is curved. Unlike a draped cable which can take loads only in one direction, a hyper-tensioned cable can take loads in all directions. The difficulty is that the load deflection behaviour is non-linear, and the stiffness, especially at small deflections, is not very great. The tension under a uniform load can be calculated by hand by considering the equation of the elastic stretch under tension to achieve a certain dip and comparing this to the geometrical extension for the same dip. If a circular arc is assumed we get the following equations.

Span	=	S
Load	=	W/unit length,
Stiffness	=	EA
Pre-tension	=	T_o
Tension under load	=	T
Stressed length	=	L
Length at T_o	=	L_o

Equilibrium equation $\quad T = RW \quad$ or $R = T/W$ (1)

New length $\quad L = 2R\sin^{-1}(S/2R) = 2T/W\sin^{-1}(WS/2T)$ (2)

Strain $\quad e = (L - L_o)/L_o = (T - T_o)/EA$ (3)

or New length from strain $\quad L = L_o(T - T_o)/EA + L_o$ (4)

Substituting for L from 2 into 4 and rearranging.

$$L_O (T - T_0) /EA = 2T/W (\sin^{-1} WS/2T) - L_O \quad (6)$$

This equation can be solved quite readily for the new tension T and the results compare well with a solution using a non-linear computer program.

Hyper-tensioned cables have been used for a few cable and fabric roof structures, the most notable being the Millennium Dome. As happens in the structural world, they also appeared almost at the same time in bridges in the form of the Millennium bridge in London. Dallard (2001). In roof structures the main concern is the risk of ponding under snow or rain loading. This can be dealt with by choosing a design where there is sufficient slope in the surface to enable the water to drain away. There can also be a problem if the roof is subject to wind turbulence, when it will move up and down rather disturbingly if the pre-tension is not high enough. With the bridge the problem was vibrations under pedestrian loading.

5.1 Basic Properties of the Millennium Bridge

Centre span = 144.2 mm
Self weight = 2 × 747 Kg/m
Cable dip = 2.290 m vertically and 1.735m horizontally.
Total dip = 2.873 m Hence Radius R= 906 m
Cable tension/side = 10,435 kN,

Cable properties
Metallic area = 10100 mm^2 , E modulus= 160 +/- 10 kN/mm^2
MBL/cable = 15,800 kN
Cable EA (4 cables) = 6.528 10^6 kN

Cable extension	T/EA	= 0.230 m
Geometrical extra length	$2R \times \sin^{-1}(S/2R)$	= 0.153 m
Initial Tension with zero weight		= 3523 kN

Natural frequency as tensioned cable = $\sqrt{(T/M)} /2L = 0.410$ Hz

Measured natural frequency horizontally =0.45 Hz. This is higher than the theoretical because of the additional stiffnesses from the cable clamps and deck structure.

Problems occurred with the second mode or the first harmonic of lateral vibrations where the natural frequency was around 0.9 Hz. As is now well known, if the bridge started to vibrate at this frequency with a certain critical amplitude the pedestrians have to adjust their walking behaviour to keep balance and this manner of walking added energy into the vibration mode. With enough pedestrians on the bridge the lateral oscillations would suddenly increase in amplitude. To guard against this the engineers retro fitted many dampers.

6. SPIDER WEBS

Comparisons have often been made between developments in tension structures and spider's webs. That there is a similarity is not surprising since the form of both is generated by the same physical forces, that of prestressed or hyper-tensioned cables. An unusual feature of spiders is that they produce two different kinds of thread. Lin (1998) A structural thread that has a fairly linear stress strain relationship until failure at 1200 MPa at a strain of 40% and a capture thread that has a very non-linear stress strain curve going up to a similar stress at nearly 500%. This trick is achieved by coating the capture thread in a hygroscopic glue. This absorbs water and then forms into a string of droplets that gather in the surplus thread. So the initial part of the stress strain curve is the surface tension of the glue and is constant but very low. The Youngs modulus of the structural thread is ten times greater than that of the capture thread.

The structural part of a spider's web consists of the frame and anchor threads and the radial threads that converge on a central point. These are pre-stressed to a force of 60 – 150μN. It would be interesting to discover how the spider achieves this pre-stress. It could be done by the spider tensioning each thread as it spins it into place. However if the threads were elastic, tensioning the subsequent threads would cause the ones installed earlier to become either slack or over tensioned depending on the arrangement of the boundary supports. This problem arose in the tensioning of the Millennium Dome structure, where the radial cables had to be tensioned under dimensional control to maintain the shape and arrive at uniform forces.

If we calculate the ratios of tensions between the curved boundary threads and the radial threads and the capture threads we find that it is of the order of 10 : 5 : 1. The boundary threads would be installed first and could be installed at a higher tension and could be reinforced by the spider doubling or trebling the threads. It could also happen that the radial threads are installed at a lower tension or the stress strain curve is flatter at the lower end than has been measured.

7. INSPECTION AND MAINTENANCE

There are clear structural advantages of using tension elements, but there is a constant anxiety about the unseen deterioration and potential failure of tensile tendons. This might come from fatigue, creep failure of polymer materials or corrosion of steel. The question for design engineers is, how do you inspect them and how do you maintain or replace them? In prestressed concrete bridges we have seen the change from grouted in cables to external cables that can be inspected and, if necessary, changed. Although hangar cables in suspension bridges can be replaced the main cables generally cannot. In old bridges this lead to the use of flat plate eyebars for the chains or the over-sizing of wire rope so that the damage from fatique or corrosion would be spotted well before the time they became critical. In buildings owners are reluctant to carryout maintenance; so solid rods are often preferred to stranded wire.

8. CONCLUSIONS

While there are very-high-strength tensile materials available they are usually utilised at relatively low stress levels compared to steel. The factor of safety for fibre ropes has changed from 8 in the 19th century to 3 – 4 today. The reason for the relatively low stress is often stiffness, energy storage which can result in violent failures, weakening of the cables because of non-uniform loading of the fibres or wires, damage at end terminations, and fatigue or flex resistance. In spite of this tensile elements and materials are frequently used in temporary, lightweight and deployable structures. In traditional applications with natural fibre ropes the design allowed for replacement when the material degraded. The same philosophy still applies with structures today although the time scales are longer.

The move to cable stayed bridges in preference to suspension bridges and to external pre-stressing tendons rather than grouted in tendons is driven in part by the need for the reliability that comes from inspection and maintenance.

While spiders webs may be a poor example for technical structures one feature that we can learn from is structural re-cycling. Certain web building spiders will eat up their silk and re-make their webs the next day.

References

P. Dallard et al (2001) The London Millennium Footbridge, *The Structural Engineer*, 79(22)
Gourok, (1950) *Gourock Ropework Book* (Ca 1950), Gourock Ropes and Canvas Ltd.

L. H. Lin, W. Sobek, (1998) Structural Hierarchy in Spider webs and Spiderweb-type Systems. *The Structural Engineer* 76(4) pp 59-64

W. I. Liddell, (1988), Structural Fabrics and Foils, *1st Oleg Kerensky Memorial Conference*, Institution of Structural Engineers.

W. J. Macquorn Rankine et al, (1866) *Shipbuilding, Theoretical and Practical.* William MacKenzie, London.

C. Mason ed, (1975), *The Best of Sail Trim,* Adlard Coles Nautical, London.

R. M. Mayrbaurl, J. Good, (1988) Re-Anchoring a Main Cable on the Manhattan Bridge, *1st Oleg Kerensky Memorial Conference*, Institution of Structural Engineers.

W. B. Parsons, (1939) *Engineers and Engineering in the Renaissance*, MIT Press

R. J. M. Sutherland. (1988) Some Aspects of the History of Suspension Bridges, *1st Oleg Kerensky Memorial Conference*, Institution of Structural Engineers.

ELASTIC MINIMUM-WEIGHT DESIGN: AN ENCOUNTER WITH ALAN TURING

R. K. Livesley
Churchill College, Storey's Way, Cambridge, CB3 0DS, UK

Abstract Alan Turing and the Author worked in the Computing Laboratory at Manchester University from 1951 to 1954. In the autumn of 1953 the Author asked Turing for help in proving a general theorem on the optimum design of elastic pin-jointed frames. The manuscript he produced in response to this request appears to have been his only venture into the field of engineering structures: because of his tragic death a year later it was never published. The general proof resulting from this collaboration has long since been superseded. It is now natural to formulate the design problem in linear programming terms: this makes the proof of the theorem entirely straightforward. However, Turing's present reputation may lend some interest to a brief historical account of the way in which the ideas developed. The first two sections of this paper describe the computing facilities at Manchester in the early 1950's and the computing work carried out there by the Author – work which eventually led to the collaboration with Turing. The remainder of the paper looks at the theorem and its proof from a present-day point of view.

Keywords: Pin-jointed frames, Optimum design, History of computing.

1. COMPUTING AT MANCHESTER UNIVERSITY IN THE EARLY 1950'S

The first computer available for general use at Manchester University was the Ferranti Mark 1, a production version of earlier experimental machines built in the Electrical Engineering Department. It was commissioned in July 1951, occupying most of the ground floor of a purpose-built two-storey red-brick building next to the old Electrical Engineering Laboratory. The building was plain and rectangular and its furnishings were utilitarian, but at least it was warm – what else could it be, seeing that it housed a machine with a power consumption of nearly 40 kW!

My contact with the Mark 1 began in October 1951. In 1949 I had been appointed as a University Demonstrator, the post being shared between the Departments of Mathematics and Engineering: my job was to provide

H.R. Drew and S. Pellegrino (eds.),
New Approaches to Structural Mechanics, Shells and Biological Structures, 155–163.

lectures in mathematics for engineering undergraduates. For the first two years of this appointment I spent my spare working time extending my mathematical knowledge. At the start of the third year, when I felt it was time to try my hand at research, the Professor of Civil Engineering, Louis Matheson, was looking for someone to explore ways in which the new computer could be used in engineering analysis and design. With some trepidation I agreed to register for a Ph.D. with this as my research topic. In view of the inter-disciplinary nature of this research I was assigned a second supervisor, Professor M. J. Lighthill, who was then Head of the Department of Mathematics.

From 1951 to 1954 I was one of the small group of regular users of the Mark 1. This group, numbering about 20, included individuals from outside the University (some of whom had previous programming experience on the EDSAC at Cambridge) and staff and research students from various Departments. In 1952 I gave up my teaching post to become a Research Assistant in the Computing Laboratory: this brought the programming staff of the Laboratory up to four. My official duties included extending the library of subroutines, teaching programming to new users and giving advice on appropriate numerical techniques. Although I had enjoyed lecturing, the change of post had the advantage that all my academic activities were now centred on the Mark 1.

Alan Turing had been appointed to a Readership in Mathematics in 1948, in the expectation that he would play a major role in the development of Computing within the University. He was closely involved in the specification of the overall architecture of both the prototype and production versions of the Mark 1, and wrote the first programming manual for the production machine – he was not, perhaps, the ideal author for the only manual available! During my time in the Laboratory he was a heavy user of the Mark 1, booking it regularly for two nights a week, but his research work on morphogenesis produced relatively little in the way of new ideas for software development. Nor did he show much interest in the work on autocodes which was beginning in the laboratory at that time. To quote the biography by Andrew Hodges (Hodges, 1983) "The development that could have absorbed both his understanding of symbolism and his willingness to do the donkey-work when necessary was that of designing programming languages, a development he had described as 'obvious' in 1947. But this was precisely what he did not do...The opportunities open to him were rich and his neglect of them very striking."

Programming the Mark 1 was certainly 'machine-code programming'. Each machine instruction consisted of 20 binary digits. An instruction was fed into the machine as four rows of holes/blanks on 5-track teleprinter tape, each row corresponding to a character on a teleprinter keyboard. Thus the written form of a program consisted of a sequence of 'words', each of four teleprinter characters. Unfortunately the characters corresponding to the 32 binary numbers 00000 . . . . 11111 were arranged in the arbitrary sequence /E@A:SIU½DRJNFCKTZLWHYPQOBG"MXV£. Anyone who used the machine regularly ended up knowing this sequence by heart – fifty years later I can still repeat it from memory! And this bizarre programming code was not the only complication facing a user. The Mark 1 had a C.R.T. store, and in this type of store the trace always goes from left to right. So all numbers were stored and processed with the most significant digit on the right. (I imagine Turing could have changed this in the original specification of the machine, but he probably had no difficulty in doing arithmetic backwards himself and couldn't imagine it being a problem for anyone else.) And of course all numbers were stored in fixed-point form: there was no hardware for floating-point arithmetic.

Those who were users of other early computers will doubtless have similar memories of the difficulties and frustrations which surrounded programming in those days, and the corresponding feeling of excitement when a program actually worked. A good account of the early history of computing at Manchester, including a description of the programming system developed for the Mark 1, has been given by Campbell-Kelly (1980).

2. A THEOREM ON THE ELASTIC MINIMUM-WEIGHT DESIGN OF PIN-JOINTED FRAMES

At the suggestion of Professor Matheson I began work on the Mark 1 by implementing some ideas on the elastic analysis of rigid-jointed structural frames put forward by J. M. Bennett, who had analysed some simple frames on the EDSAC. Although by present standards the Mark 1 had a very small working store (512 20-bit registers) it had an advantage over the EDSAC in that it possessed a magnetic drum backing-store. This made it possible to construct a program for the elastic analysis of plane frames which was capable of doing serious commercial work, and in 1953 T. M. Charlton and I carried out the analysis of a 20-joint frame for a new power station. As far as I know this was the first computer analysis of a commercial civil engineering structure in this country, although similar (and probably more sophisticated) work was being done in the field of aircraft structures at the same time by Argyris and his team at Imperial College.

The success of the frame analysis program opened up the possibility of a variety of extensions. I managed to include linear stability functions in the program before I left Manchester, but other developments – the analysis of grillages and space frames, the effects of bowing and elasto-plasticity, etc., I left for later users of more powerful machines. Instead I decided to abandon analysis for a while and look at structural design.

My training as a mechanical engineer meant that I had virtually no experience of the structural design process. However, a suggestion by J. M. Bennett in one of the first books on computing (Bowden, 1953) seemed to fit in well with what I had done so far. Bennett proposed an iterative scheme for the optimum (i.e. least-weight) design of an elastic structure of given form and loading. "An initial design is chosen and the stresses in the members evaluated. New member sections are then selected so as to give stresses closer to the maximum allowable and the analysis repeated. The design to which this process converges is clearly an optimum one." This seemed to me a sensible procedure and, moreover, one which could easily be programmed. Accordingly I asked an M.Sc. student I was supervising to set the existing program for elastic frame analysis (modified to deal with plane pin-jointed frames) within an iterative loop in which member cross-sectional areas were repeatedly modified until all stresses took specified limiting values.

With hindsight it is not surprising that this procedure failed to converge, except when applied to a statically determinate frame. As I should have realised from the start, an attempt to prescribe arbitrary stresses in all the bars of a statically indeterminate elastic frame violates the compatibility conditions which must be satisfied by the bar strains and joint displacements. It was clearly time to stop computing for a while and do some thinking. (Turing was fond of saying "An ounce of mathematics is worth a ton of computing".) Studying the results produced by the iterative design program I began to suspect that the optimum elastic design for a pin-jointed frame of specified geometry and loading is always statically determinate. Suspicion gradually turned to conviction and conviction to proof, at least in the simple case of a plane frame with one joint and up to three bars. My argument was as follows.

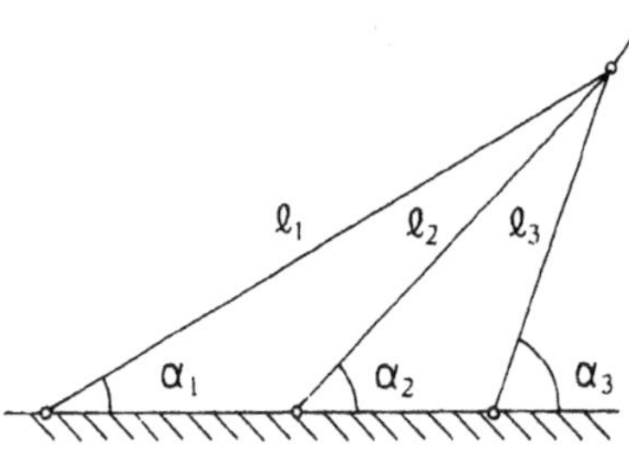

Figure 1. A simple frame.

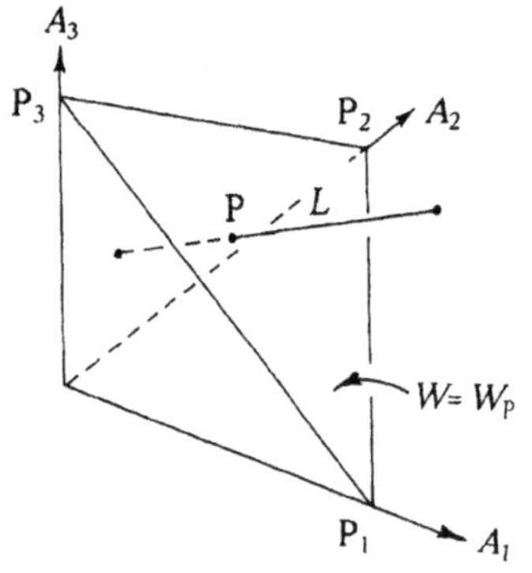

Figure 2. Its design space.

Consider the optimum design of the plane pin-jointed frame shown in Fig. 1. The geometry of the unloaded frame is fixed and the load $\boldsymbol{F}$ is specified in both magnitude and direction. The bars are of uniform cross-section and are made of material of density ρ. The bar strains produced by the load are assumed to be small and the bar stresses σ_1, σ_2, σ_3 are single-valued functions of these strains. Since the geometry of the frame and the loading are fixed, the only design variables are the cross-sectional areas A_1, A_2, A_3. A specific design can therefore be represented by a point in the three-dimensional 'design space' with axes A_1, A_2, A_3 shown in Fig. 2. A 'safe' design is one in which the stresses in the bars produced by the load $\boldsymbol{F}$ lie within prescribed limits: for each bar these limits are assumed to be independent of the bar's cross-sectional area. The safe design of least weight is the one which minimises the linear function of the cross-sectional areas

$$W = \rho(A_1\ell_1 + A_2\ell_2 + A_3\ell_3) \tag{1}$$

while satisfying the limits on the stresses.

Consider an arbitrary statically *indeterminate* safe design, with all cross-sectional areas strictly positive, represented by the point P in Fig. 2. It is clear that P lies *within* the triangle $P_1P_2P_3$ formed by the intersection of the plane of constant weight $W = W_p$ with the design space.

The equilibrium equations for the loaded joint are

$$\left.\begin{aligned} F_x &= \sigma_1 A_1 \cos\alpha_1 + \sigma_2 A_2 \cos\alpha_2 + \sigma_3 A_3 \cos\alpha_3 \\ F_y &= \sigma_1 A_1 \sin\alpha_1 + \sigma_2 A_2 \sin\alpha_2 + \sigma_3 A_3 \sin\alpha_3 \end{aligned}\right\} \tag{2}$$

Let S be the set of all the designs for which the stresses σ_1, σ_2, σ_3 produced by the load $\boldsymbol{F}$ are the same as those which occur in the design represented by the point P. The points in the design space which correspond to the set S lie on a straight line L passing through P, the equation of L being obtained by inserting the appropriate stresses in equations (2). Since all the points on L correspond to designs with the same stresses they represent safe designs. It is clear from Fig. 2 that if L does not lie in the plane $W = W_p$ there is a range of safe designs for which $W < W_p$, including a statically indeterminate design associated with the point where L intersects one of the bounding planes of the design space. [If the determinant formed from the coefficients of A_1, A_2, A_3 in equations (1) and (2) is zero then the line L lies in the plane $W = W_p$. In this special case there are two statically determinate safe designs, each of weight W_p, corresponding to the points at which L intersects the boundary of the triangle $P_1P_2P_3$.] It follows that, given an arbitrary statically indeterminate safe frame, it is always possible to find a statically determinate safe frame of less or equal weight. Consequently the overall least-weight frame must be either statically determinate or a combination of two statically determinate frames of equal weight having the same bar stresses.

This completes the proof for the special case of a three-bar plane frame. Note that the assumption that the limits on the bar stresses are independent of the cross-sectional areas ignores the possibility of a bar buckling. However, there is no assumption that the material of the frame has a linear stress/strain relationship.

The form of the proof in this simple case convinced me that the theorem held generally for a plane or space frame with n displacement degrees of freedom and up to m bars. However, it was clear that the generalisation required a change from a three-dimensional design space to an m-dimensional one, the lines and planes of the simple proof being replaced by linear sub-spaces of appropriate dimensions. Not being convinced of my own ability to get the details of the generalisation right I consulted my supervisor, who suggested that I show my work to Turing and ask for his help. To my surprise and pleasure he took up the problem with almost childish enthusiasm. I remember that he spent some time exploring the possibility of transposing it into a problem in game theory. He certainly treated it as a 'new' problem (it seems that neither of us had read the classic papers by Maxwell and Michell on the optimum design of pin-jointed frames).

After a long weekend Turing returned to the laboratory with a general proof of the theorem which ran to ten pages of highly impressive mathematics. While he certainly achieved the required result his reasoning appeared to follow a completely different line from my own. In the normal course of events the two pieces of work might have formed the basis of a joint paper in an appropriate journal. However, because of his tragic death the following year and my own interest in other problems his typescript was consigned to a file containing various other items of unfinished work: it remained in that file for nearly fifty years. His proof is too long to reproduce here: the original typescript is now in the collection of his papers held in the Archives centre at King's College, Cambridge.

Intimidated by Turing's reputation as a mathematician I gave up my own attempt at a general proof. It was only when I began to prepare the present paper that I realised that an extension of the simple geometrical argument set out above is entirely straightforward. For consider an arbitrary elastic statically indeterminate frame with n displacement degrees of freedom having m bars with strictly positive cross-sectional areas A_j ($j = 1 \,..\, m$), where $m > n$, and bar stresses σ_j lying within the stress limits. This frame corresponds to a point within the m-dimensional design space $A_j \geq 0$. The set S of safe designs having the same stresses σ_j as this arbitrary frame form a linear subspace of $m - n$ dimensions within the design space: this subspace is defined by n joint equilibrium equations which, like equations (2), are linear in the A_j's. Since the weight $W = \rho A_j \ell_j$ is a linear function of the A_j's it cannot have a minimum at an interior point of the subspace: the minimum must occur at a vertex. Now a vertex is defined by m equalities – the n equations of joint equilibrium and $m - n$ equations of the form $A_j = 0$. The remaining n A_j's must be strictly positive and therefore define a statically determinate frame. [As with the simple 3-bar frame, it may happen that the weight function has the same value at several vertices, in which case there are several statically determinate frames of equal weight.] It follows that for any safe statically indeterminate frame there is always a safe statically determinate frame of less or equal weight. Consequently the overall least-weight frame must be either statically determinate or a combination of several statically determinate frames of equal weight having the same bar stresses.

The simplicity of this argument naturally raises the question: how did someone of Turing's undoubted ability miss what now seems obvious? Perhaps the biography by Hodges quoted earlier provides a clue. "People did not have a second chance with Turing. If they tuned into a Turing wavelength they would receive hours of attention full blast, with an almost embarrassing intensity. . . . All the same, he was no world-standard mathematician and what he did *not* know was often more amazing to the professional mathematician than what he did. He never approached Von Neumann's status in breadth of knowledge: indeed, he had read very little mathematics since 1938."

A few years after Turing's death I came across a paper by Drymael (1942), which proved the general result using a very different approach.

3. THE THEOREM IN RETROSPECT

In the first two sections of this paper I have tried to set our work on minimum weight design in the context of 1950's computing facilities and 1950's thinking. During that period several papers pointed out that all the bars in a statically indeterminate pin-jointed frame can be made to carry maximum stress by giving them an appropriate amount of pre-stress. One of these papers proposed a procedure for optimum design which started with a statically determinate frame, reduced its weight by adding pre-stressed members and concluded that the resulting statically indeterminate frame was the frame of least weight. The analysis presented in the previous section shows that the result of this procedure is not the true optimum and that the procedure only works at all if the 'wrong' statically determinate frame is chosen to start with.

Our work, like much of the early work on optimum structural design, was based on the assumptions of a design philosophy based on elastic behaviour. The whole field changed rapidly as the concept of ultimate load design, developed by Baker and his team at Cambridge (among others), became accepted by civil engineers. This approach removes the need to consider elastic behaviour during the optimizing process. It has the effect of reducing the optimum design problem considered in Section 2 to what is now a standard problem in linear algebra – the linear programming problem. A paper by Foulkes (1954) is typical of the Cambridge approach. In it he considers the problem of the optimum design of a rigid-jointed plane frame and sets up the weight function and linear constraints for a simple example. However, he does not use the term 'linear programming' and does not mention any of the general computational procedures developed in operations research during the 1940's.

For a pin-jointed plane or space frame the formulation as a linear programming problem is as follows. Consider a pin-jointed frame of given geometry with n displacement degrees of freedom having up to m bars with cross-sectional areas A_j ($j = 1 \,..\, m$), where $m \geq n$, and stresses lying between specified limits σ^- and σ^+. Let a specified set of loads f_i ($i = 1 \,..\, n$) produce tensions t_j in these bars: these m tensions must satisfy the n linear equations of joint equilibrium and the m linear inequalities $\sigma^- A_j \leq t_j \leq \sigma^+ A_j$. These equations and inequalities define a permissible region of $2m - n$ dimensions within the space of $2m$ dimensions associated with the variables A_j and t_j. The design of least weight is the set A_j (where $A_j \geq 0$) which minimises the weight function $W = \rho A_j \ell_j$ over this region.

This formulation not only provides a means of computing the optimum design in any particular case: it also provides a simple proof of the general theorem. Since the weight function is linear in the A_j's the least-weight design must correspond to a vertex of the permissible region. This vertex is defined by $2m$ equalities – the n equations of joint equilibrium and $2m - n$ equalities selected from the inequalities $A_j \geq 0$ and $\sigma^- A_j \leq t_j \leq \sigma^+ A_j$. If the least-weight frame has p bars with positive cross-sectional areas then $m - p$ of the A_j's and $m - p$ of the t_j's are zero: this defines $2(m - p)$ of the required $2m - n$ equalities. The remaining $2m - n - 2(m - p) = 2p - n$ equalities must come from the p inequalities $\sigma^- A_j \leq t_j \leq \sigma^+ A_j$ associated with bars having positive cross-sectional areas. But for each of these bars only one of the pair of inequalities can be satisfied as an equality. It follows that $2p - n \leq p$, i.e. $p \leq n$. But if $p < n$ the bars form a mechanism, not a structure: hence $p = n$ and the least-weight frame is statically determinate. [As with the simple frame discussed in the previous section, there may be two or more vertices of the permissible region for which the weight function has the same value. In such cases the least-weight frame may be a combination of several statically determinate frames of equal weight.]

Note that in the general proof of the theorem set out in Section 2 the design space has m dimensions, while the linear-programming approach involves a space of $2m$ dimensions and a permissible region which includes points associated with sets of bar tensions for which the associated elastic strains do not necessarily satisfy the displacement compatibility conditions.

Although Foulkes's paper was significant from a theoretical point of view its practical value was limited. At that time no general-purpose computer programs were available for solving the linear programming problem (indeed, it was virtually impossible to transfer a program from one machine to another) and most of the literature was written in the language of operations research, which made it difficult to understand. The situation at the end of the 1950's is well illustrated in a paper by Calladine (1958) on the related problem of computing the plastic collapse load of a rigid-jointed frame. Although Calladine identifies the problem as a linear programming one he does not follow this identification with an account of one of the standard linear programming procedures. Instead he develops an ingenious analogy between the calculation of the limit load of a rigid-jointed frame and the determination of the maximum flow of fluid through a network – a problem for which a special-purpose computational procedure had been suggested by a contemporary researcher.

Fifty years on the optimization problem discussed in this paper looks very limited and academic. A review of the current state of optimum structural design will be found in a recent textbook by Renton (1999). An account of the relevant mathematical optimization techniques will be found in Whittle (1971).

4. POSTSCRIPT

Alan Turing died in June 1954 and I left Manchester at the end of that year. The Mark 1 finally went out of service in 1958 and the building was demolished a few years later. The last time I visited the site I found a smooth lawn and a small brass plate. In the middle of the lawn stood two well-established trees. I had expected the lawn, but not the trees!

References

Bowden, B.V. (Ed.) (1953) *Faster than Thought.* Pitman, London, 223-233.

Calladine, C. R. (1958) *A direct method for the calculation of the plastic collapse load of framed structures.* Essay awarded the John Wimbolt prize, Cambridge University Engineering Department. Manuscript room, Cambridge University Library.

Campbell-Kelly, M. (1980) *Programming the Mark 1: Early programming activity at the University of Manchester.* Annals of the History of Computing, **2**, No. 2, 130-188.

Drymael, J. (1942) *Design of trusses and its influence on weight and stiffness.* Journ. Roy. Aero. Soc., **46**, 297-308.

Foulkes, J. (1954) *The minimum-weight design of structural frames.* Proc. Roy. Soc., Ser. A, **223**, 482-494.

Hodges, A. (1983) *Alan Turing, the Enigma.* Hutchinson, Ch. 7.

Renton, J. D. (1999) *Elastic Beams and Frames.* Camford, Ch. 14.

Whittle, P. (1971) *Optimization under Constraints.* Wiley Interscience.

ENGINEERING AND EDUCATION

P. G. Lowe,
Department of Civil and Environmental Engineering, University of Auckland, Private Bag 92019, Auckland, New Zealand.
lowepj@bigpond.com

Abstract A brief survey of some engineering and education topics is made through the contributions of selected individuals, over a three hundred year period. The final section relates to the future. Novel construction systems are considered, with *simplicity* as the primary feature.

Keywords: Biographical details, History, New construction technology.

1. INTRODUCTION

It is with much pleasure, and a range of thoughts, that this contribution to Prof. Christopher Reuben Calladine's (CRC) Festschrift is being written. The title includes the education of the engineer: this has been a primary theme in CRC's career for over forty years, along with his prodigious research output. The themes engineering and education will be discussed in a loose historical perspective and will range over various aspects that have interested me over this forty year period. Some of the themes have direct links to our esteemed colleague.

2. THE PAST

Mechanics has been a central study for CRC and is a suitable starting point here. Cambridge and the Newtonian mechanical revolution are intimately bound together and have been much studied. There were other, earlier seminal contributions, by Galileo, Stevin and especially Wallis .

Wallis's contributions to mechanics have received relatively little attention. For the non-specialist the essay in the *Dictionary of Scientific Biography* is a suitable starting point for study of Wallis. John Wallis (1616-

H.R. Drew and S. Pellegrino (eds.),
New Approaches to Structural Mechanics, Shells and Biological Structures, 165–174.

1703) was Cambridge trained and in many respects was a key prior contributor to two subjects of prime importance to engineers, with which later Newton was particularly associated: namely the calculus and mechanics. It is Wallis's *Mechanica*, published in three volumes in 1670-71, that interests us here. Recall that this was about 16 years before Newton's celebrated *Principia* appeared in 1687.

In the first two volumes, Wallis discusses the laws of mechanics and a range of applications. He was attempting to formulate the laws of motion, and came close to enunciating these. Our interest is however in the restricted field of statics. This had been mastered earlier. In volume three Wallis applies the concepts of force and moment equilibrium, with real ease and style, to formulating a series of problems. In particular he examines several problems relating to the forces and moments induced in plate-like, floor support systems under transverse loading.

Typically these *plate bending* problems consist of assemblies of short (wooden) beam elements which were arranged in modules of three, four or more members. These modules are themselves connected to form larger assemblies. The primary feature of both the modules and the assemblies is that they are statically determinate. Wallis proceeded to write down all the equilibrium equations for the unknown forces of interaction. He then went on to solve the sets of thirty or more algebraic equations, exactly! The solutions are expressed as ratios of two whole numbers. The only discrepancy I have found in the many solutions he worked out in detail to similar problems is a one digit type-set error in one of the many numbers. The solutions are entirely correct and could not be improved upon even today, given all our available computing power. This degree of sophistication at such any early time is staggering. I regard these solutions as the earliest finite element solutions! Wallis probably knew of other proposals for building up floor support systems using short timbers, and he states where one actual construction could be seen.

Similar systems are described by some of the earliest architecture authors, such as Sebastian Serlio in the sixteenth century. Serlio's major work was first published in an English edition in 1611 and illustrates this type of floor support system. But these earlier proposals were sketches only, were probably indeterminate and no analysis of the forces was attempted. Similar topics continued to be studied and described in most of the major architectural and building-related works almost to the present day. Rondelet, Krafft and Tredgold all describe such systems, though they offer no analysis of the forces in the manner that Wallis does.

Over the next three hundred years or more many quite superb developments in mechanics as applied to structures can be found in the literature. There are some in Euler's prodigious output, including his analysis of the column instability problem in the appendix to his *Methodus Inveniendi..'* (1744) on the calculus of variations. Although probably none of us have worked through Euler's treatment at first hand, the column analysis still features in some form in most present-day courses in structural theory. What a durable piece of modelling this has proved to be! The nineteenth century saw many advances in engineering, especially in thermodynamics, and later in electrical engineering. There was also the laying of the foundations for the developments in IT at the present day. Babbage, Boole and many other names could be mentioned. On the smaller canvas of structural mechanics, many of the leading contributors on the larger canvas, such as Maxwell and Rankine, made important advances.

One lesser known engineering educator of the period is Henry Moseley (1801-1872). He graduated seventh Wrangler in the Mathematical Tripos in 1827, but this placing was not sufficient to secure him a college fellowship and hence a livelihood. He was ordained and went off to parish duties. The clerical connection was probably important to his advancing in an academic career as the foundation Professor of Natural Philosophy and Astronomy at King's College in the Strand in London. Despite the description of his chair, he seems to have spent most of his time lecturing and writing on engineering mechanics, and more especially structural mechanics. He wrote several textbooks, the most important being *The Mechanical Principles of Engineering and Architecture*, published in 1843. This is a substantial work based on wide reading and investigation by the author. Unlike most other English texts on structural mechanics of the day, Moseley's book discusses the latest European knowledge, especially of the French academic engineers who led the world at the time. The book was also adopted as a text by the US Military Academy at West Point, and can be assumed to have exerted some influence on their methods and outlook. The Great Exhibition of 1851 contributed to the realisation that American industry was becoming competitive with British and European industry. Moseley produced a revised and enlarged edition of his book in 1855. The Great Exhibition profits, which were very substantial, were largely ear-marked for promotion of education and particularly technical education, with the strong support of Albert, the Prince Consort. Towards the end of the century the main fruits were to be seen as the Imperial College and the nearby Museums. Clerk Maxwell, who was a later occupant of the Kings College London Chair, is recorded by his biographers as having studied Moseley's *..Engineering and Architecture* while preparing for the Tripos in the early 1850's.

A tradition in Cambridge teaching that had a long period of dominance, and a considerable degree of success, was that of the coach or personal tutor. These were the self-employed teachers who prepared their paying pupils for the Senate House Examinations. The coaches provided the bulk of the teaching available at the time. One of the best documented is the coaching career of Edward J. Routh (1831- 1907) in the Mathematical Tripos during the second half of the nineteenth century. Canadian born, his career dominance as a coach followed that of Peterhouse colleague and his own former coach, William Hopkins. Routh first came to prominence as Senior Wrangler for 1854 when he edged Clerk Maxwell into the Second Wrangler spot. He began coaching as the result of the illness and early death of another Peterhouse colleague, W. J. Steele, who was Second Wrangler to P. G. Tait as Senior Wrangler two years earlier. EJR continued with this career choice and went on to become the most successful coach of his era. In mid career, Routh was a candidate for the post of first Professor of Mechanism and Applied Mechanics in 1875, but he was unsuccessful. James Stuart (1843-1913), a former pupil of Routh's and twelve years his junior, was appointed instead. Seen from the long perspective of history this seems like a good example of a doubtful decision. Routh already had an established body of academic achievement including the 1855 book on Newtonian mechanics with Lord Brougham, his *Rigid Dynamics* of 1860, FRS in 1872 and later *Stability of Motion*, his 1877 Adams' prize essay. Stuart's achievement was in other areas. Routh married and hence had to give up his Fellowship at Peterhouse. His wife, Hilda, was the daughter of Sir George Airy (of the stress function) who was best known as Astronomer Royal.

Much later, during the Inglis' tenure of what had become the Mechanical Sciences chair, in the 1920's and -30's, coaches were still a prominent feature of the teaching in Engineering. A coach of particular interest to me is H.(Herbert) A. Webb (1882-1961) who taught mathematics to many engineers of the period, despite not holding a faculty post. In retirement he collaborated with the late Dr. D.(Derek) G. Ashwell on a slim volume titled *A Mathematical Toolkit for Engineers*. This collection evolved through at last three editions and contained the essential packages of mathematical manipulation that Webb had taught to his coached pupils for many years. The Macaulay bracket, as a device for organising the integration of the beam bending equation, was published in 1919. Many textbook writers, starting with John Case in 1925, publicised the method, and it is discussed in extended form in the later editions of the *...Toolkit*. The essentials of the method had been employed by Clebsch in his celebrated *Theorie der Elasticitat Fester Korper* in 1862, but this had not been followed up by later workers. In terms of practical use it was taught by Webb and Inglis, and I presume Macaulay, earlier than 1919. This is clear from surviving correspondence between Webb and Ashwell, (Ashwell, 1970).

One of Webb's last pupils was the late Prof. W. (Bill) H. Wittrick. In 1965 WHW published a paper in the AIAA Journal, on extensions of the Macaulay method. There he includes a tribute to Webb as coach, teacher and initiator of these ideas. This paper was the catalyst for my own involvement with yet further extensions of the approach (Lowe, 1971). I had met WHW earlier. We shared an interest in the Russian engineering mechanics developments that were surfacing in the West in the 1950's. He was the Professor of Aeronautical Engineering in Sydney, and I was a graduate student in Civil Engineering, supervised by Prof. J.(Jack) W. Roderick (1913-1990), who earlier had been J.F.Baker's first 'plastic theory' research assistant.

W.H. Macaulay (1853-1936) trained through the Mathematical Tripos and could perhaps best be described as a geometer. His book on solid geometry, written very late in life, has some useful content. In career terms he spent most of his adult life as a Fellow of Kings, and had some teaching duties in the Engineering Faculty but no Faculty post. He and Charles Inglis no doubt meet regularly in King's College, and Macaulay and Webb, though nearly thirty years different in age, were friends.

C.E. Inglis (later Sir Charles) (1875-1952) was John Baker's predecessor as the Professor of Mechanical Sciences. One of his primary interests was in mechanical vibrations. Vibrations had been a topic of study in the Engineering Laboratory before Inglis and it seems a connection can be traced back to the second holder of the Mechanism Chair, J.(later Sir James) A. Ewing (1855-1935). Ewing's interest in vibrations had been kindled after experiencing earthquakes in Japan when he was a professor in Tokyo very early in his career.

Prof. S.(Stephen) P. Timoshenko (1878- 1972) published his *History of Strength of Materials* in 1953 when he was seventy-five years old. It is a comprehensive and valuable work with many useful comments in what some regard as the most important of all his books. He explains in the preface that he had lectured the topic for at least twenty-five years. This equates to lectures first given in 1928, or possibly earlier. His first academic post in the USA was at University of Michigan, Ann Arbor, beginning in 1928. His final career move was to Stanford in 1936. I have two copies of the book. One is the copy the author gave to R.V. (later Sir Richard) Southwell (1888-1970). This copy is annotated in pencil, as was Southwell's habit with most of his books. The other copy belonged to Dr. R.D.Davies, whose illness and early retirement created the staff vacancy in the Cambridge Engineering Laboratory which I filled in 1963. This led to my meeting CRC and now my contributing to this Festschrift.

Sir Richard Southwell, of the *Southwell plot* and *Relaxation Methods*, was Cambridge-trained in Mathematics and Mechanical Sciences, and moved to the Engineering Science Chair at Oxford after Fleeming Jenkin's son, C. F. Jenkin, retired, c 1930. During World War 2 he became Rector of Imperial College, but retired early. Thereafter he lived in Trumpington and died there in 1970. He was one of the few English workers in engineering mechanics of the day who had a working relationship with Timoshenko, and had attended summer schools at Ann Arbor during Timoshenko's time there. Lady Southwell's maiden name was Wingate: later the family name was changed to Pemberton. Her father was a GP in Cambridge. The family home was Scroope House in Trumpington Street, the site of the Engineering Laboratory. The Wingate's lease had many years remaining but by agreement with the freeholders, Gonville and Caius College, the site was bought by the University in 1920 and was developed to accommodate the move of the Engineering Department from Free School Lane after WW1. This was an initiative taken by Charles Inglis soon after his appointment as Professor. The fine residence served for some years as staff offices and the Library. It was eventually demolished to make way for extensions to the Baker Building c.1960. Wingate Way in Trumpington is one of the associations with her family. Southwell's tenure of the Engineering Science Chair at Oxford during the 1930's meant that many Australian and New Zealand engineers came and worked with him when they migrated to Oxford as Rhodes Scholars. Some of them returned to their places of origin and taught the next generation of engineers (Lowe, 2000).

3. THE PRESENT

I shall characterise 'the present' in the field of engineering and education by the wide ramifications of the use of computers. My own expertise is in only small parts of the whole scene, and other contributions to this volume cover the period. I mention only a few of the developments. Some of the earliest successful applications of computers in engineering, and in education in particular, were the frame analysis programmes written by Dr. R.K.(Ken) Livesley and others. The finite-element era followed. From about the mid-twentieth century, and stimulated by the political and social upheavals of WW2, came the very important advances in theoretical plasticity studies, limit analysis and associated mechanics. But the emphasis on computation will probably be seen to be the iconic activity. There have been many other significant developments including bio-mechanics with our featured colleague Calladine a prominent contributor. There has been much experimentation, and computation is seldom far away.

It seems to me that there has been an over-emphasis on computation, and this has contributed to a comparative neglect of design and particularly construction as the skills and expertise which must be enhanced if the final product of civil engineering endeavour is to be the best possible. An unrelated, unresolved educational question for me is what role can there be for the study of the history of our engineering discipline in the regular course structure? Judging by the almost total absence of such courses at present, the current climate seems to say that there is no place to be found. The field of engineering mechanics has, in comparative terms, shrunk somewhat, as other engineering disciplines have grown-up and acquired their place in the sun. There is still much of value being achieved. But what of the future?

4. THE FUTURE

In my view there needs to be a shift in the focus of structural engineering. Perhaps a new name is required. The primary need in the future I argue should be a much greater emphasis on the physical realisation, the actual construction process, of the structures and infrastructure which are the end products of the theoretical study and design.

These pages are being written in the weeks following the 11th September 2001 destruction of the World Trade Centre. This event serves to reinforce a belief that many of our buildings and other infrastructure are not as well constructed as they could or should be to cope with extreme events. The more common extreme events are natural disasters such as damaging earthquakes, but clearly premeditated disasters must also be considered.

The middle decades of the twentieth century saw the flowering of the plastic theory and limit analysis themes in engineering mechanics. There are many Cambridge associations. CRC made important contributions. J.(John) F. Baker (1901-1985) (Lord Baker of Windrush) was the Head of the Engineering Faculty in Cambridge when CRC studied for the Mechanical Sciences Tripos and during his early years on the Faculty. JFB presided over the Department from 1942 until 1968. His advocacy for plastic theory and the energies he channelled into finding and encouraging the many young graduate students and staff during this twenty-six year period, and earlier at Cardiff and Bristol, were instrumental in the development of this important body of engineering knowledge. Despite the very substantial output of research programmes and the publicity which plastic theory and design generated during this period, and the award to him of many honours including a Royal Medal, JFB was a disappointed person at the end of his life. His disappointment arose from what he perceived as an inadequate response from the profession and industry toward these plastically based principles and the potential benefits they offered.

The legacy from this period is substantial, and is more to do with appreciation of concepts such as ductility and global characteristics of the structures, than the specifics of plastic theory and design. This is an important legacy and must now be carried into other avenues for the structures of the future.

My own interests in structural mechanics, and especially many of the historical aspects, have benefited over the years from conversations with and reading the publications of a Peterhouse colleague of CRC's, Professor Jacques Heyman. There is much that is special about personal associations. The Peterhouse connection continues back in time with distinguished contributors to engineering related studies, including Routh, Tait, Kelvin and many more who studied or taught there.

We need to have clear ideas for the future of engineering, and of the needs of society which engineers might help provide. Then we can think more effectively about what appropriate educational requirements should be put in place for the engineers of the future. Engineering has broadened immensely during our lifetimes, and no doubt this trend will continue. In the first half of the twentieth century mechanics in various forms was a dominant study in the Engineering Laboratory in Cambridge and elsewhere. That dominance has passed and will not return, but the fundamental importance and usefulness of this body of knowledge remains.

There are many challenges for society in the future, none more so than in the building and construction industry since the events of *'9/11'*. Collapse of the Twin Towers has produced world-changing knock-on effects. Insurance markets around the world, which have considerable influence on all our futures, have already been grossly affected, particularly in relation to building. We argue that unless progress can be made with radically new building methods, many of society's legitimate aspirations for building in the future will not be met. The scenario discussed here does not call for new materials to be developed. Instead it points to substantial benefits that can be achieved in the final built infrastructure when the present dominant materials, the steel and concrete we know already, are better utilised. We shall further argue that there are significant educational challenges embedded in such a programme.

A major challenge for the future is to find new ways to deal with wastes. There is the volume, make-up and environmental impact of the waste to be considered. Building waste is a significant contributor, as has been observed with the 'Ground Zero' clean-up. Despite huge effort, expense and sophistication involved in the promulgation of Standards to extend the life of buildings, probably a majority of today's new construction will not remain economically viable for more than some notional minimum of tens of years. Rebuilding or adaptation will be necessary, and huge quantities of waste will be generated in the process. If in the future the production of substantially less building waste is not just an aim but a requirement, then we must expect that present methods of building will not achieve substantial improvements.

The choice of building material is important. Here we assume that steel and concrete will continue to dominate as they have done for the past one hundred-and-fifty years. Other composites incorporating fibres in various forms have a place though my own hunch is that they will at best occupy a significant niche but will never displace steel and concrete globally. There are also many sheet building materials. These are particularly prone to producing unwanted waste when rebuilding becomes necessary. Much thought will be needed to devise suitable alternatives that can be recycled economically.

When we examine the use of these traditional materials in construction, it is clear that there is relatively little core innovation being worked on of any scale. It should be a given that innovation is needed in all major industries at all times. The kind of innovation I am suggesting is in the method of use rather than any technical demand on the quality or specification of the material itself. In this sense the problem is clear cut and there may be several solutions possible.

So what of the future? My contention is that what have for more than a hundred years been separate, competing industries, namely the steel and concrete (cement) industries, must in some sense be merged before all the benefits possible with steel/concrete composites can be realised. A radical shift needs to be made towards a steel/concrete composite technology that offers prospects for substantial further progress on many fronts. These include : i) reduction of building-related waste produced either by new construction or later rebuilding/refurbishment, ii) utilisation of the many attractive technical features of hollow steel components, iii) permanent formwork for the insitu concrete, iv) superior strength and ductility of a truly plastic material and v) cost-effectiveness. And these are only some of the desires for the future! Most of all, the emphasis should be to *simplify* the processes, particularly fabrication and final construction of the component or structure. Part of the simplification will depend upon all of the individual processes being capable of being automated. These end points cannot be reached if bar reinforcement is retained.

Thus, implicit in this proposed scenario is the need to abandon bar reinforcement as the steel content in our new steel-concrete composite of the future. It seems quite clear that there is little prospect of wholly automating the bar fixing parts of the construction cycle, for example. Instead what we propose is that the steel be employed as sheet, but now fabricated rather than rolled into hollow members of box-shaped cross-section. One version of such a technology has been given a name, *Externally Reinforced Concrete (ERC)*, and can be shown to offer scope to make progress on all the above features, and more (Lowe, 1997). The result is a highly adaptable, strong, ductile and cost effective steel/ concrete composite.

What are the implications for the educational process? The track record has been that innovation and substantial improvement of the building process has been traditionally slow, and some of this slowness can probably legitimately be attributed to the nature and content of the courses on offer in degree programmes, and the other means of education and training. Computation has a place in the course structure for this sort of applied structural mechanics, and other skills of visualisation and judgement also need to be fostered. Recent experience of offering courses on the subject of ERC has shown that there is considerable scope to theorise about, design and construct structures in the classroom, the laboratory and in the market place.

All the achievements of plastic theory can be readily applied in the ERC environment. Ductility and strength of the composite steel/concrete material are both easily achieved and impressive. New avenues open to achieve simpler and more effective structures. These could be some of the educational and engineering aims in the future!

5. CONCLUSIONS

This gathering celebrates the achievements of an esteemed colleague. His combination of many skills and talents is evidenced by the great variety and depth of the tributes here being paid to him. May he long continue to delight us with his clear thinking and the broad range of his activities.

Acknowledgements

It is my pleasure to record the assistance I have received from friends, A-P Robert Wheen, the HOD, and A-P Stuart Reid, of the Department of Civil Engineering , University of Sydney.

References

Ashwell, D.G. (1970). Private communication, including copies of some correspondence with H.A. Webb.

Lowe, P.G. (1971) *Classical Theory of Structures*, Cambridge University Press.

Lowe, P. G. (1997) Externally reinforced concrete composites, *Third Canmet/ACI International Symposium on Advances in Concrete Technology, Auckland*, ACI Special Publication SP 171, paper 26, p. 551-568.

Lowe, P. G. (2000) Some Aspects of Australasian Academic Engineering - from Rankine to Southwell, *Proceedings of the Second Australasian Conference on Engineering Heritage, 2000*, National Committee for Engineering Heritage, Institution of Professional Engineers New Zealand, Auckland. p. 175 - 181.

PREDICTION OF BUCKLING LOAD FROM VIBRATION MEASUREMENTS

P. Mandal
Manchester Centre for Civil & Construction Engineering, UMIST,
PO Box 88, Manchester, M60 1QD, U.K
P.Mandal@umist.ac.uk

Abstract The linear relationship between buckling load and the square of the frequency of a structure is limited to the cases in which the fundamental vibration mode and the lowest buckling mode are the same. For cases where the two modes are different researchers in the past have suggested some empirical equations. In this study (mainly numerical) it is shown that the linear relationship is reasonably valid when the modes are approximately close to each other. However, for a simply supported rectangular plate of aspect ratio two or more, the fundamental vibration mode and the lowest buckling mode are usually different to each other. It is observed that the apparent non-linear curve in this situation consists of a few linear segments depending on the aspect ratio. The buckling load could be accurately predicted by measuring the first few frequencies, instead of just one.

Key words: Buckling load, Frequency, Rectangular plates.

1. INTRODUCTION AND BACKGROUND

Buckling is a common failure mode in a wide range of structures, including civil, aerospace, marine and offshore structures. Increasing use of high-strength materials in construction and emphasis on optimisation in design often make structures vulnerable to buckling. Experiments on buckling are usually destructive in nature as large displacements occur at buckling. Therefore, the experiments are normally done on scaled models of the actual structures. This may not necessarily represent the reality. In most

H.R. Drew and S. Pellegrino (eds.),
New Approaches to Structural Mechanics, Shells and Biological Structures, 175–188.

cases, numerical results are the only guidelines available, although it is difficult to incorporate appropriate boundary conditions and accurate geometry in the numerical modelling. Moreover, the theory of buckling is somewhat imperfectly understood for certain kinds of structures, e.g., cylindrical shells under axial compression. Therefore much emphasis is put on the actual experimental data.

Several studies have been done in the past on the use of non-destructive frequency measurements to predict the buckling load of a structure. The motivation comes from the fact that both the linear buckling load and the frequency of a structure can be evaluated from an eigenvalue problem, although buckling is associated with the conventional structural stiffness and geometric stiffness while frequency is associated with conventional structural stiffness and mass. It is not exactly known who first tackled the problem. Melan in 1917 published an approximate formula as below (although not exactly in this form) for a simply-supported bar with a concentrated mass at mid-point.

$$\frac{P}{P_{cr}}+\left(\frac{f_m}{f_0}\right)^2=1 \tag{1}$$

Here P is the applied load (negative for a tensile force), P_{cr} is the buckling (or critical) load, f_0 is the fundamental frequency without any applied load, and f_m is the frequency at load P. The above equation indicates that when the buckling condition is reached (i.e., $P= P_{cr}$) the frequency of the loaded structure (f_m) becomes zero. It was established later that (1) is valid for greater-than-or-equal-to inequality and the equality holds only for the case when the vibration mode and the buckling mode are the same. For a proof see Tarnai (1995).

It is evident that if one knows the frequencies for two different load levels then the buckling load can be easily estimated by (1). These two frequencies can be measured experimentally for two different values of load. In the case of an existing structure, say the measured frequency is f_1. Now for an increase in load on the structure by ΔP, the new measured frequency will be altered to f_2. If the load ΔP is increased proportionally, the structure will buckle at

$$(\Delta \mathrm{P})_{cr}=\Delta \mathrm{P}\frac{f_1^2}{f_1^2-f_2^2} \tag{2}$$

In other words the buckling load is predicted as a multiple of the load increment, ΔP. The ratio between these two can be considered as a factor of safety against buckling.

Lurie (1952) conducted one of the earliest experimental studies on the relationship between buckling load and frequency. He did tests on a rectangular frame, a rigidly jointed truss and a flat plate. The predictions and measurements agreed well except for the flat plate. He concluded that the square of the frequency is in exact linear relationship to the end thrust of a column when the mode of vibration is identical to the buckling mode. Galef (1967) also came up with the same formula for a column from an energy-approximation. He reported that the relationship is exact if the vibrating mode shape corresponds to the buckling mode shape. This was the case at least for the investigated hinged-hinged column. In general, the buckling and vibrating mode shapes are sufficiently similar to use Equation (1) for an excellent approximation.

Some references to earlier works can be found in Lee *et al.* (1965), who conducted experiments on columns of non-uniform cross-section with various end-conditions. It was concluded that the use of vibration data to predict buckling loads can be used for elastic buckling, but it will be unsafe in situations where the column fails inelastically. Jubb *et al.* (1975) extended the relationship to include the determination of residual stresses in welded plates. Their conclusions were based on experiments on welded box-columns. They showed that when the vibration mode is similar to the elastic buckling mode, the natural frequency approaches zero as the axial load tends to the elastic buckling load. Similar numerical observations for frame structures have been reported by Ovunc (1980).

In some situations, axial stresses are present only in a certain portion of a structure while the rest of the structure is stress-free, e.g., a multistage rocket. Shastry and Rao (1986) presented the frequencies for partially stressed columns with different boundary conditions. Bokian (1988) conducted a theoretical study of a uniform single-span beam under a constant compressive axial load. He considered four boundary conditions at a particular end (i.e., free, sliding, pinned and fixed), and thus ten different combinations for both ends. The results showed that, except for free-free and pinned-free beams, Equation (1) is valid. However, for higher modes of vibration all the beams follow (1).

Singer (1982) carried out experiments on cylindrical shells and used measured frequencies at low levels of axial load to predict buckling loads. In most cases, the predicted values were higher than the actual buckling loads. Therefore, he stressed the need for a lower-bound approach, so that in conjunction they provide a range within which the buckling load would lie. Plaut and Virgin (1990) attempted to address this issue by changing the power of frequency ratio in (1): e.g., a fourth power of frequency ratio will give a lower bound of the prediction. However, their proposition was purely empirical, without a rational argument behind it. Souza and Assaid (1991) prescribed the use of a 'Hermite curve' for better accuracy instead of the linear relationship between axial load and squared frequency. They demonstrated it for the buckling of cylindrical shells.

On the assumption that the linear relationship between axial load and squared frequency exists, Sundararajan (1992) developed a method of computing fundamental frequencies of axially loaded structure with non-uniform mass or stiffness distribution. He suggested that it could be used in the optimisation of structures.

Costley *et al.* (1991) also applied the linear relationship to predict the buckling load of cylindrical shells. Their idea was to carefully monitor the modal frequencies due to random excitation. If the external load is increased, the modal frequencies will decrease, and the buckling load and corresponding shape can be predicted by extrapolating the frequencies. However, their attempt was unsuccessful, as they found it difficult to isolate experimentally the closely spaced modes. There were also difficulties in identification of the modes. Therefore, they concluded that mode identification and tracking was not a feasible method for determining the buckling load of a thin-walled, axially loaded cylinder.

Recently Rao (2001) has done an analysis to predict the fundamental frequency of initially stressed square plates. The result derived is identical to Equation (1).

These previous researches have been limited to the verification of (1), or to provide similar empirical equations. In almost all the cases only the fundamental frequency has been measured and used for the prediction of buckling loads. The present study aims to provide a better understanding of the relationship between the vibration frequency and buckling load, in particular when the fundamental mode shapes are completely different from each other.

2. EXPERIMENTS AND ANALYSES

2.1 Columns

A simple experiment has been carried out to establish the relationship between the buckling load of an axially loaded column with its frequency at different levels of axial load. The column was made of mild steel and had uniform rectangular cross-section (width 25.4 mm, thickness 3.27 mm, and length 700 mm). It had simply supported boundary conditions. Axial load was applied at one end, and the frequencies for the loaded column were measured by a portable structural analyser (DI-2203). The results are plotted in Fig. 1. Here, the buckling load, P_{cr}, is calculated by Euler's formula. It seems that Equation (1) works quite well for the tested specimen. The n-th buckling and vibration mode can be written as

$$y_n(x) = A \sin\frac{n\pi x}{L} \tag{3}$$

where A is an arbitrary constant, x is the distance along the length, and L is the total length of the column. As the fundamental ($n = 1$) buckling mode and vibration mode are essentially the same, the assumptions in (1) are valid. Therefore, it is not surprising that the test data could predict the buckling load with a reasonable accuracy.

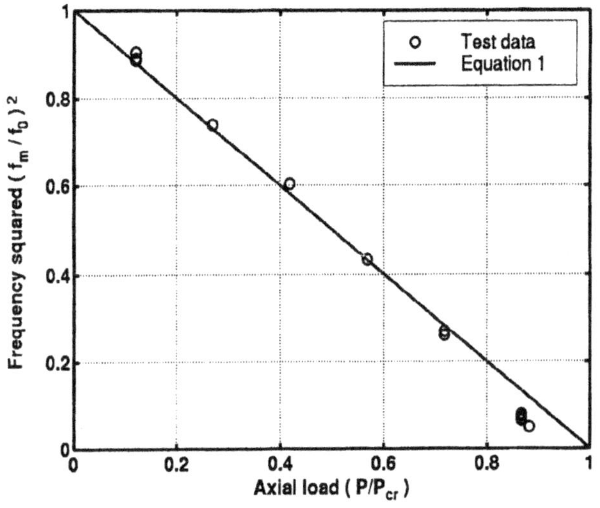

Figure 1. Dimensionless plot of measured frequency versus axial load of a slender column.

However, the situation may be different for columns with boundary conditions other than simple supports. The n-th buckling mode for a cantilever column loaded by a concentrated load at its free end can be written as

$$y_n(x) = A\left(1-\cos\frac{(2n-1)\pi x}{2L}\right) \tag{4}$$

where A is an arbitrary constant, and L is the length of the column. On the other hand, the n-th vibration mode is

$$y_n(x) = A[(\sin \beta_n L - \sinh \beta_n L)(\sin \beta_n x - \sinh \beta_n x) + (\cos \beta_n L + \cosh \beta_n L)(\cos \beta_n x - \cosh \beta_n x)] \quad (5)$$

where A is an arbitrary constant, and $\beta_1 L = 1.875$, $\beta_2 L = 4.694$, $\beta_3 L = 7.855$, etc.

Evidently, the fundamental vibration and buckling modes for a cantilever column do not match. However, they are close enough for (1) to be reasonably valid, as shown in Fig. 2, see Column A. This figure shows the numerical frequencies obtained from finite element analysis, using ABAQUS (a commercial, general purpose, finite-element package). Column D is similar to A, but it is not fully restrained against rotation at the base. Instead a rotational spring is attached to provide a finite amount of rotational flexibility. Equation (1) is also shown to be applicable in this case, which reflects a real-life situation in terms of foundations on soils. In Column E, the axial load is distributed along the length, like gravity loading in a vertical column. The analysis is done with the *nlgeom* (non-linear geometry) option in ABAQUS. Essentially this allows one to extract the eigenvalue frequency of a pre-loaded structure. For all the numerical analyses in this study, the material is considered as linear elastic, with Young's modulus of elasticity 200 GPa, Poisson's ratio 0.3, and mass density 7850 kg/m^3. P_{cr} denotes the linear eigenvalue buckling load.

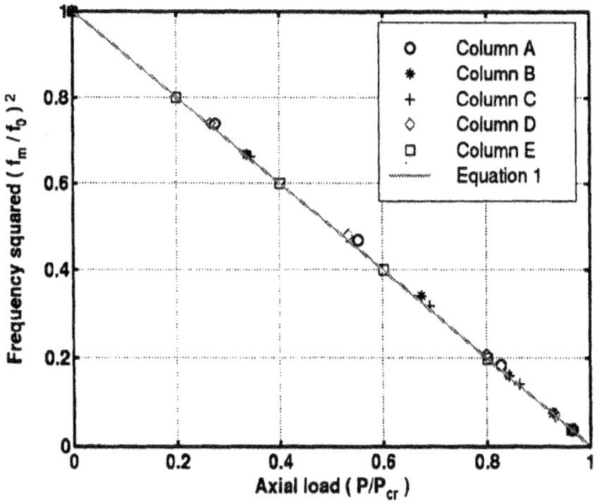

Figure 2. Frequency of pre-loaded columns with various boundary conditions. For buckling, a vertically downward load is applied at the top for Columns A-D, whereas Column E is under self-weight only. The boundary conditions for Columns A to E at the bottom and the top are Fixed-Free, Fixed-Pinned, Fixed-Sway allowed, Rotational spring-Free, Fixed-Free, respectively. All the columns are 700 mm long, 3.27 mm thick and 25.4 mm wide.

2.2 Two dimensional frames

In the next step, a 2-dimensional frame, shown in Fig. 3(a), is analysed by ABAQUS. The cross-section of the frame is rectangular, and the supports are encastred. The lowest buckling mode for two vertical concentrated loads applied at Points 2 and 3 are shown in Part (b) of the figure, while Part (c) shows the fundamental vibration mode. It is observed that they are very similar. Therefore, the non-dimensional plot of frequency versus load will closely follow Equation (1). If the loads are uniformly applied over the beam 2-3, the base configuration for the eigenvalue frequency analysis will change with the load. However, the lowest buckling mode under this uniformly distributed load is the same as for concentrated loads at two corners. The frequency plot is shown in Fig. 3(d).

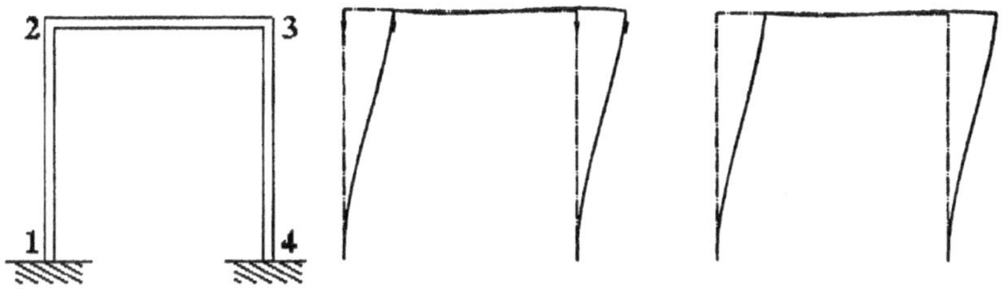

(a) (b) Lowest buckling mode (c) Fundamental vibration mode

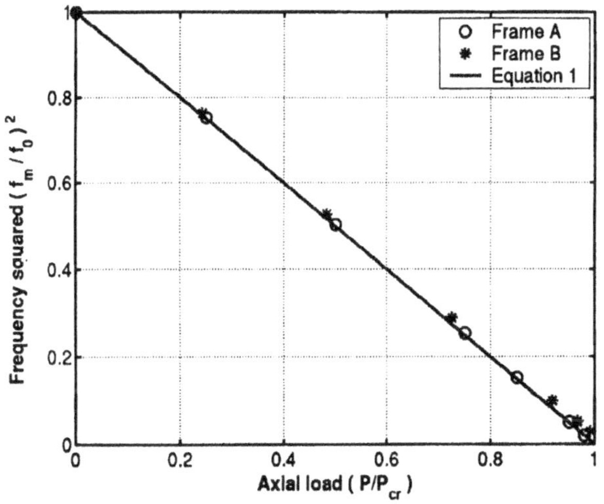

(d)

Figure 3. Frame A is the frame shown in (a) with two concentrated loads of magnitude P acting vertically downwards at Points 2 and 3. In Frame B the load is uniformly distributed over the beam 2-3. All the beams and columns are 300 mm long, 3.2 mm thick and 12.7 mm wide.

2.3 Three dimensional frames

(a) (b) Fundamental vibration modes for Frames C, D, E (c) Lowest buckling mode for Frame C (d) Lowest buckling mode for Frame D (e) Lowest buckling mode for Frame E

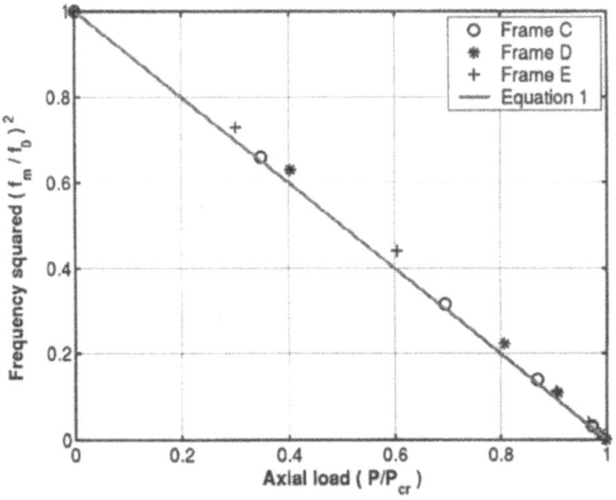

(f)

Figure 4. Frame C is the frame shown in (a) with four concentrated loads of magnitude P acting vertically downward at points 9 to 12. In Frame D the load P is uniformly distributed over the eight beams 5-6, 6-7, etc., and in Frame E the load is uniformly distributed on two floors 5-6-7-8, and 9-10-11-12. All the beams and columns are 300 mm long, 3.2 mm thick and 12.7 mm wide. The slabs are 2 mm thick.

The analysis is extended to include 3-dimensional frames. A one bay, two storey frame is chosen for this purpose, as shown in Fig. 4(a). The loads are applied either as concentrated loads at the corners (Frame C), or distributed on the beams (Frame D) or on the slabs (Frame E). The fundamental vibration mode, which will be same for all three frames as the mass distribution and geometry are the same, is shown in Fig. 4(b). The lowest buckling modes for three frames are shown in Fig. 4(c)-(e). Part (f) of the same figure shows the dimensionless plot of computed frequencies versus applied load. The buckling mode for Frame C is similar to the vibration mode, therefore Equation (1) is closely followed. However, the buckling modes for Frames D and E are different from the vibration modes and the points do not lie so close to the line. There is almost no deformation on the upper storey columns (see Fig. 4(b) and (c)). This is because the loads are applied at both floor levels. Therefore the axial loads in the lower storey columns are higher than those in the upper storey. But for concentrated loads at the top (Frame C), both the upper and lower storey columns are subjected to same amount of loads.

2.4 Plates

From the preceding analyses, it is observed that Equation (1) works quite well for regular type of structures. In all the cases reported above, the fundamental vibration and buckling mode shapes are close to each other, if not exactly identical. Therefore, the frequency data (experimental or computed) seems to satisfy the linear relationship with the buckling load. However, the fundamental vibration and buckling modes for some types of structures could be completely different, e.g., a rectangular plate. For a simply supported plate of aspect ratio two, the fundamental vibration mode shape is characterised by one half-wave in each direction. On the other hand, the lowest buckling mode shape under in-plane axial compressive load applied to the shorter edges, is characterised by one half-wave in the shorter direction and two half-waves in the longer direction. For plates of aspect ratio three, the fundamental vibration mode remains the same, but the buckling mode changes to three half-waves in the longer direction. The modes are shown in Figure 5(b)-(e). The computed plot of frequency against buckling load is shown in Figure 6 after normalisation.

From Fig. 6 it is observed that although Equation (1) accurately predicts the behaviour of a square plate, in general it over-estimates the buckling load of a rectangular plate. The reason is that the fundamental vibration and buckling modes are similar for a square plate, but very different for rectangular plates, as discussed above. On closer examination of the rectangular plate of aspect ratio 3, it is observed that the fundamental vibration mode changes shape with the level of applied axial load, and shows mode shapes similar to the buckling mode when P approaches to P_{cr}. This is shown in Fig. 7. Similarly, a plate of aspect ratio 2 is characterised by a bilinear curve: see Fig. 6.

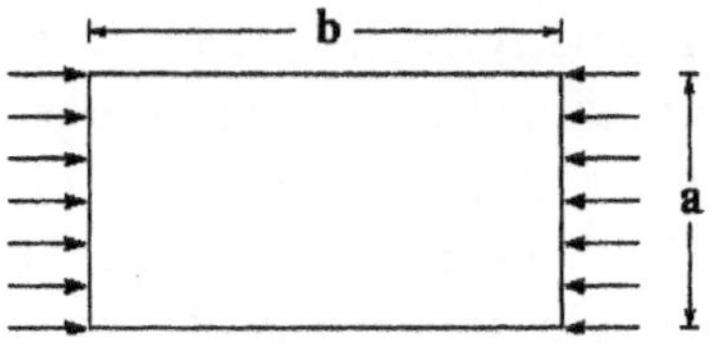

(a) Aspect ratio b/a

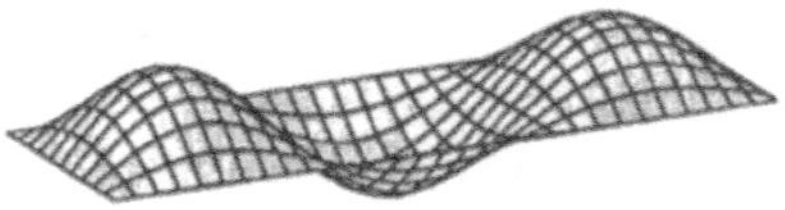

(b) Lowest buckling mode for aspect ratio 3

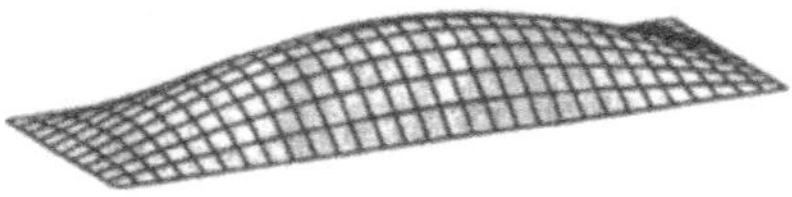

(c) Fundamental vibration mode for aspect ratio 3

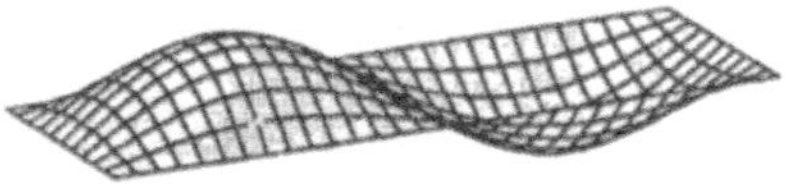

(d) Second vibration mode

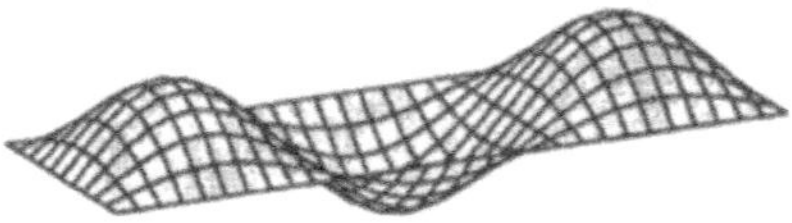

(d) Third vibration mode

Figure 5. Mode shapes for plates of aspect ratio 3. The plate is 900 mm × 300 mm × 2 mm. The edges are simply supported.

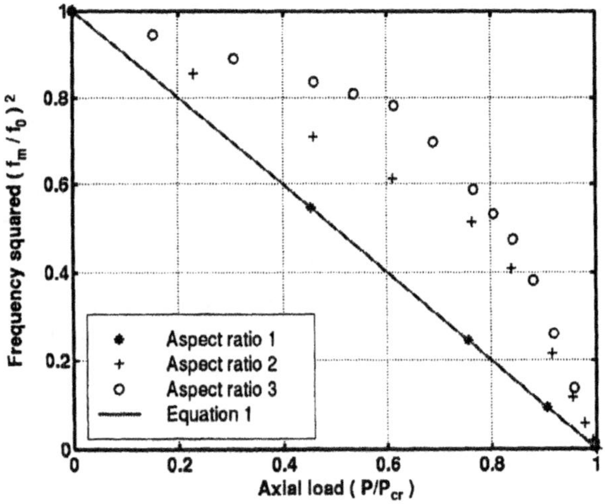

Figure 6. Non-dimensional plot of computed frequency versus in-plane axial load of rectangular plates of different aspect ratios. Aspect ratio 1 indicates a square plate. P_{cr} is the buckling load for respective plates. The width and thickness of the plate are 300 mm and 2 mm respectively.

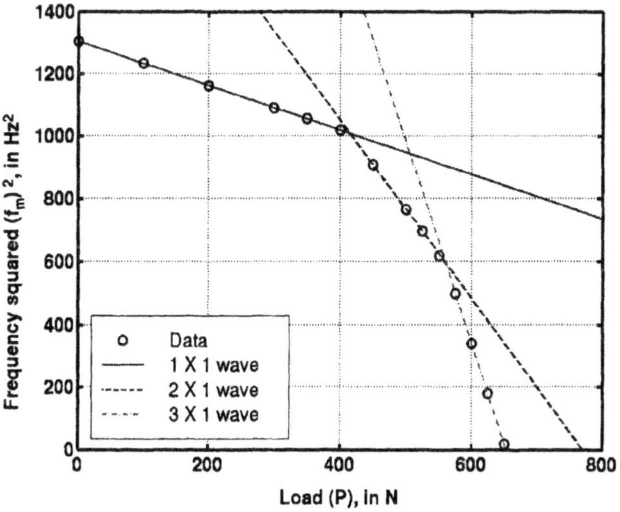

Figure 7. Plot of computed frequency versus in-plane axial load of a rectangular plate of aspect ratio 3, with lines corresponding to Equation (1) for the different vibration modes.

3. DISCUSSION

In the previous section it was evident that if the buckling and vibration modes of a structure were close to each other, formulas like (1) or (2) could be used to predict the buckling load. Frequencies of a structure are generally easy to measure and could be used as a non-destructive technique to predict buckling loads. This also takes into account the effect of real-life boundary conditions. For structures like a cantilever column the fundamental mode shapes (4) and (5) do not match, but are close enough that (1) is applicable. Essentially the problem is that of a *two-parameter eigenvalue* problem.

$$([K]-\lambda[K_G]-\mu[M])\{u\}=0 \tag{6}$$

Here $[K]$, $[K_G]$ and $[M]$ are the structure stiffness matrix, the geometric stiffness matrix, and the mass matrix, respectively. λ and μ are the eigenvalue estimates of buckling load and frequency. $\{u\}$ is the eigenvector. Normally, two separate analyses are carried out, i.e., eigenvalue buckling analysis (7) and eigenvalue frequency analysis (8):

$$([K]-\lambda[K_G])\{v\}=0 \tag{7}$$

$$([K]-\mu[M])\{w\}=0 \tag{8}$$

where $\{v\}$ is the buckling mode and $\{w\}$ the vibration mode. These are in general different. If these two vectors are equal, a relationship like Equation (1) could be easily derived from (6). If they are different, and the structure is loaded incrementally up to buckling and the frequencies and mode shapes are observed at each increment, the fundamental vibration mode $\{w_1\}$, will approach the lowest buckling mode $\{v_1\}$. At or near buckling these two modes will be exactly equal. The angle between the lowest buckling mode vector and the fundamental vibration mode vector for different axial loads is plotted in Fig. 8. It might be worth mentioning here that the set of vibration modes for a cantilever column is an orthogonal set, whereas the buckling mode set, although independent, is not orthogonal.

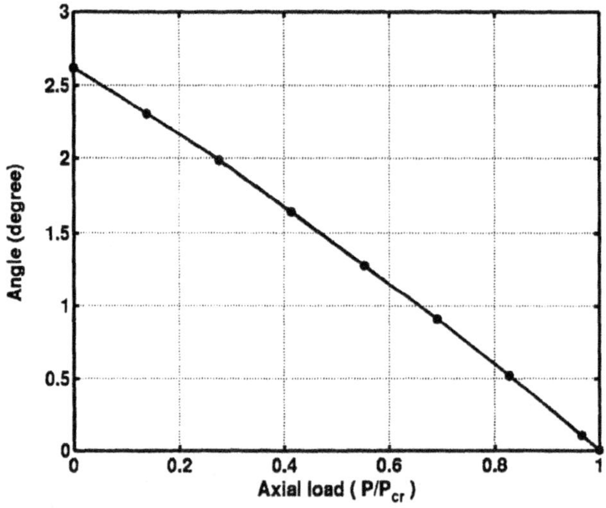

Figure 8. Plot of angle between the lowest buckling mode vector, and the fundamental vibration mode vector for different axial loads on a cantilever column.

In the case of rectangular plates both vibration and buckling mode sets are from the same orthogonal set. The ordering of modes differs with the level of applied loads. Therefore, if one measures n frequencies, instead of measuring just one, the extrapolation will predict a reasonable value of the buckling load. The number n depends on the aspect ratio of the plate in this case. As a general guide one should continue to include more and more of frequencies until the extrapolation leads to lower buckling loads. Referring to Fig. 7 it could be observed that for load less than 400 N, one needs to monitor at least three frequencies. For loads between 400 N to 525 N at least two frequencies will be necessary. After that just the fundamental frequency will be sufficient. Essentially the idea is to track the lowest buckling mode in the set of vibration modes.

The results discussed so far have assumed linear elastic behaviour of structures. The phenomenon of buckling is very often accompanied by material inelasticity. Moreover, the buckling load and frequency of a structure will be different if there is any geometric imperfection, or residual stresses. In these cases the linear relationship between the frequency and the buckling load may not be valid. The buckling load for cylindrical shells under axial compression, for example, is highly imperfection-sensitive. The buckling modes are very closely spaced as well. Although the fundamental mode shapes for free vibration and classical linear buckling for cylindrical shells under uniform axial compressions are identical (Leissa 1993), the application of the present method may prove to be a futile exercise.

4. CONCLUSIONS

- The simple formula, Equation (1) is fairly accurate if the fundamental vibrating mode shape is close to the lowest buckling mode shape. The relationship is exact if the mode shapes are exactly similar.
- If the fundamental vibration mode shape is completely different from the lowest buckling mode shape, the relationship constitutes a convex curve to the origin. For rectangular plates with aspect ratio three, it is bounded by three straight lines.
- A reasonable estimate of buckling load could be made by monitoring the first n (≥ 2) frequencies, where n should be big enough to include the lowest extrapolated value of the buckling load.
- For low values of the P/P_{cr} ratio, the predicted buckling load is much higher than the actual one – a non-conservative estimate. The prediction is much better for higher values of P/P_{cr} and almost exact when $P \rightarrow P_{cr}$.

References

Bokian, A. (1988) Natural frequencies of beams under compressive axial loads, *Journal of Sound and Vibration*, **126**(1), 49-65.

Costley, A. C., Mourad, S. A., Kazanjy, R. P. and Pardoen, G. C. (1991) Determination of critical buckling load of thin-walled cylinders using modal analysis, *Proc. Ninth International Modal Analysis Conference*, Florence, IT Union College, N.Y., Pt 1, 542-548.

Galef, A. E. (1967) Bending frequencies of compressed beams, *The Journal of the Acoustical Society of America*, **44**(8), 643.

Jubb, J. E. M., Philips, I. G. and Becker, H. (1975) Interrelation of structural stability, stiffness, residual stress and natural frequency, *Journal of Sound and Vibration*, **39**(1), 121-134.

Lee, J. A. N., Hope, B. B. and MacGowan, J. P. (1965) *The Prediction of the Buckling Load of Columns by Non-destructive Testing Methods*, Ontario Joint Highway Research Programme, Report No. 41; Department of Civil Engineering (C.E. Report no. 50), Queen's University, Kingston, Ontario, Canada.

Leissa, A. (1993) *Vibration of shells*, Acoustical Society of America.

Lurie, H. (1952) Lateral vibrations as related to structural stability, *Journal of Applied Mechanics*,**19**, 195-204.

Melan, H. (1917) Kritische Drehzahlen von Wellen mit Langsbelastung, *Zeitschrift der Österr Ingenieur- und Architekten- Vereines*, **69**, 610-612, 619-621.

Ovunc, B. A. (1980) Effect of axial force on framework dynamics, *Computers and Structures*, **11**, 389-395.

Plaut, R. H. and Virgin, L. N. (1990) Use of frequency data to predict buckling, *Journal of Engineering Mechanics*, **116**(10), 2330-2335.

Rao, G. V. (2001) A simple formula to predict the fundamental frequency of initially stressed square plates, *Journal of Sound and Vibration*, **246**(1), 185-189.

Shastry, B. P. and Rao, G. V. (1986) Vibration of partially stressed beam, *Journal of Vibration, Acoustics, Stress and Reliability in Design*, **108**, 474-475.

Singer, J. (1982) Vibration correlation techniques for improved buckling predictions of imperfect stiffened shells, in *Buckling of Shells in Off-shore Structures*, edited by Harding J. E., Dowling P. J. and Agelidis N., Granada Publishing, London, England, 285-329.

Souza, M. A. and Assaid, L. M. B. (1991) A new technique for the prediction of buckling loads from nondestructive vibration tests, *Experimental Mechanics*, **31**(2), 93-97.

Sundararajan, C. (1992) Frequency analysis of axially loaded structures, *AIAA Journal*, **30**(4), 1139-1141.

Tarnai, T. (1995) The Southwell and the Dunkerley theorems, in *Summation Theorems in Structural Stability*, edited by Tarnai, T., CISM Courses and Lectures no. 354, Springer-Verlag, New York.

NON-COLLINEARLY LOADED LAMINAE

M.N. Pavlović
Department of Civil Engineering, Imperial College
London SW7 2BU, UK

N. Tahan
Noble Denton Consultants, 14701 St. Mary's Lane,
Suite 750, Houston, Texas 77079, USA

M.D. Kotsovos
Faculty of Civil Engineering, National Technical University of Athens
42 Patission Street, Athens 106 82, Greece

Abstract The paper illustrates how symbolic computations can restore some of the earlier significance of certain analytical techniques. This is shown by reference to the two-dimensional elasticity problem in which rectangular plates (both isotropic and orthotropic) are subjected to the action of non-collinear but parallel compressive forces (as in a beam subjected to 'four-point' loading of its upper and lower surfaces). The work focuses on the case where the out-of-alignment of these actions is small, thus generating a local departure from the shear-stress distribution predicted by the simplified engineering beam theory. Parametric studies readily ensue, aided by a single symbolic matrix inversion (thus rendering repetitive equation solving unnecessary) and by the use of Fejér's summation of the Fourier-series solution (which greatly reduces the number of terms required for convergence).

Keywords: Plane-stress elasticity, Beams, Non-collinear loading, Isotropy, Orthotropy

1. INTRODUCTION

The power of currently-available numerical techniques has largely displaced classical analytical methods in engineering design. While this is understandable since modern computational tools, such as the finite-element method, enable extremely complex problems to be analysed even by engineers with relatively modest background in mechanics, a good design team will not automatically ignore the wealth of information available from

H.R. Drew and S. Pellegrino (eds.),
New Approaches to Structural Mechanics, Shells and Biological Structures, 189–202.

classical-solution sources. After all, engineers with a knowledge of these sources will often recognize either the ready-availability of an existing closed-form solution or the possibility of attaining such a solution through some analytical means, either of which could at least offer a good first approximation to the problem in hand. As much of Calladine's work shows, the great advantage of analytical methods is that they provide not only an understanding of the relevant physical phenomena (Calladine, 1983) but also permit the identification of, and the classification into, different regimes of behaviour (Calladine, 1982); thus, such techniques enable the designer to attain, through simple parametric studies, a quick understanding of the effect of the variables involved. An efficient and rational approach to engineering analysis, therefore, should retain both numerical and analytical options, especially as the latter can be enhanced when combined with symbolic computer techniques. One such instance is explored in the present article.

The Levy-type solution, extensively used for a wide range of problems, belongs to that type of computational technique which could be described as semi-analytical, for it straddles both analytical and numerical domains. Thus, the general solution consists of a single Fourier series, each term of which involves constants which can only be obtained through the solution of a set of linear simultaneous equations. By employing symbolic computer techniques, however, the latter need be carried out only once in non-numerical fashion, thus not only avoiding repetitive equation-solving but also yielding a truly closed-form solution (Pavlović and Poulton, 1985; Pavlović and Sapountzakis, 1986; Pavlović *et al*, 1998a, 1998b; Tahan, 1991; Tahan *et al*, 1993a, 1993b, 1995a, 1995b, 1997). Such an approach, then, leads to simple addition of Fourier terms, the number of which needed for convergence can frequently be reduced quite spectacularly by employing Fejér's summation instead of the ordinary series summation (Tahan *et al*, 1993a). We shall now proceed to apply this symbolic approach to the two-dimensional problem of a thin, rectangular, elastic plate (akin to a simply-supported beam) which is acted upon by two non-collinear (equal and opposite) forces that are relatively close to each other, as depicted in Fig. 1: both isotropic and orthotropic cases will be considered.

2. ISOTROPIC CASE

As is well known, when the lever arm e between the non-collinear forces in Fig. 1 is larger than a certain value, the distribution of shear stresses across the middle plane Y-Y resembles the parabolic distribution given by the elementary theory of bending of elastic beams; otherwise, for "small" values of e, deviations from the parabolic distribution occur. Our present aim is to determine the limiting distance beyond which the well-known beam-theory results apply (studying the parameters affecting this distance) and to investigate the stress distributions for distances smaller than these "limiting" ones. Although the study is restricted to the dominant shear stresses in this localized region, the method provides a truly closed-form solution to all three stress components throughout the plane-stress system of Fig. 1, including regions in the vicinity of the applied forces.

The application of Levy's method to the biharmonic equation for (isotropic) plane stress combines the Airy stress function Φ (for all positive integers m), i.e.

$$\Phi = \Sigma\, Y_m (y) \sin a_m x \tag{1}$$

with the explicit solution for $Y(y)$ obtained by substituting each term of (1) into the governing (biharmonic) equation, namely

$$Y(y) = C_1 \cosh \alpha y + C_2 \sinh \alpha y + C_3\, y \cosh \alpha y + C_4\, y \sinh \alpha y \tag{2}$$

and $\alpha = m\pi/a$ (henceforth, the subscript m is implicit in α, Y and the constants C). The constants are obtained by prescribing, along the longitudinal edges, the boundary conditions, the latter conveniently expressed in the form of a half-range Fourier series of period $2a$. For our problem (but also, more generally, for arbitrary normal loadings) this implies

$$\sigma_y = -\,\sigma_u \sin ax \ \text{ at } y = 0,\ \sigma_y = -\,\sigma_l \sin ax \ \text{ at } y = b,\ \tau_{xy} = 0 \text{ at } y = 0, b \tag{3a-d}$$

where the Fourier coefficients of the applied loading are denoted by the subscripts u and l referring to the "upper" and " lower" edges of the plate. With the use of the standard expressions for the stress tensor in terms of the second derivatives of Φ (Timoshenko and Goodier, 1968), the generic system of simultaneous equations stemming from (3) is

$$\begin{bmatrix} \alpha & 0 & 0 & 0 \\ 0 & \alpha & 1 & 0 \\ \alpha\beta & \alpha\gamma & \alpha b\beta & \alpha b\gamma \\ \alpha\gamma & \alpha\beta & \beta + \alpha b\gamma & \gamma + \alpha b\beta \end{bmatrix} \begin{bmatrix} C_1 \\ C_2 \\ C_3 \\ C_4 \end{bmatrix} = \begin{bmatrix} \sigma_u/\alpha \\ 0 \\ \sigma_l/\alpha \\ 0 \end{bmatrix} \tag{4}$$

and $\beta = \cosh \alpha b$, $\gamma = \sinh \alpha b$. Once the system of equations has been set up, the computer language ALTRAN (ALgebra TRANlator) can be used to invert symbolically the matrix of the left-hand side of (4), a technique introduced in earlier works (Pavlović and Poulton, 1985; Pavlović and Sapountzakis, 1986). Then, the C constants are obtained in explicit form :

$$\begin{bmatrix} C_1 \\ C_2 \\ C_3 \\ C_4 \end{bmatrix} = \begin{bmatrix} \dfrac{1}{\alpha} & 0 \\ \dfrac{\alpha b + \beta\gamma}{\alpha(\alpha^2 b^2 - \gamma^2)} & -\dfrac{\alpha b\beta + \gamma}{\alpha(\alpha^2 b^2 - \gamma^2)} \\ -\dfrac{\alpha b + \beta\gamma}{\alpha^2 b^2 - \gamma^2} & \dfrac{\alpha b\beta + \gamma}{\alpha^2 b^2 - \gamma^2} \\ \dfrac{\gamma^2}{\alpha^2 b^2 - \gamma^2} & \dfrac{-\alpha b\gamma}{\alpha^2 b^2 - \gamma^2} \end{bmatrix} \begin{bmatrix} \sigma_u/\alpha \\ \sigma_l/\alpha \end{bmatrix} \tag{5}$$

For each m, the stress distribution due to the mth term of the applied surface forces can now be obtained. The full solution follows by superposition until convergence is achieved; the latter is dramatically improved by using Fejér's summation which eliminates an oscillatory convergence for the shear stresses worse than that reported for collinear loading (Tahan *et al*, 1993a). Thus, a relatively small number of terms is usually sufficient for the requisite accuracy (although m increases with a and decreasing e/b (Tahan, 1991)).

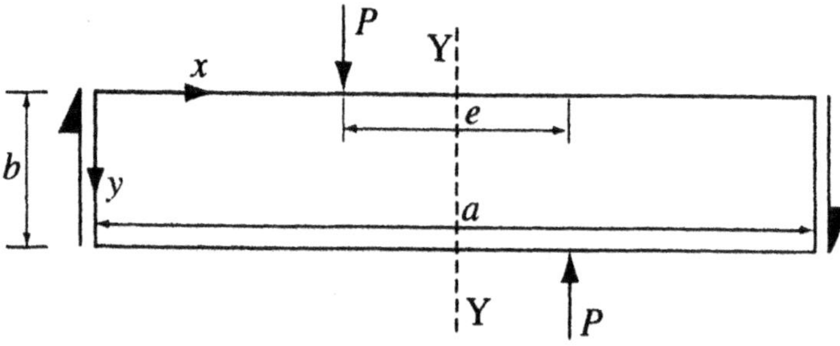

Figure 1. Lamina under non-collinear (equal and opposite) point loads.

According to Filon (1903), who studied the limiting case of an infinitely-wide lamina (i.e. $a \rightarrow \infty$), the profile of the shear-stress distribution across the section Y-Y is only dependent on the ratio e/b. This finding is certainly true when the length of the plate is very large; however, no previous study has addressed the case where the dimension a of the plate is small (or finite), and this can easily be done with the present symbolic approach by reference to an arbitrary aspect ratio defined as $r(=a/b)$. Filon pointed out that the parabolic distribution of shear, which is characteristic of Saint-Venant's type of flexure, does not begin to be approximated until e/b is above unity. Figure 2 shows that, for $e/b = 0.5$, the variation of the shear stress is approximately as expected (from simple beam theory) near the edges but it becomes almost flat in the middle. As e/b becomes smaller, a trough symmetrical around $y = 0.5b$ (the "neutral" axis) is formed and the U-shaped character of the curve becomes more and more accentuated until two sharp peaks form near the top and bottom boundaries, the shear stress falling off to quite a low between these.

After a considerable parametric study, in which different aspect ratios were studied, it was found that the shapes of the distribution curves are similar for all aspect ratios for a given e/b value. Although, as the aspect ratio r increases, the peak becomes sharper (reaching its sharpest at $r = \infty$), there is little difference between these various plates as can be seen in Fig. 3; in fact, this small difference in the peak values may just be seen between $r = 1$ and $r = 3$, while, beyond $r = 3$, there is practically no change.

The effect of non-collinear patch loads has also been studied (Tahan, 1991). Although the magnitude of the static shear force across the middle section Y-Y ($= P(1-e/a)$) remains constant for a given e/a irrespective of the loaded width (w), the maximum shear stress that develops near the top and bottom faces of the beam decreases with increase in w. As the loads spread nearer to each other, the maximum stress moves towards the outer faces with a slight reduction in its magnitude. The peaky profile still persists as the two loads just overlap ($w = e$), and even somewhat beyond this transitional point (i.e. $w > e$), but then the shear-stress distribution begins to revert towards the parabola-like shape associated with ordinary beam theory, as the shear force becomes negligible since the loading tends to collinearity. However, as soon as overlapping takes place, the peak values begin to move from the edges. This behaviour is illustrated in Fig. 4.

3. ORTHOTROPIC CASE

We now turn our attention to the system of Fig. 1 for the case where the material of the plate conforms to the elastic laws of orthogonal anisotropy. Unlike the isotropic case, this problem does not seem to have received much attention in the literature, despite its practical relevance (whether one is dealing with proper material orthotropy, or in instances where "smeared" geometric orthotropy can be invoked such as, for example, stiffened plating (Tahan, 1991)).

The expression for the governing equation for orthotropic plane-stress (Timoshenko and Goodier, 1968) is now dependent on three material coefficients (unlike the isotropic case, where a single material coefficient, namely Young's modulus, is common to all three terms of the equation, so that the ensuing stress distribution is independent of this material constant).

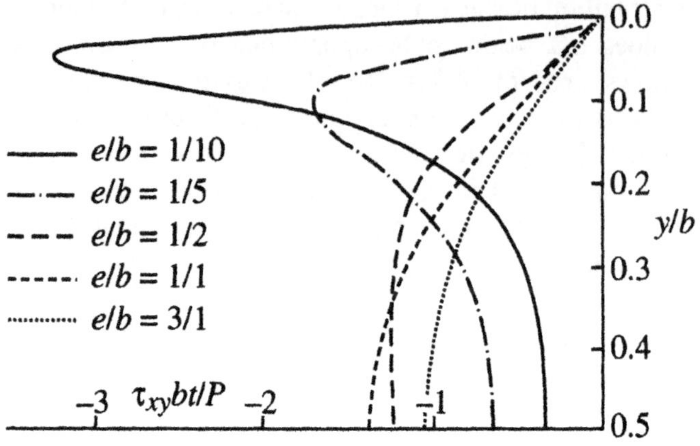

Figure 2. Isotropic plate with $r = 10$: $\tau_{xy}bt/P$ at Y-Y for various e/b.

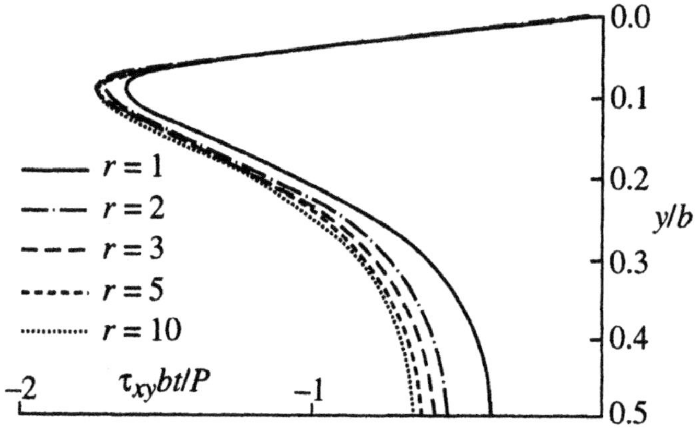

Figure 3. Isotropic plate with $e/b = 1/5$: $\tau_{x}ybt/P$ at Y-Y for various r.

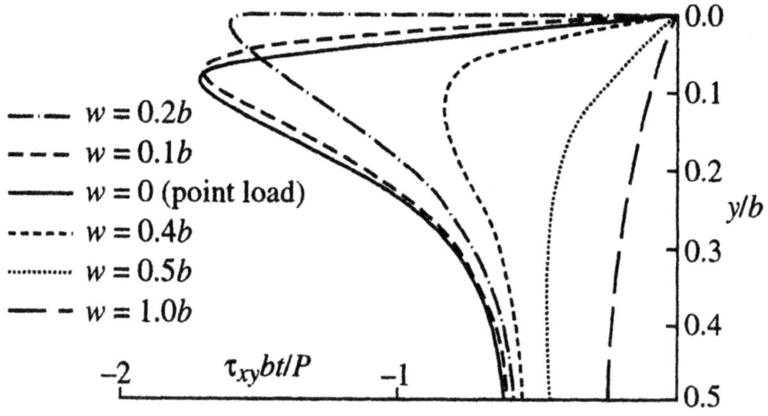

Figure 4. Isotropic plate ($w > 0$) with $r = 5$ and $e/b = 1/5$: $\tau_{xy}bt/P$ at Y-Y for various w.

Nevertheless, the solution procedure described in the previous section can be adopted except that the form of expression (2) now changes to

$$Y(y) = C_1 \cosh \lambda_1 y + C_2 \sinh \lambda_1 y + C_3 \cosh \lambda_2 y + + C_4 \sinh \lambda_2 y \qquad (6)$$

$$\lambda_1 = \alpha(A+B)^{1/2}, \ \lambda_2 = \alpha(A-B)^{1/2} ; \ A = H/D_y, \ B = (H^2/D_y^2 - D_x / D_y)^{1/2} \qquad (7a\text{-}d)$$

and, as before, the subscript m is implicit in α, λ, C. The three material coefficients appearing in (7c,d) are functions of the four orthotropic material constants, namely

$$D_x = E_x G, \ D_y = E_y G, \ 2H = E_x E_y - 2 \nu_x E_x G \qquad (8a\text{-}c)$$

where the subscripts x, y indicate Young's moduli along these co-ordinate axes, G is the shear modulus and, of the two Poisson's ratios ν (i.e. along x, y), only one appears as the other is related to it through the reciprocity theorem. Subject to the boundary conditions (3), the inversion of the relevant matrix yields the following explicit expressions for the constants defining a typical term m of the series:

$$\begin{bmatrix} C_1 \\ C_2 \\ C_3 \\ C_4 \end{bmatrix} = \frac{1}{\varepsilon} \begin{bmatrix} \lambda_2(\beta_1\lambda_1\beta_2 - \gamma_1\gamma_2\lambda_2 - \lambda_1) & \lambda_1\lambda_2(\beta_2 - \beta_1) \\ \lambda_2(\beta_1\gamma_2\lambda_2 - \gamma_1\lambda_1\beta_2) & \lambda_2(\gamma_1\lambda_1 - \gamma_2\lambda_2) \\ (\beta_1\beta_2\lambda_2 - \gamma_1\lambda_1\gamma_2 - \lambda_2) & -\lambda_1\lambda_2(\beta_2 - \beta_1) \\ -\lambda_1(\beta_1\gamma_2\lambda_2 - \gamma_1\lambda_1\beta_2) & -\lambda_1(\gamma_1\lambda_1 - \gamma_2\lambda_2) \end{bmatrix} \begin{bmatrix} \sigma_u \\ \sigma_l \end{bmatrix} \qquad (9)$$

$$\beta_1 = \cosh \lambda_1 b, \ \beta_2 = \cosh \lambda_2 b, \ \gamma_1 = \sinh \lambda_1 b, \ \gamma_2 = \sinh \lambda_2 b,$$

$$\varepsilon = \alpha^2 (2\beta_1\lambda_1\beta_2\lambda_2 - \gamma_1\lambda_1^2 \gamma_2 - \gamma_1\gamma_2\lambda_2^2 - 2\lambda_1\lambda_2) \qquad (10a\text{-}e)$$

The above implies that expression (6) holds, which is true provided the material orthotropy corresponds to the case

$$\mu = H/(D_x D_y)^{1/2} > 1 \qquad (11)$$

This is almost invariably the case as few orthotropic materials fall into the category $\mu < 1$ (for example, a ± 45°angle-ply laminate was found to be one such instance – but just, as its μ value is 0.993 (Tahan, 1991)). Thus, the case $\mu < 1$ will not be pursued here (although its C constants have also been obtained by using the appropriate expression for $Y(y)$ (Tahan 1991)). On the other hand, the case $\mu = 1$ (often imposed arbitrarily – for convenience – in the design of stiffened steel plating and/or bridge decks) corresponds essentially to the solution for the isotropic case (with minor changes (Tahan 1991)), which was considered earlier.

In view of the above, we shall concentrate our present study on materials with $\mu > 1$. A useful exercise is a comparison between rectangular plates made, respectively, of pine wood with

$E_x = 10.00$ GPa, $E_y = 0.42$ GPa, $G = 0.75$ GPa; $\nu_x = 0.238$

and 5-ply maple with

$E_x = 12.89$ GPa, $E_y = 4.14$ GPa, $G = 1.10$ GPa; $\nu_x = 0.122$.

These materials have been chosen as they represent two typical cases on the orthotropic spectrum, in the sense that one (pine wood) is highly unidirectional while the other (5-ply maple) is slightly unidirectional (about 8 times less). Therefore, by studying these two cases, with reference also to the isotropic one (which, as will be seen, turns out to be an intermediate case "bounded" by the other two), some insight into the stress distribution in orthotropic plates can be achieved. This insight is further aided by considering also the effect of rotating the principal axes of orthotropy by 90° so that the subscripts x and y in the two Young's moduli and the two Poisson's ratios are interchanged, with the shear modulus G unchanged. Indices I and II will be used henceforth to indicate whether the major orthotropy axis (e.g. along the grain in the pine) is aligned parallel to the longitudinal x–axis or rotated by 90°, respectively.

As already mentioned, when the ratio e/b in isotropic plates decreases below unity, the shear-stress distribution starts to deviate from the parabolic distribution of the simple theory and a trough-like shape begins to appear, reproduced in Fig. 5(a) as a reference for comparison. Similar distributions of the shear stress are also found in plates with orthotropic properties, as will be shown below. Results of an extensive study indicate that, for any given constant e/b and r, the U-shaped character of the curve varies with varying elastic properties of the orthotropic plates, while maintaining the general feature of developing maxima near the edges and dropping to low values at the midsurface level. This is illustrated by the shear stress distribution across the plate depth at the midspan section for gradually decreasing e/b ratios in plates of pine wood and 5-ply maple plates, as depicted in Figs 5(b) to 5(e). The same scale of values has been taken for all cases in order to show clearly the effect of the orthotropic characteristics of the plate on the stress distribution with varying plate materials. It can be seen that pine wood (I), i.e. where the major orthotropy axis is aligned parallel to the x-axis, produces significantly more peaky distributions when compared to the isotropic case or the other orthotropic materials. This may be attributed to the highly unidirectional nature of pine wood (I).

Let us determine the factor f, defined as the ratio of the maximum shear stress to that which develops at the midsurface of the plate (so that, when the

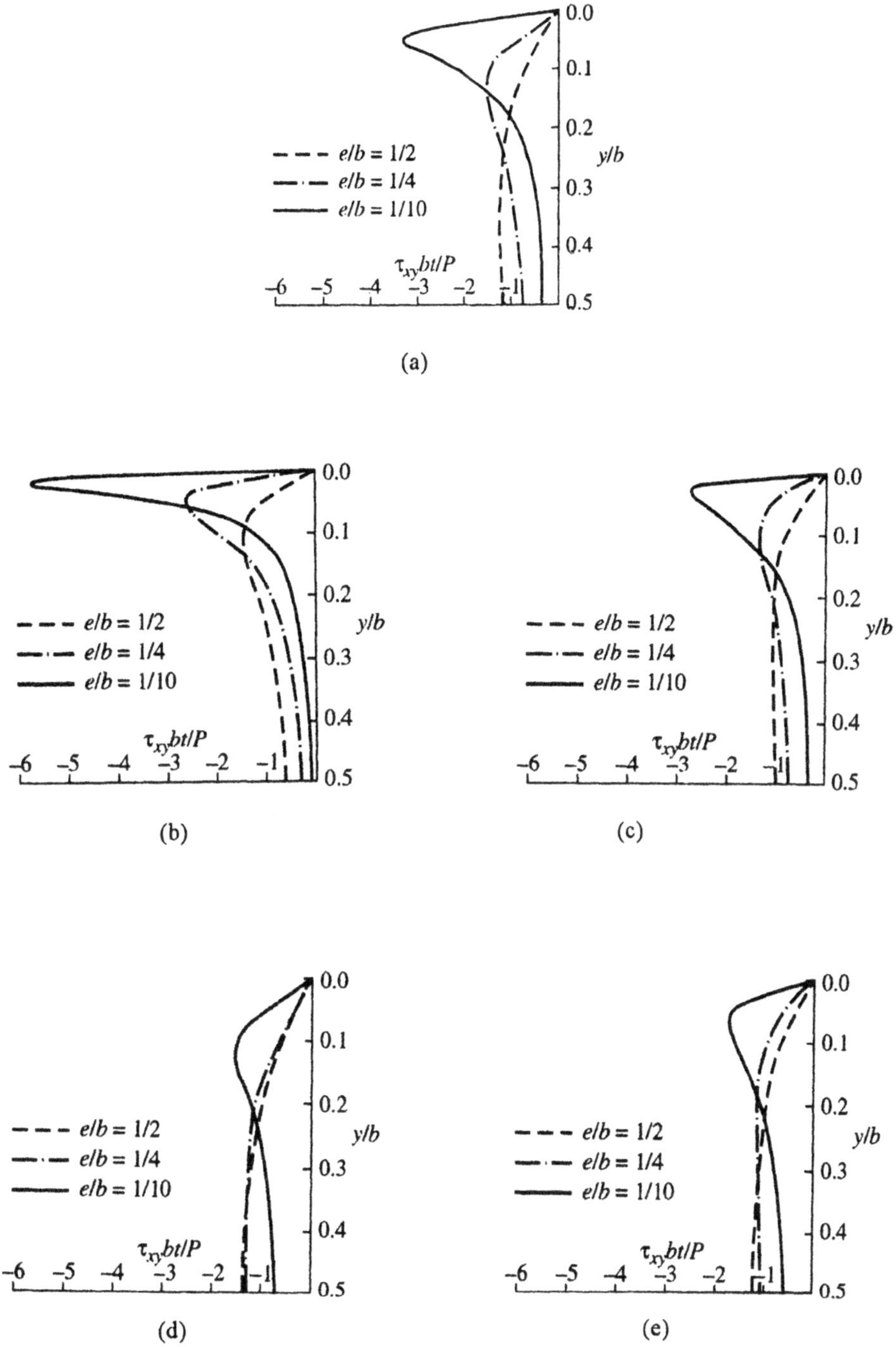

Figure 5. Plates with $r = 5$: $\tau_{xy}bt/P$ at Y-Y for varying degrees of orthotropy and various e/b: (a) isotropy; (b) pine wood (I); (c) 5-ply maple (I); (d) pine wood (II); (e) 5-ply maple (II).

maximum develops at the centroid, $f = 1$). The increase in the maximum stress near the top and bottom edges is accompanied by a decrease in the minimum stress which develops, in most orthotropic plates, at midsurface for e/b ratios less than unity; this leads to an increase in the factor f. Therefore, f could be regarded as a measure of the variation in the shear stress distribution against varying orthotropies as well as e/b ratios.

As may be seen in Fig. 5, a general feature of the variation of the shear stress distribution, common to all $\mu > 1$ orthotropic materials, is that a trough-like curve starts developing at a certain e/b ratio; this U-shaped curve continues deepening as e/b decreases. However, the threshold of the trough-like shape and the rate of its variation with e/b varies with the elastic constants of the orthotropic plates. For example, while the shear stress distribution in isotropic plates for $e/b = 0.5$ still exhibits a parabolic-like distribution ($f = 1$), that for pine wood (I) develops a peaky distribution ($f \sim 2.05$). Note that, regardless of the properties of the material, the area under the curve that depicts the shear stress distribution is constant (as required by statics) for a given e/b ratio, independently of the elastic constants of the material.

As the axes of orthotropy in pine wood are rotated by 90° the shear stress distribution in the rotated plate (II) shows a smoother distribution than even the isotropic case, as may be seen in Fig. 5(d), and the parabolic shear stress distribution is obtained at lower e/b ratios than the unrotated one or even the isotropic case. This behaviour is similarly found for the 5-ply maple materials (I) and (II) (Figs 5(c) and 5(e), where the shear stress distribution is smoother in the rotated position than in the original one. From the above, one might be tempted to conclude that, the higher the ratio ρ of Young's moduli (along x and y respectively), the peakier the stress distribution at the midspan. This conclusion, however, is undermined by the fact that the isotropic material produces peakier distributions than that of the 5-ply maple (I), as may be seen by comparing the solid lines of Figs 5(a) and 5(c), which correspond to the ratio e/b of 0.1: in fact, $f \sim 11$ for $\rho = 1$ (isotropy), compared to $f \sim 7$ for $\rho = 3.1$ (5-ply maple). Therefore, one may only conclude that ρ has a strong effect on the shear stress distribution.

The effect of spreading the load over some width (i.e. patch-load case) has also been studied. Figure 6 shows the shear stress distribution across half the plate for a fixed $e/b = 0.5$ in isotropic material and pine wood (in its two orthogonal configurations I and II) plates for gradually varying patch-load widths $w = 0$ (point load), $0.25b$, $0.50b$ and $0.75b$. It may be seen that the isotropic distribution provides an intermediate situation between the two (unrotated and rotated) pine-wood cases for all load widths. At the same time, a close examination of Figs 6(a) to 6(d) shows that, as the load width increases from 0 to the point at which the loads overlap (i.e. $w = e/b = 0.5$), all curves tend to become more and more peaky, with the location of the peaks nearing the outer edges. Once the loads overlap, however, the pattern reverses and the curves become less and less peaky with increasing load widths (as can be seen in Fig. 6(d)), and the position of the peak moves further away from the edges.

The effect of the aspect ratio r on the shear stress distribution at the centre has also been investigated. It has been found that, for a given e/b ratio in a certain orthotropic rectangular plate, the distribution is hardly affected by the variation in r; this conclusion may be drawn from the fact that f remains practically constant as r varies. Note that the absolute values, and thereby the area under the shear-stress distribution curve, increases with increasing aspect ratio (as required by statics).

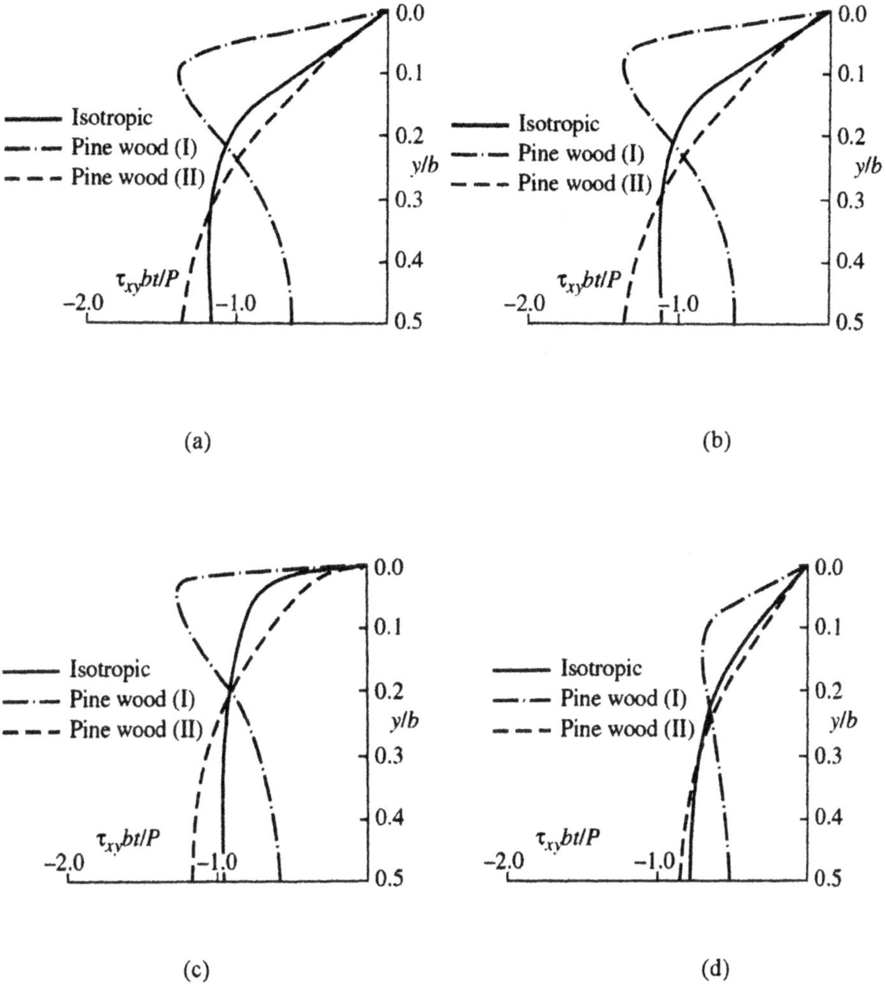

Figure 6. Plates ($w > 0$) with $r = 5$ and $e/b = 1/2$: $\tau_{xy}bt/P$ at Y-Y for varying degrees of orthotropy and various w: (a) $w = 0$; (b) $w = 0.25b$; (c) $w = 0.50b$; (d) $w = 0.75b$.

4. CONCLUDING REMARKS

The article provides an example of how symbolic computation can "modernize" certain types of classical techniques: for the problem considered, such an approach is more attractive than either numerical methods or much more complex closed-form analytical solutions (Baker *et al*, 1993; Zhao, 1996). First, there is the advantage of eliminating a potentially large cycle of numerical equation solving, which takes on added significance when larger problems (involving 8X8, 12X12 or even bigger matrices) have to be tackled (Pavlović and Sapountzakis, 1986). Secondly, while the present 4X4 matrix may be inverted by hand with relative ease in the case of isotropy, the orthotropic case not only requires very considerable effort but the lengthy and cumbersome equations do not allow the easy application of simplifying functional identities (Tahan, 1991): by using ALTRAN's symbolic inversion, however, full benefit was made of these identities and factorization was systematically optimized. Thirdly, a truly closed-form solution is achieved, which aids both mechanical understanding and the ready obtainment of parametric studies. The whole procedure is further enhanced by the adoption of Fejér's summation, as this not only reduces the number of terms needed for the series' convergence, but also avoids potential excesses as regards current numerical computation facilities in the case of very large/small numbers stemming from the exponential functions in the solution as m becomes very large (Tahan *et al*, 1993a).

As regards the actual problem considered, this has numerous applications in various fields of structural engineering. One such instance can be found in the area of offshore structures. Some of these, weighing in excess of 2,000 tonnes, are installed using heavy lift vessels (with capacities of up to 14,000 tonnes) in a single lift. Lifting aids such as slings and shackles connect the hook of a crane barge to strong points on the structure to be lifted. These strong points (referred to as lift points) are often made of fabricated steel plates welded to the legs of the structure at four corners, as shown in Fig. 7. The size of these plates, and of the welds connecting them to the legs of the structure, depends on the magnitude of the applied external force, herein referred to as "applied sling force". The latter has vertical and horizontal components (Fig. 7). For present purposes, attention is focused on the vertical component. It is usual design practice to calculate the shear stress on the basis of this vertical component divided by the depth of the plate times its thickness and compare it with an allowable value, typically 0.4 times the yield strength of the plate material. In other words, the shear-stress distribution is assumed to be uniform. And yet, our article has demonstrated that this common design approach is rather crude and does not reflect the true stress distribution which depends on e/b (in this case e/b = 385/1970 ~ 0.2, thus being well outside the "engineering theory" range).

This points to further work that is required to provide the designer with "shape factors" to be applied to the nominal shear stress so that the resulting design is not unconservative, nor is the safety factor uneconomical.

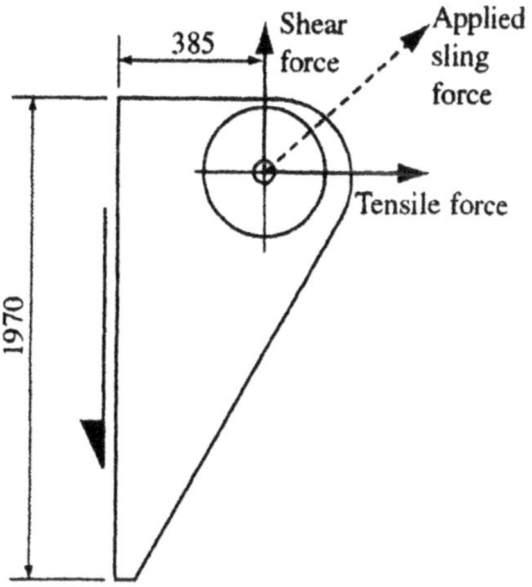

Figure 7. Typical lift point: applied sling force.

References

Baker, G., Pavlović, M.N. and Tahan, N. (1993) An exact solution to the two-dimensional elasticity problem with rectangular boundaries under arbitrary edge forces, *Philosophical Transactions of the Royal Society (London), Series A* **343**, 307-336.

Calladine, C.R. (1982) The theory of shell structures : Aims and methods, *International Journal of Mechanical Sciences* **24**, 219-230.

Calladine, C.R. (1983) *Theory of Shell Structures*, Cambridge University Press, Cambridge.

Filon, L.N.G. (1903) On an approximate solution for the bending of a beam of rectangular cross-section under any system of load, with special reference to points of concentrated or discontinuous loading, *Transactions of the Royal Society (London), Series A* **201**, 63-155.

Pavlović, M.N. and Poulton, S.M. (1985) On the computation of slab effective widths, *Journal of Structural Engineering, American Society of Civil Engineers* **111**, 363-377.

Pavlović, M.N. and Sapountzakis, E. (1986) Computers and structures : Non-numerical applications, *Computers and Structures* **24**, 455-474.

Pavlović, M.N., Tahan, N and Kotsovos M.D. (1998a) Shear lag and effective breadth in rectangular plates with material orthotropy. Part 1: Analytical formulation, *Thin-Walled Structures* **30**, 199-213.

Pavlović, M.N., Tahan, N. and Kotsovos, M.D. (1998b) Shear lag and effective breadth in rectangular plates with material orthotropy. Part 2: Typical results of parametric studies, *Thin-Walled Structures* **30**, 215-237.

Tahan, N. (1991) *Single Fourier Series for Two-Dimensional Analysis of Plates and Box Girders*, Ph.D Thesis, University of London.

Tahan, N., Pavlović, M.N. and Kotsovos, M.D. (1993a) Single Fourier series solutions for rectangular plates under in-plane forces, with particular reference to the basic problem of colinear compression. Part 1: Closed form solution and convergence study, *Thin-Walled Structures* **15**, 291-303.

Tahan, N., Pavlović, M.N. and Kotsovos, M.D. (1993b) Single Fourier series solutions for rectangular plates under in-plane forces, with particular reference to the basic problem of colinear compression. Part 2: Stress distribution, *Thin-Walled Structures* **17**, 1-26.

Tahan, N., Pavlović, M.N. and Kotsovos, M.D. (1995a) Orthotropic rectangular plates under in-plane loading part 1: Closed-form solutions for stresses, *Composite Structures* **33**, 35-48.

Tahan, N., Pavlović, M.N. and Kotsovos, M.D. (1995b) Orthotropic rectangular plates under in-plane loading part 2: Application of the series solution by reference to the problem of colinear compression, *Composite Structures* **33**, 49-61.

Tahan, N., Pavlović, M.N. and Kotsovos, M.D. (1997) Shear-lag revisited: The use of single Fourier series for determining the effective breadth in plated structures, *Computers and Structures* **63**, 759-767.

Timoshenko, S.P. and Goodier, J.N. (1968) *Theory of Elasticity*, third edition, McGraw-Hill, New York.

Zhao, Y.Q. (1996) An exact solution to the plane orthotropic elasticity problem with rectangular boundaries under arbitrary edge forces, *Philosophical Transactions of the Royal Society (London), Series A* **354**, 2469-2511.

LINEAR MATCHING METHODS FOR SHAKEDOWN ANALYSIS

Alan R. S. Ponter and Haofeng Chen
Department of Engineering, University of Leicester, University Road, Leicester LE1 7RH

Abstract: The paper discusses a programming method, the Linear Matching Method, that provides a particularly simple and convenient means of evaluation of the shakedown limits for structures subjected to variable load and temperature. The method possesses a general theoretical structure that allows a broad range of problems to be solved, involving a range of yield conditions and structural problem types.

Keywords: Plasticity theory, Shakedown, Computational methods

1. INTRODUCTION

Design analysis methodology, where inelastic strains are important, varies significantly these days between applications. At the two extremes, many problems are solvable only through detailed finite element solutions derived from realistic constitutive equations. At the other extreme there is still significant reliance upon traditional design rules and related methods that derive from plasticity theory concepts, such as limit and shakedown load, applications of Neuber's rule, estimates of Elastic Follow-up Factors and other such methods. Between these two extremes there has been steady progress in the development of computational methods that may be viewed as the inverse of analysis, of which computational shakedown analysis is the best known example (see, for example Weichert and Maier (2000)). Such methods generally rely upon the application of linear and non-linear programming methods to the optimisation of upper and lower bound load parameters derived from the upper and lower bound shakedown limits. The rather specialist nature of such methods has confined them to specialist

H.R. Drew and S. Pellegrino (eds.),
New Approaches to Structural Mechanics, Shells and Biological Structures, 203–213.

computer codes used largely in a research environment. There have, however, been alternative methods that attempt to simulate the shakedown state by matching a linear description to the non-linear problem through a spatially varying linear moduli. The most developed of such methods is the Elastic Compensation method (Mackenzie and Boyle 1993) which has its origins in simple lower bound methods used in industry for some time.

The purpose of this paper is to demonstrate that such methods are, in fact, derivable from a programming technique of a particularly simple and convenient form. Effectively, the non-linear minimisation problem posed by the upper bound shakedown theorem may be sequentially replaced by a quadratic minimisation problem that may be solved as the solution of linear equations. The linear problem may be interpreted in terms of familiar physical variables. By suitable adaptation of user routines such solutions may be solved using a standard commercial finite element code. There seems, therefore to be a very efficient and convenient method that brings shakedown analysis into the mainstream of the type of computations that are usually carried out in industry.

The paper gives an account of the theory behind the method and an example that demonstrates the application of the method particularly well, the shakedown of a half-space subjected to a moving Hertzian contact repeatedly traversing a line of travel.

2. THE UPPER BOUND SHAKEDOWN THEOREM

Consider a body composed of an elastic-perfectly plastic solid which satisfies the yield condition;

$$f(\sigma_{ij}) \leq 0 \tag{1}$$

The associated flow rule for the plastic strain rate $\dot{\varepsilon}_{ij}^{p}$ is given by

$$\dot{\varepsilon}_{ij}^{p} = \dot{\alpha}\frac{\partial f}{\partial \sigma_{ij}} \tag{2}$$

where $\dot{\alpha}$ is a plastic multiplier. We assume that the yield function is convex and that the maximum work principle holds,

$$(\sigma_{ij}^{p} - \sigma_{ij}^{*})\dot{\varepsilon}_{ij} \geq 0 \tag{3}$$

where σ_{ij}^{p} is the stress associated with plastic strain rate $\dot{\varepsilon}_{ij}^{p}$ at yield and σ_{ij}^{*} is any state of stress which satisfies the yield condition (2.1).

Consider the following problem. A body of volume V and surface S is subjected to a cyclic history of loading with cycle time Δt. The history consists of mechanical loads $\lambda P_i(x_j,t)$ over S_T, part of S , and a temperature history $\lambda \theta(x_j,t)$ within V, where λ is a positive load parameter. On the remainder of S , namely S_U, the displacement rate, $\dot{u}_i = 0$. The linear elastic solution to the problem, corresponding to $\dot{\varepsilon}_{ij}^p = 0$, is denoted by $\lambda \hat{\sigma}_{ij}$. The elastic material properties are assumed to be independent of temperature.

We define a class of kinematically admissible strain rate histories $\dot{\varepsilon}_{ij}^c$ with a corresponding displacement increment fields Δu_i^c and associated compatible strain increment,

$$\Delta \varepsilon_{ij}^c = \tfrac{1}{2}\left(\frac{\partial \Delta u_i^c}{\partial x_j} + \frac{\partial \Delta u_j^c}{\partial x_i}\right) \tag{4}$$

The strain-rate history $\dot{\varepsilon}_{ij}^c$, which need not be compatible, satisfies the condition that,

$$\int_0^{\Delta t} \dot{\varepsilon}_{ij}^c dt = \Delta \varepsilon_{ij}^c \tag{5}$$

In terms of such a history of strain rate an upper bound on the shakedown limit is given by (see Koiter, (1960); Gokhfeld and Cherniavsky (1980); Konig (1987)),

$$\int_V \int_0^{\Delta t} \sigma_{ij}^c \dot{\varepsilon}_{ij}^c dt dV = \lambda_{UB}^c \int_V \int_0^{\Delta t} \hat{\sigma}_{ij} \dot{\varepsilon}_{ij}^c dt dV \tag{6}$$

where $\lambda_{UB}^c \geq \lambda_s$, the exact shakedown limit and σ_{ij}^c denotes the stress at yield associated with $\dot{\varepsilon}_{ij}^c$. We wish to discover solution methods which involve a sequence of kinematically admissible strain increment fields that give a reducing sequence of upper bounds. The sequence should converge to the shakedown limit λ_s or the least upper bound associated with the class of displacement fields and strain rate histories chosen.

We may express the upper bound (6) in the following alternative form. Consider the functional

$$I(\dot{\varepsilon}_{ij}^c, \lambda) = \int_V \int_0^{\Delta t} (\sigma_{ij}^c - \lambda \hat{\sigma}_{ij}) \dot{\varepsilon}_{ij}^c dt dV \tag{7}$$

then (6) may be expressed as

$$I(\dot{\varepsilon}_{ij}^{c},\lambda_{UB}^{c})=0\,,\quad I(\dot{\varepsilon}_{ij}^{c},\lambda_{s})\geq 0 \tag{8}$$

The methodology described in the next section, the Linear Matching Method, is essentially concerned with the generation of kinematically admissible strain-rate histories that reduce functionals of the general form of $I(\dot{\varepsilon}_{ij}^{c},\lambda)$ for a prescribed value of λ. In the case of shakedown analysis, equation (8) allows the process to be translated into the generation of reducing upper bounds.

3. THE LINEAR MATCHING METHOD

Suppose we have an initial kinematically admissible strain rate history $\dot{\varepsilon}_{ij}^{i}$ with a corresponding upper bound λ_{UB}^{i}. We now derive an algorithm that generates a new strain rate history $\dot{\varepsilon}_{ij}^{f}$ such that $\lambda_{UB}^{f}\leq\lambda_{UB}^{i}$ with equality only if $\dot{\varepsilon}_{ij}^{f}\equiv\dot{\varepsilon}_{ij}^{i}$. The new solution $\dot{\varepsilon}_{ij}^{f}$ is generated from the solution of a linear problem. The material parameters of this problem are derived from a matching process with the yield condition and associated flow rule. In essence the method is very simple and the following is a summary of its main components.

First consider the problem of generating $\dot{\varepsilon}_{ij}^{f}$ so that $I(\dot{\varepsilon}_{ij}^{f},\lambda)\leq I(\dot{\varepsilon}_{ij}^{i},\lambda)$ for any value of λ. We define a linear material with a quadratic dissipation potential $U(\dot{\varepsilon}_{ij})$ with corresponding stress state σ_{ij}^{L};

$$\sigma_{ij}^{L}=\frac{\partial U}{\partial\dot{\varepsilon}_{ij}} \tag{9}$$

In its most general form U may be expressed as;

$$U=\tfrac{1}{2}C_{ijkl}(\dot{\varepsilon}_{ij}-\dot{r}_{ij})(\dot{\varepsilon}_{kl}-\dot{r}_{kl}) \tag{10}$$

where C_{ijkl} and $\dot{r}_{ij}$ are material tensors to be determined. Generally they will both be functions of position x_i and time t.

The components of these tensors are defined by two conditions, a matching condition and an inequality. The matching condition requires that the stresses defined by the associated flow rule and the linear material are identical for the history $\dot{\varepsilon}_{ij}^{i}$, i.e.

$$\sigma_{ij}^{Li}=\frac{\partial U(\dot{\varepsilon}_{ij}^{i})}{\partial\dot{\varepsilon}_{ij}^{i}}=\sigma_{ij}^{pi} \tag{11}$$

where σ_{ij}^{pi} is the stress state, at yield, associated with $\dot{\varepsilon}_{ij}^{p}=\dot{\varepsilon}_{ij}^{i}$. The inequality condition involves a comparison between the matching strain rate $\dot{\varepsilon}_{ij}^{i}$ and, potentially, any other strain rate $\dot{\varepsilon}_{ij}^{f}$;

$$U(\dot{\varepsilon}_{ij}^{f})-U(\dot{\varepsilon}_{ij}^{i})\geq\sigma_{ij}^{f}\dot{\varepsilon}_{ij}^{f}-\sigma_{ij}^{i}\dot{\varepsilon}_{ij}^{i} \tag{12}$$

or, equivalently, taking into account the matching condition (11),

$$U(\dot{\varepsilon}_{ij}^{f})-U(\dot{\varepsilon}_{ij}^{i})-\frac{\partial U}{\partial\dot{\varepsilon}_{ij}^{i}}(\dot{\varepsilon}_{ij}^{f}-\dot{\varepsilon}_{ij}^{i})\geq(\sigma_{ij}^{pf}-\sigma_{ij}^{pi})\dot{\varepsilon}_{ij}^{f}\geq 0 \tag{13}$$

For any U that satisfies these conditions we may define a functional

$$I_{L}(\dot{\varepsilon}_{ij}^{c},\lambda)=\int_{V}\int_{0}^{\Delta t}\{U(\dot{\varepsilon}_{ij}^{c})-\lambda\hat{\sigma}_{ij}\dot{\varepsilon}_{ij}^{c}\}dtdV \tag{14}$$

the linearised equivalent to $I(\dot{\varepsilon}_{ij}^{c},\lambda)$. From inequality (12) we immediately see that the functionals (7) and (14) are related by;

$$I(\dot{\varepsilon}_{ij}^{f},\lambda)-I(\dot{\varepsilon}_{ij}^{i},\lambda)\leq I_{L}(\dot{\varepsilon}_{ij}^{f},\lambda)-I_{L}(\dot{\varepsilon}_{ij}^{i},\lambda) \tag{15}$$

Hence if we choose $\dot{\varepsilon}_{ij}^{f}$ so that $I_{L}(\dot{\varepsilon}_{ij}^{c},\lambda)$ is minimised amongst the class of kinematically admissible strain rate histories, then

$$I(\dot{\varepsilon}_{ij}^{f},\lambda)-I(\dot{\varepsilon}_{ij}^{i},\lambda)\leq I_{L}(\dot{\varepsilon}_{ij}^{f},\lambda)-I_{L}(\dot{\varepsilon}_{ij}^{i},\lambda)\leq 0 \tag{16}$$

and hence $I(\dot{\varepsilon}_{ij}^{f},\lambda)\leq I(\dot{\varepsilon}_{ij}^{i},\lambda)$. In particular, if $\lambda=\lambda_{UB}^{i}$ then, from the definition of the upper bound (6) and (11) it follows that $\lambda_{UB}^{f}\leq\lambda_{UB}^{i}$. In summary, if we solve for the minimum of $I_{L}(\dot{\varepsilon}_{ij}^{c},\lambda_{UB}^{i})$ we obtain a kinematically admissible strain rate field that reduces the upper bound to the shakedown limit. The absolute minimum of $I_{L}(\dot{\varepsilon}_{ij}^{c},\lambda_{UB}^{i})$ is provided by the solution of the following linear problem;

$$\frac{\partial U}{\partial\dot{\varepsilon}_{ij}^{c}}=\lambda_{UB}^{i}\hat{\sigma}_{ij}+\bar{\rho}_{ij}\ ,\ \int_{0}^{\Delta t}\dot{\varepsilon}_{ij}^{c}dt=\Delta\varepsilon_{ij}^{c} \tag{17}$$

where $\bar{\rho}_{ij}$ is a constant residual stress field, in equilibrium with zero surface tractions on S_{T}. The equations (17) may be integrated over $0\leq t\leq\Delta t$ to produce a single linear problem for $\bar{\rho}_{ij}$ and $\Delta\varepsilon_{ij}^{c}=\Delta\varepsilon_{ij}^{f}$.

This single stage of generating a reduced upper bound as the result of the solution of a linear problem, may be recognised as typical of a programming technique. Provided equality in (12) and (13) only occurs when $\dot{\varepsilon}_{ij}^{i} = \dot{\varepsilon}_{ij}^{f}$ then repeated application of the method will generate a sequence of reducing upper bounds that converge to the least upper bound. When the strain field $\Delta\varepsilon_{ij}^{c}$ is expressed in terms of a class of displacement fields Δu_{i}^{c} in the form, for example, of a finite element mesh and element type, the condition of equilibrium for $\bar{\rho}_{ij}$ may be represented by the Galerkin condition,

$$\int_V \bar{\rho}_{ij}\Delta\varepsilon_{ij}^{c}dV = 0 \tag{18}$$

for all members of the class of strain increment fields $\Delta\varepsilon_{ij}^{c}$. In this case λ_{UB} converges to the least upper bound associated with the chosen class.

In summary, this approach provides a very straightforward programming method for minimising a certain class of functional by replacing the original non-linear functional by an equivalent quadratic function. The method may either be understood as a programming method or a method of constructing linear solutions that increasingly accurately mimic the shakedown state.

4. EXAMPLES OF APPLICATIONS

4.1 Von Mises Yield Condition

In this case the matching condition (9) and inequality (12) are satisfied if U is chosen as a function of the effective strain rate $\bar{\dot{\varepsilon}} = \sqrt{\tfrac{2}{3}\dot{\varepsilon}_{ij}\dot{\varepsilon}_{ij}}$,

$$U = \frac{1}{2E}\bar{\dot{\varepsilon}}^{2} \tag{19}$$

where E is a modulus. This corresponds to a general isotropic incompressible linear viscous material. The conditions (11) and (12) become

$$E\bar{\dot{\varepsilon}}^{i} = \sigma_y \tag{20}$$

and

$$U(\bar{\dot{\varepsilon}}^{f}) - U(\bar{\dot{\varepsilon}}^{i}) \geq \sigma_y(\bar{\dot{\varepsilon}}^{f} - \bar{\dot{\varepsilon}}^{i}) \tag{21}$$

Inequality (21) is always satisfied for $E > 0$. The linear problem (17) then becomes;

$$\dot{\varepsilon}_{ij}^{f'} = \frac{1}{\mu(t)}(\lambda_{ub}^{i}\hat{\sigma}_{ij} + \overline{\rho}_{ij}^{f})', \quad \dot{\varepsilon}_{kk}^{f} = 0, \text{ and } E = \frac{3}{2}\mu(t) = \frac{\sigma_y}{\overline{\dot{\varepsilon}}^{i}} \tag{22}$$

Integration of (22) over the cycle of loading produces the following linear relationship between $\Delta\varepsilon_{ij}^{f}$ and $\overline{\rho}_{ij}$,

$$\Delta\varepsilon_{ij}^{f'} = \frac{1}{\overline{\mu}}\left(\overline{\rho}_{ij}^{\prime f} + \sigma_{ij}^{\prime in}\right), \quad \Delta\varepsilon_{kk}^{f} = 0 \tag{23}$$

where $\sigma_{ij}^{\prime in} = \overline{\mu}\left\{\int_0^{\Delta t} \frac{1}{\mu(t)}\lambda_{UB}^{i}\hat{\sigma}_{ij}'(t)dt\right\}$ and $\frac{1}{\overline{\mu}} = \int_0^{\Delta t}\frac{1}{\mu(t)}dt$.

The repeated application of this method produces a monotonically reducing sequence upper bounds, with corresponding distributions of linear modulus, E_k, k=1,2.... The initial choice E_1 is arbitrary but is usual chosen as a constant. Note that the values of two consecutive values, E_k and E_{k+1} are given, from (20) by;

$$E_{k+1}\overline{\dot{\varepsilon}}^{k} = \sigma_y \text{ and } E_k\overline{\dot{\varepsilon}}^{k} = \overline{\sigma}^{k} \quad \text{i.e.} \quad E_{k+1} = E_k\frac{\sigma_y}{\overline{\sigma}^{k}} \tag{24}$$

This can be recognised as the standard relationship for the Elastic Compensation method (Mackenzie and Boyle 1993). Of course, the advantage of the present approach is that it may be applied in a systematic way to provide convergent solutions for any yield condition. The characteristics of such applications are described for a particular case in the following section.

4.2 Yield Conditions for Geotechnical Materials

In this case the yield condition depends upon both the deviatoric stress and the hydrostatic pressure. The only additional complication in this case is ensuring that the inequality (12) is always satisfied. It can be shown (see Ponter, Fuschi and Engelhardt (2000)) that this condition requires a restriction on the matching linear material. This is best expressed in terms of the complementary dissipation rate $\overline{U}(\sigma_{ij})$ for the linear material. As

$$\dot{\varepsilon}_{ij} = \frac{\partial\overline{U}}{\partial\sigma_{ij}} \tag{25}$$

the matching condition (12) ensures that, in stress space the surface of constant $\overline{U}$ that passes through the yield point at the matching strain rate will be tangential to the yield condition or lie between adjacent tangents at a corner. If this surface always lies outside the yield surface, i.e. the constant $\overline{U}$ surface surrounds the yield surface, then inequality (12) is satisfied. Examples of applications of the method to yield conditions of the Prager-Drucker type have been given by Ponter, Fuschi and Engelhardt (2000) for limit analysis and by Parrinello (2000) and Parrinello and Ponter (2001) for shakedown analysis.

5. EXAMPLE: ROLLING CONTACT OF A HALF SPACE FOR A VON MISES YIELD CONDITION

The problem shown in Figure (1) is identical in form to the problem solved by Ponter, Hearle and Johnson (1985) using a semi-analytical upper bound method. A circular Herzian contact region repeatedly traverses an elastic-perfectly plastic half space. The solution method assumes that the deformation field in independent of *x*, the axis in the direction of travel. As a result, for any slice of material perpendicular to the direction of travel the displacements on the opposing surfaces of the slice are identical. Hence equation (23) is solved with this assumption, where the history of elastic stress corresponds to positions of the contact at various points along the path of travel relative to the slice. The equations were solved by finite element methods, using the commercial code ABAQUS via a method described by Engelhardt (1999).

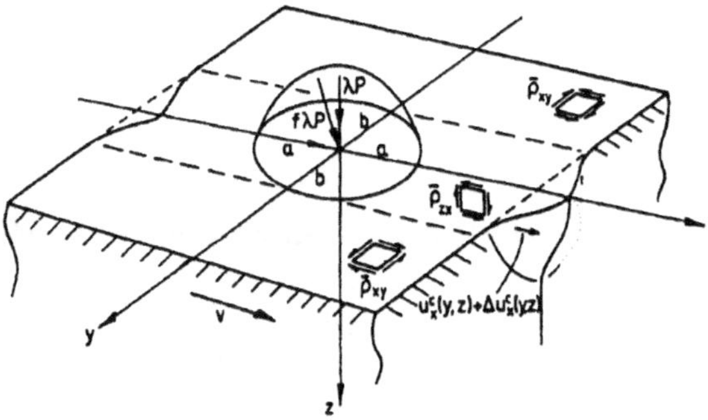

Figure 1. Rolling contact problem considered by Ponter, Hearle and Johnson (1985).

The shakedown boundary is shown in Figure (2), composed from a number of shakedown calculations, where the axes are given by the non-dimensionalised normal force P and the net tangential force fP where f is a coefficient of friction. Here k is the shear yield stress and a, the radius of the contact region. Whereas in Figure (2) the frictional force is shown as acting in the direction of travel, in the calculations fP was assumed to act at an angle α to the direction of travel, where $\alpha = \tan^{-1}(F/Q)$ with F and Q the components of frictional force acting along and perpendicular to the direction of travel.

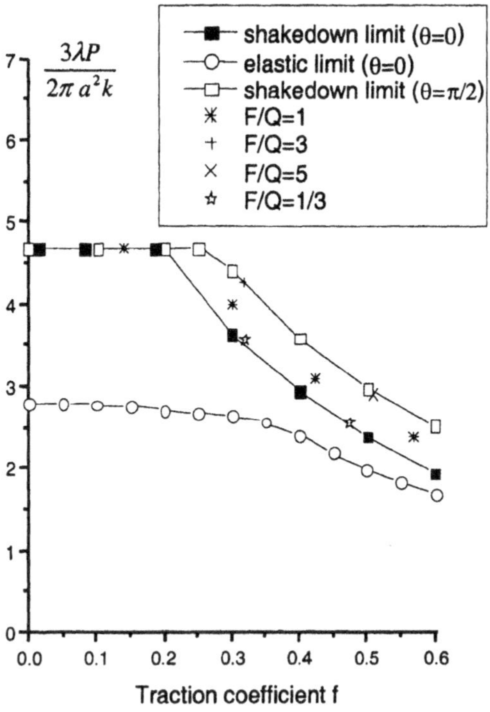

Figure 2. Shakedown and elastic limits of a circular loaded region for different normal and tangential loads.

The solution for the case $Q = 0$ is very close to the solution discussed by Ponter, Hearle and Johnson (1985). For positive Q the shakedown boundary lies outside the boundary for $Q = 0$. Figure (3) shows the convergence of the load parameter λ for three finite element meshes of increasing element density. The solutions in Figure 2 were generated with model 3. It can be seen that convergence occurs in about 30 cycles, i.e as a result of the solution of 30 linear problems. The least upper bound reduces with increasing mesh density. In addition, the rate of convergence is independent of the mesh density, i.e. the number of degrees of freedom. This contrasts with most programming techniques, including linear programming, where the number of iterations to convergence increases with the number of degrees of freedom. This appears, from experience, to be a general property of the method. A detailed discussion of this problem is given by Chen and Ponter (2001).

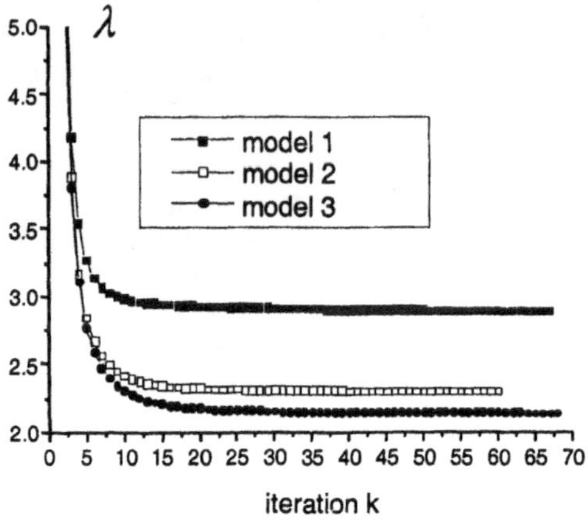

Figure 3. Convergence conditions for finite element models with increasing density of elements.

6. CONCLUSIONS

The Linear Matching Method is a non-linear programming method that has the characteristic of sequentially replacing the original non-linear problem by an equivalent quadratic programming problem that yields a linear problem as its solution. As a result the linear problem retains most of the characteristic of the original problem. In the case discussed in this paper, the evaluation of an optimal upper bound shakedown load parameter, the linear problems are of a form that may be solved by a standard finite element code. For the case of a Von Mises yield condition we recover the standard iterative equation of the Elastic Compensation method. This method is based upon the idea of mimicking the non-linear problem with linear problems with spatially varying linear moduli. Hence the theory described here provides a method that has the characteristics of being both a programming method, for which convergence may be guaranteed, and also a method that mimics the non-linear problem. The method is currently being applied to a range of inversion problems involving high temperature creep, Ponter (2000). Other applications to plasticity problems have been given by Chen and Ponter (2001a,b), Ponter and Engelhardt (2000) and Ponter and Chen (2001c).

Acknowledgement

The authors wish to acknowledge the grant support of the Engineering and Physical Sciences Research Council during the course of the work described in this paper.

References

Chen H. and Ponter A. R. S. (2001a). A method for the evaluation of a ratchet limit and the amplitude of plastic strain for bodies subjected to cyclic loading. *European Journal of Mechanics, A/Solids* 20:555-571.

Chen H. and Ponter A. R. S. (2001b). The 3-D shakedown and limit load using the elastic compensation method, *Int Journal of Pressure Vessel & Piping*, to appear.

Chen H. and Ponter A. R. S. (2001c), The Linear Matching Method for the shakedown analysis of rolling and sliding point contact problems, Department of Engineering, University of Leicester, Report October 2001.

Engelhardt M. (1999), Computational modelling of shakedown, PhD thesis, University of Leicester.

Koiter, W.T., (1960) General theorems of elastic-plastic solids. In Sneddon J.N. and Hill R., eds., *Progress in Solid Mechanics*, 1:167-221.

Konig, J.A., (1987) *Shakedown of Elastic-Plastic Structures,* PWN-Polish Scientific Publishers, Warsaw and Elsevier, Amsterdam.

Mackenzie D. and Boyle J.T. (1993) A simple method of estimating shakedown loads for complex structures *Proc. ASME Pressure Vessel and Piping Conference*, Denver.

Parrinello F. and Ponter A. R. S. (2001) Shakedown limits based on linear solutions, for a hydrostatic pressure dependent material. *Proc. 2nd European Conference on Computational Mechanics*, Cracow, Poland, June 2001, Abstracts, ISBN83-85688-68-4, 718-719, Institute of Computer Methods in Civil Engineering, Cracow University of Technology, Poland.

Parrinello F, (2000) PhD thesis, University of Palermo, Italy.

Ponter, A.R.S., Hearle, A.D. and Johnson, K.L., Application of the kinematical shakedown theorem to rolling and sliding point contacts, *Journal of the Mechanics and Physics of Solids*, Vol. 33(4), pp.339-362, 1985.

Ponter A. R. S., Fuschi P. and Engelhardt M. (2000) Limit analysis for a general class of yield conditions. *European Journal of Mechanics, A/Solids*. 19:401-421.

Ponter A. R. S. and Engelhardt, M. (2000) Shakedown limits for a general yield condition. *European Journal of Mechanics, A/Solids* 19:423-445.

Ponter A. R. S. and Chen H. (2001) A minimum theorem for cyclic load in excess of shakedown, with applications to the evaluation of a ratchet limit. *European Journal of Mechanics, A/Solids*, 20:539-553.

Ponter A. R. S. (2001), Minimum theorems and iterative solution methods for creep cyclic loading problems, *Meccanica*, 296-8:1-11.

Weichert D. and Maier G (eds.),(2000) Inelastic Analysis of Structures Under Variable Loads, *Theory and Applications*; Kluwer Academic Publishers, ISBN 0-7923-6645-X.

ANALYSIS OF COILED PIEZOELECTRIC STRUCTURES

K. A. Seffen
Department of Engineering, University of Cambridge
Trumpington Street, Cambridge, CB2 1PZ, U.K.
kas14@eng.cam.ac.uk

Abstract This study is concerned with piezoelectric bimorph structures of novel shape. A bimorph is a thin strip with two layers of actuating material that enable the strip to bend and to extend independently of the applied loads under electrical control; the associated generalised Hooke's law is derived by simple means. This approach complements the analysis of, first, a straight helix and, second, a coiled-coil, which have been recently proposed as actuators of relatively large displacements. Two specific quantities are obtained in closed form as measures of overall performance for each actuator; the freely actuated displacement and the load required to restore this displacement to zero. It is shown that a coiled-coil greatly magnifies small actuation strains at the expense of a low load-carrying capacity.

Keywords: piezoelectric, bimorph, helix, coiled-coil.

1. INTRODUCTION AND REVIEW

"Supercoiled" structures belong to a class of structures of general helical form. A typical example is a straight bar or rod that is first coiled into a straight helix, which is then wound into a second, larger helix: this is relatively straightforward to show by wrapping a telephone handset cord around a cylinder or, indeed, one's forearm. The resulting shape of the smaller, "minor" helix and the larger, "major" helix is more commonly described as being a *coiled-coil* and is depicted in Fig. 1(a). Fletcher *et al.* (2001) precisely define the structure as a *second-order hyperhelix*, where the zeroth- and first-order hyperhelices are a straight rod and a straight helix, respectively; another, singular description is a *superhelix*.

Supercoiled and coiled-coil structures are peculiar to several phenomena. For example, consider a long piece of twisted string held taut and

H.R. Drew and S. Pellegrino (eds.),
New Approaches to Structural Mechanics, Shells and Biological Structures, 215–227.

straight between its ends. If the ends of the string are brought closer together, the string buckles into a slowly modulating helix: reducing the axial tension further causes the helix to collapse suddenly into a relatively large supercoiled "knot". Although this behaviour is an unlikely source of failure in most engineering structures, it cannot be discounted if a structure is unsupported over great lengths and if the end loads are not precisely known, *e.g.* in a fast-spinning oil-rig drill string or during the installation of an undersea cable. Hence, the study of post-buckling of a simple, flexible rod into a supercoil is a worthy starting point for more complex, practical systems; see Thompson and Champneys (1996) and the references therein for more detail.

Another, important example is from biology. The DNA molecule in each living cell is a long chain comprising two juxtaposed but interconnected helical nucleotide strands. During the initiation of protein synthesis, the DNA chain writhes in space, leading to the supercoiling of its helices. A similar scenario occurs when a cell divides; its DNA unravels into a pair of helices, in order to make identical copies of the parent DNA for each new cell. Deeper insight into these biochemical reactions can be gained by ascribing the kinematics of DNA to a deforming supercoil, and pioneering research in this area has been led by Calladine and co-workers (Calladine and Drew, 1992).

Studies of the above kind deal with the *transition* of initially straight structures into supercoiled forms. On the other hand, this investigation introduces a novel, artificial *piezoelectric* coiled-coil when undeformed, see Fig. 1(b): manufacture of this intricate shape has only been possible with recent advances in ceramic processing technology (Pearce, 1999). In general, piezoelectric material experiences strain in response to an applied electric field, and *vice versa*. It is commonly used in sensors to measure strains and pressures, and in actuators to generate forces and displacements. The recent advent of "smart" structures, which automatically monitor and respond to environmental loads, has increased both the awareness and use of piezoelectric as an actuating material. However, it is well known that the maximum levels of raw actuation strain are small (Huber *et al.*, 1997), and the resulting direct displacements are also small. Here, the coiled-coil increases the arc-length of material within a given volume of structure and, therefore, the path over which the free strains integrate to yield relatively large displacements. Thus, it is proposed as one solution for large-stroke displacement actuators. A drawback is that its open structure gives rise to a low stiffness, which limits the maximum load it may carry.

A key aim of this study is to quantify the actuation response of a piezoelectric coiled-coil, and this is imperative for proper actuator de-

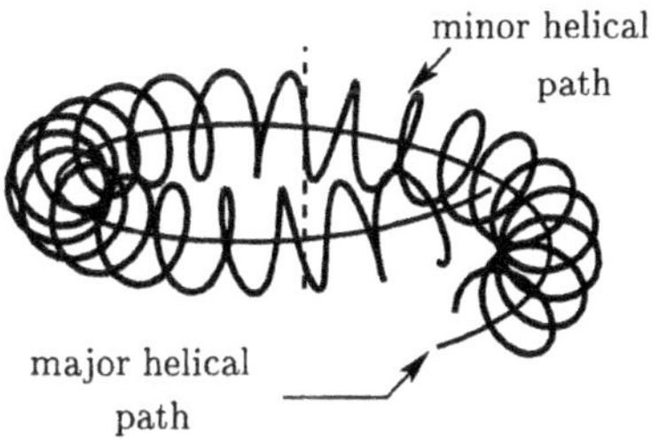

(a) A coiled-coil in which a small helix locally winds along a larger helical path.

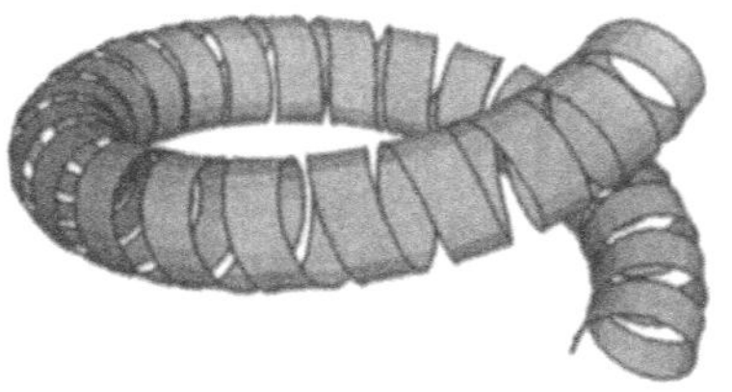

(b) Schematic piezoelectric coiled-coil of one major turn with ribbon-like cross-section.

Figure 1. Types of coiled-coil.

sign. Its response is most readily expressed by a generalised Hooke's law in terms of the induced displacements, the stiffness of the structure, the external loads and the degree of actuation. However, rather than beginning with the analysis of a coiled-coil, the procedure is made more transparent by first dealing with simpler structures. In the next section, the generalised Hooke's law is determined for a special type of straight piezoelectric beam known as a "bimorph" that bends during actuation. This is followed by the study of a coiled helical bimorph whose behaviour helps to elucidate the performance of a coiled-coil. A comparison of structures concludes the investigation.

2. ANALYSIS

2.1 A Piezoelectric Beam: the "Bimorph"

Figure 2(a.i) shows a straight, thin strip of piezoelectric material of length L. The uniform, rectangular cross-section has width w. External voltage potentials are applied to the top and bottom surface electrodes; the electric field is the ratio of the voltage difference, V, to the distance, t, between electrodes. If this field is positive in the direction of through-thickness polarisation, the strip freely expands in the same direction and contracts longitudinally, and *vice versa*, see Fig. 2(a.ii).

During actuation, the unconstrained or *free* longitudinal strain, Λ, is expressed in terms of a material constant, d_{31}, and the electric field as follows (Spearitt and Asokanthan, 1996)

$$\Lambda = d_{31}\frac{V}{t}. \tag{1}$$

Consider now an axial, tensile force, P, applied to the strip, as shown. The total strain, ϵ, is the sum of the elastic component and the free strain, *i.e.*

$$\epsilon = \frac{\sigma}{E} + \Lambda, \tag{2}$$

where σ is the normal stress at any section and E is the Young's modulus. Substituting for $P = \sigma A$, with the cross-sectional area A equal to wt, and re-arranging, gives the generalised Hooke's law

$$P = EA\left(\epsilon - \Lambda\right). \tag{3}$$

Equation 3 indicates that a pair of independent variables may be used to *control* the remaining third parameter. For example, the force necessary to "block" any displacements is obtained by setting ϵ equal to zero and, clearly, $P = -EA\Lambda$. This quantity includes the axial stiffness, EA, and the degree of free strain, which is governed by the electric field in Eqn 1. Conversely, if $P = 0$, then ϵ is now equal to Λ and the expansion/contraction of the strip is independent of the axial stiffness.

The above is rather elementary but it illustrates the analysis route. Note that the choice of an axial force was not incidental: if equal and opposite end couples, M, are applied to the strip, it is easy to verify that

$$M = EI\kappa; \quad I = \frac{wt^3}{12}, \tag{4}$$

where κ is the curvature of the neutral axis and I is the second moment of area of cross-section. There is no term involving the actuation strain because of its symmetry about the neutral axis and bending effects cannot be actively controlled in this simple structure.

An improvement is gained if two identical strips are connected together but with total thickness t, as shown in Fig. 2(b.i). Independent electric fields can be applied between each pair of electrodes with actuation strain $\Lambda_t = d_{31}V_t/(t/2)$ in the top strip and $\Lambda_b = d_{31}V_b/(t/2)$ in the bottom strip. This is known as a "bimorph" structure for its dual actuation capability: if Λ_t is not equal to Λ_b, each strip attempts to expand by different amounts, causing the structure to bend as well as to stretch. This deformation can be characterised by the curvature, κ, and the axial strain, ϵ, of the bimorph neutral axis; equal and opposite end couples M are now suffered along with P.

The composite action of the bimorph is more simply determined from each strip as follows. The neutral axes in the top and bottom strips have strains $\epsilon + (t/4)\kappa$ and $\epsilon - (t/4)\kappa$, respectively, see Fig. 2(c). Returning to Fig. 2(b.ii), let there be corresponding axial forces P_t and P_b on each

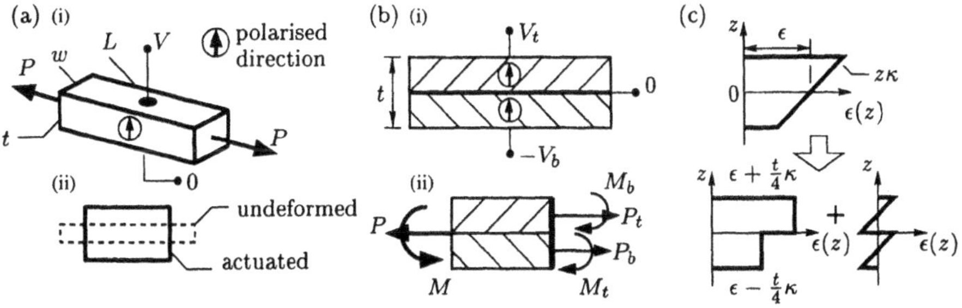

Figure 2. Piezoelectric strip structures: (a.i) single strip and associated geometry; (a.ii) freely actuated deformation; (b.i) two perfectly connected strips form a bimorph structure; (b.ii) external loads P and M, and internal stress resultants for each strip; (c) linear strain distribution $\epsilon + z\kappa$ for bimorph and decomposition of profile in each strip into components of axial extension and pure bending.

half section, and bending moments M_t and M_b. Making use of Eqns 3 and 4

$$P_t = Ew(t/2)\left[\epsilon + (t/4)\kappa - \Lambda_t\right];$$
$$P_b = Ew(t/2)\left[\epsilon - t/4\kappa - \Lambda_b\right];$$
$$M_t = M_b = E\frac{w(t/2)^3}{12}\kappa, \qquad (5)$$

and ensuring equilibrium with the external loads yields

$$P = P_t + P_b; \quad M = M_t + M_b + (t/4)\left[P_t - P_b\right]. \qquad (6)$$

Substituting for Eqns 5 and tidying up, finally gives

$$P = Ewt\left[\epsilon - \frac{1}{2}(\Lambda_t + \Lambda_b)\right] \ \text{(a)}; \quad M = E\frac{wt^3}{12}\left[\kappa - \frac{3}{2t}(\Lambda_t - \Lambda_b)\right] \ \text{(b)}. \qquad (7)$$

These are the generalised Hooke's laws for axial loading and bending of the bimorph. A simple example that demonstrates the difference between these two modes of actuation is given in Fig. 3. A bimorph cantilever is subject to, first, a vertical tip load. Since axial effects may be neglected, it is appropriate to have maximum bending actuation, as given by equal and opposite free strains in each layer. In practice, this is achieved by setting V_t equal to $-V_b$. Hence, substituting for $\Lambda_t = \Lambda$ and $\Lambda_b = -\Lambda$ into Eqn 7b

$$M = EI\left[\kappa - \frac{3\Lambda}{t}\right]. \qquad (8)$$

Second, an axial force is applied to the tip of an identical strip and the actuation strains are the same; bending effects are not induced and Eqn 3 is recovered as for the previous monolithic strip.

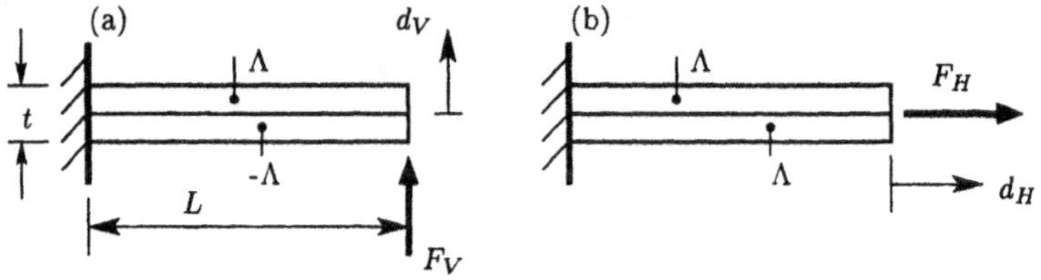

Figure 3. A bimorph cantilever subject to two actuation-load cases: (a) bending actuation with a vertical tip force, F_V; (b) axial behaviour under F_H. The analysis for case (a) equates the bending moment at any section to Eqn 8 with κ as the second derivative of the small transverse displacements. It can be shown that $F_V = (3EI/L^3)[d_V + (3\Lambda L^2/2t)]$. Case (b) is described by Eqn 3 with $d_H = \epsilon L$ and $F_H = P$.

It is left to the reader to verify that, first, the ratio of absolute tip displacements without loads is $3L/2t$ in favour of bending actuation. For a slender strip, the curvature change more efficiently amplifies the small actuation strains, and Seffen and Butler (2002) perform a more detailed examination. However, it can be shown that the ratio of blocking forces is $3t/8L$, and a much greater force is needed to restore axially induced displacements to zero.

Whilst analysis of the bimorph is not the primary focus of study, it is essential since the following helical structures are, fundamentally, curved bimorphs. Even though the generalised Hooke's laws in Eqns 7 have been developed for initially straight strips, they are appropriate for curved strips provided the local radius of initial curvature is larger than the thickness by an order of magnitude, see Seffen (2001). To simplify matters, only pure bending actuation and specific load cases are addressed.

2.2 A Piezoelectric Helical Bender

One turn of a typical right-handed helix is shown in Fig. 4(a). It is a curved bimorph with cylindrical radius R_1 to mid-thickness. A tangent to the helix at any point makes an angle of pitch, β_1, relative to a radial plane through this point. Equal and opposite axial twisting couples, C, are applied at each end.

The behaviour of the bimorph helix may be approximated to a helical line endowed with the properties of the strip bimorph from before. The curvature, κ_1, and the rate of twist, τ_1, of this line are

$$\kappa_1 = \frac{\cos^2 \beta_1}{R_1}; \quad \tau_1 = \frac{\sin \beta_1 \cos \beta_1}{R_1}. \tag{9}$$

Note that τ_1 is the *spatial* twist-rate; Thompson and Champneys (1996) introduce an extra intrinsic or *kinematic* twist-rate to account for the highly-wound nature of their buckled rods. By comparison, the rotations here are small and the kinematic twist-rate may be neglected.

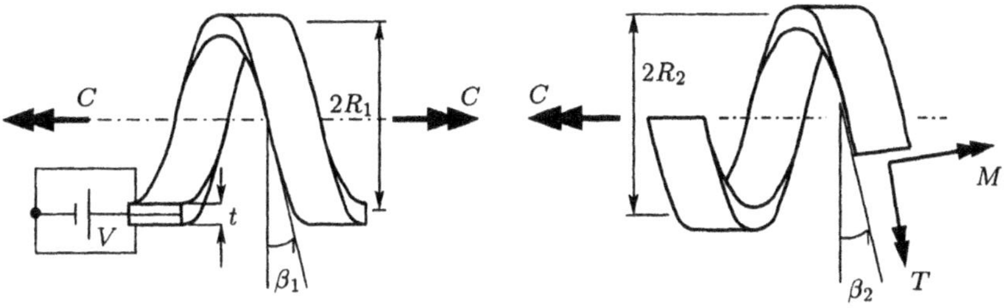

(a) Single turn of piezoelectric bimorph helix. The cylindrical radius to mid-thickness is R_1 and the pitch angle is β_1. Equal and opposite electric fields give pure bending actuation along the wrapped length of section.

(b) Free-body section of the deformed helix. The "cut" exposes a non-zero bending moment, $M = EI(\kappa_2 - \kappa_1 - 3\Lambda/t)$, and torque, $T = GJ(\tau_2 - \tau_1)$, in equilibrium with the applied axial couple, C.

Figure 4. Piezoelectric bimorph helical bender.

The deformed helix must adopt uniform R_2 and β_2 by symmetry. A free-body section in Fig. 4(b) has a bending moment, M, about the minor axis of cross-section and a normal torque, T, on the exposed internal face; these are the only non-zero stress resultants from simple equilibrium considerations. M is equivalently described by Eqn 8 and κ now equals $\kappa_2 - \kappa_1$ with $\kappa_2 = \cos^2 \beta_2 / R_2$. The torque is $GJ[\tau_2 - \tau_1]$ where GJ is the torsional stiffness of the cross-section and $\tau_2 = \sin \beta_2 \cos \beta_2 / R_2$.

Resolving moments radially and axially in Fig. 4(b) results in the governing equations of deformation

$$EI\left[\kappa_2 - \kappa_1 - \frac{3\Lambda}{t}\right]\cos \beta_2 + GJ\left[\tau_2 - \tau_1\right]\sin \beta_2 = 0;$$

$$EI\left[\kappa_2 - \kappa_1 - \frac{3\Lambda}{t}\right]\sin \beta_2 - GJ\left[\tau_2 - \tau_1\right]\cos \beta_2 = C. \tag{10}$$

Equations 10 are more compactly stated by defining $\Gamma = \kappa_1 + 3\Lambda/t$, $\xi = GJ/EI$ and $\bar{C} = C/EI$. Writing the deformed curvature and rate of twist in terms of R_2 and β_2, and eliminating R_2 between the above equations produces

$$\xi\tau_1\cos^2\beta_2 - \Gamma\xi\cos\beta_2\sin\beta_2 = \bar{C}\,(\xi-1)\cos^2\beta_2\sin\beta_2. \tag{11}$$

This non-linear equation can be numerically solved for β_2; a solution in closed form does become available if the pitch angle is assumed small, as in a closely-coiled device. Therefore, setting $\sin\beta_2 \approx \beta_2$ and $\cos\beta_2 \approx 1$, solving for β_2 and inserting the result into either of Eqns 10, in order to extract R_2, yields

$$\beta_2 = \frac{\xi\tau_1}{\Gamma\xi + \bar{C}\,(\xi-1)}\ \text{(a)};\quad R_2 = \frac{\left(\Gamma\xi + \bar{C}\,[\xi-1]\right)^2}{\left[\left(\Gamma\xi + \bar{C}\,[\xi-1]\right)^2 + \xi^2\tau_1^2\right]}\frac{1}{\left(\bar{C}+\Gamma\right)}\ \text{(b)}, \tag{12}$$

with $\tau_1 = \beta_1/R_1$ and $\Gamma = 1/R_1 + 3\Lambda/t$. The first quotient term on the right-hand side of Eqn 12b is approximately equal to unity and R_2 is simplified to

$$R_2 \approx \frac{R_1}{1 + R_1\left(\bar{C} + 3\Lambda/t\right)}. \tag{13}$$

Furthermore, since the applied load is an axial couple, it is appropriate to express the generalised Hooke's law in terms of the relative rotation between the ends. This rotation is estimated by assuming that the arc-length of helix, s, does not change during actuation as there is no net axial strain. If the undeformed helix subtends a total angle, θ_1, then $s_1 = R_1\theta_1/\cos\beta_1$ by simple geometry. The new shape subtends angle θ_2 and $s_2 = R_2\theta_2/\cos\beta_2$. The relative end rotation, $\Delta\theta = \theta_2 - \theta_1$, is found by equating s_1 to s_2 and, for small pitch angles, it can be shown

$$\frac{\Delta\theta}{\theta_1} \approx \frac{R_1}{R_2} - 1. \tag{14}$$

Replacing R_2 using Eqn 13, and tidying up, results in the generalised Hooke's law

$$C = EI\left[\frac{1}{R_1\theta_1}\Delta\theta - \frac{3\Lambda}{t}\right]. \tag{15}$$

The small pitch implies that the external couple is carried, primarily, in bending.

Finally, it is noted that, when $C = 0$ and without assuming small angles, Eqn 11 is *exactly* solved for β_2 and, subsequently, either of Eqns 10 for R_2 as

$$\tan\beta_2 = \frac{\tau_1}{\Gamma};\quad R_2 = \frac{\Gamma}{\tau_1^2 + \Gamma^2}. \tag{16}$$

Conserving arc-length and substituting for R_2 and β_2 from Eqns 16, re-arranging, and neglecting small terms compared to unity, presents

$$\frac{\Delta\theta}{\theta_1} = \frac{3\Lambda}{t} R_1. \tag{17}$$

This more accurate solution has the same free rotation as Eqn 15 with C equal to zero and has been confirmed by experiments (Seffen, 2001).

2.3 A Piezoelectric Coiled-Coil

The geometry of a coiled-coil can be described by its major and minor helical paths, recall Fig. 1(a); these helices form a right-handed pair. Practical devices are compact, with a closely-coiled major path and, for simplicity, its angle of pitch may be taken as zero; the major radius is denoted as b. Relative to the major helix, the minor helix has a radius, a, and a small but non-zero angle of pitch, β. During actuation, the structure is loaded by equal and opposite axial forces, F, see Fig. 5(a.i), and the corresponding displacement between the ends is d.

One approach towards a generalised Hooke's law mimics the procedure in the previous section, and this requires the curvature and twist-rate of the coiled-coil to be obtained in terms of the above parameters. From Gray (1993), these expressions depend upon the position along the arc-length of coil; also, they are extremely lengthy in spite of the major pitch angle being zero, rendering algebraic manipulation difficult. An alternative route must make simplifying assumptions; naturally, some accuracy is lost but it is a good starting point, as is now detailed.

The major radius tends to be much greater than the minor radius in practice. Therefore, any portion of the coiled-coil has, locally, negligible toroidal curvature: for some analytical purposes, the axis of the minor helix may be treated as being straight. The major radius is also assumed to remain constant during deformation. From overall moment equilibrium with the axial load, a couple, $C = Fb$, is carried everywhere along the major helical path and normal to every radial plane of the minor helix, see Fig. 5(a.ii).

With these assumptions, a generalised Hooke's law for the coiled-coil can be adapted from Eqn 15 for a straight helix, if the only stress resultant on the major helix axis is the couple, C: but it is clear from Fig. 5(a.ii) that there is also a vertical shear force, F, and the problem is similar to conventional close-coiled spring theory with the minor coil replacing the helical wire. However, because the minor helix is an open structure, this force is carried by a combination of bending and twisting, as well as shearing everywhere along the arc-length of coil. A rigorous analysis of this behaviour is not reported here, suffice to say that the extra deformation due to F cannot be neglected in comparison to the axial displacement from the twisting couple alone. Hence, using only Eqn 15, the axial stiffness of the coiled-coil will be over-estimated; the prediction of the actuated axial displacement may be too small. Nonetheless, the study proceeds in this manner, in order to appreciate the levels of performance between structures: a more accurate analysis will be given elsewhere.

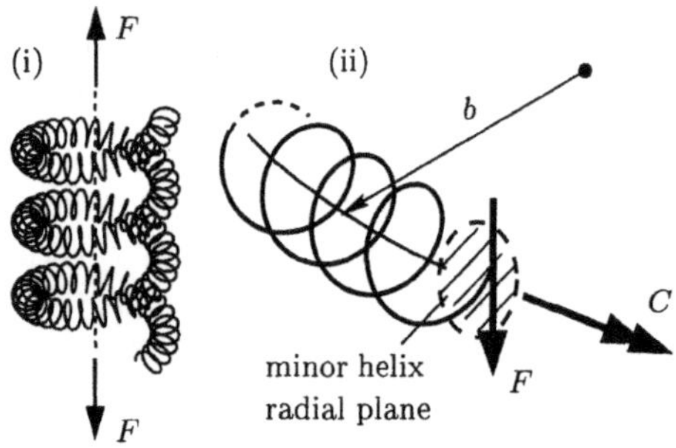

(a) External axial forces, F, are applied to the coiled-coil, which is schematically shown as a wire-frame. Provided the major helix is closely coiled, there is a vertical shear force, F, and a normal twisting couple, $C = Fb$, at every radial plane of the minor helix.

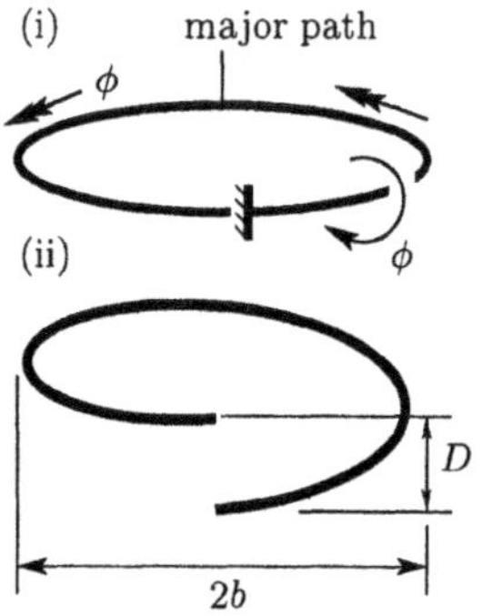

(b) One full turn of the major helical path subject to a twist rate, ϕ, along its length. The vertical displacement (ii) between the start and end of the loop due to ϕ is D.

Figure 5. Equilibrium and compatibility considerations for coiled-coil.

Using Eqn 15, C is replaced by Fb and R_1 by a. $\Delta\theta$ in this equation now refers to the rotation of the minor helix along its axis, and is related to d by the compatibility argument illustrated in Fig. 5(b). A twist rate, ϕ, is imposed along a planar curved rod of radius b, causing it to displace out of plane, as shown. Integrating the product of twist rate and radial lever arm along the circumference and assuming small displacements, the resulting vertical displacement between the ends of one turn, D, is found to be

$$D = 2\pi b^2 \phi. \tag{18}$$

Obviously, $\Delta\theta$ is $2\pi b\phi$ and, thus, $D = b\Delta\theta$. If there are N major turns of coiled-coil, d is analogously equal to ND and

$$\Delta\theta = \frac{d}{Nb}. \tag{19}$$

Correspondingly, θ_1 in Eqn 15 is now the total angle subtended by the minor helix along one major turn, and simple geometry gives

$$\theta_1 = \frac{2\pi b}{a \tan\beta} \approx \frac{2\pi b}{a\beta}. \tag{20}$$

Equations 19 and 20 are now substituted into Eqn 15 and, after re-arrangement, the approximate generalised Hooke's law is

$$F = EA\left[\frac{\beta t^2}{24\pi N b^3} d - \frac{\Lambda t}{4b}\right]; \quad A = wt. \tag{21}$$

The coiled-coil can now be compared to the other structures of this study. The performance measures are the freely actuated change of shape and the blocking load under pure bending actuation. Therefore, a meaningful comparison deals only with linear displacements and forces, and the strip bimorph with the transverse tip load from Fig. 3(a) is chosen; the torsional response of the straight helix is not appropriate.

3. COMPARISON OF STRUCTURES

From the caption in Fig. 3(a), the blocking force, F_{str}, and the free displacement of the tip, d_{str}, are calculated as

$$F_{\mathrm{str}} = \frac{3EA\Lambda t}{8L} \text{ (a)}; \quad d_{\mathrm{str}} = \frac{3\Lambda L^2}{2t} \text{ (b)}, \tag{22}$$

where the subscript "str" denotes "strip"; note that the above terms are absolute quantities. Using Eqn 21, the magnitudes of the equivalent parameters for the coiled-coil ("cc") are

$$F_{\mathrm{cc}} = \frac{EA\Lambda t}{4b} \text{ (a)}; \quad d_{\mathrm{cc}} = \frac{6\Lambda\pi N b^2}{\beta t} \text{ (b)}. \tag{23}$$

Consider the ratio of blocking forces; in particular, the cross-sectional properties and the levels of actuation are assumed to be identical between structures. Thus, the terms EA, Λ and t in Eqns 22 and 23 are respectively equal to one another, leading to

$$\frac{F_{\mathrm{cc}}}{F_{\mathrm{str}}} = \frac{4}{3}\left[\frac{L}{2b}\right], \tag{24}$$

and the ratio of free displacements is

$$\frac{d_{\mathrm{cc}}}{d_{\mathrm{str}}} = \frac{\pi N}{\beta}\left[\frac{2b}{L}\right]^2. \tag{25}$$

Moreover, if the major diameter of the coiled-coil, $2b$, is equal to the length of the strip, L, the devices are similar in size and

$$\left[\frac{F_{\rm cc}}{F_{\rm str}}\right]_{L=2b} = \frac{4}{3} \text{ (a);} \quad \left[\frac{d_{\rm cc}}{d_{\rm str}}\right]_{L=2b} = \frac{\pi N}{\beta} \text{ (b).} \tag{26}$$

Equation 26(a) suggests that the blocking force of the coiled-coil is 33% more than that of the strip. However, in neglecting the effects of shear force in the previous section, this disparity may decrease in practice but needs to be confirmed by experiments. Conversely, actuated displacements in the coiled-coil are significantly greater; for example, the ratio in Eqn 26(b) is greater than five for one turn of major helix, $N = 1$, with β equal to 30°, which is a typical value in actual coils. In absolute terms, initial experiments have indicated a displacement per turn equal to $0.1b$ for $b/t = 30$ under very small ($\approx 0.01\%$) actuation strains.

In closing, it is noted that comparisons can be made according to other metrics, for example, the arc-length of structure rather than the major dimension: such exercises are left to future work.

4. CONCLUSION

The actuation responses of novel coiled piezoelectric structures have been determined by compact analytical models, which may be used in the design of actuator systems. It has been shown that a coiled-coil shape efficiently amplifies the naturally small actuation strains of piezoelectric material.

Acknowledgments

The piezoelectric coiled-coil is registered under the trade-name "Helimorph" by its inventors, 1...Ltd, St John's Innovation Centre, Cambridge, UK. The author is grateful to 1...Ltd for kind permission to write this paper.

References

Fletcher, N H, Tarnopolskaya, T & de Hoog, F R (2001) Wave propagation on helices and hyperhelices: a fractal regression, *Proc. Roy. Soc. A, Lond.* **457**, 33-43.

Thompson, J M T & Champneys, A R (1996) From helix to localized writhing in the post-buckling of elastic rods, *Proc. Roy. Soc. A, Lond.* **452**, 117-138.

Calladine, C R & Drew, H R (1992) *Understanding DNA: the Molecule & How it Works*, Academic Press.

Pearce, D H (1999) Piezoelectrics spring into action, *Materials World* **7**, 748-750.

Huber, J E, Fleck, N A & Ashby, M F (1996) The selection of mechanical actuators based on performance indices, *Proc. Roy. Soc. A, Lond.* **453**, 2185-2205.

Spearitt, D J & Asokanthan, S F (1996) Torsional vibration control of a flexible beam using laminated PVDF actuators, *J. Sound. Vib.* **193**, 941-955.

Seffen, K A & Butler, D J (2002) An exercise in elementary smart structures, *Int. J. Mech. Eng. Education* **30**, 59-72.

Seffen, K A (2001) *Through-Thickness Effects in Curved Piezoelectric Beams*, technical report UMIST/ME/AM/15.05.01/CURV1, UMIST.

Seffen, K A (2001) *Analysis of Transducers for Digital Loudspeakers and Phased Array Antennae*, technical report UMIST/ME/AM/09.05.01/PADS2, UMIST.

Gray, A (1993) *Modern Differential Geometry of Curves and Surfaces*, CRC Press.

FRICTIONAL IMPACT IN MECHANISMS

W.J. Stronge
Department of Engineering, University of Cambridge
Cambridge CB2-1PZ, UK
wjs@eng.cam.ac.uk

Abstract This paper obtains smooth (continuous) dynamics for impact between two rough (frictional) mechanisms. The mechanisms are composed of rigid bodies joined together by ideal non-dissipative pinned joints. The system has a configuation described in terms of generalized coordinates q_i and time t, and it has kinetic energy $T(\dot{q}_i, q_i, t)$. Generalized momentum of the system is defined as a vector, $\partial T / \partial \dot{q}_i$. During collision the system is subject to a set of constraint and friction forces that give another vector — the differential of generalized impulse $\mathrm{d}\Pi_i$. If the applied forces act impulsively, then the differentials of generalized momentum and generalized impulse are equal, $\mathrm{d}(\partial T / \partial \dot{q}_i) = \mathrm{d}\Pi_i$.

When applied to impact between systems of hard bodies where there is slip that changes direction during contact, this differential relation is required. If the direction of slip is constant, however, it is more convenient to use an integrated form of this generalized impulse-momentum relation. In either case, terminal reaction impulse is obtained from the energetic coefficient of restitution.

Key words: Rigid-body dynamics, Impact, Energetic coefficient of restitution, Friction

1. INTRODUCTION

Impact on a mechanism (a system of linked rigid bodies) induces reaction forces at the joints or connections between the bodies. If joint compliance is small, the reactions are generated by kinematic constraints and methods of analytical mechanics apply to this multi-body impact problem.

Synge and Griffith (1959), Wittenberg (1977), Drazetic et al. (1996) etc. have presented methods for analysing impulse response of systems of rigid

H.R. Drew and S. Pellegrino (eds.),
New Approaches to Structural Mechanics, Shells and Biological Structures, 229–244.

bodies that are joined by frictionless pinned joints (workless constraints). These methods equate changes in a generalized momentum to a generalized impulse, where generalized momentum and impulse are obtained from the principle of virtual work. When these methods are applied to a collision, where two systems come together at an impact point, the analytical method can be applied directly only if the ratio of normal to tangential components of force at the impact point remains constant throughout the indefinitely small period of contact. Where this ratio of components of force remains constant, the terminal impulse acting at the impact point can be obtained from the normal impulse for compression (Souchet, 1993 or Zhao, 1999). The ratio of forces is constant however only if: (i) friction is negligible or (ii) the incident tangential relative velocity is negligible or (iii) motion is constrained to be planar and the incident tangential relative velocity is sufficiently large so that sliding does not vanish before separation (Batlle, 1996). This paper obtains equations of motion that equate differentials of generalized momentum and impulse — this formulation is required whenever the ratios of components of contact force are not constant.

2. EQUATIONS OF MOTION BASED ON GENERALIZED IMPULSE AND MOMENTUM

Equations of motion are developed first for a set of particles. These are applicable as well for a set of rigid bodies where the configuration and properties of a body are obtained by defining fixed distances between the particles which comprise the body; i.e. there are a set of constraint equations (holonomic and/or nonholonomic) which define both the relative positions of particles within each rigid body and kinematic relationships between bodies.

Let S be a set of N particles with the jth particle located at a position vector $\mathbf{r}_j$, $j = 1,\cdots,N$; each particle is subject to an external force $\mathbf{F}_j$. These forces give differential impulses $\mathrm{d}\mathbf{p}_j$ that act in the very brief period of impact, $\mathrm{d}\mathbf{p}_j = \mathbf{F}_j\,\mathrm{d}t$. The velocity $\mathbf{V}_j$ of the jth particle is the rate-of-change of the position vector and this is a function of the impulse $\mathbf{p}_j$; i.e. $\mathbf{V}_j(\mathbf{p}_j) \equiv \mathrm{d}\mathbf{r}_j(\mathbf{p}_j)/\mathrm{d}t$. Suppose the particle velocities are subject to $3N-n$ holonomic constraint equations; e.g. a fixed distance separates the jth and kth particles. Then the particle velocities can be expressed in terms of generalized coordinates q_i, generalized speeds $\dot{q}_i$, and time t; i.e. $\mathbf{V}_j = \mathbf{V}_j(\dot{q}_i, q_i, t)$, $i = 1, \ldots, n$. These expressions for particle velocities are consistent with $3N-n$ holonomic constraints. In addition, suppose there are $n-m$ nonintegrable equations (nonholonomic constraints) of the form $0 = \sum_{i=1}^{n}(a_{si}\dot{q}_i + b_s)$ $s = 1, \ldots, n-m$ with coefficients a_{si} and b_s. This nonholonomic system has m degrees of freedom.

Virtual displacements $\delta \mathbf{r}_j$ are any displacement field that is compatible with the displacement constraints. Similarly virtual velocities $\delta \mathbf{V}_j$ are compatible with the velocity constraints of the system. The virtual velocity of the *j*th particle can be expressed in terms of generalized speeds $\dot{q}_i$ as

$$\delta \mathbf{V}_j = \sum_{i=1}^{n} (\partial \mathbf{V}_j / \partial \dot{q}_i) \delta \dot{q}_i \tag{1}$$

During impact on a mechanism, the virtual differential of work $\delta(\mathrm{d}W)$ done by external forces $\mathrm{d}\mathbf{p}_j / \mathrm{d}t$ is used to define a *differential of generalized impulse* $\mathrm{d}\Pi_i$,

$$\delta(\mathrm{d}W) = \sum_{j=1}^{N} \sum_{i=1}^{n} \mathrm{d}\mathbf{p}_j \cdot (\partial \mathbf{V}_j / \partial \dot{q}_i) \delta \dot{q}_i = \sum_{i=1}^{n} (\mathrm{d}\Pi_i) \delta \dot{q}_i \tag{2}$$

where $\mathrm{d}\Pi_i \equiv \sum_{j=1}^{N} \mathrm{d}\mathbf{p}_j \cdot (\partial \mathbf{V}_j / \partial \dot{q}_i)$. Generalized active forces $F_i = \sum_{j=1}^{N} (\mathrm{d}\mathbf{p}_j / \mathrm{d}t \cdot \partial \mathbf{V}_j / \partial \dot{q}_i)$ are the only forces that contribute to the differential of generalized impulse; other forces, which do not contribute, include any equal but opposite forces of interaction at rigid (i.e. non-compliant) constraints and external body forces or pressures which remain constant during impact.

A system of N particles connected by $3N - n$ holonomic velocity constraints has kinetic energy T which varies with applied impulse. Kinetic energy can be expressed in either a global coordinate system or as a function of the generalized speeds,

$$T = \frac{1}{2} \sum_{j=1}^{N} M_j (\mathbf{V}_j \cdot \mathbf{V}_j) = \frac{1}{2} \sum_{i=1}^{n} \sum_{s=1}^{n} m_{is} \dot{q}_i \dot{q}_s \tag{3}$$

where M_j is the mass of the *j*th particle and m_{is} is an inertia matrix obtained from kinetic energy of the constrained motion.

For a set of N particles subjected to an external impulse $\mathbf{p}_j$, the differential form of Newton's second law can be expressed as

$$\mathrm{d}\left(\sum_{j=1}^{N} M_j \mathbf{V}_j \right) = \mathrm{d}\left(M \hat{\mathbf{V}} \right) = \sum_{j=1}^{N} \mathrm{d}\mathbf{p}_j \tag{4}$$

where the system has mass $M = \sum_{j=1}^{N} M_j$ and the velocity of the centre of mass is $\hat{\mathbf{V}} = M^{-1} \sum_{j=1}^{N} M_j \mathbf{V}_j$. Taking the scalar product of this equation and the set of virtual velocity coefficients $\partial \mathbf{V}_j / \partial \dot{q}_i = \partial \mathbf{r}_j / \partial q_i$ gives

$$\sum_{j=1}^{N} M_j \frac{\partial \mathbf{V}_j}{\partial \dot{q}_i} \cdot \mathrm{d}\mathbf{V}_j = \mathrm{d}\left(\frac{\partial T}{\partial \dot{q}_i} \right) - \sum_{j=1}^{N} M_j \mathbf{V}_j \cdot \mathrm{d}\left(\frac{\partial \mathbf{V}_j}{\partial \dot{q}_i} \right) = \mathrm{d}\Pi_i - \sum_{j=1}^{N} \mathbf{p}_j \cdot \mathrm{d}\left(\frac{\partial \mathbf{V}_j}{\partial \dot{q}_i} \right) \tag{5}$$

where $d(\partial T / \partial \dot{q}_i)$ is termed a *differential of generalized momentum*. Assuming that the configuration is invarient during collision, $d(\partial \mathbf{V}_j / \partial \dot{q}_i) = 0$; hence in Eq (5) the second terms vanish.

For a system subject to velocity constraints, the equations of motion in terms of generalized speeds $\dot{q}_i$ are obtained directly from the kinetic energy T and the differential of generalized impulse $d\Pi_i$.

> Theorem — *If an impulsively loaded system can be represented by* n *generalized coordinates* q_i *and* $n - m$ *nonholonomic velocity constraints* $\sum_{i=1}^{n}(a_{si}\dot{q}_i + b_s) = 0, \quad s = 1, \ldots, n-m$ *and this system is subject to a differential of generalized impulse* $d\Pi_i$, *then the equations of motion in terms of generalized speeds* $\dot{q}_i$ *are obtained as,*
>
> $$d(\partial T / \partial \dot{q}_i) = d\Pi_i$$

The differential of generalized momentum can be expressed in terms of generalized speeds as $d(\partial T / \partial \dot{q}_i) = m_{is} d\dot{q}_s$ where m_{is} are inertia coefficients of the system.

3. IMPACT PROCESS

Impact initiates when two colliding bodies B and B′ first come into contact at C, an initial point of contact. Each body has a point of contact, C or C′, and at incidence these points have velocities $\mathbf{V}_C(0)$ and $\mathbf{V}_{C'}(0)$, respectively. Between the contact points there is a relative velocity $\mathbf{v}(t)$ defined as $\mathbf{v}(t) \equiv \mathbf{V}_C(t) - \mathbf{V}_{C'}(t)$. If at least one of the bodies is smooth in a neighbourhood of C, there is a common tangent plane (c.t.p.) and perpendicular to this plane there is a normal direction $\mathbf{n}_3$. At incidence the bodies come together with a negative relative velocity at C; i.e. $\mathbf{n}_3 \cdot \mathbf{v}(0) < 0$.

Analyses of ‘rigid’ body impact can follow the process of velocity change at the contact point by introducing an infinitesimal deformable particle between the bodies at C — this particle simulates a small deforming region. The deforming region is assumed to have negligible mass because it is small in comparison with the size of either body; consequently the contact forces on either side of the particle are equal but opposite. Note that the normal contact force F_3 is always compressive so that the normal component of impulse $p_3(t)$ is a monotonously increasing function of time; thus normal impulse $p \equiv p_3$ can replace time as an independent variable.

4. TERMINAL IMPULSE

The key to calculating changes in velocity during impact is to find a means of evaluating the terminal impulse p_f at separation. The theory of rigid body impact can have broader applicability if the terminal impulse can be based on physical considerations. Here we relate the terminal impulse to the *energetic coefficient of restitution*; this represents dissipation of (kinetic) energy due to inelastic deformation in the region surrounding the contact point.

4.1 Compression and Restitution Phases of Collision

After the colliding bodies first touch, the contact force $F(t)$ rises as the deformable particle is compressed; this causes an indentation δ. If compliance is rate-independent, the maximum indentation and maximum force occur simultaneously when the normal component of relative velocity vanishes. Figure 1(b) illustrates the normal contact force as a function of indentation δ while Fig. 1(a) shows this force as a function of time. The latter figure shows the separation of the contact period into an initial phase of approach or *compression* and a subsequent phase of *restitution.*

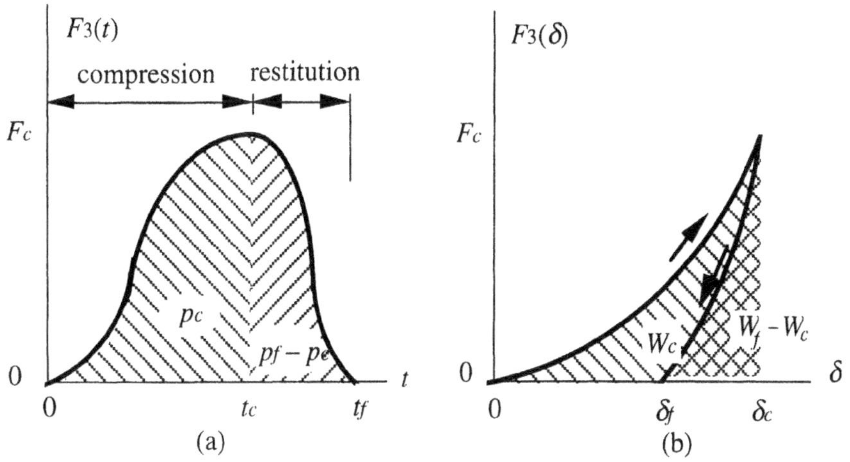

Figure 1. Variation of contact force during impact.

At any time t after incidence, the normal component of contact force F has an impulse p which equals the area under the curve of force shown in Fig. 1(a). Let the instant when indentation changes from compression to restitution be t_c. The colliding bodies have a relative velocity between contact points that vanishes at the end of compression, $v(t_c) = 0$. The reaction impulse p_c which brings the two bodies to a common speed is termed the *normal impulse for compression*; this characteristic impulse is useful for analyzing collision processes.

4.2 Coefficient of Restitution and Kinetic Energy Absorbed in Collision

Dissipation of energy during collision results in smaller compliance during unloading (restitution) than was present during loading (compression); i.e. the force-deflection curve given in Fig. 1(b) exhibits hysteresis. The kinetic energy of normal relative motion that is transformed to internal energy of deformation during loading equals the area under the loading curve in Fig. 1(b); this area is denoted by $W_c \equiv W_3(p_c)$. On the other hand, the area under the unloading curve equals the elastic strain energy released from the deforming region during restitution; in Fig. 1(b) this is denoted by $W_f - W_c \equiv W_3(p_f) - W_3(p_c)$. This work is easily calculated for separate phases of compression and restitution if changes in relative velocity are obtained as a function of normal impulse as illustrated in Fig. 2(b).

During compression the impulse of the normal contact force does work $W_3(p_c)$ on the rigid bodies that surround the small deforming region — this work equals the internal energy of deformation absorbed in compressing the deformable region. If the relative velocity at C is expressed as a function of normal impulse p,

$$W_c \equiv W_3(p_c) = \int_0^{p_c} v(p)\,\mathrm{d}p \tag{7}$$

This is just the kinetic energy of normal relative motion that is transformed during compression. During the succeeding phase of restitution the rigid bodies regain some of this kinetic energy of normal relative motion due to the work $W_3(p_f) - W_3(p_c)$ done by contact forces.

$$W_f - W_c \equiv W_3(p_f) - W_3(p_c) = \int_{p_c}^{p_f} v(p)\,\mathrm{d}p \tag{8}$$

These expressions for work done by the contact force during separate parts of the collision (period) are used to express the part of the initial kinetic energy of normal relative motion that is dissipated. Expressions (7) and (8) give the part of this transformed energy that is irreversible and this can be used to define an *energetic coefficient of restitution*, e_*.

Definition — *The square of the coefficient of restitution* e_*^2 *is the negative of the ratio of the elastic strain energy released during restitution to the internal energy of deformation absorbed during compression,*

$$e_*^2 = -\frac{W_3(p_f) - W_3(p_c)}{W_3(p_c)} \tag{9}$$

This coefficient has values in the range $0 \le e_* \le 1$ where 0 implies a perfectly plastic collision (i.e. no final separation so that none of the initial kinetic energy of normal relative motion is recovered) while a value of 1 implies a perfectly elastic collision (i.e. no loss of kinetic energy of normal relative motion), Stronge (1990).

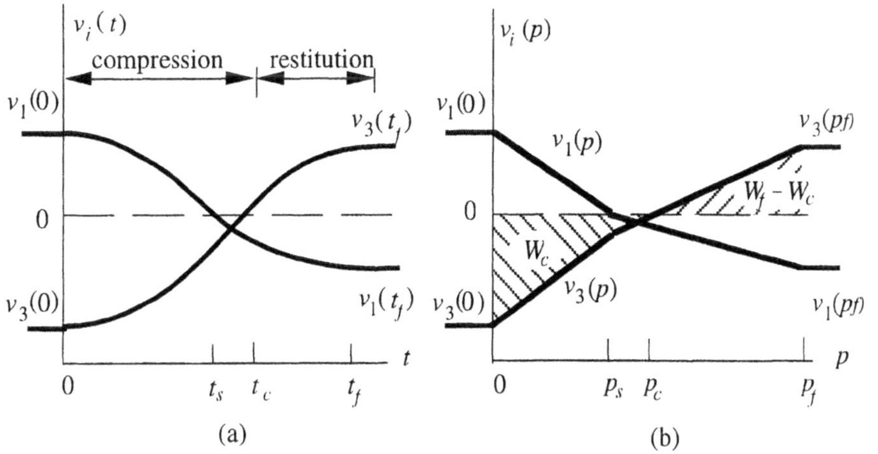

Figure 2. Variation of components of relative velocity at contact point during impact with small initial speed of sliding.

5. KINETIC ENERGY OF SLENDER UNIFORM BAR

A slender uniform rigid bar of mass M with ends A and B and velocity $\hat{\mathbf{V}}$ of the center of mass, has kinetic energy T. This can be decomposed into part T_V that is translational and another part T_ω that is rotational;

$$T_V = \frac{1}{2} M \hat{\mathbf{V}} \cdot \hat{\mathbf{V}} = \frac{M}{8}\left(\mathbf{V}_\mathrm{A} \cdot \mathbf{V}_\mathrm{A} + 2\mathbf{V}_\mathrm{A} \cdot \mathbf{V}_\mathrm{B} + \mathbf{V}_\mathrm{B} \cdot \mathbf{V}_\mathrm{B}\right) \tag{10}$$

$$T_\omega = \frac{1}{2} \boldsymbol{\omega} \cdot \hat{\mathbf{I}} \cdot \boldsymbol{\omega} = \frac{M}{24}\left(\mathbf{V}_\mathrm{A} \cdot \mathbf{V}_\mathrm{A} - 2\mathbf{V}_\mathrm{A} \cdot \mathbf{V}_\mathrm{B} + \mathbf{V}_\mathrm{B} \cdot \mathbf{V}_\mathrm{B}\right) \tag{11}$$

where $\boldsymbol{\omega}$ is the angular velocity of the bar and $\hat{\mathbf{I}}$ is the inertia dyadic for the center of mass. Thus the kinetic energy of the bar $T = T_V + T_\omega$ is given by the following expression;

$$T = \frac{M}{6}\left(\mathbf{V}_\mathrm{A} \cdot \mathbf{V}_\mathrm{A} + \mathbf{V}_\mathrm{A} \cdot \mathbf{V}_\mathrm{B} + \mathbf{V}_\mathrm{B} \cdot \mathbf{V}_\mathrm{B}\right) \tag{12}$$

The useful expression (12) was presented by Bahar (1994).

6. EXAMPLE PROBLEMS

Example 1: Normal impulse applied to pin-jointed mechanism

A regular hexagon is constructed of 6 bars, each of length L and mass M. Each end of each bar is joined to an adjacent bar by a frictionless pin. The hexagon lies at rest on a smooth level plane before application of a radial impulse $\mathbf{p}$ at one corner. Figure 3(a) shows half of this symmetric impulse applied at joint 1 on the axis of symmetry of the regular hexagonal mechanism.

Solution

The velocity of the individual nodes or joints can be expressed as

$$\begin{aligned}
\mathbf{V}_1 &= V_1\mathbf{n}_1 = \dot{q}_i\mathbf{n}_1 \\
\mathbf{V}_2 &= \mathbf{V}_1 + 0.5L\omega_{12}\left(-\sqrt{3}\mathbf{n}_1 + \mathbf{n}_2\right) = 0.5\left(2\dot{q}_1 - \sqrt{3}\dot{q}_2\right)\mathbf{n}_1 + 0.5\dot{q}_2\mathbf{n}_2 \\
\mathbf{V}_3 &= \mathbf{V}_2 + L\omega_{23}\mathbf{n}_2 = 0.5\left(2\dot{q}_1 - \sqrt{3\dot{q}_2}\right)\mathbf{n}_1 + 0.5\left(\dot{q}_2 + 2\dot{q}_3\right)\mathbf{n}_2 \\
\mathbf{V}_4 &= \mathbf{V}_3 + 0.5L\,\omega_{34}\left(\sqrt{3}\mathbf{n}_1 + \mathbf{n}_2\right) \\
&= 0.5\left(2\dot{q}_1 - \sqrt{3}\dot{q}_2 + \sqrt{3\dot{q}_4}\right)\mathbf{n}_1 + 0.5\left(\dot{q}_2 + 2\dot{q}_3 + \dot{q}_4\right)\mathbf{n}_2
\end{aligned}$$

where generalized speeds $\dot{q}_2$ have been defined as follows:

$$\dot{q}_1 = V_1, \qquad \dot{q}_2 = L\omega_{12}, \qquad \dot{q}_3 = L\omega_{23}, \qquad \dot{q}_4 = L\omega_{34}$$

Symmetric deformations

If the deformation field is symmetric, points on the axis of symmetry have no velocity component transverse to the axis. This gives $0 = \mathbf{n}_1 \cdot \mathbf{V}_4 = \dot{q}_2 + 2\dot{q}_3 + \dot{q}_4$. After substituting for $\dot{q}_4$ and employing Eq. (12), one can obtain an expression for the kinetic energy T of each half of the hexagon,

$$T = \frac{M}{6}\left\{9\dot{q}_1^2 - 9\sqrt{3}\dot{q}_1\dot{q}_2 + (19/2)\dot{q}_2^2 + 10\dot{q}_2\dot{q}_3 + 5\dot{q}_3^2 - 3\dot{q}_1\dot{q}_3\right\}$$

This kinetic energy is differentiated with respect to $\dot{q}_i$ in order to obtain the generalized momentum $\partial T / \partial \dot{q}_i$.

Equation of relative motion

An external impulse $\mathbf{p} = p\mathbf{n}_1$ is applied at point 1; this point of application has velocity $\mathbf{V}_1 = \dot{q}_1\mathbf{n}_1$. Hence one obtains the components of the generalized impulse $\Pi_1 = 0.5p\mathbf{n}_1 \cdot \mathbf{n}_1 = 0.5p$, $\Pi_2 = \Pi_3 = 0$. Equating the change in generalized momentum to the generalized impulse results in the equations of motion,

$$\frac{M}{6}\begin{bmatrix} 18 & -9\sqrt{3} & -3\sqrt{3} \\ -9\sqrt{3} & 19 & 10 \\ -3\sqrt{3} & 10 & 10 \end{bmatrix}\begin{Bmatrix} \dot{q}_1 \\ \dot{q}_2 \\ \dot{q}_3 \end{Bmatrix} = \frac{p}{2}\begin{Bmatrix} 1 \\ 0 \\ 0 \end{Bmatrix}$$

which can be solved for the change in generalized speeds $\dot{q}_i$ due to the applied impulsive load,

$$\dot{q}_1 = 0.909\, p/M\,, \qquad \dot{q}_2 = 1.050\, p/M\,, \qquad \dot{q}_3 = -0.577\, p/M\,,$$
$$\dot{q}_4 = 0.105\, p/M$$

After application of the impulse, the velocity $\hat{\mathbf{V}}$ of the center-of-mass of the system is $\hat{\mathbf{V}} = 0.1667\ p\mathbf{n}_1 / M$, so that relative to the center-of-mass, the joints have relative velocities $\mathbf{v}_i$ where $\mathbf{v}_i = \mathbf{V}_i - \hat{\mathbf{V}}$. The terminal relative velocities of joints $\mathbf{v}_i$ and angular velocities of bars $\omega_{ij}\,\mathbf{n}_3$ are

$$\mathbf{v}_1 = (0.742\mathbf{n}_1 + 0.0\mathbf{n}_2)p/M \qquad \omega_{12} = 1.050 p/M$$
$$\mathbf{v}_2 = (-0.1667\mathbf{n}_1 + 0.525\mathbf{n}_2)p/M \qquad \omega_{23} = -0.577 p/M$$
$$\mathbf{v}_3 = (-0.1667\mathbf{n}_1 - 0.0523\,\mathbf{n}_2)p/M \qquad \omega_{34} = 0.104 p/M$$
$$\mathbf{v}_4 = (-0.0764\mathbf{n}_1 + 0.0\,\mathbf{n}_2)p/M$$

This relative velocity field is illustrated in Fig. 3(b). The result of the normal impulse acting on a symmetric hexagonal ring is to develop a heart shaped configuration. Johnson (1991) speculated that a normal impulse acting on a flexible circular ring (negligible bending stiffness) would cause the ring to move towards a heart shaped configuration. The present geometrically crude approximation for a flexible ring indicates that his conjecture is correct.

The previous problem involved an impulsive load acting on a multi-bar mechanism of rigid elements. A somewhat different problem arises in a system where two 'rigid' bodies collide; i.e. they come together with some relative velocity at a contact point C. In this case the active impulse at C is unknown; this impulse must be obtained from kinematic and energetic considerations.

The more complex problem of impact at the tip of a double compound pendulum was proposed by Kane & Levinson (1985). The double pendulum is swinging when the tip strikes against a rough inelastic half-space. This problem has generated renewed interest in analytical methods for representing impact with friction (see Hurmuzlu & Marghitu, (1994) or Pereira & Nikravesh, (1996).

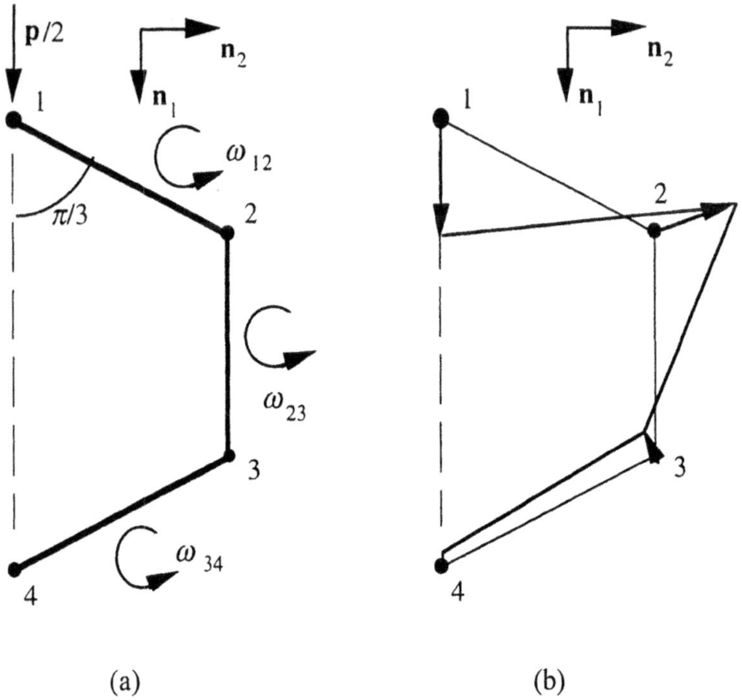

Figure 3. (a) Initial configuration and (b) changes in relative velocities of joints in pin-jointed hexagon subject to normal impulse at a joint.

Example 2: Double pendulum colliding against rough, inelastic half-space.

Two identical uniform rods OB and BC are joined at ends B by a frictionless joint in order to form a double pendulum; the other end of OB is suspended from a frictionless hinge at O as shown in Fig. 4. When the free end C of rod BC strikes against a rough half-space, the rods have angles of inclination from vertical denoted by θ_1 and θ_2 and angular speeds $\dot{\theta}_1$ and $\dot{\theta}_2$ respectively. Denote the coefficient of friction between C and the half-space by μ and the energetic coefficient of restitution at the same location by e_*. Assume the motion is planar.

Solution

After defining generalized speeds, $\dot{q}_1 = L\dot{\theta}_1$, $\dot{q}_2 = L\dot{\theta}_2 / 2$, the kinetic energy of system can be expressed as;

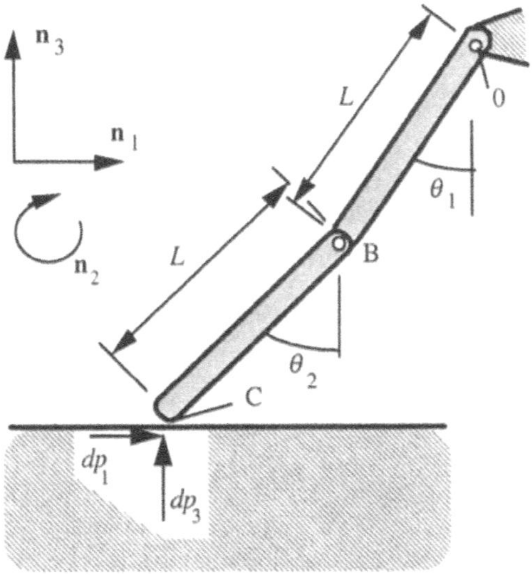

Figure 4. Double pendulum pivoting about 0 that collides at C against rough, inelastic half-space.

$$T = (ML^2/6)[4\dot{\theta}_1^2 + \dot{\theta}_2^2 + 3\dot{\theta}_1\dot{\theta}_2\cos(\theta_2 - \theta_1)]$$
$$= (M/3)[2\dot{q}_1^2 + 2\dot{q}_2^2 + 3\dot{q}_1\dot{q}_2\cos(\theta_2 - \theta_1)]$$

Velocity of contact point $\mathbf{V}_C$ using contracted notation, $s\theta_i \equiv \sin\theta_i$ and $c\theta_i \equiv \cos\theta_i$;

$$\mathbf{V}_C = L\dot{\theta}_1(-c\theta_1\mathbf{n}_1 + s\theta_1\mathbf{n}_3) + L\dot{\theta}_2(-c\theta_2\mathbf{n}_1 + s\theta_2\mathbf{n}_3)$$
$$= -(\dot{q}_1 c\theta_1 + 2\dot{q}_2 c\theta_2)\mathbf{n}_1 + (\dot{q}_1 s\theta_1 + 2\dot{q}_2 s\theta_2)\mathbf{n}_3$$

The impulse $d\mathbf{p} = dp_1\mathbf{n}_1 + dp_3\mathbf{n}_3$ results in differentials of generalized impulse $d\Pi_1 = -dp_1 c\theta_1 + dp_3 s\theta_1$ and $d\Pi_2 = -2dp_1\, c\theta_2 + 2dp_3\, s\theta_2$.

Initially slip is in direction $\mathbf{n}_1$ so Coulomb's law of friction gives $dp_1 = -\mu dp_3 \equiv -\mu dp$ and Eq. (9) results in equations of motion;

$$\begin{bmatrix} 4 & 3c(\theta_2 - \theta_1) \\ 3c(\theta_2 - \theta_1) & 4 \end{bmatrix} \begin{Bmatrix} d\dot{q}_i / dp \\ d\dot{q}_2 / dp \end{Bmatrix} = \frac{3}{M} \begin{Bmatrix} \mu c\theta_1 + s\theta_1 \\ 2\mu c\theta_2 + 2s\theta_2 \end{Bmatrix} \equiv \begin{Bmatrix} b_1 \\ b_2 \end{Bmatrix}$$

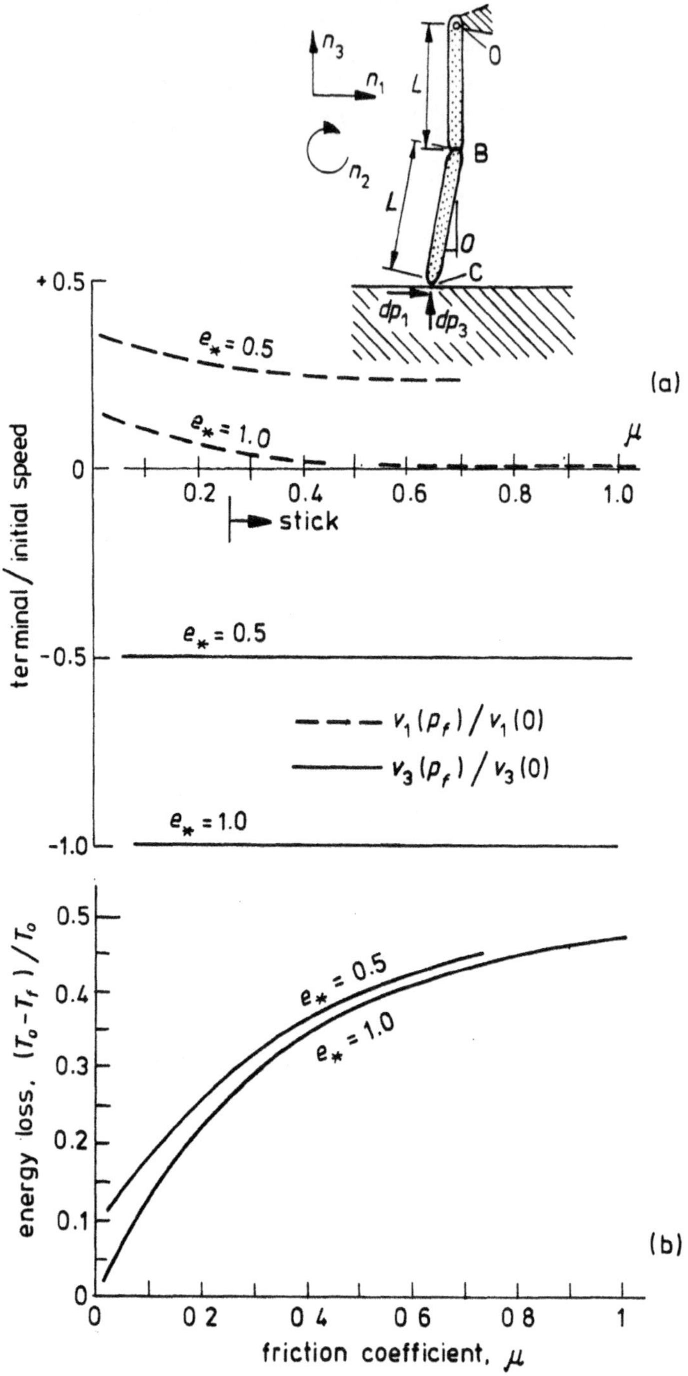

Figure 5. Effect of friction coefficient μ on (a) components of terminal velocity at C and (b) energy loss $(T_0 - T_f)/T_0$. Double pendulum striking half-space in configuration $\theta_1 = 0$, $\theta_2 = \pi/15$; initial speeds $\dot{\theta}_1(0) = -0.1\,\mathrm{rad/s}$, $\dot{\theta}_2(0) = -0.2\,\mathrm{rad/s}$.

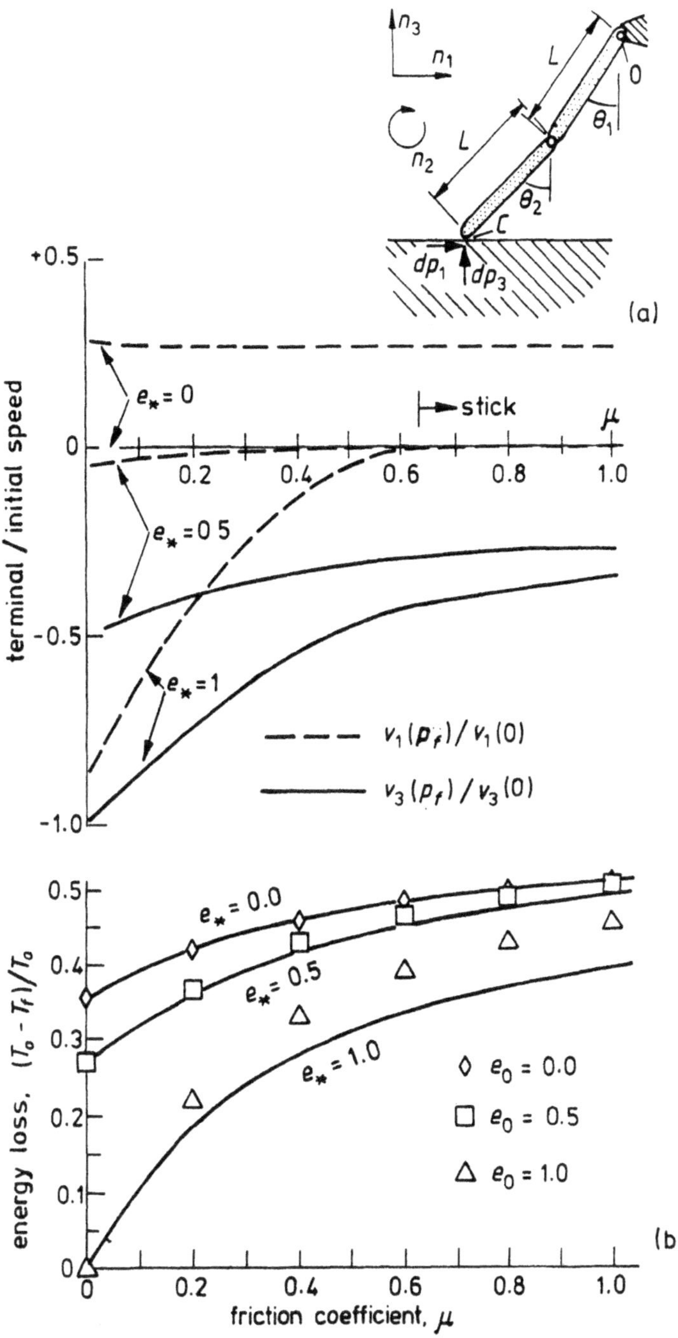

Figure 6. Effect of friction coefficient μ on (a) components of terminal velocity at C and (b) energy loss $(T_0 - T_f)/T_0$. Double pendulum striking half-space in configuration $\theta_1 = \pi/10$, $\theta_2 = \pi/6$; initial speeds $\dot{\theta}_1(0) = -0.1\,\text{rad/s}$, $\dot{\theta}_2(0) = -0.2\,\text{rad/s}$.

After solving for the differentials and then integrating with initial conditions $\dot{q}_i(0)$,

$$\begin{Bmatrix} \dot{q}_1 \\ \dot{q}_2 \end{Bmatrix} = \begin{Bmatrix} \dot{q}_1(0) \\ \dot{q}_2(0) \end{Bmatrix} + \frac{1}{D} \begin{Bmatrix} 4b_1 - 3b_2\cos(\theta_2 - \theta_1) \\ 4b_2 - 3b_1\cos(\theta_2 - \theta_1) \end{Bmatrix} p, \qquad p < p_s$$

$$D = 16 - 9\cos^2(\theta_2 - \theta_1)$$

where at impulse p_s the tangential speed vanishes $\mathbf{V}_C \cdot \mathbf{n}_1 = \dot{q}_1 c\theta_1 + 2\dot{q}_2 c\theta_2 = 0$.

$$p_s = \frac{-[\dot{q}_1(0)c\theta_1 + 2\dot{q}_2(0)c\theta_2]D}{[4b_1 - 3b_2 c(\theta_2 - \theta_1)]c\theta_1 + 2[4b_2 - 3b_1 c(\theta_2 - \theta_1)]c\theta_2}$$

while at impulse p_c the normal component of relative velocity vanishes, $\mathbf{V}_C \cdot \mathbf{n}_3 = \dot{q}_1 s\theta_1 + 2\dot{q}_2 s\theta_2 = 0$.

$$p_c = \frac{-[\dot{q}_1(0)s\theta_1 + 2\dot{q}_2(0)s\theta_2]D}{[4b_1 - 3b_2 c(\theta_2 - \theta_1)]s\theta_1 + 2[4b_2 - 3b_1 c(\theta_2 - \theta_1)]s\theta_2}$$

If $p_s > p_c$ the contact point C slides in the initial (positive) direction throughout the period of compression.

After initial sliding is brought to a halt, either the direction of sliding reverses or friction is large enough to cause stick. If sliding resumes it occurs in direction $-\mathbf{n}_1$ so that coefficients b_1 and b_2 transform to

$$\begin{Bmatrix} \bar{b}_1 \\ \bar{b}_2 \end{Bmatrix} \equiv \frac{3}{M} \begin{Bmatrix} -\hat{\mu}\, c\theta_1 + s\theta_1 \\ -2\hat{\mu}\, c\theta_2 + 2s\theta_2 \end{Bmatrix}$$

where $\hat{\mu} = \min(\mu, \bar{\mu})$ and $\bar{\mu}$ is the *coefficient of friction for stick* during $p_s < p < p_f$. This coefficient $\bar{\mu}$ gives the impulse that is sufficient to prevent the resumption of sliding; it depends solely on the impact configuration and is obtained from $\mathbf{V}_C \cdot \mathbf{n}_1 = \dot{q}_1 c\theta_1 + 2\dot{q}_2 c\theta_2 = 0$ with $\dot{q}_i(b_1, b_2)$,

$$\bar{\mu} = \frac{8s\theta_2 c\theta_2 + 2s\theta_1 c\theta_1 - 3s(\theta_2 + \theta_1)c(\theta_2 - \theta_1)}{8c^2\theta_2 + 2c^2\theta_1 - 6c\theta_1 c\theta_2 c(\theta_2 - \theta_1)}$$

For specific initial angular speeds, $\dot{\theta}_1 = -0.1$ rad s^{-1}, $\dot{\theta}_2 = -0.2$ rad s^{-1}, Figs. 5 and 6 show the components of terminal velocity and the energy loss for this double pendulum in two different initial configurations. These calculations were obtained with an energetic coefficient of restitution at C, $e_* = 0, 0.5, 1.0$. For the configuration of Fig. 5 where the upper arm is vertical, the direction of sliding does not change during collision so the various definitions of the coefficient of restitution all give the same result. For the configuration of Fig. 6 however, the direction of sliding reverses during collision if $e_* \geq 0.3$. Consequently, calculations of the terminal state that are based on the kinematic (Poisson) coefficient of restitution e_0 give different results from those based on the energetic coefficient of restitution e_*. These calculations show that the kinematic coefficient of restitution e_0 (Ahmed et al., 1999) overestimates the loss of energy in those collisions where the direction of slip reverses during the collision. According to Gauss's principle of least constraint, the energy loss is a minimum consistent with kinematic constraints on the system.

7. CONCLUSION

Analyses of 'rigid' body impact problems with friction and possible changes in direction of slip during contact require differential equations of relative motion in order to correctly account for changes in tangential components of contact forces. If the tangential velocity is relatively small and the impact configuration is non-collinear (or unbalanced), these tangential forces are not proportional to the change in the normal component of force. The formulation developed in this paper is obtained from the principle of virtual power; this formulation is efficient for analyzing multi-body systems with perfect or workless constraints. If the connections between bodies have compliances that are similar in magnitude to that at the impact point however, it is necessary to explicitly include these compliances in modelling the dynamic system.

References

Ahmed, S., Lankarani, H.M., & Pereira, M.F.O.S. (1999) Frictional impact analysis in open-loop multibody mechanical systems, ASME J. Mech. Design **121**, 119-127.

Batlle, J.A. (1996) Rough balanced collisions, *ASME J. Appl. Mech.* **63**, 168-172.

Bahar, L. (1994) On use of quasi-velocities in impulsive motion, *Int. J. Engr. Sci.* **32**, 1669-1686.

Calladine, C.R. (1990) Teaching of some aspects of the theory of inelastic collisions, *Int. J. Mech. Engng. Ed.* **18**(4), 301-310.

Drazetic, P., Level, P., Canaple, B. & Mongenie, P. (1996) Impact on planar kinematic chain of rigid bodies: application to movement of anthropomorphic dummy in crash, *Int. J. Impact Engng.* **18**(5), 505-516.

Hurmuzlu, Y. & Marghitu, D. (1994) Rigid body collisions of planar kinematic chains with multiple contact points, *Int. J. Robotics Research* **13**(1), 82-92.

Johnson, W. (1991) Simple model of jack-knifing of a train of coaches and Samuel Vince (1749-1821) *Int. J. Mech Engng. Ed.* **19** (3), 159-170.

Kane, T. & Levinson, D.A. (1985) *Dynamics: Theory and Application*, McGraw-Hill, New York.

Pereira, M.S. & Nikravesh, P. (1996) Impact dynamics of systems with frictional contact using joint coordinates and canonical equations of motion, *Nonlinear Dynamics* **9**, 53-71.

Souchet, R. (1993) Analytical dynamics of rigid body impulsive motions. *Int. J. Engng. Sci.* **31**, 85-92.

Stronge, W.J. (1990) Rigid body collisions with friction, *Proc. Roy. Soc. Lond.* A**431**, 169-181.

Synge, J.L. & Griffith, B.A. (1959) *Principles of Mechanics*. McGraw-Hill, New York.

Wittenburg, J. (1977) *Dynamics of System of Rigid Bodies*, Teubner, Stuttgart.

Zhao, W. (1999) Kinetostatics and analysis methods for the impact problem, *Eur. J. Mech. A/Solids* **18**, 319-329.

RIGIDITY AND STABILITY OF PRESTRESSED INFINITESIMAL MECHANISMS

T. Tarnai and J. Szabó
Department of Structural Mechanics, Budapest University of Technology and Economics
Budapest, Műegyetem rkp. 3. , H-1521 Hungary
tarnai@ ep-mech.me.bme.hu

Abstract This paper is concerned with both statically and kinematically indeterminate bar-and-joint assemblies. It investigates whether the kinematic indeterminacy appears in the form of a first-order or a higher-order infinitesimal mechanism. It is shown that the method by Calladine and Pellegrino and the method by Kuznetsov can be synthesized in a unified approach based on the Hellinger-Reissner principle. In the case of a single state of self-stress, examples are presented.

Keywords: bar-and-joint assembly, infinitesimal mechanism, self-stress, rigidity, stability

1. INTRODUCTION

Consider a bar-and-joint assembly which is both statically and kinematically indeterminate. Such an assembly can have one or more states of self-stress and one or more inextensional infinitesimal or finite mechanisms. If pre-stress is introduced into this assembly, the question arises whether pre-stress can make the assembly rigid. This question is interesting in particular for assemblies having fewer bars than necessary to be simply stiff (Calladine, 1978). The answer to this question is in connection with the order of infinitesimal mechanisms.

Koiter (1984) has defined "*an infinitesimal mechanism of the first order by its property that any infinitesimal displacement of the mechanism is accompanied by second order elongations of at least some of the bars. An infinitesimal mechanism is called of second (or higher) order, if there exists an infinitesimal motion such that no bar undergoes an elongation of lower than the third (or higher) order.*" From the point of view of the stiffening effect of pre-stress, first-order infinitesimal mechanisms play an important

H.R. Drew and S. Pellegrino (eds.),
New Approaches to Structural Mechanics, Shells and Biological Structures, 245–256.

role. Thus, a basic problem is to distinguish between first-order infinitesimal mechanisms and higher-order infinitesimal or finite mechanisms.

Using different techniques, many authors such as Kötter (1912), Levi-Civita and Amaldi (1930), Besseling (1979), Tanaka and Hangai (1986), Tarnai (1989), Salerno (1992), Connelly and Whiteley (1996), Vassart *et al.* (2000) have contributed to this field. Answering the question of first-order infinitesimal mechanisms (Tarnai, 1980; Koiter, 1984), by using matrix methods, two rival theories have been proposed. In the first, Pellegrino and Calladine (1986) showed that positive stiffness produced by self-stress is the criterion. In the second, Kuznetsov (1989), using kinematic constraint functions and virtual displacements, emphasized stability as a criterion. In this paper we will show that these two different approaches can be synthesized in a unified theory based on the Hellinger-Reissner principle. We will deal only with assemblies having a single state of self-stress.

2. POSITIVITY OF WORK OF THE PRODUCT FORCE

Let an unloaded assembly be characterized by Q_i ($i = 1,\ldots,m$) joint coordinates and l_k ($k = 1,\ldots,n$) bar lengths. Let us write the equilibrium equations of the joints in terms of coordinate differences and tension coefficients Λ_k ($\Lambda_k = S_k/l_k$ where S_k is the force in the kth bar). The coefficient matrix of this set of homogeneous equations is the $m \times n$ equilibrium matrix $\mathbf{J}^{\mathrm{T}}$ whose rank is r. The assembly has $s = n - r$ independent states of self-stress with n-dimensional tension coefficient vectors $\mathbf{\Lambda}_1,\ldots,\mathbf{\Lambda}_s$ (in our case $s = 1$), and $p = m - r$ independent inextensional mechanisms with m-dimensional displacement vectors $\mathbf{d}_1,\ldots,\mathbf{d}_p$.

Pellegrino and Calladine (1986) set up an algorithm by which they determined the so-called product forces, which arise at the joints of the pre-stressed assembly when the mechanisms are excited. If the assembly is under a state of self-stress $\mathbf{\Lambda}_1$ and the mechanisms are excited with displacement vectors $\mathbf{d}_1,\ldots,\mathbf{d}_p$, then the respective product forces can be formed as m-dimensional vectors $\mathbf{p}_1,\ldots,\mathbf{p}_p$. An $m \times m$ matrix is composed of the independent columns of $\mathbf{J}^{\mathrm{T}}$ and the product force vectors. If this matrix has full rank, then the infinitesimal mechanism is of the first order. They remark that this matrix test should be supplemented by a check that the scalar product of all linear combinations of vectors $\mathbf{d}_1,\ldots,\mathbf{d}_p$ with the corresponding product forces are positive. This shows that activation of the pre-stressed mechanism requires positive work.

Later, Calladine and Pellegrino (1991) replaced this requirement by the following one. Compose the $m \times p$ matrices $\mathbf{D} = [\mathbf{d}_1 \ldots \mathbf{d}_p]$ and $\mathbf{P} = [\mathbf{p}_1 \ldots \mathbf{p}_p]$, and then form the $p \times p$ matrix $\mathbf{D}^{\mathrm{T}} \bullet \mathbf{P}$. We have an infinitesimal mechanism of the first order if $\mathbf{D}^{\mathrm{T}} \bullet \mathbf{P}$ is a positive definite (more correctly, sign definite) matrix.

3. STABILITY OF PRE-STRESS

Following Levi-Civita and Amaldi (1930), Kuznetsov (1989) characterized the assembly with a compatible set of constraint equations of bars:

$$F_k(Q_1,\ldots,Q_i,\ldots,Q_m) = 0, \quad k = 1, 2,\ldots, n. \tag{1}$$

In actual 3-dimensional applications, the constraint functions have the form

$$F_k = (Q_{i1} - Q_{j1})^2 + (Q_{i2} - Q_{j2})^2 + (Q_{i3} - Q_{j3})^2 - l_k^2\,. \tag{2}$$

Power series expansion of F_k at the solution point $Q_i = Q_i^\circ$, $i = 1, 2,\ldots, m$, results in the constraint variation

$$f_k = \sum_i \left.\frac{\partial F_k}{\partial Q_i}\right|_\circ dQ_i + \frac{1}{2!}\sum_{i,j} \left.\frac{\partial^2 F_k}{\partial Q_i \partial Q_j}\right|_\circ dQ_i dQ_j + \ldots = 0\,. \tag{3}$$

The linear form in (3) is a linearized constraint function, which for $k = 1, 2,\ldots, n$ results in the set of linearized compatibility equations

$$\sum \left.\frac{\partial F_k}{\partial Q_i}\right|_\circ dQ_i = 0, \quad k = 1, 2,\ldots, n \tag{4}$$

whose coefficient matrix is the $n\times m$ Jacobian matrix of the constraint functions, that is,

$$\mathbf{J} = \left[\frac{\partial F_k}{\partial Q_i}\right]$$

where the matrix is represented by a general element. Since dQ_i, $i = 1, 2,\ldots, m$, can be considered as virtual displacement components, from (4), the theorem of virtual work yields homogeneous equilibrium equations for the constraint reactions Λ_k (which in practice are usually tension coefficients):

$$\sum \Lambda_k \left.\frac{\partial F_k}{\partial Q_i}\right|_\circ = 0, \quad i = 1, 2,\ldots, m \tag{5}$$

where the coefficient matrix is $\mathbf{J}^{\mathrm{T}}$. Matrix $\mathbf{J}$ has rank r. From (4) and (5) it follows that the assembly has $p = m - r$ independent inextensional mechanisms, and $s = n - r$ independent states of self-stress (in our case, $s = 1$). Maxwell (1864) anticipated that stiffness could be produced for assemblies having fewer bars than necessary, if one or more bars have maximum or minimum length. Levi-Civita and Amaldi (1930) considered the linear combination of f_k ($k = 1, 2,\ldots, m$) as an implicit function of bar lengths, and sought its maximum or minimum. Kuznetsov (1989) introduced constraint reactions as constants in the linear combination

$$\sum_k \Lambda_k \left(\sum_i \frac{\partial F_k}{\partial Q_i}\bigg|_{\circ} dQ_i + \frac{1}{2!}\sum_{i,j} \frac{\partial^2 F_k}{\partial Q_i \partial Q_j}\bigg|_{\circ} dQ_i dQ_j + \ldots \right). \tag{6}$$

In this way, the constraint reactions play the role of Lagrange multipliers of the constraint functions. The necessary condition for an extremum is satisfied because the linear terms in (6) are missing due to (4). The quadratic form

$$\sum_k \Lambda_k \left(\frac{1}{2!}\sum_{i,j} \frac{\partial^2 F_k}{\partial Q_i \partial Q_j}\bigg|_{\circ} dQ_i dQ_j \right) \tag{7}$$

needs to be examined. By substituting the solutions of equations (4) and (5) into (7), a quadratic form depending on p variables is obtained. It can be formulated also by means of a $p \times p$ symmetric matrix $\mathbf{H}_p$. If the quadratic form is sign definite (indefinite) then the assembly is an infinitesimal mechanism (finite mechanism). Since this quadratic form is related to the stability of equilibrium, it gives information about the stability of self-stress of the assembly as well.

Later Kuznetsov (1991a) formulated his criterion by using the $(m+n)\times(m+n)$ matrix

$$\begin{bmatrix} \mathbf{H} & \mathbf{J}^{\mathrm{T}} \\ \mathbf{J} & 0 \end{bmatrix},$$

where $\mathbf{H}$ is the Hessian matrix of $\Sigma\Lambda_k F_k$, to which the quadratic form (7) is associated. Additionally, Kuznetsov (1991b) showed the equivalence of the different criteria given by him.

4. THE HELLINGER-REISSNER PRINCIPLE

Although the two methods in Sections 2 and 3 have some common points, they follow two different lines; so, it is not at all obvious whether the two methods are equivalent. In Section 2, it is not easy to see that the product of two rectangular matrices $\mathbf{Q}$ and $\mathbf{P}$ results in a symmetrical quadratic matrix. In Section 3, stability of self-stress is not immediately understandable, because the investigation is based on virtual work and not on potential energy. Therefore, it will be useful to consider a different approach to the problem. In the following analysis, we use potential energy (Tarnai, 1988). Its actual form comes from the Hellinger-Reissner principle

(Washizu, 1982), where the total potential energy is considered with auxiliary conditions related to constraints for the bars. That is, constraint functions with Lagrange multipliers are added to the total potential energy.

4.1 Assembly with Elastic Bars

Consider an assembly composed of elastic bars and frictionless joints, and loaded at its joints. Let P_i and Q_i $(i = 1, 2,..., m)$ be load components and node coordinates, $\widetilde{F}_k$ and Λ_k $(k = 1, 2,..., n)$ be kinematical constraint functions of bars and the respective Lagrange multipliers. The constraint functions are given in the form:

$$\widetilde{F}_k = \sqrt{(Q_{i1} - Q_{j1})^2 + (Q_{i2} - Q_{j2})^2 + (Q_{i3} - Q_{j3})^2} - l_k - e_k - t_k \tag{8}$$

where l_k, e_k and t_k denote the length, the elastic elongation and a given elongation of the kth bar, respectively. If EA_k denotes the tensional rigidity of the kth bar, then the Hellinger-Reissner functional of the assembly is:

$$\Pi_R = -\sum_{i=1}^{m} P_i(Q_i - Q_i^\circ) + \frac{1}{2}\sum_{k=1}^{n} \frac{EA_k}{l_k} e_k^2 + \sum_{k=1}^{n} \Lambda_k \widetilde{F}_k \, . \tag{9}$$

where Q_i° $(i = 1, 2,..., m)$ are reference coordinates (node coordinates of the investigated configuration). In this context, the Lagrange multiplier Λ_k means force in the kth bar (Lanczos, 1986). Therefore, the elongations can be expressed as $e_k = \Lambda_k l_k/(EA_k)$. Substituting this into (8) and (9) we obtain

$$\Pi_R = -\sum_{i=1}^{m} P_i(Q_i - Q_i^\circ) - \frac{1}{2}\sum_{k=1}^{n} \Lambda_k^2 \frac{l_k}{EA_k} + \sum_{k=1}^{n} \Lambda_k F_k \tag{10}$$

where

$$F_k = \sqrt{(Q_{i1} - Q_{j1})^2 + (Q_{i2} - Q_{j2})^2 + (Q_{i3} - Q_{j3})^2} - l_k - t_k \, . \tag{11}$$

Stationarity of the Hellinger-Reissner functional Π_R requires that its first variation be equal to zero:

$$\delta\Pi_R = \left[\left[\frac{\partial \Pi_R}{\partial Q_i}\right] \left[\frac{\partial \Pi_R}{\partial \Lambda_k}\right]\right] \bullet \begin{bmatrix} [\delta Q_i] \\ [\delta \Lambda_k] \end{bmatrix} = 0 \tag{12}$$

that yields the equilibrium equations

$$\sum_{k=1}^{n} \Lambda_k \frac{\partial F_k}{\partial Q_i} - P_i = 0 \, , \quad i = 1, 2,..., m \tag{13}$$

and compatibility equations

$$\tilde{F}_k = 0, \quad k = 1, 2, \ldots, n, \tag{14}$$

whose linearized form is:

$$\sum \frac{\partial F_k}{\partial Q_i} dQ_i - \frac{\Lambda_k l_k}{EA_k} - t_k = 0, \quad k = 1, 2, \ldots, n. \tag{15}$$

Now consider the second variation of Π_R:

$$\delta^2 \Pi_R = \begin{bmatrix} [\delta Q_i]^{\mathrm{T}} & [\delta \Lambda_k]^{\mathrm{T}} \end{bmatrix} \bullet \begin{bmatrix} \left[\sum_{l=1}^{n} \Lambda_l \frac{\partial^2 F_l}{\partial Q_i \partial Q_j} \right] & \left[\frac{\partial F_i}{\partial Q_k} \right] \\ \left[\frac{\partial F_k}{\partial Q_i} \right] & \left[-\frac{l_k}{EA_k} \right] \end{bmatrix} \bullet \begin{bmatrix} [\delta Q_j] \\ [\delta \Lambda_k] \end{bmatrix}. \tag{16}$$

We take all matrices and equations at the given, compatible configuration of the assembly. Let us use the second variation (16) to formulate the incremental technique, and let us apply the following notation

$\mathbf{H} = \left[\sum_{l=1}^{n} \Lambda_l \frac{\partial^2 F_l}{\partial Q_i \partial Q_j} \right]_{m \times m}$ the Hessian matrix of $\sum_{k=1}^{n} \Lambda_k F_k$

$\mathbf{J} = \left[\frac{\partial F_k}{\partial Q_i} \right]_{n \times m}$ the Jacobian matrix of F_k $(k = 1, 2, \ldots, n)$

$\mathbf{F} = \left[-\frac{l_k}{EA_k} \right]_{n \times n}$ the negative of the flexibility matrix

$d\mathbf{Q} = [\delta Q_i]_{m \times 1}$ increment of nodal displacement

$d\Lambda = [\delta \Lambda_k]_{n \times 1}$ increment of bar forces

$d\mathbf{P} = [\delta P_j]_{m \times 1}$ increment of nodal load

$d\mathbf{t} = [\delta t_k]_{n \times 1}$ increment of kinematic load

Let us write (16) in the form

$$\delta^2 \Pi_R = \begin{bmatrix} d\mathbf{Q}^{\mathrm{T}} & d\Lambda^{\mathrm{T}} \end{bmatrix} \bullet \begin{bmatrix} d\mathbf{P} \\ d\mathbf{t} \end{bmatrix}$$

that defines the nodal load increments and the kinematic load increments:

$$\begin{bmatrix} \mathbf{H} & \mathbf{J}^{\mathrm{T}} \\ \mathbf{J} & \mathbf{F} \end{bmatrix} \bullet \begin{bmatrix} d\mathbf{Q} \\ d\Lambda \end{bmatrix} = \begin{bmatrix} d\mathbf{P} \\ dt \end{bmatrix}. \qquad (17)$$

Eq. (17) is the well-known differential equation of change of state of bar structures (Szabó and Roller, 1971). With negative sign, **H** is known as complementary stiffness matrix, **J** is called the compatibility matrix, and $\mathbf{J}^{\mathrm{T}}$ is called the equilibrium matrix.

4.2 Assembly with Rigid Bars

Consider now an unloaded assembly composed of rigid bars. Suppose that there are no prescribed elongations of the bars, that is the kinematic loads are zero. In this case, the relationships in Subsection 4.1 are significantly reduced.

The constraint functions take the form

$$F_k = \sqrt{(Q_{i1} - Q_{j1})^2 + (Q_{i2} - Q_{j2})^2 + (Q_{i3} - Q_{j3})^2} - l_k, \quad k = 1, 2, ..., n, \qquad (18)$$

and the Hellinger-Reissner functional reads

$$\Pi_R = \sum_{k=1}^{n} \Lambda_k F_k \,. \qquad (19)$$

The equilibrium equations (13) and the linear compatibility equations (15), set up for a compatible configuration of the assembly, become homogeneous. In matrix-vector form they are:

$$\mathbf{J}^{\mathrm{T}} \bullet \mathbf{\Lambda} = \mathbf{0}, \qquad (20)$$

$$\mathbf{J} \bullet d\mathbf{Q} = \mathbf{0}. \qquad (21)$$

The assembly that is being investigated is both statically and kinematically indeterminate. Therefore, **J** has rank r, and the assembly has $s = n - r$ independent states of self-stress (in our case $s = 1$) and $p = m - r$ independent mechanisms. That is, the equilibrium equation (20) has one linearly independent solution $\mathbf{\Lambda}_1$ and the compatibility equation (21) has p linearly independent solutions $d\mathbf{Q}_1,...,d\mathbf{Q}_p$ as displacements of the p independent mechanisms.

Consider the quadratic form defined by the second variation of the energy functional Π_R, which is:

$$\begin{bmatrix} d\mathbf{Q}^{\mathrm{T}} & d\Lambda^{\mathrm{T}} \end{bmatrix} \bullet \begin{bmatrix} \mathbf{H} & \mathbf{J}^{\mathrm{T}} \\ \mathbf{J} & \mathbf{0} \end{bmatrix} \bullet \begin{bmatrix} d\mathbf{Q} \\ d\Lambda \end{bmatrix}. \qquad (22)$$

If this quadratic form is positive definite then the potential energy has a minimum, and so the state of self-stress is stable. Here, however, the components of the displacement vector and those of the force vector are not varied arbitrarily. Variation can be made only for the self-stress and the independent displacements of the mechanisms. If the parameter of the self-stress is varied, then we always have a force system satisfying (20). Therefore, the cross terms in (22) are equal to zero. Therefore, it is enough to investigate the positive definiteness of the quadratic form $d\mathbf{Q}^{\mathrm{T}}\bullet\mathbf{H}\bullet d\mathbf{Q}$. Here, vectors $d\mathbf{Q}$ should lie in the subspace spanned by the independent solutions $d\mathbf{Q}_1,\ldots,d\mathbf{Q}_p$. This requirement reduces an m-dimensional problem to a p-dimensional one. If we substitute the solutions of (21) into $d\mathbf{Q}^{\mathrm{T}}\bullet\mathbf{H}\bullet d\mathbf{Q}$, then a quadratic form depending on p variables is obtained, that can be formulated also by means of a p-order reduced Hessian matrix $\mathbf{H}_p$. This is the same as Kuznetsov's analysis of the quadratic form (7). So, Kuznetsov indeed investigated self-stress stability.

An alternative way of reduction can be the following. Activation of a pre-stressed mechanism produces forces at the joints. According to (17), if $d\Lambda = 0$, this force vector is $d\mathbf{P} = \mathbf{H}\bullet d\mathbf{Q}$, called product force by Pellegrino and Calladine (1986). They made a sign check for the scalar product $d\mathbf{Q}^{\mathrm{T}}\bullet d\mathbf{P}$. If $d\mathbf{Q}_1,\ldots, d\mathbf{Q}_p$ are the independent motions of the p mechanisms, then they required that the scalar products of all linear combinations of these vectors with the corresponding product forces be positive. Let $d\mathbf{Q} = c_1 d\mathbf{Q}_1 + \ldots + c_p d\mathbf{Q}_p$ be a general linear combination, then the product $d\mathbf{Q}^{\mathrm{T}}\bullet d\mathbf{P} = d\mathbf{Q}^{\mathrm{T}}\bullet\mathbf{H}\bullet d\mathbf{Q}$ takes the form

$$(c_1 d\mathbf{Q}_1 + \ldots + c_p d\mathbf{Q}_p)^{\mathrm{T}} \bullet \mathbf{H} \bullet (c_1 d\mathbf{Q}_1 + \ldots + c_p d\mathbf{Q}_p) =$$

$$\begin{bmatrix} c_1 & \cdots & c_p \end{bmatrix} \bullet \begin{bmatrix} d\mathbf{Q}_1^{\mathrm{T}} \bullet \mathbf{H} \bullet d\mathbf{Q}_1 & \cdots & d\mathbf{Q}_1^{\mathrm{T}} \bullet \mathbf{H} \bullet d\mathbf{Q}_p \\ \vdots & & \vdots \\ d\mathbf{Q}_p^{\mathrm{T}} \bullet \mathbf{H} \bullet d\mathbf{Q}_1 & \cdots & d\mathbf{Q}_p^{\mathrm{T}} \bullet \mathbf{H} \bullet d\mathbf{Q}_p \end{bmatrix} \bullet \begin{bmatrix} c_1 \\ \vdots \\ c_p \end{bmatrix}.$$

This is a quadratic form in which the matrix can be produced as the product $\mathbf{D}^{\mathrm{T}}\bullet\mathbf{P}$ where $\mathbf{D} = [d\mathbf{Q}_1 \ldots d\mathbf{Q}_p]$ and $\mathbf{P} = [d\mathbf{P}_1 \ldots d\mathbf{P}_p]$. That was the point suggested by Calladine and Pellegrino (1991). The p-order matrix $\mathbf{D}^{\mathrm{T}}\bullet\mathbf{P}$ is obviously symmetric, because $\mathbf{P} = \mathbf{H}\bullet\mathbf{D}$ and $\mathbf{H}$ is symmetric. Sign definiteness of the matrix $\mathbf{D}^{\mathrm{T}}\bullet\mathbf{P}$, considered by Calladine and Pellegrino, is equivalent to the sign definiteness of Kuznetsov's matrix $\mathbf{H}_p$ (not necessarily $\mathbf{H}_p = \mathbf{D}^{\mathrm{T}}\bullet\mathbf{P}$).

If we use the constraint functions (2) instead of (18) then, for the same pre-stress, the Hessian matrix $\mathbf{H}$ for (2) is in general different from that for (18). From the results of numerical examples, however, we have found that the two different types of constraint functions result in the same matrix $\mathbf{D}^{\mathrm{T}}\bullet\mathbf{P}$.

5. EXAMPLES

Example 1

The space frame in Fig. 1 is the edge network of a unit cube with space diagonals, connected by six bars to a rigid foundation. This assembly has been studied previously by Kötter (1912), Calladine (1978), Connelly (1982), Pellegrino (1986). It has 8 non-foundation spherical joints, that is, $m = 3\times 8 = 24$ node coordinates, and $n = 22$ bars. The number of bars is 2 less than that required by Maxwell's (1864) rule for simple stiffness.

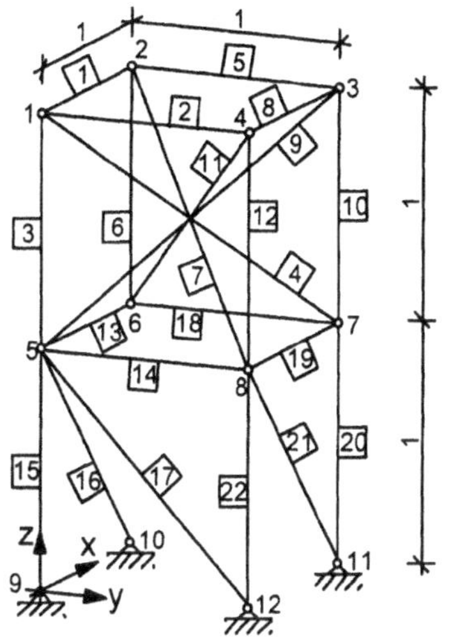

Figure 1. A frame having the form of a cube with space diagonals

In a Cartesian coordinate system x, y, z with axes parallel to the edges of the cube, we set up the 24×22 equilibrium matrix $\mathbf{J}^{\mathrm{T}}$ using constraint functions (18). It has rank $r = 21$, so the assembly has $s = 1$ state of self-stress and $p = 3$ independent inextensional mechanisms. Assuming that one of the space diagonals is in compression with force magnitude 1, results in the following vector $\mathbf{\Lambda}$ of forces in bars (in which the elements are arranged in accordance with the serial numbers of bars):

$$\Lambda^{\mathrm{T}} = \begin{bmatrix} a & a & a & -1 & a & a & -1 & a & -1 & a & -1 & a & a & a & 0 & 0 & 0 & a & a & 0 & 0 & 0 \end{bmatrix}$$

where $a = 1/\sqrt{3}$. Matrix $\mathbf{D}$ is formed from the three vectors of the linearly independent displacements of the 8 nodes. Each vector is built up from x-, y-, z-displacement-component triplets of nodes, where the triplets are arranged in accordance with the serial numbers of nodes:

$$\mathbf{D}^{\mathrm{T}}=\left[\begin{array}{ccc|ccc|ccc|ccc|ccc|ccc|ccc|ccc} 1 & 0 & 0 & 1 & 1 & 0 & -1 & 1 & 0 & -1 & 0 & 0 & 0 & 0 & 0 & 0 & 1 & 0 & 0 & 1 & 0 & 0 & 0 & 0 \\ -1 & 0 & 0 & -1 & 1 & 2 & -1 & 1 & 0 & -1 & 0 & 0 & 0 & 0 & 0 & 0 & -1 & 2 & 0 & -1 & 0 & 0 & 0 & 0 \\ -1 & 2 & 0 & -1 & 1 & 2 & -1 & 1 & 0 & -1 & 2 & 0 & 0 & 0 & 0 & 0 & 1 & 2 & 0 & 1 & 0 & 0 & 0 & 0 \end{array}\right].$$

Then, from the constraint functions (18) and the vector $\mathbf{\Lambda}$ of bar forces, we produced the 24×24 Hessian matrix $\mathbf{H}$. The product $\mathbf{H{\bullet}D}$ resulted in the following matrix $\mathbf{P}$ of product forces (with $b = 2a$):

$$\mathbf{P}^{\mathrm{T}}=b\left[\begin{array}{ccc|ccc|ccc|ccc|ccc|ccc|ccc|ccc} 1 & 0 & 0 & 1 & 0 & 0 & -1 & 0 & 0 & -1 & 0 & 0 & -1 & 0 & 0 & -1 & 0 & 0 & 1 & 0 & 0 & 1 & 0 & 0 \\ 0 & -1 & -1 & 0 & 1 & 1 & 0 & 1 & -1 & 0 & -1 & 1 & 0 & 1 & -1 & 0 & -1 & 1 & 0 & -1 & -1 & 0 & 1 & 1 \\ 0 & 1 & -1 & 0 & -1 & 1 & 0 & -1 & -1 & 0 & 1 & 1 & 0 & -1 & -1 & 0 & 1 & 1 & 0 & 1 & -1 & 0 & -1 & 1 \end{array}\right]$$

which produced the reduced Hessian matrix

$$\mathbf{D}^{\mathrm{T}} \bullet \mathbf{P} = 4b\begin{bmatrix} 1 & 0 & 0 \\ 0 & 2 & 0 \\ 0 & 0 & 2 \end{bmatrix}.$$

Since matrix $\mathbf{D}^{\mathrm{T}}{\bullet}\mathbf{P}$ is positive definite, the assembly in Fig. 1 is a first-order infinitesimal mechanism.

Example 2

Consider the planar bar-and-joint assembly in Fig. 2, introduced by Connelly and Servatius (1994). It has n = 15 bars and 8 non-foundation joints, that is, m = 16 node coordinates. It has one bar less than required for simple stiffness.

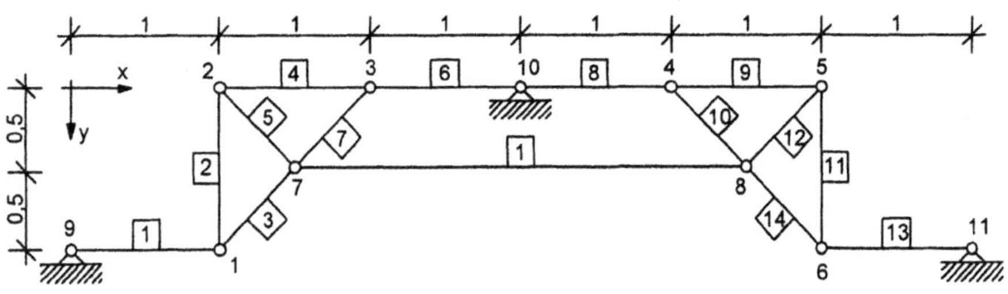

Figure 2. The mechanism by Connelly and Servatius

The constraint functions (18) result in a 16×15 equilibrium matrix $\mathbf{J}^{\mathrm{T}}$ with rank r = 14. Thus, the assembly has s = 1 state of self-stress and p = 2 independent inextensional mechanisms with bar forces $\mathbf{\Lambda}$ and node displacements $\mathbf{D}$:

$$\Lambda^{\mathrm{T}} = \begin{bmatrix} -c & c & -1 & c & -1 & c & 0 & c & c & 0 & c & -1 & -c & -1 & -2c \end{bmatrix},$$

$$\mathbf{D}^{\mathrm{T}} = \left[\begin{array}{cc|cc|cc|cc|cc|cc|cc|cc} 0 & 1 & 0 & 1 & 0 & 1 & 0 & 0 & 0 & 0 & 0 & 0 & 0 & 1 & 0 & 0 \\ 0 & 0 & 0 & 0 & 0 & 0 & 0 & 1 & 0 & 1 & 0 & 1 & 0 & 0 & 0 & 1 \end{array}\right]$$

where $c = 1/\sqrt{2}$. Constraint functions (18) and bar forces $\mathbf{\Lambda}$ give the 16×16 Hessian matrix **H**, from which, by **H•D**, product forces are obtained:

$$\mathbf{P}^{\mathrm{T}} = \left[\begin{array}{cc|cc|cc|cc|cc|cc|cc|cc} 0 & -c & 0 & 0 & 0 & c & 0 & 0 & 0 & 0 & 0 & 0 & 0 & -d & 0 & d \\ 0 & 0 & 0 & 0 & 0 & 0 & 0 & c & 0 & 0 & 0 & -c & 0 & d & 0 & -d \end{array}\right]$$

where $d = 0.4714045$. This yields the reduced Hessian matrix

$$\mathbf{D}^{\mathrm{T}} \bullet \mathbf{P} = d\begin{bmatrix} -1 & 1 \\ 1 & -1 \end{bmatrix}$$

which is not sign definite (its eigenvalues are: $-2d$ and 0). Therefore, the assembly is a mechanism of higher-than-first order. It has been shown that the assembly in Fig. 2 in fact is a finite mechanism (Connelly and Servatius, 1994; Gáspár and Tarnai, 1994; Tarnai and Szabó, 2000).

6. CONCLUSIONS

We have shown that the positive stiffness principle introduced by Calladine and Pellegrino is practically the same as the pre-stress stability principle considered by Kuznetsov. The results from both authors are in agreement with the results obtained from graph theory (Connelly and Whiteley, 1992, 1996).

It was found that the Hellinger-Reissner principle is a proper tool to decide whether a bar-and-joint assembly that is both statically and kinematically indeterminate forms a first-order infinitesimal mechanism. It gave a deeper insight into the problem of infinitesimal mechanisms in the case where the assembly had a single state of self-stress. It would be of interest to know how the Hellinger-Reissner principle works for multi-state-of-self-stress infinitesimal mechanisms.

Acknowledgements. The research reported here was partially supported by OTKA grant No. T031931. We thank Prof. Zs. Gáspár for comments and Zs. Hortobágyi for help with the figures.

References

Besseling, J.F. (1979) Finite element methods, in *Trends in Solid Mechanics* (ed. Besseling, J.F. and van der Heiden, M.A.), Noordoff, Leyden, pp. 53-78.

Calladine, C.R. (1978) Buckminster Fuller's "tensegrity" structures and Clerk Maxwell's rules for the construction of stiff frames, *Int. J. Solids Structures* **14**, 161-172.

Calladine, C.R. and Pellegrino, S. (1991) First-order infinitesimal mechanisms, *Int. J. Solids Structures* **27**, 505-515.

Connelly, R. (1982) Rigidity and energy, *Invent. Math.* **66**, 11-33.

Connelly, R. and Servatius, H. (1994) Higher-order rigidity – What is the proper definition? *Discrete Comput. Geom.* **11**, 193-200.

Connelly, R. and Whiteley, W. (1992) Remarks on Underconstrained Structural Systems, *Preprint*, Department of Mathematics, Cornell University.

Connelly, R. and Whiteley, W. (1996) Second-order rigidity and prestress stability for tensegrity frameworks, *SIAM J. Discrete Math.* **9**, 453-491.

Gáspár, Zs. and Tarnai, T. (1994) Finite mechanisms have no higher-order rigidity. *Acta Technica Acad. Sci. Hung.* **106**, 119-125.

Koiter, W.T. (1984) On Tarnai's conjecture with reference to both statically and kinematically indeterminate structures, *Lab. Report No.* **788**, Delft Univ. Tech.

Kötter, E. (1912) Über die Möglichkeit, *n* Punkte in der Ebene oder im Raume durch weniger als $2n - 3$ oder $3n - 6$ Stäbe von ganz unveränderlicher Länge unverschieblich miteinander zu verbinden, in *Festschrift Heinrich Müller-Breslau*, Kröner, Leipzig, pp. 61-80.

Kuznetsov, E.N. (1989) On immobile kinematic chains and a fallacious matrix analysis, *Trans. ASME, J. Appl. Mech.* **56**, 222-224.

Kuznetsov, E.N. (1991a) Systems with infinitesimal mobility: Part I – Matrix analysis and first-order infinitesimal mobility, *Trans ASME, J. Appl. Mech.* **58**, 513-519.

Kuznetsov, E.N. (1991b) *Underconstrained Structural Systems*, Springer, New York.

Lanczos, C. (1986) *The Variational Principles of Mechanics*, 4th edn., Dover, New York.

Levi-Civita, T. and Amaldi, U. (1930) *Lezioni di Meccanica Razionale*, Vol. 2, Zanichelli, Bologna.

Maxwell, J.C. (1864) On the calculation of the equilibrium and stiffness of frames, *Phil. Mag.* Ser. 4, **27**, 294-299 (also in *The Scientific Papers of J.C. Maxwell*, Vol. 1, University Press, Cambridge, 1980, pp. 598-604).

Pellegrino, S. (1986) *Mechanics of Kinematically Indeterminate Structures*, PhD Dissertation, University of Cambridge.

Pellegrino, S. and Calladine, C.R. (1986) Matrix analysis of statically and kinematically indeterminate frameworks, *International Journal of Solids and Structures* **22**, 409-428.

Salerno, G. (1992) How to recognize the order of infinitesimal mechanisms: a numerical approach, *Int. J. Num. Meth. Engng* **35**, 1351-1395.

Szabó, J. and Roller, B. (1971) *Theory and analysis of bar structures* (in Hungarian), Műszaki Könyvkiadó, Budapest (in German: Akadémiai Kiadó, Budapest, 1978).

Tanaka, H. and Hangai, Y. (1986) Rigid body displacement and stabilization conditions of unstable truss structures, in *Shells, Membranes and Space Frames*, Proc. IASS Symp., Osaka (ed. Heki, K.), Elsevier Sci., Amsterdam, pp. 55-62.

Tarnai, T. (1980) Problems concerning spherical polyhedra and structural rigidity, *Structural Topology* **4**, 61-66.

Tarnai, T. (1988) Bifurcations and singularities of compatibility, in *Abstracts, XVIIth Int. Cong. of Theor. Appl. Mech.*, Grenoble, p. 179.

Tarnai, T. (1989) Higher-order infinitesimal mechanisms, *Acta Technica Acad. Sci. Hung.* **102**, 363-378.

Tarnai, T. and Szabó, J. (2000) On the exact equation of inextensional, kinematically indeterminate assemblies. *Computers and Structures* **75**, 145-155.

Vassart, N., Laporte, R. and Motro, R. (2000) Determination of mechanism's order for kinematically and statically indetermined systems. *International Journal of Solids and Structures* **37**, 3807-3839.

Washizu, K. (1982) *Variational Methods in Elasticity and Plasticity*, 3rd edn, Pergamon, Oxford.

AMPLITUDE OF WRINKLES IN THIN MEMBRANES

Y.W. Wong and S. Pellegrino
Department of Engineering, University of Cambridge
Trumpington Street, Cambridge, CB2 1PZ, U.K.
pellegrino@eng.cam.ac.uk

Abstract This paper presents a theory for predicting the wavelength and amplitude of the wrinkles that form in a membrane in pure shear. Predictions from this theory are compared to experimental measurements and a finite element simulation and are found to be very accurate.

Keywords: Membrane, Wrinkles, Gossamer spacecraft

1. INTRODUCTION

A number of space missions will require high-accuracy membrane structures acting as sunshields, solar arrays and solar sails, radars, and inflatable antennas. These prestressed membrane structures will have to remain partially wrinkled in their operational configuration. Removing the wrinkles would require a biaxially tensile stress state, thus significantly increasing the loads transmitted to the edge deployable structure that supports the membrane.

Wrinkles can reduce the performance of reflectors and sunshields, or cause difficulties in maneuvering solar sails. Hence, it is now important to predict details of the wrinkles, such as wavelength and amplitude, in order to determine if the membrane structures meet the requirements of each particular application.

The wrinkling of membranes has attracted much interest in the past, starting from the observation that the thin-walled shear web of a beam can carry loads well above the initial buckling value; which prompted the development of tension field theory by Wagner (1929). Simpler formulations of this theory by Reissner (1938), Stein and Hedgepeth (1961), and Mansfield (1969, 1989) made it possible to find solutions to problems in which the tension-lines are non parallel. Further generalizations were

H.R. Drew and S. Pellegrino (eds.),
New Approaches to Structural Mechanics, Shells and Biological Structures, 257–270.

proposed by Pipkin (1986) and, more recently, Epstein and Forcinito (2001).

A premise common to all of these formulations, and also to the accompanying numerical solutions that have been pursued alongside, is that the membrane is modelled as a *no-compression, two-dimensional continuum with negligible bending stiffness.* See Jenkins and Leonard (1991) for an extensive list of references and Adler (2000) for a more recent perspective; see also Liu et al. (2000) for a numerical study in which wrinkle amplitudes were predicted. Hence, it is assumed in effect that an infinite number of wrinkles of infinitesimally small amplitude will form. This is not, of course, what actually happens and it is well known that, although the stress fields produced by these theories are essentially correct, the out-of-plane displacements are significantly different from those observed in practice. This is not a problem in the design of membranes that are purely load-carrying elements, but it is not acceptable in the present context.

The only study, to our knowledge, which has taken into account the role played by the membrane bending stiffness is Rimrott and Cverko's (1986) study of the "hanging blanket" problem. These authors considered a vertical membrane subject to in-plane gravity loading and supported on two level points, a problem for which the tension field solution in terms of cosine-shaped wrinkles had already been determined by Mansfield (1981). However, it can be readily verified —e.g. with a bath towel— that in practice there is only a small number of tension strips, which are in equilibrium in a strongly three-dimensional configuration. Rimrott and Cverko showed that the introduction of a critical, *compressive stress perpendicular to the wrinkles* produced trends matching the experimentally observed variation of the number of tension strips with the density of the membrane.

In this paper we return to the simple configuration considered in the earliest study of wrinkled membranes, namely the rectangular membrane in shear. For this problem we compute a series of reference solutions, by means of non-linear finite element analysis, where the membrane is accurately modelled with a fine mesh of thin-shell elements. These solutions provide an insight into the stress distribution in the wrinkled membrane, on which we then base a simple analytical solution inspired by Rimrott and Cverko's approach to the hanging blanket. Thus, we obtain predictions for the wrinkle wavelength and amplitude as functions of the angle of shear and of the geometric and material properties of the membrane.

This theory has three key components. First, it is assumed that the tension lines, i.e. the direction of the wrinkles, are determined by a standard tension field approach. Second, the (small) compressive stress in the direction perpendicular to the wrinkles is set equal to a critical buckling stress which is a known function of the wavelength of the wrinkles. Third, the stress components along and perpendicular to each wrinkle —modelled as a doubly-curved shell— are in equilibrium in the out-of-plane direction. Results from this model are compared to both direct experimental measurements and finite element simulations, and are found to be in good agreement.

2. PRELIMINARY CONSIDERATIONS

Figure 1 shows a photograph of an aluminized Kapton membrane, of length $L = 380$ mm, height $H = 128$ mm, and thickness $t = 25$ μm, that is attached to rigid blocks along the top and bottom edges and is free along the sides. A relative horizontal displacement $\delta = 3$ mm has been imposed between the two blocks without changing their distance, thus imposing a geometric shear strain $\gamma_g = 2.3\%$ on the membrane.

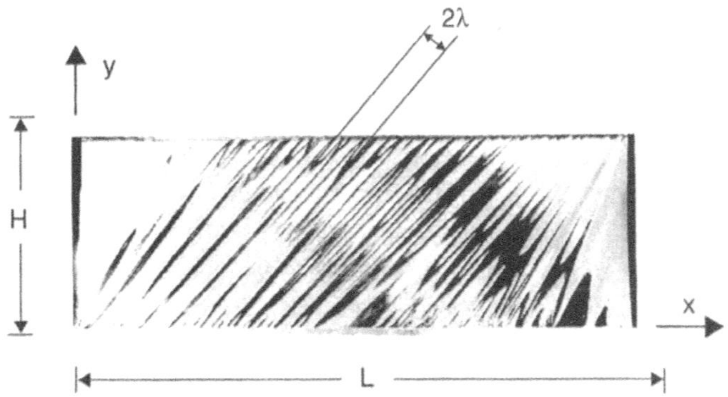

Figure 1. Sheared Kapton sheet, showing extensive wrinkling.

The photograph shows that, instead of deforming purely in two-dimensional shear, the membrane has buckled out of plane, forming a series of approximately straight wrinkles. Excluding the wrinkles originating near the bottom-left and top-right corners, the wrinkles in the central part of the membrane are parallel and uniformly spaced. The wrinkle wavelength, i.e. the distance between two consecutive crests, or troughs, is denoted by 2λ.

We have carried out a finite-element simulation of the shearing process, using the finite-element package ABAQUS (2000) where the membrane was modelled by four-node thin shell elements (type S4R5). The

Table 1. Kapton membrane properties.

Length, L (mm)	380
Height, H (mm)	128
Thickness, t (mm)	0.025
Young's Modulus, E (N/mm^2)	3530
Poisson's ratio, ν	0.3

numerical singularities that occur at the beginning of the analysis, due to the small bending stiffness of the shell elements, were avoided by applying a small, uniform initial stress in the y-direction to the membrane and by carrying out an initial, geometrically non-linear step. This was followed by a prediction of the possible wrinkling modes, made by running an eigenvalue buckling step. These modes were superposed and amplified, and applied as initial geometric imperfections, before running the final step. The final analysis of the wrinkled membrane was facilitated by using the automated "stabilize" option in ABAQUS, which introduces dynamic pseudo-inertia and pseudo-damping forces when an instability is detected, thus triggering a pseudo-time integration of the dynamic snap of the system. The default ABAQUS values for the amount of damping that is introduced in the calculation were tuned down to achieve the greatest possible accuracy.

A single type of membrane, made of Kapton, was used for all the studies presented in this paper (except that photographed in Figure 1); its properties are given in Table 1.

Figure 2 is a plot of the principal stresses in a heavily wrinkled membrane, predicted by ABAQUS. For each element we have plotted two vectors aligned with the principal stress directions and of length proportional to the principal stresses. Note that: (i) only the major stress σ_1 is visible in the plot, as the minor stress σ_2 is very small; (ii) in the central part of the membrane the major principal stress is inclined at 45°; and (iii) there are stress concentrations by a factor of up to 2.5 in the top-left and bottom-right corners.

The same results are also found from tension field theory (Mansfield, 1969) and it is interesting to compare the loci of the crests and troughs predicted with ABAQUS, Figure 3(a), with the tension lines for a semi-infinite strip. Thus, Figure 3(b) shows the tension lines at 5° intervals as well as the first six wrinkles (a-f) predicted by ABAQUS. It can be seen that wrinkles (a-d), forming respectively at 63°, 56°, 51°, and 46° to the horizontal, practically coincide with the corresponding tension lines.

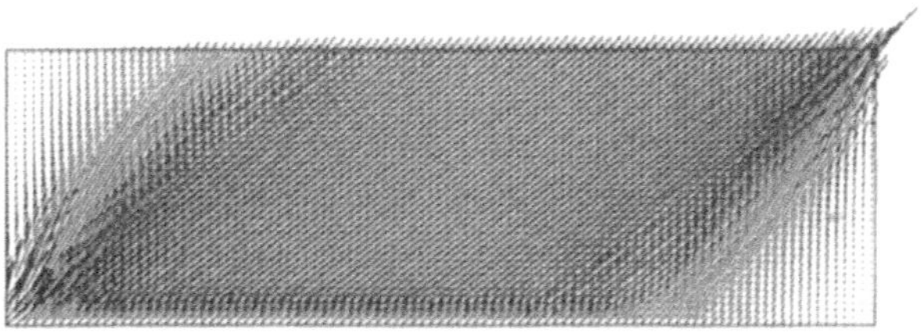

Figure 2. Stress trajectories from ABAQUS solution.

The 45° tension line is at infinity; but the actual wrinkles are inclined at practically 45° at only a short distance from the corner.

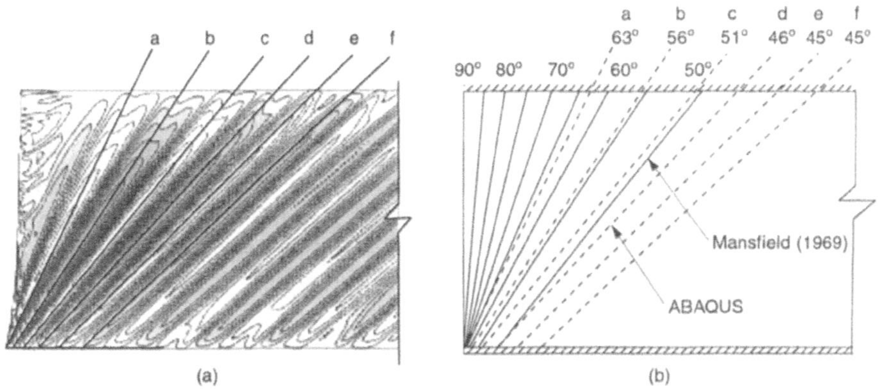

Figure 3. Comparison between (a) ABAQUS solution and (b) tension field.

Figure 4 is a plot of the distribution of the principal mid-surface stresses across the centre of the membrane, for two different shear displacements. Near the free edges, σ_1 rapidly increases to an approximately uniform, positive value. σ_2 is also essentially uniform in the central region and, although it appears to be zero from this plot, in fact it is negative, albeit small.

The value of the minor principal mid-surface stress at a representative point in the middle of the membrane has been plotted in Figure 5 as a function of the shear displacement δ. The behaviour shown here is characteristic of the post-wrinkled response of the membrane, and it provides a series of important clues. The fact that the mid-surface stress —which of course does not include bending effects— is compressive in

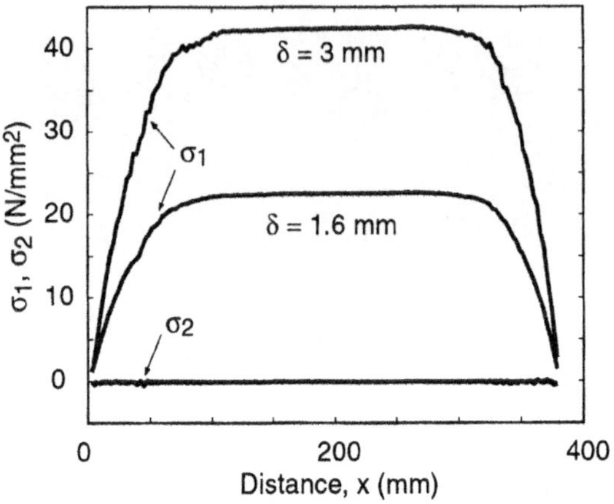

Figure 4. Mid-surface principal stresses through horizontal section at mid-height.

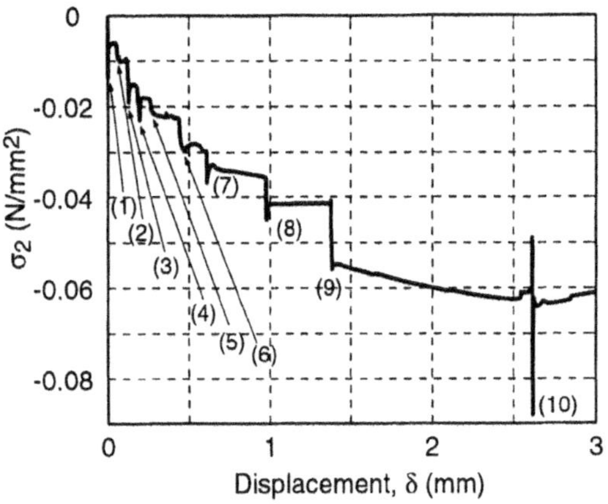

Figure 5. Variation of minor principal stress with shear displacement, at the centre of the membrane.

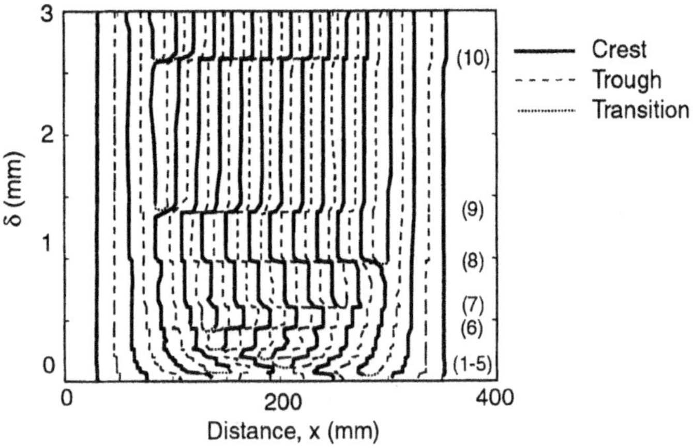

Figure 6. Trajectories of maximum and minimum out-of-plane displacement at mid-height, for increasing shear displacement.

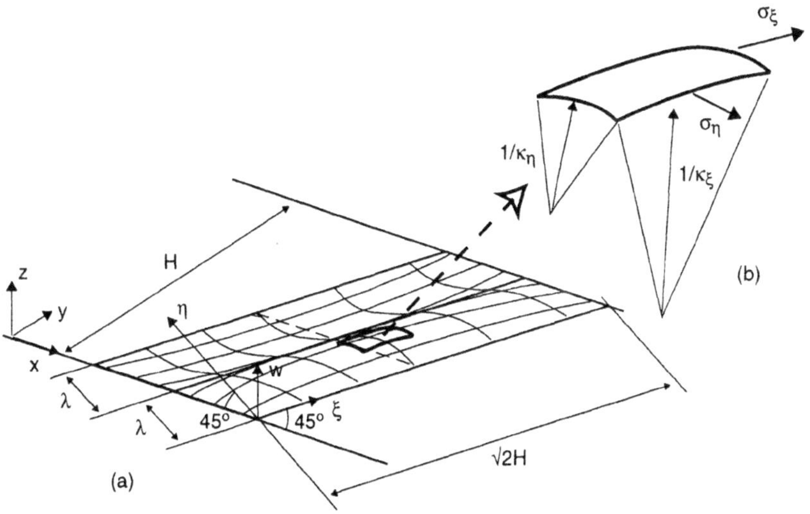

Figure 7. (a) Perspective view of a single wrinkle and (b) infinitesimal free body.

the direction perpendicular to the wrinkles is particularly significant. It is also significant that its value increases with the shear displacement.

This variation has an underlying, square-root type trend, superposed onto a series of sharp peaks and jumps, which correspond to sudden increases in the numbers of wrinkles and a rearrangement of the shape of the central part of the membrane. The complete sequence of the formation of wrinkles at mid-height is shown in Figure 6, where each bifurcation point —and the corresponding "transition" in the plot— corresponds to a jump in Figure 5. Note that the side wrinkles do not move, as they are "pinned" by the corner supports.

3. ANALYTICAL MODEL

Consider one of the wrinkles in the central, uniformly wrinkled part of the membrane. From standard tension field theory, its direction is found to be at 45° to the edges of the membrane, as shown in Figure 7(a).

The initially flat membrane has deformed into a doubly-curved shape lying alternately above and below the original, xy plane of the membrane. This wrinkled surface intersects the xy plane at a regular half-wave distance λ and it is observed in practice that these intersections are along straight lines.

A simple mode-shape describing the wrinkled surface can be readily set up using the coordinate system ξ, η shown in Figure 7(a). Note that ξ is parallel to the wrinkle direction, and η is perpendicular to it. The out-of-plane deflection is

$$w = A \sin \frac{\pi(\xi + \eta)}{\sqrt{2}H} \sin \frac{\pi\eta}{\lambda} \tag{1}$$

and the in-plane deflection is small. Here, A is the amplitude and λ is the half-wavelength, both unknown. Since the wrinkles are long and narrow, $\eta << \xi$ apart from a small region near the origin. Hence, the mode-shape can be simplified to

$$w = A \sin \frac{\pi\xi}{\sqrt{2}H} \sin \frac{\pi\eta}{\lambda} \tag{2}$$

The compressive stress σ_η acting perpendicular to the wrinkles is set equal to the stress required to buckle a simply-supported, infinitely wide plate of length λ which is loaded length-wise. Hence, by Euler's formula (Calladine, 1983)

$$\sigma_\eta = -\frac{\pi^2 E t^2}{12(1-\nu^2)\lambda^2} \tag{3}$$

This is very small in comparison with the longitudinal (tensile) stress σ_ξ and hence the stress-strain relationship for a uniaxial stress state can be assumed to be valid

$$\sigma_\xi = E\epsilon_\xi \tag{4}$$

Since the wrinkles are at 45° to the x and y axes

$$\epsilon_\xi = \gamma/2 \tag{5}$$

where $\gamma = \delta/H$ if the strain/stress variation across the wrinkle is neglected. Substituting Equation 5 into Equation 4

$$\sigma_\xi = E\gamma/2 \tag{6}$$

For equilibrium of the membrane in the normal direction

$$\sigma_\xi \kappa_\xi + \sigma_\eta \kappa_\eta = 0 \tag{7}$$

where the principal curvatures are given by

$$\kappa_\xi = -\frac{\partial^2 w}{\partial \xi^2} = \frac{\pi^2 A}{2H^2} \sin \frac{\pi\xi}{\sqrt{2}H} \sin \frac{\pi\eta}{\lambda} \tag{8}$$

$$\kappa_\eta = -\frac{\partial^2 w}{\partial \eta^2} = \frac{\pi^2 A}{\lambda^2} \sin \frac{\pi\xi}{\sqrt{2}H} \sin \frac{\pi\eta}{\lambda} \tag{9}$$

Substituting Equations 3, 6, 8 and 9 into Equation 7, simplifying and rearranging we obtain the following expression for the half-wavelength λ

$$\lambda = \frac{\sqrt{\pi}}{[3(1-\nu^2)]^{\frac{1}{4}}} \frac{\sqrt{Ht}}{\gamma^{\frac{1}{4}}} \tag{10}$$

Next, we obtain an expression for the amplitude A of the wrinkle. This must be such that the contraction of the membrane in the η-direction

$$\epsilon_\eta = -\gamma/2 \tag{11}$$

matches the sum of the strain in the membrane

$$\epsilon_{\eta M} = -\frac{\nu}{E}\sigma_\xi \tag{12}$$

plus the geometric strain produced by the out-of-plane deformation of the membrane, which at mid-height can be shown to have the expression

$$\epsilon_{\eta G} \approx -\frac{\pi^2 A^2}{4\lambda^2} \tag{13}$$

Hence,

$$-\frac{\gamma}{2} = -\frac{\nu}{E}\sigma_\xi - \frac{\pi^2 A^2}{4\lambda^2} \tag{14}$$

Substituting Equation 6 and solving for A gives

$$A = \frac{\sqrt{2(1-\nu)\gamma}}{\pi}\lambda \tag{15}$$

from which λ can be eliminated using Equation 10, to find

$$A = \frac{\sqrt{2(1-\nu)}}{\sqrt{\pi}[3(1-\nu^2)]^{\frac{1}{4}}}\sqrt{Ht}\,\gamma^{\frac{1}{4}} \tag{16}$$

3.1 Alternative Formulation

An alternative approach is to find λ by minimizing the strain energy in the wrinkled membrane, modelled as a thin plate stretched in the ξ-direction and wrinkled in the η-direction.

The general expression for the bending strain energy per unit area of an initially flat plate that is bent into a cylindrical shape is

$$U_b = \frac{Et^3}{24(1-\nu^2)}\kappa_\eta^2 \tag{17}$$

As κ_η is not constant, see Equation 9, the average strain energy, $\overline{U_b}$, per unit area is obtained from

$$\overline{U_b} = \frac{Et^3}{24(1-\nu^2)}\left(\frac{1}{\sqrt{2}H\lambda}\int_0^{\sqrt{2}H}\int_0^{\lambda}\kappa_\eta^2 d\xi d\eta\right) = \frac{Et^3}{24(1-\nu^2)}\frac{\pi^4 A^2}{4\lambda^4} \tag{18}$$

The strain energy for stretching in the η-direction is negligible, hence a general expression for the stretching strain energy per unit area is given by

$$U_s = \frac{1}{2} E t \epsilon_\xi^2 \tag{19}$$

Here, ϵ_ξ is the sum of the strain due to the in-plane shear, Equation 5, plus that due to the out-of-plane deflection due to wrinkling

$$\epsilon_\xi = \frac{\gamma}{2} + \frac{1}{2}\left(\frac{\partial w}{\partial \xi}\right)^2 \tag{20}$$

Thus, ϵ_ξ also is not constant, and hence the average strain energy of stretching, $\overline{U_s}$, over a wrinkle is given by

$$\begin{aligned} \overline{U_s} &= \frac{1}{2} E t \left(\frac{1}{\sqrt{2} H \lambda} \int_0^{\sqrt{2}H} \int_0^{\lambda} \epsilon_\xi^2 d\xi d\eta \right) \\ &\approx \frac{Et}{2}\frac{\gamma^2}{4} + \frac{Et}{2}\frac{\pi^2 A^2 \gamma}{16 H^2} \end{aligned} \tag{21}$$

The first of these terms is independent of the wrinkle amplitude A, and hence will not be carried through to the next stage of the analysis.

Hence, the total mean strain energy of wrinkling per unit area (neglecting the term without A) is given by

$$\overline{U} = \overline{U_b} + \overline{U_s} = \frac{Et^3}{24(1-\nu^2)}\frac{\pi^4 A^2}{4\lambda^4} + \frac{Et}{2}\frac{\gamma^2}{4} + \frac{Et}{2}\frac{\pi^2 A^2 \gamma}{16 H^2} \tag{22}$$

Next, we substitute Equation 15 —thus expressing A in terms of λ— and obtain

$$\overline{U} = \frac{(1-\nu)Et\gamma}{2}\left(\frac{\pi^2 t^2}{24(1-\nu^2)\lambda^2} + \frac{\gamma \lambda^2}{8H^2}\right) \tag{23}$$

which is minimised by differentiating with respect to λ. This gives

$$\lambda^4 = \frac{\pi^2 t^2 H^2}{3(1-\nu^2)\gamma} \tag{24}$$

Equations 10 and 24 are equivalent, which shows that the simple equilibrium formulation with an assumed stress σ_η captures the same effects as the energy formulation.

4. VALIDATION OF THEORY

We have tested the accuracy of our theory against (i) an ABAQUS simulation of the wrinkling process, and (ii) an experiment on a 0.025 mm

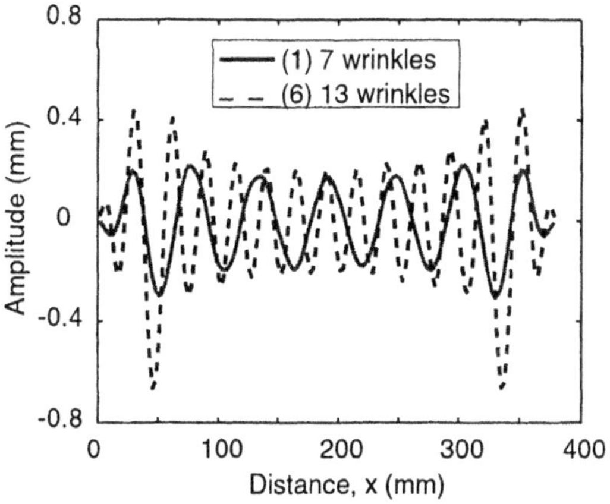

Figure 8. Wrinkle profiles from ABAQUS.

thick Kapton sheet mounted in a steel shear frame. The out-of-plane deflections of this membrane were measured with a non-contact device for monotonically increasing shear angles, both positive and negative. The smallest value of γ for which measurements could be obtained was 0.00195, i.e. $\delta = 0.25$ mm.

For any given shear displacement, both the simulation and the experiment produce a characteristic pattern with two large side-wrinkles —corresponding to wrinkle b in Figure 3— and a series of wrinkles of approximately uniform amplitude in between. Figure 8 shows the cross-section of the membrane at mid-height, for the configurations immediately following jumps 1 and 6, defined in Figures 5 and 6. Note that the distance between successive peaks in Figure 8 has to be divided by $\sqrt{2}$ to obtain the wrinkle wavelength. The side-wrinkles, which are only beginning to show after jump 1, are clearly visible after jump 6.

After carrying out the simulation and experiment, the average wrinkle wavelengths and amplitudes were determined after excluding the region of the edge wrinkles. Figures 9 and 10 show plots of these two parameters, together with the predictions from our theory; note that the experimental values plotted were obtained by averaging the values measured for positive and negative shear angles. Finally, Figure 11 compares the average, mid-plane, minor principal stress at mid-height of the membrane, predicted by ABAQUS, with predictions obtained by substituting Equation 10 into Equation 3.

It can be seen that the wavelengths obtained from ABAQUS and the experiment differ by less than 2 mm (note that the numbers on the ordinate of Figure 9 have to be doubled), i.e. about 12% of the values.

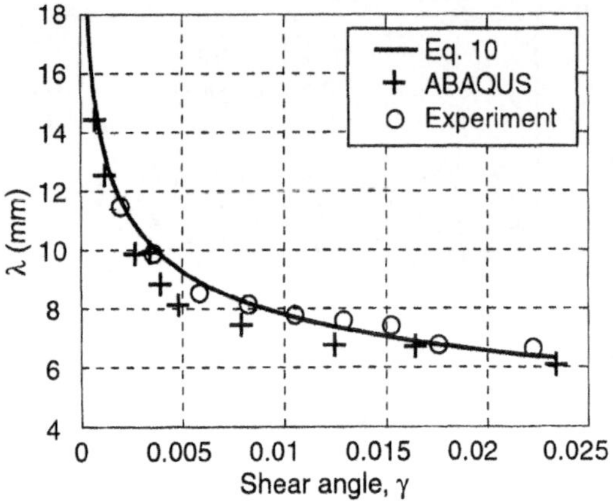

Figure 9. Comparison of wrinkle half-wavelengths.

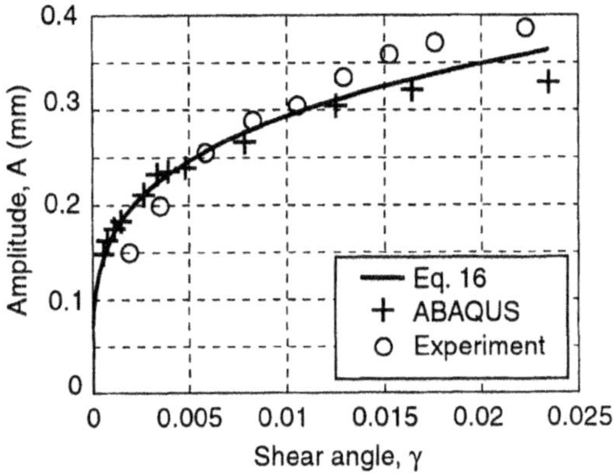

Figure 10. Comparison of wrinkle amplitudes.

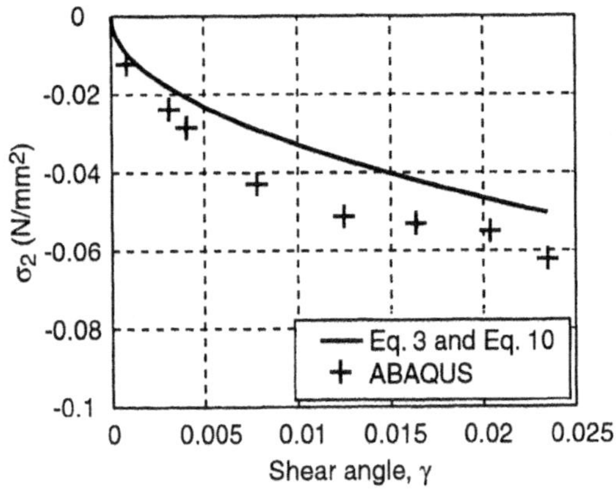

Figure 11. Comparison of compressive stress at mid-height.

The theoretically predicted amplitudes follow both the experiment and the ABAQUS predictions very closely up to $\gamma \approx 1\%$; the Kapton remains elastic within this range. For larger values of γ some plastic deformation occurs; it is remarkable, and perhaps fortuitous, that in Figure 10 the theoretical predictions fall right in the middle of the range of the ABAQUS predictions and experimental measurements.

5. DISCUSSION AND CONCLUSION

It has been shown that the wrinkle half-wavelength, λ, of a membrane in simple shear is accurately predicted by Equation 10 and the amplitude, A, by Equation 16. Thus, λ is *inversely proportional* and A is *directly proportional* to the fourth root of the shear angle; and both of them are *directly proportional* to the square roots of the width and thickness of the membrane. They are both independent of the Young's Modulus. Therefore, in the sheared-panel problem the wrinkle amplitude varies quite slowly with the angle of shear.

The equilibrium formulation of our theory can be readily extended to membranes with other boundary conditions; it will be interesting to see if these conclusions remain valid. However, it should be noted that the assumed mode-shape allows the required geometric strain only in the middle of the membrane. A more accurate description of the wrinkled surface may be required, in future.

To conclude, we note that the spread in our numerical and experimental results is due, in part at least, to the fact that the number of wrinkles corresponding to a particular shear angle depends on the exact number of bifurcations that have taken place, e.g. in Figure 6, which is sensitive to small imperfections. The irregularities in the variation of the number of wrinkles with the shear angle originate from the difference between the response predicted upon increasing or decreasing the angle of shear, coupled with the fact that —instead of using a path-following algorithm— we have used displacement incrementation.

Acknowledgments

We thank Chris Calladine for many useful discussions and for suggesting the approach presented in Section 3.1. Financial support for Y.W.W. from the Cambridge Commonwealth Trust is gratefully acknowledged.

References

ABAQUS (2000) *Theory and User's Manual Version 6.1.* Hibbit Karlsson & Sorensen Inc.

Adler, A. (2000) *Finite Element Approaches for Static and Dynamic Analysis of Partially Wrinkled Membrane Structures.* Ph.D. Dissertation, The University of Colorado at Boulder.

Calladine, C.R. (1983) *Theory of Shell Structures*, Cambridge University Press, Cambridge.

Epstein, M. and Forcinito, M.A. (2001) Anisotropic membrane wrinkling: theory and analysis. *Int. J. Solids Structures* **38**, 5253-5272.

Jenkins, C. and Leonard, J.W. (1991) Nonlinear dynamic response of membranes: State of the art. *ASME Appl. Mech. Reviews*, **44**, 319–328.

Liu, X., Jenkins, C.H. and Schur, W.W. (2000) Fine scale analysis of wrinkled membranes. *Int. J. Computational Eng. Sci.* **1**, 281–298.

Mansfield, E.H. (1969) Tension field theory a new approach which shows its duality with inextensional theory. *Proc. 12th Int. Cong. Appl. Mech.*, 305–320.

Mansfield, E.H. (1981) Gravity-induced wrinkle lines in vertical membranes. *Proc. Roy. Soc. Lond. A* **375**, 307-325.

Mansfield, E.H. (1989) *The Bending and Stretching of Plates*, second edition, Cambridge University Press, Cambridge.

Pipkin, A.C. (1986) The relaxed energy density for isotropic elastic membranes. *IMA J. Appl. Math.*, **36** 85–99.

Reissner, E. (1938) On tension field theory. *Proc. 5th Int. Cong. Appl. Mech.*, 88–92.

Rimrott, F.P.J. and Cvercko, M. (1985) Wrinkling in thin plates due to in-plane body forces. In: *Inelastic Behaviour of Plates and Shells*, (Edited by L. Bevilacqua, R. Feijoo and R. Valid), 19–48. Springer-Verlag.

Stein, M. and Hedgepeth, J.M. (1961). *Analysis of Partly Wrinkled Membranes.* NASA Langley Research Center, NASA TN D-813.

Wagner, H. (1929) Flat sheet metal girder with very thin metal web, *Zeitschrift für Flugtechnik Motorlurftschiffahrt.* **20**, 200–207, 227–233, 256–262, 279-284.

Wong, Y.W. (2000) *Analysis of Wrinkle Patterns in Prestressed Membrane Structures.* M.Phil. Dissertation, University of Cambridge.

ON A HIGH-FIDELITY HIERARCHICAL APPROACH TO BUCKLING LOAD CALCULATIONS

Johann Arbocz
Delft University of Technology, The Netherlands

James H. Starnes
NASA Langley Research Center, Hampton, VA USA

Abstract A step towards developing a new design philosophy for buckling-critical thin-walled shells is described. This new design philosophy is intended to advance thin-walled shell design technology from the traditional empirical design approach used today towards a science-based design-technology approach. This science-based design-technology approach is based on the hierarchical "high-fidelity-analysis" approach to buckling load calculations proposed by Arbocz et al. [1] where the uncertainties involved in a design are simulated by refined and accurate numerical methods. This hierarchical analysis approach includes three levels of shell buckling analyses that range from classical linear bifurcation buckling analysis (Level-1 analysis) to nonlinear finite element collapse analysis (Level-3 analysis) to provide an accurate prediction of the critical buckling load of a given shell structure. The critical buckling load and the estimated imperfection sensitivity of the shell are verified using a sufficiently refined finite element model with current generation two-dimensional shell analysis codes that include both geometric and material nonlinearities. The new approach is demonstrated for a quasi-isotropic composite shell [2].

Keywords: Shell, Buckling

1. INTRODUCTION

The development of future advanced space transportation systems that successfully achieve desired performance goals making full and efficient use of the technical expertise accumulated in the past 50 years or so, and combining it with the computational capability now available. The performance goals associated with weight critical space applications can only be achieved by using a "high fidelity analysis" approach, where the uncertainties involved in a design are simulated by refined and accurate numerical models. The use of "high fidelity" numerical simulations should also lead to overall cost reduction, since the analysis and design phases can be completed faster and it is only necessary to verify the reliability of the final configuration by structural testing.

H.R. Drew and S. Pellegrino (eds.),
New Approaches to Structural Mechanics, Shells and Biological Structures, 271–292.

The light-weight shell structures used in aerospace applications are often buckling critical. The buckling load calculations are usually conducted by one of the many currently available finite element based computer codes [e.g., 3,4]. In order to reduce computer execution time, buckling analyses are often done using only the small displacement stiffness matrix K_0. This approach is used, despite the fact that the "initial stability problem" so formulated can only give physically meaningful answers if the elastic solutions based on K_0 (at least approximately) are identically equal to zero [5].

When the qualitative nature of the expected behavior is completely unknown, the stability of the structure should be investigated using the full tangent stiffness matrix K_T in order to guarantee accurate and reliable buckling load and buckling mode predictions. To discover the load level at which K_T ceases to be positive definite (that is, the load level when buckling occurs), a step-by-step analysis procedure is needed.

It is imperative (though often completely neglected), that the accuracy of the discrete model used in a stability investigation should be checked with available analytical or semi-analytical results. This step is part of a study needed to establish the dependence of the buckling load predictions on the mesh distribution used. Furthermore, if one carries out imperfection sensitivity investigations which involve an extension of the solution into the postbuckling response region, further mesh refinement may be needed since the wavelength of the dominant large deformation pattern may often decrease significantly (e.g. Byskov [6]).

Finally, when conducting a shell stability analysis, it is especially important to be aware of the possible detrimental effects of a series of factors that have been investigated extensively in the late 1960s and the early 1970s. Thus, for an accurate and reliable prediction of the critical buckling load of a real structure, one must not only account for the influence of initial imperfections [e.g., 7,8] and of the boundary conditions [e.g., 9], but one must also consider the effects of stiffener and load eccentricity [e.g.,10] and the prebuckling deformations caused by the edge restraints [e.g., 11,12].

The results of a quasi-isotropic composite shell experiment conducted by Waters [2] is used to illustrate how such a hierarchical approach to buckling load calculations can be conducted. The platform for the multi-level computations, needed for an accurate prediction of the critical buckling loads and a reliable estimation of their imperfection sensitivity, is provided by the **D**elft **I**nteractive **S**hell **DE**sign **CO**de or DISDECO [13]. With this open ended, hierarchical, interactive computer code the user can access a succession of programs of increasing complexity and fidelity from an engineering workstation.

2. SOLUTION OF THE BUCKLING PROBLEM

In the following it will be shown that, with the help of DISDECO, the shell designer can study the buckling behavior of a specified shell, accurately calculate its critical buckling load, and make a reliable prediction of the expected degree of imperfection sensitivity of the critical buckling load. The proposed procedure consists of a hierarchical approach, where the analyst proceeds step-by-step from the simpler (Level-1) methods used by the early investigators to the more sophisticated analytical and numerical (Level-2 and Level-3) methods used presently.

Level-1 Perfect Shell Buckling Analysis

The geometric and material properties of the 8-ply, quasi-isotropic composite shell of Ref. [2] are listed in Table 1.

Table 1. Geometric properties of $[\pm 45/0/90]_S$ composite shell [2].

$t_{total}(=h)$	$=0.039976$	in	(= 1.01539 mm)
L	$=14.0$	in	(= 355.600 mm)
R	$=7.99945$	in	(= 203.18603 mm)
E_{11}	$=18.5111\times 10^6$	psi	$(=12.7629\times 10^4\ N/mm^2)$
E_{22}	$=1.64\times 10^6$	psi	$(=1.13074\times 10^4\ N/mm^2)$
G_{12}	$=0.8706\times 10^6$	psi	$(=6.00257\times 10^3\ N/mm^2)$
ν_{12}	$=0.300235$		

Note: Symmetrical lay-up with 8 plies of equal thicknesses (= 0.004997 in).

Assuming a perfect shell with initial imperfection amplitude equal to zero, $(\overline{W}=0)$ and the following membrane prebuckling state

$$w^{(o)} = hW_v = h\overline{A}^*_{12}\frac{\lambda}{c} \tag{1}$$

$$F^{(o)} = -\frac{Eh^2}{cR}\frac{1}{2}\lambda y^2$$

where $\lambda = \dfrac{\sigma}{\sigma_{c\ell}} = \dfrac{N_x}{N_{c\ell}}$; $\sigma_{c\ell} = \dfrac{Eh}{cR}$; $N_{c\ell} = \sigma_{c\ell}h$ and $c=\sqrt{3(1-\nu^2)}$

then the nonlinear equations governing the prebuckling state are identically satisfied and the linearized stability equations reduce to a set of equations with constant coefficients. It has been shown [14] that, by assuming an asymmetric bifurcation mode of the form

$$W^{(1)} = h \sin m\pi \frac{x}{L} \cos \frac{n}{R}(y - \tau_K x) \tag{2}$$

where

$m = k$ = number of axial half waves
$n = \ell$ = number of circumferential full waves
τ_K = Khot's skewedness parameter

The solution of the linearized stability equations can be reduced to an algebraic eigenvalue problem. The eigenvalue $\lambda_{mn\tau}$ depends on the wave numbers m and n, and on Khot's skewedness parameter τ_K, a real number. The critical load parameter λ_c is the lowest of all possible eigenvalues. Thus, finding λ_c involves not only a search over the integer valued wave numbers m and n, but the search has to be repeated over a range of possible positive and negative real numbers for τ_K. Using the Level-1 computational module AXBIF [14] in DISDECO, a search over integer valued axial half-wave numbers m and over a range of possible positive and negative real numbers τ_K yielded the lowest eigenvalues listed in Table 2 for the specified circumferential wave numbers n.

In addition to the absolute minimum of $\lambda_c^m = 0.365992$ at n = 7 there is a local minimum of $\lambda_c^m = 0.369089$ at n = 12. The results of the search for the critical (lowest) buckling load λ_c can be displayed in a contour map as shown in Fig. 1. Using the membrane prebuckling analysis results, the critical eigenvalue is

$$\lambda_c^m = 0.365992$$

with m=1 half-waves in the axial direction and n=7 full waves in the circumferential direction. To provide an overview of the distribution of eigenvalues, the values displayed in the contour plot are re-normalized, and the following re-normalized eigenvalues are plotted in Fig. 1

$$\rho_c^m = \lambda_{mn\tau} / 0.365992$$

The critical buckling load can be calculated using a simple multiplication operation

$$N_c^m = \lambda_c^m N_{c\ell} = 0.365992(-2238.325) = -819.209 \text{ lb/in}$$

Table 2. Buckling loads of a quasi-isotropic composite shell [2].
Buckling load map for the perfect shell using AXBIF [14] ($N_{c\ell}$ = - 2238.325 lb/in).

$n=4$	$\lambda_c^m=0.371613$	$(m=7, \tau_K=3.320)$
$n=5$	$\lambda_c^m=0.370302$	$(m=7, \tau_K=2.680)$
$n=6$	$\lambda_c^m=0.370029$	$(m=7, \tau_K=2.178)$
$n=7$	$\lambda_c^m=0.365992$	$(m=1, \tau_K=0.011)$
$n=8$	$\lambda_c^m=0.370131$	$(m=6, \tau_K=1.681)$
$n=9$	$\lambda_c^m=0.371980$	$(m=6, \tau_K=1.488)$
$n=10$	$\lambda_c^m=0.372880$	$(m=5, \tau_K=1.345)$
$n=11$	$\lambda_c^m=0.375261$	$(m=4, \tau_K=1.214)$
$n=12$	$\lambda_c^m=0.369089$	$(m=1, \tau_K=0.579)$
$n=13$	$\lambda_c^m=0.370076$	$(m=1, \tau_K=0.653)$
$n=14$	$\lambda_c^m=0.376748$	$(m=1, \tau_K=0.734)$

Level-2 Perfect Shell Buckling Analysis

To investigate the effects of edge constraint and of different boundary conditions on the critical buckling load of the perfect shell ($\overline{W}=0$), one has to switch to the Level-2 computational module ANILISA [15] in DISDECO. In this module the axisymmetric prebuckling state is represented by

$$W^{(o)} = hW_v + hw_o(x) \tag{3}$$
$$F^{(o)} = \frac{Eh^2}{cR}\left[-\frac{1}{2}\lambda y^2 + R^2 f_o(x)\right]$$

It has been shown [15] that with these assumptions the prebuckling problem is reduced to the solution of a single fourth order ordinary differential equation with constant coefficients. Solutions for simply supported and clamped boundary conditions have been published in the literature [16].

For anisotropic shells, the linearized stability equations admit separable solutions of the form

$$W^{(1)} = h[w_1(x)\cos n\theta + w_2(x)\sin n\theta] \tag{4}$$
$$F^{(1)} = \frac{ERh^2}{c}[f_1(x)\cos n\theta + f_2(x)\sin n\theta]$$

where $\theta = \dfrac{y}{R}$.

Using a generalization of Stodola's method [17] first published by Cohen [18] the resulting nonlinear eigenvalue problem is reduced to a sequence of linearized eigenvalue problems. The resulting ordinary differential equations are solved numerically by a technique known as "parallel shooting over N-intervals" [19]. With this approach the effect of edge restraint and the specific boundary conditions are satisfied rigorously. To find the critical load parameter λ_c, an n-search, or search over a range of values of n, is carried out. In this n-search one must be careful to find the absolute minimum and not a local minimum. As can be seen from the results presented in Table 3, the n-search using membrane prebuckling analysis results and a rigorous satisfaction of SS-3 ($N_x = v = w = M_x = 0$) boundary conditions for the stability problem now yields a local minimum of $\lambda_c = 0.364715$ at n = 7 and an absolute minimum of $\lambda_c = 0.364370$ at n = 11.

The most accurate Level-2 solutions are obtained when one employs a rigorous nonlinear prebuckling analysis. For this particular shell (see Table 3), the critical buckling loads with nonlinear prebuckling results are always lower than the corresponding results obtained using membrane prebuckling analysis results. Specifically, the local minimum of $\lambda_c^{nl} = 0.337088$ at $n = 7$ is about 8% lower, and the absolute minimum of $\lambda_c^{nl} = 0.328594$ at $n = 11$ is about 11% lower. The critical load N_c can be calculated easily by multiplying the lowest eigenvalue λ_c by the normalizing factor $N_{c\ell} = -2238.325$ lb/in yielding

$$N_c = \lambda_c N_{c\ell} = -735.500 \text{ lb/in } (n = 11)$$

The critical buckling modes using membrane and rigorous nonlinear prebuckling results are shown in Fig. 2. The solutions with nonlinear prebuckling analysis results differ significantly from the ones obtained using membrane prebuckling analysis results, especially at n = 11 where a typical edge buckling type behavior occurs.

Table 3. Buckling loads of a quasi-isotropic composite shell [2] ($N_{c\ell} = -2238.325$ lb/in)
Buckling load map for the perfect shell using ANILISA [15]
(B.C. $N_x = v = w = M_x = 0$)

Prebuckling	Membrane	Nonlinear
n = 6	$\lambda_c^m = 0.371920$	$\lambda_c^{nl} = 0.347378$
n = 7	$\lambda_c^m = 0.364715$	$\lambda_c^{nl} = 0.337088$
n = 8	$\lambda_c^m = 0.372592$	$\lambda_c^{nl} = 0.339701$
n = 9	$\lambda_c^m = 0.371460$	$\lambda_c^{nl} = 0.330957$
n = 10	$\lambda_c^m = 0.367479$	$\lambda_c^{nl} = 0.329163$
n = 11	$\lambda_c^m = 0.364370$	$\lambda_c^{nl} = 0.328594$
n = 12	$\lambda_c^m = 0.364551$	$\lambda_c^{nl} = 0.330271$
n = 13	$\lambda_c^m = 0.368528$	$\lambda_c^{nl} = 0.334044$
n = 14	$\lambda_c^m = 0.376314$	$\lambda_c^{nl} = 0.339388$

Level-3 Perfect Shell Buckling Analysis

The finite difference version [20] of the shell analysis code STAGS [21] has been used to verify the Level-1 and 2 predictions. Due to the slightly skewed buckling pattern predicted by the Level-1 and Level-2 computations it is necessary to model the whole shell.

A convergence study has been conducted to establish the mesh size needed to model accurately the buckling behavior of the shell. For this purpose, the analysis option to determine asymmetric bifurcation from a nonlinear prebuckling path was used, and the previous results obtained with the Level-2 module ANILISA listed in Table 3 serve as a reference.

In the convergence study, a fixed number of mesh points was first used in the axial direction (NR = 161) and the number of mesh points in the circumferential direction (NC) was increased until the bifurcation load approached a horizontal tangent. As shown in Fig. 3, the results converge to a limiting value from below at about NC = 201. Next, a fixed number of mesh points was used in the circumferential direction (NC = 201) and the number of rows (NR) was varied. For this case convergence is from above and, as shown in Fig. 3, a horizontal tangent is reached at about NR = 201.

Using a mesh of 161 rows and 201 columns and the SS-3 boundary conditions $(N_x = -N_o, \quad v = W = M_x = 0)$, the following three lowest eigenvalues were obtained:

$$\lambda_c^{(1)} = 0.327759 \ \ (n = 11) \rightarrow \quad N_c^{(1)} = \lambda_c^{(1)} N_{c\ell} = -733.630 \text{ lb/in}$$

$$\lambda_c^{(2)} = 0.328652 \ \ (n = 10) \rightarrow \quad N_c^{(2)} = \lambda_c^{(2)} N_{c\ell} = -735.631 \text{ lb/in}$$

$$\lambda_c^{(3)} = 0.329421 \ \ (n = 12) \rightarrow \quad N_c^{(3)} = \lambda_c^{(3)} N_{c\ell} = -737.352 \text{ lb/in}$$

The critical buckling mode is shown in Fig. 4. The sequence of the three lowest buckling loads and the corresponding buckling modes agree closely with the predictions obtained with the Level-2 DISDECO module ANILISA (see also Table 3).

3. IMPERFECTION SENSITIVITY STUDY

It is well known that initial imperfections may decrease the load carrying capacity of a thin-walled shell structure. However, in order to calculate the effect of initial imperfections, one must know the shape and amplitude of these initial imperfections, which is information that is rarely available. In the absence of initial imperfection measurements, one must establish whether a given loading condition is imperfection sensitive, and, if it is, one must estimate how much certain characteristic imperfection shapes can reduce the buckling load.

Single Axisymmetric Imperfection

The simplest imperfection model is based on Koiter's pioneering work [7,22] and consists of a single axisymmetric imperfection

$$\bar{W} = h\bar{\xi}_1 \cos i\pi \frac{x}{L} \tag{5}$$

where i is an integer denoting the number of half-waves in the axial direction and $\bar{\xi}_1$ is the amplitude of the axisymmetric imperfection normalized by the shell wall thickness h. If one assumes that both the axial load and the boundary conditions are independent of the circumferential coordinate, then the prebuckling solution will also be axisymmetric, which simplifies the solution considerably.

Level-1 Analysis of Axisymmetric Imperfection

Neglecting the effect of the prebuckling boundary conditions, the nonlinear equations governing the prebuckling state admit the following axisymmetric solutions

$$W^{(o)} = hW_v + hw_o(x) \tag{6}$$

$$F^{(o)} = -\frac{Eh^2}{cR}\frac{1}{2}\lambda y^2 + f_o(x)$$

where

$$w_o(x) = \frac{\lambda}{\lambda_{c_i} - \lambda} h\bar{\xi}_1 \cos i\pi \frac{x}{L}$$

$$f_o(x) = \frac{\lambda}{\lambda_{c_i} - \lambda} \frac{(1+\bar{B}^*_{21}\alpha_i^2)}{2\alpha_i^2 \bar{A}^*_{22}} \frac{Eh^3}{c} \bar{\xi}_1 \cos i\pi \frac{x}{L} \tag{7}$$

$$\lambda_{c_i} = \frac{1}{2}\left\{\alpha_i^2 \bar{D}^*_{11} + \frac{(1+\bar{B}^*_{21}\alpha_i^2)^2}{2\alpha_i^2 \bar{A}^*_{22}}\right\}$$

The linearized stability equations now become a set of equations with variable coefficients. The reduced wave number α_i and the normalized stiffness coefficients $\bar{A}^*_{22}$, $\bar{B}^*_{21}$ and $\bar{D}^*_{11}$ are all listed in Ref. [14]. By assuming an asymmetric bifurcation mode of the form [14]

$$W^{(1)} = h \sin m\pi \frac{x}{L} \cos \frac{n}{R}(y - \tau_K x) \tag{8}$$

where

m = k = number of axial half waves
n = ℓ = number of circumferential full waves

τ_K = Khot's skewedness parameter

a Galerkin type approximate solution yields a characteristic equation in the form of a cubic polynomial for the eigenvalues λ of the problem

$$\lambda^3 - (\lambda_{mn\tau} + 2\lambda_{c_i} - \hat{C}_1\bar{\xi}_1\delta_{i=2m})\lambda^2 + \{2\lambda_{mn\tau} + \lambda_{c_i} + (\hat{C}_2 - \hat{C}_1)\bar{\xi}_1\delta_{i=2m}\}\lambda_{c_i}\lambda$$
$$-\{\lambda_{mn\tau} + \hat{C}_2\bar{\xi}_1\delta_{i=2m} + (\hat{C}_3 - \hat{C}_4\delta_{im})\bar{\xi}_1^2\}\lambda_{c_i}^2 = 0 \tag{9}$$

where

$$\delta_{i=2m} \begin{array}{ll} =1 & \text{if } i=2m \\ =0 & \text{otherwise} \end{array} \quad ; \quad \delta_{im} \begin{array}{ll} =1 & \text{if } i=m \\ =0 & \text{otherwise} \end{array}$$

and the constants $\hat{C}_1, \hat{C}_2, \ldots$ are listed in Appendix C of Ref. [14].

One will get a noticeable degrading influence on the buckling load only if the assumed axisymmetric imperfection $\bar{\xi}_1$ is negative and if the coupling condition i=2m is satisfied. The physical explanation for this observation can be found in Koiter's 1963 paper [22]. To obtain the smallest real root of Eq. (9), an n-search must be carried out for a given axisymmetric imperfection $\bar{\xi}_1$. The terms involving the Kronecker delta $\delta_{i=2m}$ are all linear in $\bar{\xi}_1$, and thus they dominate the buckling behavior of the shell with an axisymmetric imperfection.

Assuming that the most likely axisymmetric imperfection of the steel mandrel used to lay-up the quasi-isotropic composite shell [2] is given by

$$\overline{W} = h\bar{\xi}_1 \cos 2\pi\frac{x}{L} \tag{10}$$

the Level-1 DISDECO computational module AXBIF was used to generate the solid curve shown in Fig. 5. The curve is re-normalized by $\lambda_c^m = 0.366892$, which is the critical Level-1 buckling load of the perfect shell computed using AXBIF [14] with membrane prebuckling analysis results for $\tau_K = 0.0$ and n=7. An initial imperfection amplitude equal to the wall thickness of the shell ($\bar{\xi}_1 = -1.0$) generates a "knockdown factor" of $\rho_c = \lambda_c / \lambda_c^m = 0.486$, resulting in the following low buckling load

$$\lambda_c = 0.486\lambda_c^m = 0.486(0.366892) = 0.178310$$

$$N_c = \lambda_c N_{c\ell} = \quad 0.178310(-2238.325) = -399.115 \text{ lb/in}$$

with $\tau_K = 0.0$ and n = 7 full waves in the circumferential direction.

Level-2 Analysis of Axisymmetric Imperfection

Since the applied load, the boundary conditions and the assumed initial imperfection are axisymmetric, the prebuckling solution will also be axisymmetric. By assuming [23]

$$W^{(o)} = hW\nu + hw_o(x) \tag{11}$$
$$F^{(o)} = \frac{Eh^2}{cR}\left\{-\frac{1}{2}\lambda y^2 + R^2 f_o(x)\right\}$$

the solution of the nonlinear partial differential equations governing the prebuckling state can be reduced to the solution of a single fourth order ordinary differential equation with constant coefficients. For anisotropic shells, the resulting linearized stability equations admit separable solutions of the form

$$W^{(1)} = h[w_1(x)\cos n\theta + w_2(x)\sin n\theta] \tag{12}$$
$$F^{(1)} = \frac{ERh^2}{c}[f_1(x)\cos n\theta + f_2(x)\sin n\theta]$$

where $\theta = \frac{y}{R}$.

The solution proceeds as outlined in Ref. [23]. Using an updated version of the Level-2 computational module ANILISA [24] in DISDECO and SS-3 ($N_x = v = w = M_x = 0$) boundary conditions, one obtains the results presented in Table 4. A rigorous nonlinear prebuckling analysis was used and an n-search was carried out for each specified axisymmetric imperfection amplitude $\bar{\xi}_1$. The values of Table 4 are plotted as the dashed curve in Fig. 5. A comparison of the results obtained using the Level-1 module AXBIF in DISDECO (solid curve) and the Level-2 module ANILISA in DISDECO (dashed curve) shows that, in the case of axisymmetric imperfections, a rigorous pre-buckling analysis should also be used. For very small initial imperfection amplitudes ($|\bar{\xi}_1| < 0.1$), the Level-1 predictions are especially inaccurate and over-estimate the critical buckling load. Both curves have been normalized by $\lambda_c^m = 0.366892$, which is the critical Level-1 buckling load of the perfect shell computed using membrane prebuckling results by AXBIF [14] for $\tau_K = 0.0$ and $n = 7$. These results indicate that the effect of using a rigorous prebuckling analysis becomes easily discernible.

For small enough initial imperfection amplitudes ($|\bar{\xi}_1| < 0.07$, say), the critical buckling load of the shell is insensitive to the initial imperfection shape specified by Eq. (10). The critical buckling modes have $n = 11$ full waves in the circumferential direction, and a somewhat skewed buckling mode shape which is dominated by edge buckling. However, larger initial imperfections ($|\bar{\xi}_1| \geq 0.07$, say) force the shell to respond in another mode shape with $n = 7$ full waves in the circumferential direction and with practically straight nodal lines. The critical buckling load of the shell is now sensitive to the axisymmetric initial imperfection shape specified by Eq. (10) and an initial imperfection amplitude equal to the wall thickness of the shell ($\bar{\xi}_1 = -1.0$) generates a "knockdown factor" of $\rho_c = \lambda_c / \lambda_c^m = 0.418$. The analysis thus predicts the following low buckling load

$$\lambda_c = 0.418\lambda_c^m = 0.418(0.366892) = 0.153361$$

$$N_c = \lambda_c N_{c\ell} = \quad 0.153361(-2238.325) = -343.271 \text{ lb/in}$$

with a very slight skewedness of the buckling pattern and $n = 7$ full waves in the circumferential direction.

Table 4. Buckling loads of a quasi-isotropic composite shell [2] ($N_{c\ell} = -2238.325$ lb/in) Axisymmetric imperfection using ANILISA [24] (B.C.: $N_x = v = w = M_x = 0$).

$\bar{\xi}_1$	λ_c^{nl}	$\bar{\xi}_1$	λ_c^{nl}
0.	0.328594 (n=11)	- 0.2	0.299830 (n=7)
- 0.01	0.328671 (n=11)	- 0.3	0.274773 (n=7)
- 0.02	0.328745 (n=11)	- 0.4	0.251275 (n=7)
- 0.05	0.328939 (n=11)	- 0.5	0.229976 (n=7)
- 0.06	0.328994 (n=11)	- 0.6	0.210876 (n=7)
- 0.07	0.328332 (n=7)	- 0.7	0.193827 (n=7)
- 0.08	0.326652 (n=7)	- 0.8	0.178655 (n=7)
- 0.09	0.324860 (n=7)	- 0.9	0.165187 (n=7)
- 0.10	0.322960 (n=7)	- 1.0	0.153263 (n=7)

Level-3 Analysis of Axisymmetric Imperfection

Since both the axial load and the boundary conditions are independent of the circumferential coordinate, the prebuckling solution will also be axisymmetric. As a result, one can use the analysis option that predicts the asymmetric bifurcation response from a nonlinear prebuckling path. By modeling the full shell, the code can automatically choose the critical number of full waves in the circumferential direction. No n-search is needed. Using a uniformly spaced mesh of 161 rows and 201 columns and the user written subroutine option WIMP in STAGS to introduce the following axisymmetric imperfection

$$\overline{W} = 1.0 h \cos 2\pi \frac{x}{L}$$

(STAGS defines $\overline{W}$ positive outward), the following critical bifurcation load was found

$$N_c = -333.449 \text{ lb/in}$$

The critical buckling mode has $n = 7$ full waves in the circumferential direction and no visible skewedness. The nondimensional bifurcation load of the shell with axisymmetric imperfection is for $\bar{\xi}_1 = 1.0$

$$\lambda_c = \frac{N_c}{N_{c\ell}} = \frac{-333.449}{-2238.325} = 0.148973$$

The re-normalized bifurcation load is

$$\rho_c^{nl} = \frac{N_c}{\bar{N}_{c\ell}} \qquad \text{where} \qquad \bar{N}_{c\ell} = \lambda_c^{nl} N_{c\ell}$$

thus

$$\rho_c^{nl} = \frac{\lambda_c}{\lambda_c^{nl}} = \frac{0.148973}{0.327759} = 0.4545$$

The Level-3 re-normalization is done using $\lambda_c^{nl} = 0.327759$, which is the critical Level-3 buckling load of the perfect shell computed using STAGS-A with nonlinear prebuckling response and a 161×201 mesh. The Level-2 ANILISA prediction ($\rho_c^{nl} = 0.4664,\ n = 7$) agrees closely with the Level-3 STAGS-A prediction ($\rho_c^{nl} = 0.4545,\ n = 7$). The slight difference is partly due to the fact that ANILISA uses the Donnell type nonlinear shell equations, and STAGS-A uses the higher-order Marlowe-Flugge equations.

Single Asymmetric Imperfection

The effect of a single asymmetric initial imperfection can be investigated by solving the full nonlinear response problem or by using the well known Lyapunov-Schmidt-Koiter [7] reduction technique. When investigating the degrading effect of a single-mode asymmetric imperfection

$$\bar{W} = t\bar{\xi}_2 \sin m\pi\frac{x}{L}\cos n\frac{y}{R} \tag{13}$$

where m and n are integers denoting the number of axial half-waves and the number of circumferential full waves, respectively, instability occurs at the limit point of the prebuckling state in the generalized load-deformation space. Assuming that the eigenvalue problem for the critical buckling load Λ_c will yield a unique asymmetric buckling mode $W^{(1)}$, then, for an imperfect shell ($\bar{\xi}_2 \neq 0$), the shape of the generalized load-deflection curve in the vicinity of the bifurcation point $\Lambda = \Lambda_C$ is given by the following asymptotic expansion

$$(\Lambda - \Lambda_c)\xi = \Lambda_c a\xi^2 + \Lambda_c b\xi^3 + \ldots - \Lambda_c \alpha\bar{\xi}_2 - (\Lambda - \Lambda_c)\beta\bar{\xi}_2 + O(\xi\bar{\xi}_2) \tag{14}$$

Expressions for the postbuckling coefficients "a" and "b" and the imperfection form factors "α" and "β" are derived in References [25,26]. If the limit point is close to the bifurcation point, then the maximum load Λ_s that the structure can support prior to buckling can be evaluated from Eq. (14) by maximizing Λ with respect to ξ. For cases where the first postbuckling coefficient "a" is zero, this analysis yields the modified Koiter formula [26]

$$(1-\rho_s)^{3/2} = \frac{3}{2}\sqrt{-3\alpha^2 b}[1-\frac{\beta}{\alpha}(1-\rho_s)]\left|\bar{\xi}_2\right| \tag{15}$$

where $\rho_s = \Lambda_s / \Lambda_c$.

If the second postbuckling coefficient "b" is positive, Eq. (15) has no real solutions. Thus, the buckling load of the specified shell-loading combination is not sensitive to small asymmetric initial imperfections of the shape given by Eq. (13). If, however, the second postbuckling coefficient "b" is negative, the equilibrium load Λ decreases following buckling and the buckling load of the real structure Λ_s is sensitive to the asymmetric initial imperfection specified by Eq. (13).

Level-1 Analysis of Asymmetric Imperfection

For the composite shell under investigation, there are many eigenvalues only slightly higher than the critical one of $\lambda_c = 0.365992$ for m = 1, n = 7 and $\tau_K = 0.011$ [see Table 2]. Hence, strictly speaking, the proposed form of the perturbation expansion given by Eqs. (14) is not applicable, since the nonlinear interaction between the many nearly simultaneous eigenmodes is not accounted for. Thus, the following results, where one considers the eigenfunctions corresponding to certain critical eigenvalues chosen one at the time, can at best give an indication of the severity of the expected imperfection sensitivity.

Initially assuming an asymmetric imperfection affine to the critical buckling mode of the perfect quasi-isotropic composite shell [2] as computed by the Level-1 computational module AXBIF (see also Table 2)

$$\overline{W} = h\bar{\xi}_2 \sin\pi\frac{x}{L}\cos\frac{7}{R}(y-0.011x) \tag{16}$$

and using the Level-1 computational module BFACT in DISDECO to conduct the initial postbuckling analysis yields the following results

$$\lambda_c = \lambda_c^m = 0.365992(m=1,\ n=7,\ \tau_K = 0.011)$$

$$b = -0.048844 \qquad \alpha = \beta = 1.0$$

Substituting these values into Eq. (15), one can plot the degrading effect of an asymmetric imperfection of the shape given by Eq. (16) as a function of its amplitude $\bar{\xi}_2$. As shown in Fig. 6, an initial imperfection amplitude equal to the wall thickness of the shell ($\xi_2 = 1.0$) generates a "knockdown factor" of $\rho_s = \lambda_s / \lambda_c^m = 0.541$, resulting in the following low buckling load

$$\lambda_s = 0.541\lambda_c^m = 0.541(0.365992) = 0.198002$$

$$N_s = \lambda_s N_{c\ell} = 0.198002(-2238.325) = -443.192 \text{ lb/in}$$

The imperfection form factors "α" and "β" are identically equal to 1.0 because BFACT uses membrane prebuckling analysis results to calculate the necessary first and second order fields and the assumed asymmetric imperfection shape of Eq. (16) is affine to the buckling mode. The collapse load is re-normalized by $\lambda_c^m = 0.365992$ in Fig. 6, which is the critical Level-1 buckling load of the perfect shell computed using membrane prebuckling analysis results by AXBIF [14] for $\tau_K = 0.011$ and $n = 7$.

Level-2 Analysis of Asymmetric Imperfection

To investigate the effects of edge-constraint and/or different boundary conditions on the imperfection sensitivity of the critical buckling load of the quasi-isotropic composite shell [2] one has to use the Level-2 module ANILISA [15] in DISDECO and execute its postbuckling analysis option. In this module, as described earlier, the axisymmetric prebuckling state is represented by Eqs. (3), the buckling modes by Eqs. (4) and the postbuckling state by

$$W^{(2)} = h[w_\alpha(x) + w_\beta(x)\cos n\theta + w_\gamma(x)\sin n\theta]$$

$$F^{(2)} = \frac{ERh^2}{c}[f_\alpha(x) + f_\beta(x)\cos 2n\theta + f_\gamma(x)\sin 2n\theta]$$

where $\theta = y/R$. Details of the computational procedures used are reported in Refs. [15,23].

Next, assume that the specified asymmetric imperfection is affine to the critical buckling mode obtained by the rigorous Level-2 perfect shell analysis discussed previously

$$\overline{W} = h\bar{\xi}_2[w_1(x)\cos 11\,\theta + w_2(x)\sin 11\,\theta]$$

where $\theta = y/R$ and the component functions $w_1(x)$ and $w_2(x)$ are shown in Fig. 2b. Executing ANILISA with rigorous prebuckling results and SS-3 boundary conditions ($N_x = -N_o$, $v = w = M_x = 0$) yields the following results

$$\lambda_c = \lambda_c^{n\ell} = 0.328594 (n = 11)$$

$$b = -0.37605 \quad ; \quad \alpha = 0.46663 \quad ; \quad \beta = -0.22174$$

Using Eq. (15) to plot the degrading effect of the asymmetric imperfection specified by Eq. (16) as a function of its amplitude $\bar{\xi}_2$, one obtains the results shown as a solid line in Fig. 7. For an imperfection shape affine or similar to the buckling mode with an amplitude of $\bar{\xi}_2 = 1.0$, one obtains an unrealistic negative load carrying capacity.

Here one must remember that Koiter's imperfection sensitivity theory is asymptotically exact, that is, it yields accurate predictions for sufficiently small imperfections, and what is sufficiently small may vary from case to case. Also, Eq. (15) was obtained by using the perturbation expansion given by Eq. (14), where terms of order $(\xi\bar{\xi})$ are neglected. As can be seen from the dotted curve plotted in Fig. 7, one obtains more reasonable predictions by using more advanced computational modules in DISDECO, such as COLLAPSE [27], where a full nonlinear solution is used and terms up to and including order $(\xi\bar{\xi}^2)$ are kept. Up to about $\bar{\xi}_2 = 0.3$ the asymptotic predictions from ANILISA and the nonlinear results of COLLAPSE agree very closely. Thus, in this case, the range of validity of the asymptotic solution is $0 \geq \bar{\xi}_2 \geq 0.3$.

4. DISCUSSION OF THE RESULTS

When comparing and analyzing the results obtained so far, it is important to recall that all Level-1 and Level-2 solutions are based on approximate representations of the unknown functions. The Level-1 solutions use a single term double Fourier series approximation to reduce the solution of the stability problem, to algebraic eigenvalue problems. The effect of edge restraint is neglected (one uses a membrane prebuckling solution) and the assumed field functions satisfy approximately SS-3 ($N_x = -N_o$, $v = w = M_x = 0$) boundary conditions. The Level-2 solutions eliminate the y-dependence of the governing equations by a truncated Fourier decomposition in the circumferential direction. The resulting system of nonlinear ordinary differential equations are solved numerically, and both the specified boundary conditions and the effect of edge restraint are rigorously satisfied. Thus, by using this approach, the only approximation that is made is that one represents the variation of the solution in the circumferential direction by a single harmonic with n full waves, and an n-search is used to establish which wave number is the critical one. The Level-2 module ANILISA can also be used to investigate the effect of using different boundary conditions. The results of analyses with four different boundary conditions indicate that the fully clamped C4 boundary conditions have the highest critical buckling load. The results for weaker SS3 boundary conditions are about the same as for an isotropic shell with similar characteristic dimensions (i.e., the same L/R and R/t ratios) of Ref. [28].

The Level-3 solutions are based on a two-dimensional finite difference or finite element formulation. In both cases, if one uses the appropriate meshes, one can obtain rigorous solutions where all nonlinear effects are properly accounted for in the analysis. The only real problem with Level-3 type solutions is that for each problem one must establish the appropriate mesh size. Meshes that are too coarse yield inaccurate solutions, and what is meant by too coarse depends on the particular problem under investigation. Thus, for a general nonlinear solution, a convergence study must always be conducted.

As can be seen from the results presented herein, the buckling load of the quasi-isotropic composite shell analyzed is sensitive to all of the initial imperfection shapes investigated. For a more specific prediction of the final collapse load, the final goal of a "High-Fidelity Buckling Load Analysis", one would have to conduct a refined Level-3 (i.e., the highest hierarchical analysis level) analysis which includes measured values of all the significant generalized imperfections such as the traditional shell-wall imperfections, variations in the shell-end or loading surface geometry, and especially for composite shells variations in the shell-wall thickness distribution. It has been shown in Ref. [29] that such an approach yields very good agreement between the predicted collapse load and the experimental buckling load.

5. CONCLUDING REMARKS

A hierarchical high-fidelity analysis approach for predicting the buckling loads of thin-walled shells is described. This hierarchical analysis approach includes three levels of shell buckling analyses that range in sophistication from classical linear bifurcation buckling analysis to nonlinear finite element collapse analysis to provide an accurate prediction of the critical buckling load of a given shell structure. Initial imperfections are included in the hierarchical analysis approach, and the sensitivity of the stability response to initial imperfections is determined by the analysis approach.

By relying on a series of theoretical results of various degrees of sophistication, the hierarchical approach used in the present paper has resulted in a series of buckling load predictions of increasing accuracy for thin-walled shell structures. It is shown that in order to arrive at a reliable prediction of the critical buckling load of a thin-walled shell and to make an estimate of its imperfection sensitivity which can be used with confidence, one must proceed step by step from simple to more complex models and solution procedures. In order to predict the critical buckling load of a geometrically imperfect thin-walled shell accurately and to make a reliable estimate of its imperfection sensitivity, the nonlinear effects caused by the edge restraint conditions must be included in the analysis. Any solution procedure which fails to account for these effects, should be suspect of having provided incorrect results.

The most approximate of the analyses described herein, the Level-1 (i.e., the lowest hierarchical analysis level) solutions which neglect the effects of edge restraints, can still be used to great advantage to establish the approximate behavior of a shell subjected to the specified external loading conditions. However, depending on the value of the prebuckling stiffness, resulting from the different types of wall construction used, the solutions may be either conservative or non-conservative.

As can be seen from the results presented herein, the buckling load of the quasi-isotropic composite shell analyzed is sensitive to all of the initial imperfection shapes investigated. For a more specific prediction of the final collapse load, the final goal of a "High-Fidelity Buckling Load Analysis", one would have to conduct a refined Level-3 (i.e., the highest hierarchical analysis level) analysis which includes measured values of all the significant generalized imperfections such as the traditional shell-wall imperfections, variations in the shell-end or loading surface geometry, and especially for composite shells variations in the shell-wall thickness distribution.

Acknowledgement

The research reported in this paper was supported in part by NASA Grant NAG 1-2129. This aid is gratefully acknowledged.

References

1. Arbocz, J., Starnes, J.H. and Nemeth, M.P., "A Hierarchical Approach to Buckling Load Calculations," Proceedings 40th AIAA/ASME/ASCE/AHS/ASC Structures, Structural Dynamics and Materials Conference, April 12-15, 1999, St. Louis, MO, pp. 289-299.
2. Waters, W.A. Jr., "Effects of Initial Geometric Imperfections on the Behavior of Graphite-Epoxy Cylinders Loaded in Compression," M.Sc. Thesis in Engineering Mechanics, Old Dominion University, Norfolk, Virginia, December 1996.
3. Anonymous, NASTRAN, The MacNeal-Schwendler Corporation, Los Angeles, California.
4. Anonymous, ABAQUS, Hibbitt, Karlsson & Sorensen, Inc., Providence, Rhode Island.
5. Zienkiewicz, O.C., "The Finite Element Method," (third edition) McGraw-Hill Book Company (UK) Limited, London, 1977, p. 504.
6. Byskov, E., "Smooth Postbuckling Stresses by a Modified Finite Element Method," DCAMM Report No. 380, Technical University of Denmark, Lyngby, 1988.
7. Koiter, W.T., "On the Stability of Elastic Equilibrium," Ph.D. Thesis (in Dutch), TH-Delft, The Netherlands, H.J. Paris, Amsterdam, 1945, English translation NASA TTF-10, 1967, pp. 1-833.
8. Budiansky, B. and Hutchinson, J.W., "Dynamic Buckling of Imperfection Sensitive Structures," Proceedings 11th IUTAM Congress, Munich, 1964, Julius Springer Verlag, Berlin, 1966, pp. 636-651.
9. Hoff, N.J., "Buckling of Thin Shells," Proceedings of an Aerospace Symposium of Distinguished Lecturers in Honor of Theodore von Kármán on his 80th Anniversary, Institute of Aerospace Sciences, New York, 1961, pp. 1-42.
10. Stuhlman, C.E., De Luzio, A. and Almroth, B., "Influence of Stiffener Eccentricity and End Moment on Stability of Cylinders in Compression," AIAA Journal, Vol. 4, No. 5, May 1966, pp. 872-877.
11. Fischer, G., "Über den Einfluss der gelenkingen Lagerung auf die Stabilität dünnwandiger Kreiszylinderschalen unter Axiallast und Innendruck," Z.f. Flug-wissenschaften, Vol. 11, 1963, pp. 111-119.
12. Stein, M., "The Influence of Prebuckling Deformations and Stresses on the Buckling of Perfect Cylinders," NASA TR-190, February 1964.
13. Arbocz, J. and Hol, J.M.A.M., "Shell Stability Analysis in a Computer-Aided Engineering (CAE) Environment," Proceedings 34th AIAA/ASME/ASCE/ AHS/ASC Structures,

Structural Dynamics and Materials Conference, April 19-22, 1993, La Jolla, California, pp. 300-314.

14. Arbocz, J., "The Effect of Initial Imperfections on Shell Stability - An Updated Review," Report LR-695, Delft University of Technology, Faculty of Aerospace Engineering, The Netherlands, September 1992.
15. Arbocz, J. and Hol, J.M.A.M., "Koiter's Stability Theory in a Computer Aided Engineering (CAE) Environment," Int. J. Solids and Structures, Vol. 26, No. 9/10, pp. 945-973, 1990.
16. Booton, M., "Buckling of Imperfect Anisotropic Cylinders under Combined Loading," UTIAS Report No. 203, University of Toronto, 1976.
17. Von Kármán, T. and Biot, M.A., "Mathematical Methods in Engineering," McGraw-Hill Book Co., New York, London, 1940, p. 317.
18. Cohen, G.A., "Computer Analysis of Asymmetric Buckling of Ring-stiffened Orthotropic Shells of Revolution," AIAA Journal, Vol. 6, No. 1, January 1968, pp. 141-149.
19. Keller, H., "Numerical Methods of Two-Point Boundary Value Problems," Blaisdell Publishing Co., Waltham, Mass., 1968.
20. Almroth, B.O., Brogan, F.A., Miller, E., Zele, F. and Peterson, H.T., "Collapse Analysis for Shells of General Shape; II. User's Manual for the STAGS-A Computer Code," Technical Report AFFDL-TR-71-8, Air Force Flight Dynamics Laboratory, Wright-Patterson Air Force Base, Ohio, March 1973.
21. Brogan, F.A., Rankin, C.C. and Cabiness, H.D., "STAGS Access Routines - STAR Reference Manual," Report LMSC P032595, Version 2.0, Lockheed Palo Alto Research Laboratory, Palo Alto, California, June 1994.
22. Koiter, W.T., "The Effect of Axisymmetric Imperfections on the Buckling of Cylindrical Shells under Axial Compression," Koninkl. Ned. Akad. Wetenschap. Proc. B66, 1963, pp. 265-279.
23. Arbocz, J. and Hol, J.M.A.M., "ANILISA - Computational Module for Koiter's Imperfection Sensitivity Theory," Report LR-582, Delft University of Technology, Faculty of Aerospace Engineering, The Netherlands, January 1989.
24. Romkes, A., "Stability and Imperfection Sensitivity of Anisotropic Cylindrical Shells under General Boundary Conditions," Memorandum M-809, Delft University of Technology, Faculty of Aerospace Engineering, The Netherlands, July 1997.
25. Hutchinson, J.W. and Amazigo, J.C., "Imperfection Sensitivity of Eccentrically Stiffened Cylindrical Shells," AIAA Journal, Vol. 5, No. 3, March 1967, pp. 392-401.
26. Cohen, G.A., "Effect of a Nonlinear Prebuckling State on the Postbuckling Behavior and Imperfection Sensitivity of Elastic Structures," AIAA Journal, Vol. 6, No. 8, August 1968, pp. 1616-1619.
27. Arbocz, J., de Vries, J. and Hol, J.M.A.M., "On the Buckling of Imperfect Anisotropic Shells with Elastic Edge Supports under Combined Loading - Part I: Theory and Numerical Analysis," Memorandum M-849, Delft University of Technology, Faculty of Aerospace Engineering, The Netherlands, May 1997.
28. Almroth, B.O., "Influence of Edge Conditions on the Stability of Axially Compressed Cylindrical Shells," NASA CR-161, February 1965.
29. Hilburger, M.W. and Starnes, J.H. Jr., "Effects of Imperfections on the Buckling Response of Compression-Loaded Composite Shells," Proceedings 41st AIAA/ASME/ASCE/AHS/ ASC Structures, Structural Dynamics and Materials Conference, 3-6 April 2000, Atlanta, Georgia, (paper AIAA-2000-1382).

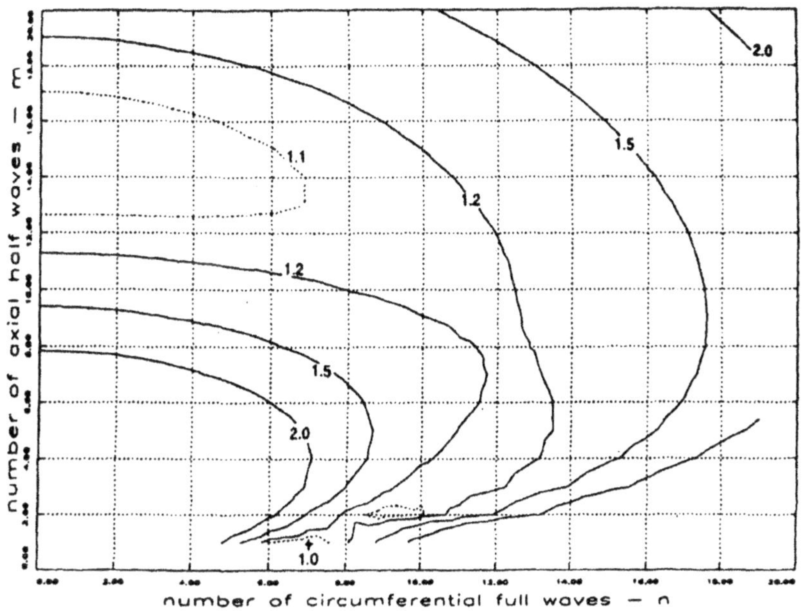

Figure 1. Distribution of buckling loads based on Level-1 membrane prebuckling analysis for a quasi-isotropic composite shell [AW-CYL-1-1] [2].

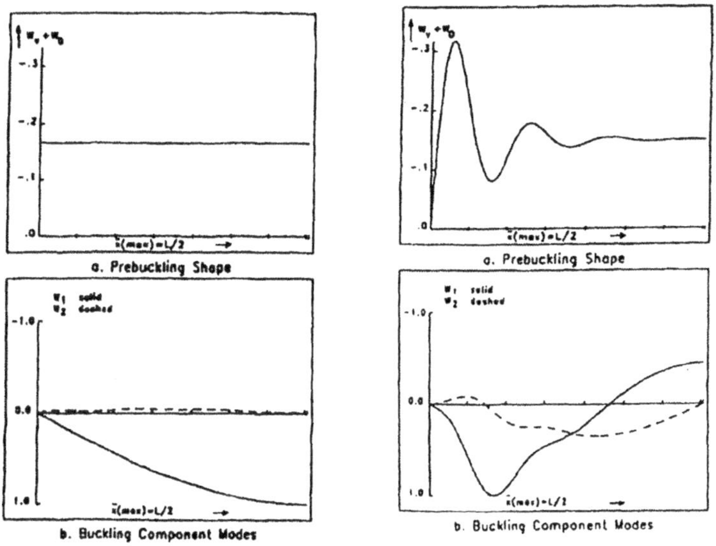

a. Membrane prebuckling, n=7

$\lambda_c^m = 0.364715$

b. Nonlinear prebuckling, n=11

$\lambda_c^{nl} = 0.328594$

Figure 2. Buckling modes for the axially compressed quasi-isotropic composite shell [2] SS-3 B.C. $N_x = -N_o$, $v = W = M_x = 0$: $N_{cl} = -2238.325$ lb/in.

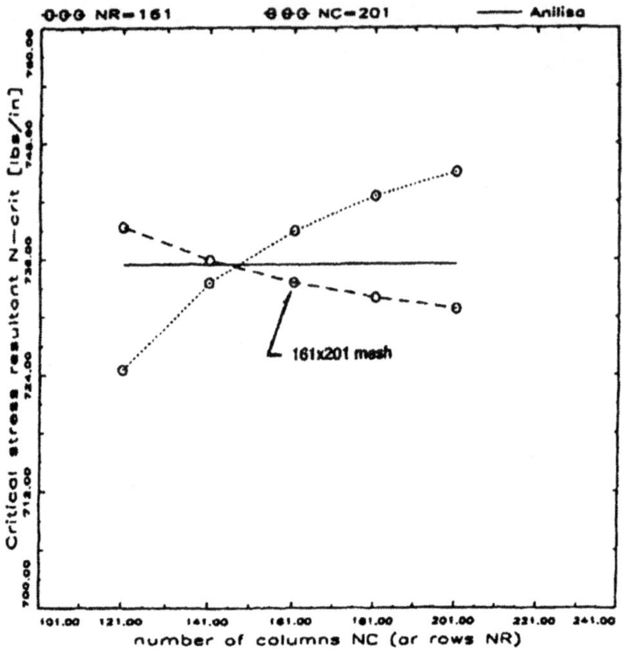

Figure 3. STAGS-A convergence study results for a quasi-isotropic composite shell [2].

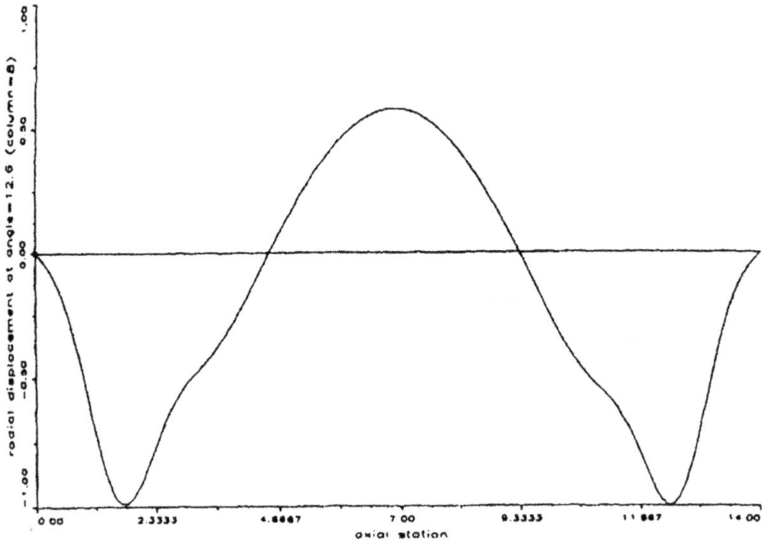

Figure 4. Buckling mode of a quasi-isotropic composite shell [2]
SS-3 B.C. $N_x = -N_o$, $v = W = M_x = 0$; $N_{c\ell} = -2238.325$ lb/in; STAGS-A results.

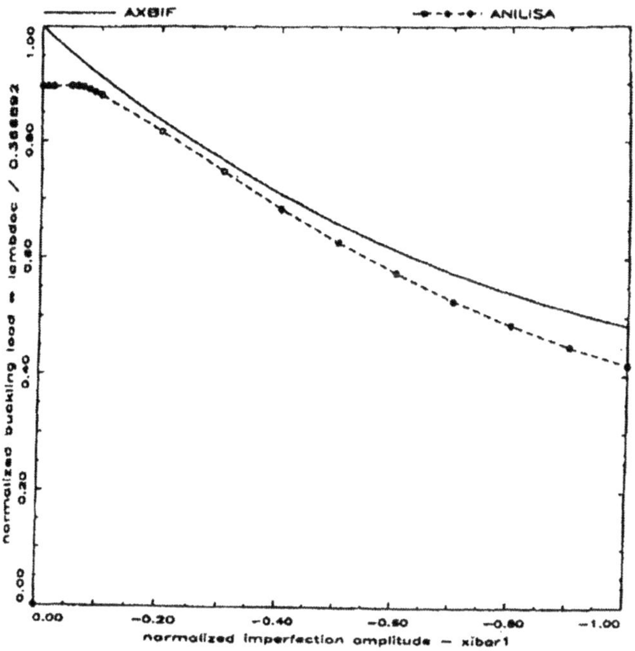

Figure 5. Imperfection sensitivity for axisymmetric imperfection (SS-3 B.C.: $N_x = -N_o$, $v = W = M_x = 0$; $N_{c\ell} = -2238.325$ lb/in).

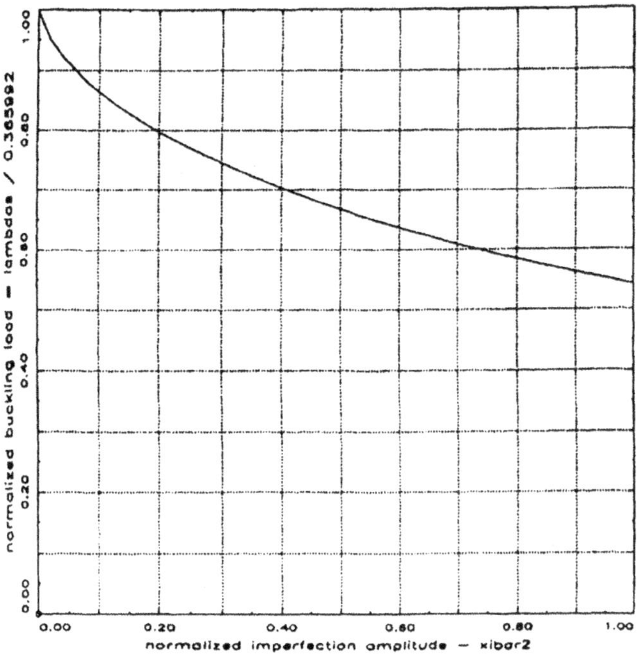

Figure 6. Level-1 imperfection sensitivity calculation using BFACT [14]

Asymptotic theory: $\bar{W} = h\xi_2 \sin \pi \frac{x}{L} \cos \frac{7}{R}(y - 0.011x)$.

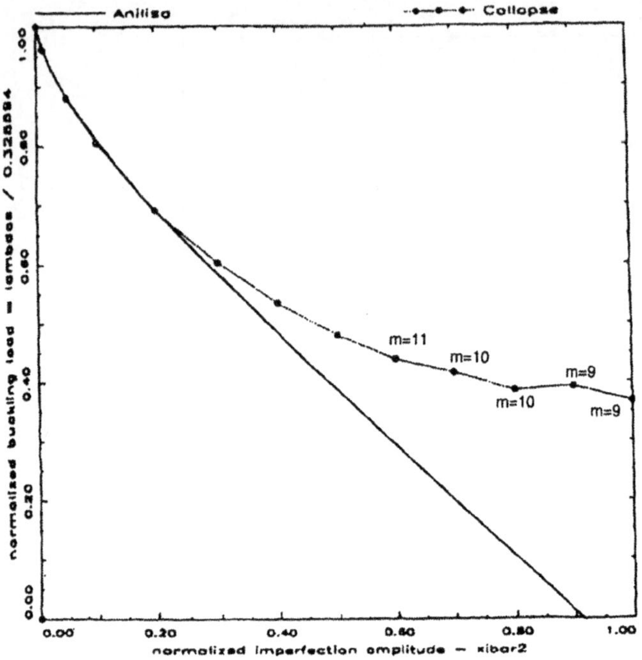

Figure 7. Imperfection sensitivity calculation for affine asymmetric imperfections. Asymptotic theory versus nonlinear collapse analysis.

CAT EARDRUM RESPONSE MECHANICS

J.P. Fay, S. Puria, and C.R. Steele
Mechanics and Computation Division
Department of Mechanical Engineering
Stanford University
Stanford, CA 94305, USA
chasst@stanford.edu

Abstract The function of the eardrum and middle ear is to resolve the acoustic impedance mismatch between the air of the outside world and the fluid of the inner ear. Without an impedance matching device, very little acoustic energy would be absorbed by the inner ear and hearing would be severely limited. Although the role of the middle ear is clear, it remains a mystery how the eardrum accomplishes this task over the audible frequency range. In the present work, a computer simulation of the cat eardrum was constructed. For the first time, the vibrations of the eardrum were fully coupled to the acoustics of the ear canal and the dynamics of the middle ear bones. The fibrous microstructure of the eardrum was taken into account. This model appears to be valid up to 20 kHz, which represents a significant improvement over previous modelling efforts.

Keywords: Tympanic membrane, Composite, High frequency, Shell-acoustic interaction

1. INTRODUCTION

Hearing is second only to sight in the amount of information humans receive from their surroundings. According to the National Institute on Deafness and Other Communication Disorders (NIDCD, 2000) , 28 million people in the US alone live with some kind of hearing disorder. The total cost of lost productivity, special education, and medical care associated with hearing loss is nearly $56 billion dollars annually (Better Hearing Institute, 2000).

A thorough understanding of middle-ear mechanics will facilitate several advances in helping the hearing-impaired. First of all, through exploring how the normal ear functions, it may be possible to pinpoint where conduction problems may arise. Second, the effectiveness of surgical procedures could be tested, reducing costly and time-consuming animal studies or clinical trials. The

H.R. Drew and S. Pellegrino (eds.),
New Approaches to Structural Mechanics, Shells and Biological Structures, 293–302.

freedom to try out many surgical procedures could lead to innovative solutions for improving hearing. Third, the design of hearing aids could be improved. Another major reason to study the middle-ear mechanics is diagnostics. An understanding of eardrum and middle-ear mechanics should help develop tests that better reveal the nature of a problem. Finally the anatomy indicates a novel mechanical design, so it is of basic scientific and engineering interest to understand the principles that have evolved.

This article contains a few highlights from the material in Fay (2001).

2. BACKGROUND

An early comprehensive paper on the eardrum and middle-ear ossicles, shown in Fig. 1, is by Helmholtz (1869). Extensive pressure measurements

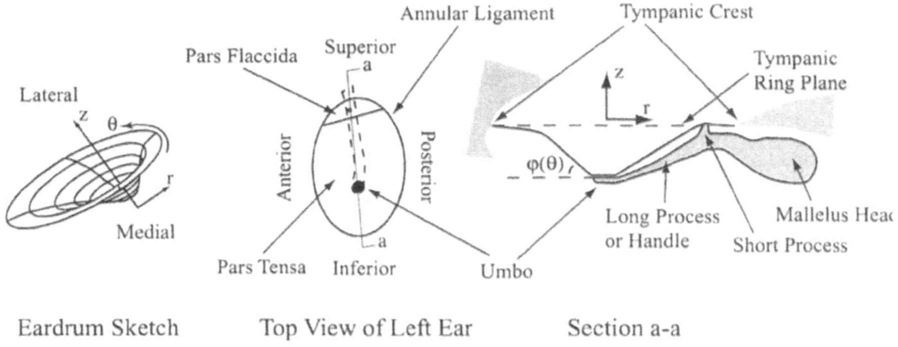

Figure 1. Sketch of a generic mammalian tympanic membrane and its relationship to the malleus. At the far left is a perspective sketch giving the generic shape of the eardrum. In the middle is a top view of a left eardrum. At the far right is a view of cross-section a-a in the middle sketch. Bone is shown in gray.

have been made, e.g. (Stinson and Khanna, 1994), while impedance measurements have been made in a number of laboratories (e.g. Hudde and Engel, 1998, Lynch et.al, 1994, Puria and Allen, 1998). A startling finding was provided by holographic measurement of the cat eardrum by Khanna and Tonndorf (1972), who found that the eardrum breaks into complex modes of vibration above about 2 kHz. Since the cat hears to 50 kHz, this means that the eardrum is not acting as a simple piston through a significant range of frequency. This has been controversial to this day, because one would expect that a mode of vibration, with different portions of the eardrum at different phases, would cause a loss of the force-transmission capability. However, the existence of the complex modes is confirmed in extensive vibration measurements (Decraemer et al., 1994).

The modeling of the middle ear has been dominated by circuit models, which convert the mechanics to an equivalent electrical circuit. The model of Zwislocki (1962) provides the basis of most of the subsequent models. The disadvantages include the fact that the physical dimensions and material properties tend to be lost in the conversion, and the eardrum is considered as a piston. Consequently, these models tend to be invalid above about 2 kHz. The first finite element model of the eardrum is by Funnell and Laszlo (1978). There is growing interest in correct mechanical modeling in several laboratories (e.g. Beer et al., 1999, Eiber, 1999, Wada et al., 1992, Williams et al., 1997). However, FEM has not yet delivered the insight desired. All the models have treated the eardrum as isotropic, have used too coarse a mesh to capture the high frequency response, have not included the acoustic-shell interaction, and have not included the inclination of the eardrum in the ear canal, see Fig. 2).

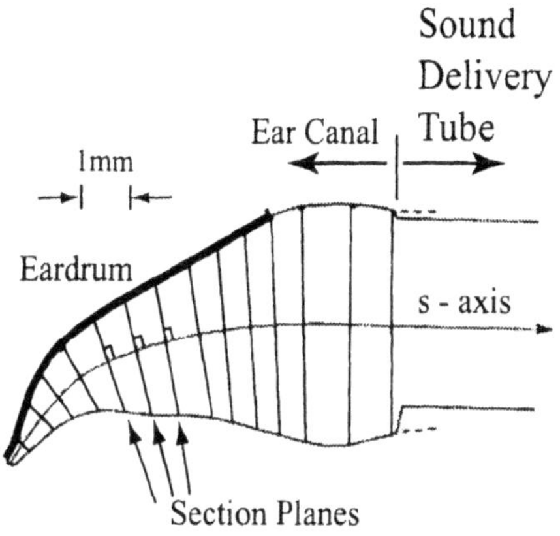

Figure 2. Sketch of cat ear canal geometry showing the sectioning planes perpendicular to the calculated midline curve. This sketch is a realistic representation of the casts that were made by Khanna and Stinson (1985).

Only a constant pressure on the eardrum has been considered. To include all features properly, it seems that well over 10^6 degrees of freedom would be required, which is a computational challange, since the responses at hundreds of frequencies are needed to construct a response curve similar to the measurements.

4.1 PRESENT WORK

3.1 Anatomy and Properties

The first task was to understand the anatomy of the eardrum from many sources, as detailed in Fay (2001). In summary, the description by Helmholtz (1869) seems to be accurate. Figure 3 shows the contour lines on the cat ear-

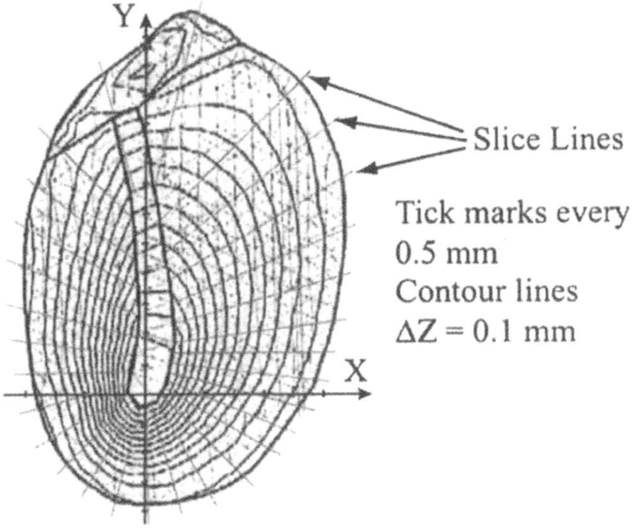

Figure 3. Contour plot of cat eardrum with slice lines used to divide up the eardrum. The shape is roughly elliptic, with the semi-major and semi-minor axes of 4.5 and 3 mm, respectively. The average thickness is around 36 microns, about half of which consists of the radial fibers.

drum. The eardrum is conical in the central region and flares out in a toroidal region at the attachment to the wall. Figure 4 shows the profile along one slice line of Fig. 3 along with the cone and toroid fits of that meridian. The eardrum is not at all isotropic. The central conical region has a relatively thick layer (avg. 17 m) of collagen fibers in the radial direction with few fibers in the circumferential direction. From various static measurements, it seems that an elastic modulus of 1 GPa for the fibers is correct. In Fig. 3 the asymmetry of the eardrum can be seen. The radial slices are much shorter on one side than the other. An automated procedure starts with the geometry, Fig. 3, and generates the local cone and toroid geometry, Fig. 4, for each slice.

The dominance of the radial fibers in the main, central portion of the eardrum provides justification for a simplified mechanical model in which the circumferential elastic coupling is neglected.

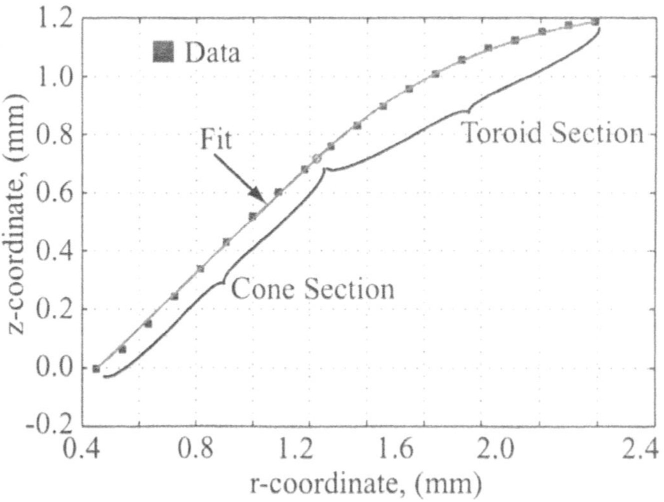

Figure 4. Best fit of a cone and toroid through a particular slice. Each slice is different, but this fit is typical.

3.2 Model

The air in the ear canal (Fig. 2) is treated by one-dimensional acoustics, in which the pressure on one plane perpendicular to the center line is assumed to be constant. The combination of the ear canal and shell is shown in Fig. 5.

3.3 Computational Procedure

A block diagram of the computational system is indicated in Fig. 6. For this preliminary work, the cochlear load is approximated by a spring and dashpot, and the ossicles are represented by a ball and stick. Each radial slice of the eardrum is divided into ten patches in the radial direction. On each patch the pressure is assumed to be constant. Because of the uncoupling of the slices, the computation for each slice is done independently. For confirmation, results for 20 patches showed no significant difference in the results.

For each frequency and for each slice, a flexibility matrix is computed, which provides the vector of volume displacements of the patches to the vector of pressures on each patch. The calculation is performed with Fast4, which is a shell-of-revolution program that combines asymptotic and numerical methods for efficiency (Steele and Shad, 1995). The pressure is assumed to be constant on a patch, but there is no restriction on the variation of the shell solution within the patch.

The next step is to obtain the transformation from the slice matrices to each acoustic region between two successive circles in Fig. 5. This is a condensation

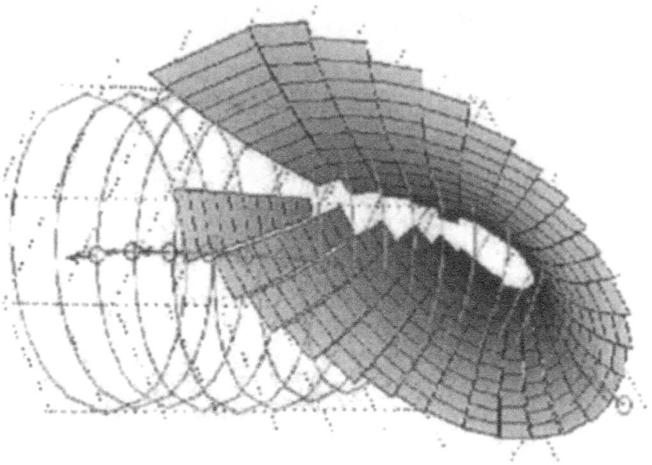

Figure 5. Interaction of tympanic membrane and air in ear canal. The solid line with small circles is the center line for the ear canal. The pressure is assumed constant on each plane perpendicular to the center line. The eardrum is cut into radial slices, and the elastic coupling between slices is neglected. However, the circles of constant air pressure cut across several slices, which provides significant acoustic-shell interaction.

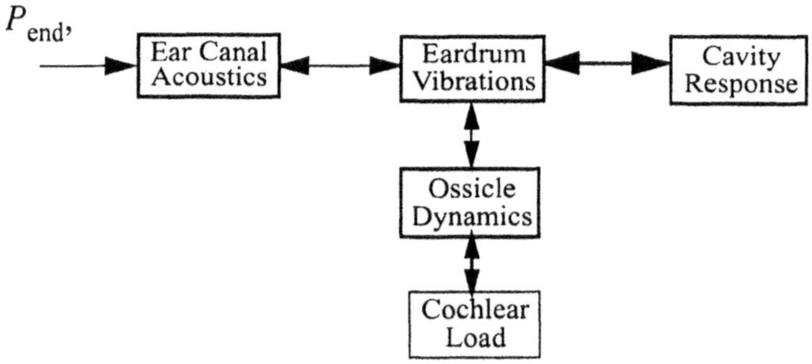

Figure 6. Flow diagram for fully coupled system

from the slice matrices to a pressure-wall displacement value for each acoustic region.

4.4 RESULTS

4.1 Cat Ear

The ear canal inward-looking impedance (with the cavities closed) is shown in Fig. 7. At low frequencies, the magnitude matches the experimental values,

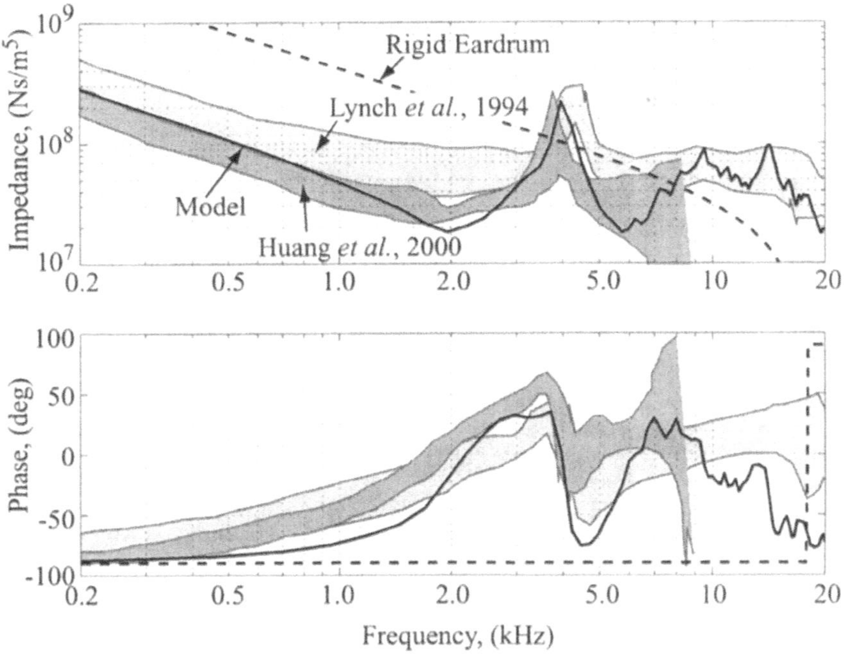

Figure 7. Inward looking ear canal impedance measured close to the eardrum. The middle-ear cavities are closed. The baseline model result is shown as the solid black lines. The lighter gray area is the range of experimental measurements reported by Lynch *et al.* 1994 . The darker gray area is the range of experimental measurements made by Huang *et al.* 2000 . The dashed black line is a calculation with a completely rigid eardrum.

but the value is somewhat sensitive to how close the measurements are made to the eardrum. This could account for how little the two experimental results overlap. At higher frequencies, the 4 kHz cavity resonance dominates the response. The model captures the effect of the cavities. Above 10 kHz, the model remains within the experimental bounds of the Lynch data. Huang does not report measurements in this frequency range. So far as we are aware, this is the first time that a structurally based model has been able to predict the impedance through the entire 200 to 20000 Hz range.

Decraemer *et al.* (1989) measured the response of four points on the eardrum. Figure 8 shows the comparison with the calculations. There is qualitative agreement in both the size of the variations and the frequency locations where the oscillations begin to occur. The rather different oscillations experienced at the different points show the presence of complex modes of vibration, that are best seen with three-dimensional animation.

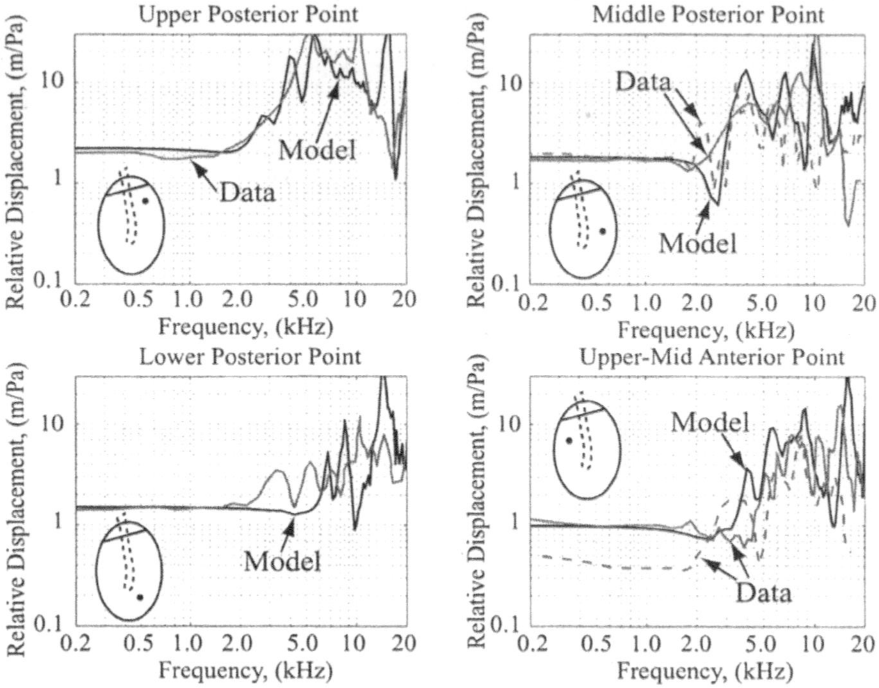

Figure 8. Displacement of four points on the eardrum relative to the umbo displacement. The position of the measurements is indicated by the black dot in the eardrum sketch inserted in each plot. Solid black lines are the model calculations. The lighter lines solid and dashed are data from Decraemer *et al.* 1989 .

4.2 Design Variation

Because the model is in reasonable agreement with the data available for various experimental conditions, there is some confidence that it is behaving much like the real system. Consequently, we may alter parameters and obtain some understanding of the present design. Some model parameters are: 1) the depth of the eardrum, 2) the angle of the eardrum within the ear canal, 3) the fiber arrangement assumed within the drum, and 4) the pre-tension within the membrane.

Here we include only the consequence of changing the depth of the eardrum on the pressure delivered to the inner ear (Fig. 9), because this transfer function determines the threshhold of hearing (Puria et al., 1997). Fig. 9 shows that flattening the eardrum decreases its ability to transfer pressures at high frequencies. By 8 kHz, the shallow eardrum is down over 20 dB compared to the others. The main reason for this is that a conical inner portion provides a better stiffness-to-mass ratio than a flat plate. A plate can only use its bending stiffness while a cone can use its bending stiffness and also its in-plane stiffness to

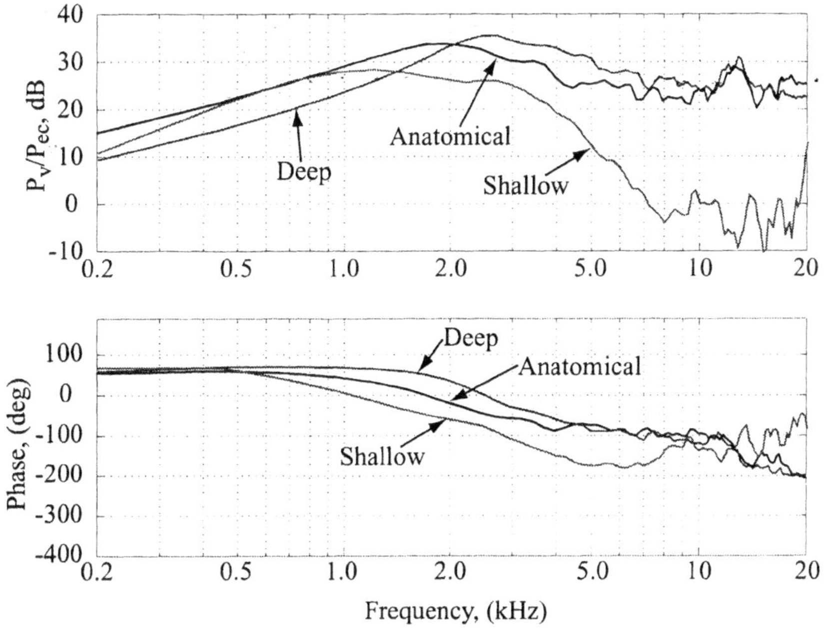

Figure 9. Pressure in the inner ear divided by the ear-canal pressure from the model for three different eardrum geometries. For each case, the shape of the drum was modified by multiplying the z-coordinate by a factor. The shallow case is with 0.1*z. The anatomical case (black line) is with 1.0*z. The deep case is with 2.0*z.

drive the umbo, which is the point of attachment to the end of the bone. Making the cone deeper has little advantage and costs more material.

5. CONCLUSION

The procedure is efficient, requiring about two minutes of computation time on a Macintosh G3 per frequency. The results in Fig. 9 and additional parameter studies make a compelling argument that the feline eardrum's shape, fiber arrangement, inclination in the ear canal, and complex modes are all the result of an evolutionary optimization to provide enhanced hearing capabilities. The orthotropic structure causes each radial slice, with different length and thickness, to have different resonances, Fig. 8. Thus with a deliberate mistuning of the individual slices, the ensemble delivers a fairly flat transfer of pressure to the inner ear for the higher frequencies, as seen in Fig. 9.

References

1. Beer, H.J., Bornitz, M., Hardtke, H.J., Schmidt, R., Hofmann, G., Vogel, U., Zahnert, T., Huttenbrink, K.B. (1999) Modelling of components of the human middle ear and simulation of their dynamic behavior, *Audiol. Neurootol.*, **4** 156-162.
2. Better Hearing Institute (2000) Facts about hearing disorders, (http://www.betterhearing.org/demograp.htm).
3. Decraemer, W.F., Khanna, S.M., Rosowski, J.J., Merchant, S.N. (2001) Complete 3-dimensional motion of the ossicular chain in a human temporal bone, ARO Midwinter Meeting, Abstract #19694.
4. Decraemer, W. F., Khanna, S. M., Funnell, W. R. J. (1989) Interferometric measurements of the amplitude and phase of tympanic membrane vibrations in cat, *Hearing Research*, **38** 1-18.
5. Eiber, A. (1999) Mechanical modeling and dynamical behavior of the human middle ear, *Audiol. Neurootol.* **4** 170-177.
6. Fay, P.G. (2001) *Cat Eardrum Mechanics*, Ph.D. thesis Stanford University.
7. Funnell, W. R. J., Laszlo, C. A. (1978) Modeling of the cat eardrum as a thin shell using the finite-element method, *J. Acoust. Soc. Am.* **63** (5) 1461-1467.
8. Helmholtz, H. (1869)The mechanism of the ossicles of the ear and membrana tympani, *Pfluger Archiv fur Physiologie*, Vol. 1.
9. Hudde, H., Engel, A. (1998) Measuring and modeling basic properties of the human middle ear and ear canal. Part III: Eardrum impedances, transfer function and model calculations, *Acta Acustica*, **84** 1091-1109.
10. Huang, G.T., Rosowski, J.J., Puria, S., Peake, W.T. (2000) A non-invasive method for estimating acoustic admittance at the tympanic membrane, *J. Acoust. Soc. Am.*, **108** (3 Pt 1) 1128-1146.
11. Khanna, S. M. and Stinson, M. R. (1985) Specification of the acoustical input to the ear at high frequencies, *J. Acoust. Soc. Am.* **77** (2) 577-589.
12. Khanna, S. M. and Tonndorf, J. (1972) Tympanic membrane vibrations in cats studied by time-averaged holography, *J. Acoust. Soc. Am.* **51** (6, Part 2)1904-1920.
13. Lynch, T. J. III, Peake, W. T., Rosowski, J. J. (1994) Measurements of the acoustic input impedance of cat ears: 10 Hz to 20 kHz, *J. Acoust. Soc. Am.* **96**(4) 2184-2209.
14. National Institute on Deafness and Other Communication Disorders (2000) Health information: hearing and balance (http://www.nih.gov/nidcd/health/hb.htm).
15. Puria, S., Peake, W.T., Rosowski, J.J. (1997) Sound-pressure measurements in the cochlear vestibule of human-cadaver ears, *J. Acoust. Soc. Am.* **101**(5) 2754-2770.
16. Puria, S. and Allen, J. B. (1998) Measurements and model of the cat middle ear: Evidence of tympanic membrane acoustic delay, *J. Acoust. Soc. Am.* **104** (6) 3463-3481.
17. Steele, C.R. and Shad, K.uR. (1995). Asymptotic-numeric solution for shells of revolution, *Applied Mechanics Reviews*, 48(11, part 2).
18. Stinson, M. R. and Khanna, S. M. (1994) Spatial distribution of sound pressure and energy flow in the ear canals of cats, *J. Acoust. Soc. Am.* **96** (1) 170-180.
19. Wada, H., Metoki, T., Kobayashi, T. (1992) Analysis of dynamic behavior of human middle ear using a finite-element method, *J. Acoust. Soc. Am.* **92** (6) 3157-3166.
20. Williams, K. R., Blayney, A. W., Lesser, T. H. J. (1997) Mode shapes of a damaged and repaired tympanic membrane as analysed by the finite element method, *Clin. Otolaryngol,* **22** 126-131.
21. Zwislocki, J. J. (1962) Analysis of the middle-ear function. Part I: Input impedance, *J. Acoust. Soc. Am.*, **34** (8, Pt. 2)1514-1523.

INVERSION OF METAL CYLINDERS

A rigid-plastic model for the first stage of forced inversion

J. J. Harrigan and S. R. Reid
Department of Mechanical, Aerospace & Manufacturing Engineering, UMIST, P.O. Box 88, Manchester, M60 1QD, U.K.

john.j.harrigan@umist.ac.uk and steve.reid@umist.ac.uk

Abstract The quasi-static behaviour of inversion tubes can provide insight into features of the load pulses generated during dynamic/impact loading of the tubes. A rigid-plastic analysis of the first stage of forced, internal inversion is provided. The assumption that the tube conforms to the profile of the shaped die is removed. The influence of the interaction surface in plastic deformation is investigated by utilising a three-dimensional Tresca yield criterion for shells.

Key words: Tube inversion, Yield criteria, Impact energy absorption

1. INTRODUCTION

Metal tubes are used extensively in impact energy absorbing systems and devices [11,14]. Tube inversion is a particularly effective and efficient deformation mechanism for applications requiring a constant load over large displacements. Internal inversion is the process of turning a tube "outside-in" thus creating a co-axial cylinder of smaller diameter than the original. Forced inversion results from the axial compression of a tube onto a suitably profiled die [2,3].

The loading arrangement and a typical load-displacement curve for quasi-static forced internal inversion are shown schematically in Fig. 1. A constraining barrel was used to prevent radial expansion and possible collapse of the part of the tube that is not constrained by the die. All results

H.R. Drew and S. Pellegrino (eds.), New Approaches to Structural Mechanics, Shells and Biological Structures, 303–314.

reported are for cold drawn steel tubes with a yield stress of 570 MN/m^2 and an outside diameter of 101.6 mm. The contact surfaces were lubricated. Good agreement between finite element analysis and experimental results was achieved for a coefficient of friction of 0.15, for which the steady-state loads predicted using finite element analysis were increased by typically 40% [8] compared with a frictionless model.

The three stages of inversion are indicated on the load-displacement plot in Fig. 1 for a tube thickness of 2 mm and a die-forming radius of 5 mm. During stage 1 only the leading edge of the tube is in contact with the forming die, as was first noted experimentally by Al-Hassani *et al.* [2]. There is often a noticeable maximum in the load during stage 1. This feature is more pronounced for increasing t_o/r and t_o/R_o ratios, where t_o and R_o are the initial thickness and radius of the tubes and r is the forming radius on the die. During stage 2 the tube conforms more fully to the die. A second unloading occurs during stage 2 at a stroke for which the leading edge has rotated so far that it is no longer in contact with the die. Reid and Harrigan [15] predicted this second unloading using rigid-plastic analysis. Under quasi-static conditions stage 3 is steady-state inversion at a constant load with a high degree of conformity between the tube and the die.

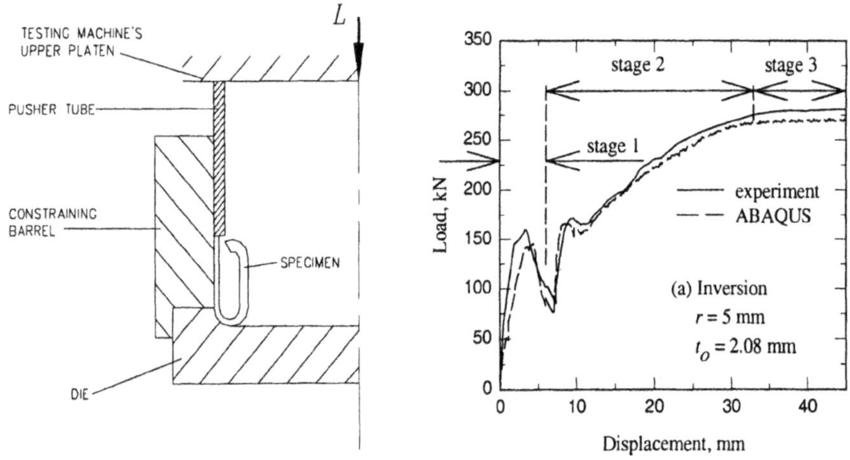

Figure 1. Quasi-static loading arrangement and typical load-displacement curve [15].

During dynamic inversion, force peaks can occur during the early stages that are far in excess of the stage 3 loads for quasi-static inversion [15]. These dynamic peaks are more prominent for geometric parameters that produce notable maxima during stage 1 of quasi-static inversion [9]. The presence of the load drop during stages 1 is therefore of extreme importance to the use of inversion tubes as impact energy absorbers.

2. BACKGROUND

Historically, in the rigid-plastic analysis of axially compressed tubes, both the assumed kinematic deformation mechanisms and the yield criterion used tend to incorporate simplifying assumptions. In this way axial crushing loads have been estimated for complex problems. Alexander's analysis of axisymmetric buckling of a cylindrical tube [1] is a good example. A non-interactive yield criterion was employed and the deformation was considered to consist of rotation at three plastic hinge circles and circumferential stretching of the material between these hinge circles. The work done by the applied load was equated to the energy dissipated in plastic deformations during the flattening of a buckle region. The characteristic length between the hinge circles was then determined by minimising the mean crushing load during this deformation process with respect to the characteristic length. The predicted length approximates well experimental observations. Several large-deformation plastic shell problems, including tube buckling, tube inversion [7] and spherical shell inversion [6], have been analysed in this manner. The attractiveness of the approach is the simplicity it can offer for extremely complicated deformation mechanisms. By quantifying the dissipation of energy in plastic zones/hinges a fundamental physical understanding of the deformations can be achieved. This physical appreciation can be lost when analysis is carried out via finite element packages. However, there is a flaw in this type of approach. This is discussed by Calladine [5], who states that "Disposable parameters in an assumed mode of propagation cannot in general be determined directly by means of a minimisation of the energy dissipated". Put simply, problems involving gross plastic deformations are history-dependent and Alexander's prediction of a critical length during axisymmetric compression owes more to good fortune than rigorous analysis. Unfortunately, the issues addressed in Ref. [5] do not appear to have been appreciated fully by the wider research community, as an examination of many engineering journals will testify.

Reddy [13] provided a first approximation to the overall load-displacement characteristic for both internal and external inversion. Stages 1 and 2 are treated as a single continuous deformation process. The outer surface of the tube wall was assumed to conform fully to the die radius, the knuckle radius was assumed to be uniform, and changes in meridional curvature occured at a travelling hinge circle only, i.e. at the point at which the tube enters the die. The rigid-plastic tube material was assumed to deform according to a criterion in which there is no interaction between bending and membrane stress resultants, although direct stress resultants are interactive according to the Tresca criterion. The rate of work of the applied load was equated to the dissipation of energy in the assumed kinematic mechanism. The assumptions relating to both the kinematics and the yield criterion simplified the problem greatly. However, certain important features of the process are overlooked as a consequence.

The accurate predictions of the quasi-static and dynamic inversion characteristics obtained using finite element analysis [15] result from allowing the changes in geometry and contact to develop naturally and progressively. The study reported here was motivated by the presence of peak loads during dynamic inversion [9,15]. Only stage 1 is considered in detail. A rigid-plastic analysis is used to facilitate a better understanding of the deformation mechanism during this stage. The model is based on solving the equilibrium equations while employing a more accurate representation of the yield condition than that in previous studies. The simplification of assuming a non-interacting yield criterion has been removed. The majority of the kinematic assumptions have been removed also. Only one kinematic restraint is imposed. This simplifies greatly the yield surface employed in the analysis.

The form of the Tresca condition for a uniform shell was first obtained by Onat and Prager [12]. The resulting four-dimensional surface is generally presented as a table of equations [10]. At a given location on the shell wall, the choice of which of the set of tabulated equations defines the stress resultant relationships depends on the relative values of three non-dimensional direction parameters. These parameters define the direction of the strain-rate vector and hence the normal to a point on the yield hypersurface. For cylindrical shells the system of equations is simplified [4], as a three-dimensional yield surface defines the fully plastic behaviour. This yield surface can be derived without reference to non-dimensional direction parameters [16]. The key step in the analysis used below is formulating the yield criteria that are valid and appropriate through the deforming region. A full discussion is given in Ref. [8]. Only the key elements will be given below.

3. ANALYSIS

In order to obtain an analytical solution for the first stage of internal inversion, for a material that yields in accordance with the Tresca criterion, the following assumptions are made:

1. Shell theory applies, so that plane sections remain plane and the through-thickness stresses are negligible compared to those in the circumferential and meridional directions.
2. Contact is considered to occur at a point.
3. At all points in the tube wall $\dot{\varepsilon}_\theta \leq 0$.

The assumption that the circumferential strain rate is negative throughout the deforming portion of the tube means that all possible combinations of stress resultants need not be considered. At the same time important effects associated with interaction in loading between the generalised stress resultants are retained and the majority of the kinematic assumptions of previous analyses are removed. In previous analyses of inversion problems it has been assumed that $\dot{\varepsilon}_\theta \leq 0$ for the first stage of internal inversion and that $\dot{\varepsilon}_\theta \geq 0$ during the first stage of external inversion. However the most restrictive of the previous kinematic assumptions is removed, *viz.* that meridional bending occurs only at the entrance to the die and that the deforming portion of the tube conforms to the die.

The direct and bending stress resultants for the meridional direction, which is denoted by the sub-script x are as follows:

$$N_x = \int_{-t/2}^{t/2} \sigma_x \, dz , \tag{1}$$

$$M_x = - \int_{-t/2}^{t/2} \sigma_x \, z \, dz , \tag{2}$$

where the z co-ordinate is measured outward from the mid-surface. M_θ and N_θ are defined similarly. The yield condition is defined using dimensionless stress resultants $n = N/\sigma_o t$ and $m = 4M/\sigma_o t^2$, where σ_o is the yield stress.

During stage 1 the force per unit circumferential length at the point of contact (C of Fig. 2) has axial and radial components. In the vicinity of the leading edge M_x is negative and increases in magnitude with increasing x. N_x is negative also. This implies that the governing yield conditions are those stated in Section 3.1.1. The deformation mechanism illustrated in Fig. 2 results from combining the kinematic assumption with the Tresca criterion. Solutions of the equations of equilibrium and testing of the yield criteria require the use of two other criteria for section II and point G.

3.1 TRESCA YIELD CONDITION FOR $\dot{\varepsilon}_\theta \leq 0$

If at any point on the tube wall there exists a known loading in terms of m_x and n_x, provided circumferential strain increments are less than or equal to zero everywhere, the values of m_θ and n_θ which correspond to fully plastic yielding can be calculated. Three distinct sets of governing equations define the three portions of the yield surface that can apply while $\dot{\varepsilon}_\theta \leq 0$. The Tresca yield criterion is plotted for stresses in Fig. 3. All points in the deforming region of the tube must lie on this surface.

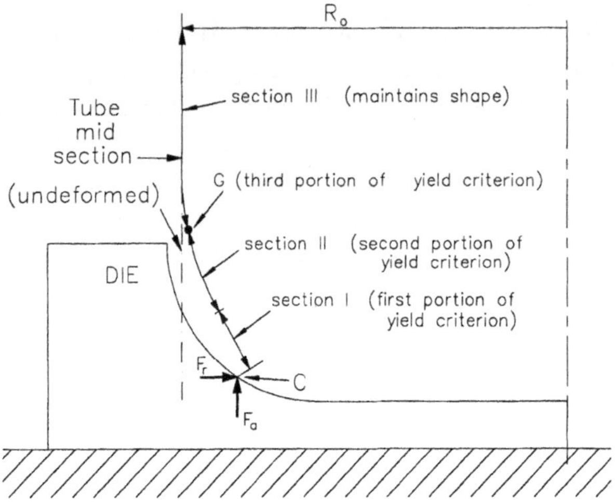

Figure2. Deformation mechanism in rigid-plastic analysis

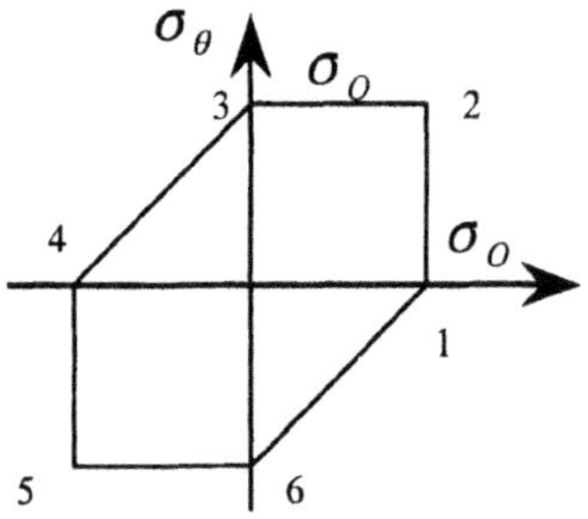

Figure 3. Tresca yield criterion.

3.1.1 Portion of yield surface appropriate to Section I

The solution procedure implies that in Section I m_x and n_x lie within that portion of the yield surface described by

$$\text{(i)} \quad |m_x| \leq 2\,|n_x|(1-|n_x|), \tag{3}$$

$$\text{(ii)} \quad -m_x + n_x^2 < 1.$$

The first limit, equation 3(i), provides the maximum magnitude of m_x for which the meridional stress can be packed into one block, see [16]. The second limit, equation 3(ii), is simply the limit provided by the two-dimensional Tresca criterion for combined bending and membrane loading in the meridional direction. Figure 4(a) illustrates the distribution of stress and strain-rate in the through-thickness direction corresponding to the range of values of m_x and n_x that lie within these limits. The stress and strain-rate distributions result from combining m_x and n_x in such a way as to satisfy the kinematic assumption. The part of the Tresca yield criterion, Fig. 3, that applies to each section of material through the thickness is indicated in Fig. 4. Stress resultants m_x and n_x can be derived from the equilibrium equations and the circumferential stress resultants can be calculated from the circumferential stress distribution, which leads to

$$\text{(i)}\quad n_\theta = -1, \qquad \text{(ii)}\quad m_\theta = 0 \tag{4}$$

3.1.2 Portion of yield surface appropriate to Section II

The through-thickness stress and strain-rate distributions for the second portion of the exact Tresca criterion that is applicable while $\dot{\varepsilon}_\theta \leq 0$ are shown in Fig. 4(b). In this portion of the yield surface the meridional stress resultants lie within the limits described by Equations 3(ii) and by

$$|m_x| > 2\,|n_x|\,(1-|n_x|)\,. \tag{5}$$

The specific meridional stress resultants can be defined in terms of two non-dimensional parameters a and b (Fig. 4(b)), where $a > b > 0$ and $a + b <$ 2, given by equations (6).

$$\begin{aligned}&\text{(i)}\quad m_x = -\,a\left(1 - \frac{a}{2}\right) - b\left(1 - \frac{b}{2}\right)\\ &\text{(ii)}\quad n_x = \frac{b-a}{2}\end{aligned} \tag{6}$$

The circumferential stress resultants can be found in terms of a and b since only one through-thickness stress distribution is consistent with the kinematic assumption:

$$\begin{aligned}&\text{(i)}\quad n_\theta = \frac{b}{2} - 1\\ &\text{(ii)}\quad m_\theta = -\,b\left(1 - \frac{b}{2}\right)\end{aligned} \tag{7}$$

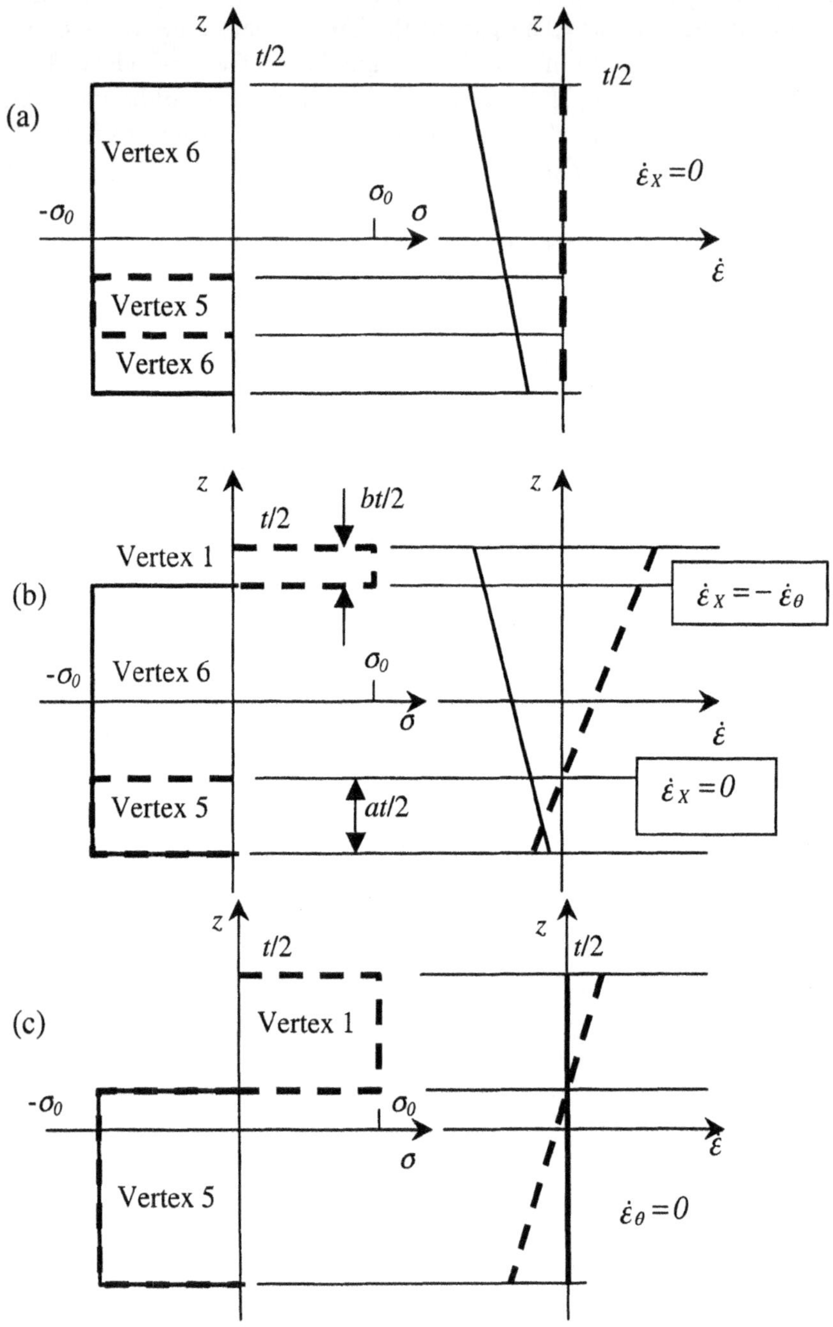

Figure 4. Stress and strain-rate distributions. Solid and dashed lines denote circumferential and meridional directions respectively.

3.1.3 Yield condition at G

The dimensionless stress resultant m_x increases through sections I and II and finally satisfies Equation 3(ii) with < replaced by =. In terms of the meridional direction it is simply the two-dimensional Tresca condition for combined bending and direct loading. The through-thickness stress and strain-rate distributions for the third portion of the exact Tresca criterion that can apply while $\dot{\varepsilon}_\theta \leq 0$ are shown in Fig. 4(c).

3.2 STAGE 1 DEFORMATION MECHANISM

For the assumptions, stated it is shown in [8] that the resulting deformation mechanism is that illustrated in Fig. 2. The mid-thickness line of the deforming tube is split into three regions. At any level of deformation there is a meridional length of tube starting at the leading edge of the deforming region (Section I) within which the normalised stress resultants correspond to those for the first portion of the Tresca criterion, Fig.4(a). Within the next meridional length (Section II of Fig. 2) the normalised stress resultants correspond to the second portion of the Tresca criterion, Fig. 4(b). At point G of Fig. 2 the normalised stress resultants correspond to those of the third portion of the Tresca criterion, Fig. 3(c). In Section III of Fig. 2 the stress resultants are below the levels required for yielding so that no deformation occurs in this section over the next incremental compression. However, as the deformation develops the geometry and loading change. Therefore some parts of Section III may previously have been deformed and during continuing deformation the lengths of Sections I and II, and subsequently the position of point G will vary. At a given displacement the values of the stress resultants at any mid-thickness location are governed by the geometry of the deforming tube at that displacement. During any incremental deformation, the normality rule and constancy of volume can be used to find the strain-rate vectors associated with the changes in curvature and lengths. The deformation is similar to that for a hinge rotation mechanism, with a pivot at point *G*, but with additional changes in shape in Sections I and II. In Section I meridional lengths remain constant, Fig. 4(a), the thickness increases and there is no change in meridional curvature. In Section II meridional lengths shorten while the thickness increases. Additionally there are changes to meridional curvature.

A discrete model has been used to obtain approximate load-displacement solutions to stage 1 of internal inversion [8] and limitations on space precludes it being described in detail here. A brief summary of the model is provided below.

3.3 THE DISCRETE MODEL

In the model, equilibrium equations are applied to the deformation mechanism described above. The mid-thickness of the tube is represented by a series of points that are separated by a small distance (typically 0.1mm). A further point monitors contact between the tube and the die which moves axially relative to the tube. The top end of the tube is fixed and far enough from the deforming region to be unaffected by plastic deformation. All stress resultants are assumed to be constant over the small elemental lengths and are calculated at the mid-thickness points only.

The solution is obtained via a series of incremental deformations of the discrete model. Initially the model represents the undeformed tube and therefore is in the form of a straight, circular cylinder. Equilibrium equations are used to find the plastic collapse load for the given geometry by an iterative procedure. Next an incremental change in geometry (corresponding to an incremental plastic deformation) is specified. This geometry change affects all the mid-thickness points and all element dimensions for the deforming portion of the tube. For all these elements the incremental strains are determined using the values of the specific stress resultants, normality and constancy of volume. This procedure is repeated for each incremental change in geometry.

The load-displacement predictions obtained are compared with experimental results in Fig. 5 for tube thicknesses ranging from 1.07 to 3.54 mm and die forming radii of 5 mm and 10 mm. A coefficient of friction of 0.15 was assumed in the analysis [15]. There is good agreement between theory and experiment. In particular the reduction in load that occurs in stage 1 is predicted by the rigid-plastic model. This suggests that the instability that exists often in stage 1 is not a consequence of elastic buckling, but is due to the rotation of a plastic bending zone and the changing contact angle at the leading edge of the tube. The deformation mechanisms are illustrated in Fig. 6 for a 2 mm thick tube inverted by dies with 5 mm and 10 mm forming radii.

4. CONCLUSIONS

Although the Stage 1 deformation mechanism is dominated by combined meridional bending and circumferential compression, the meridional bending is concentrated in a region which is some distance from the leading edge of the tube, which cannot therefore conform to the die. The unloading during this stage is a consequence of both the large geometry changes and the changing contact angle and is not in general the result of a buckling or bifurcation load being reached. The transition from stage 1 to stage 2 is caused by a "second contact" between the deforming specimen and the die. The sharp rise in load that follows the first major instability is caused by this transition. Subsequent studies have shown that under dynamic loading conditions inertial effects delay the onset of meridional bending resulting in significant enhancements of the applied force. This is critical to the use of inversion tubes as a load limiter/energy absorber component.

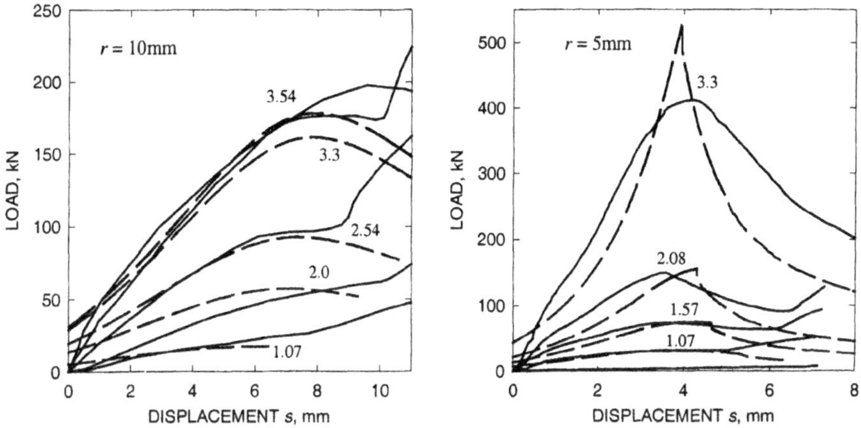

Figure 5. Experimental (solid lines) and predicted (dashed) load-displacements. Tube thicknesses indicated are in mm.

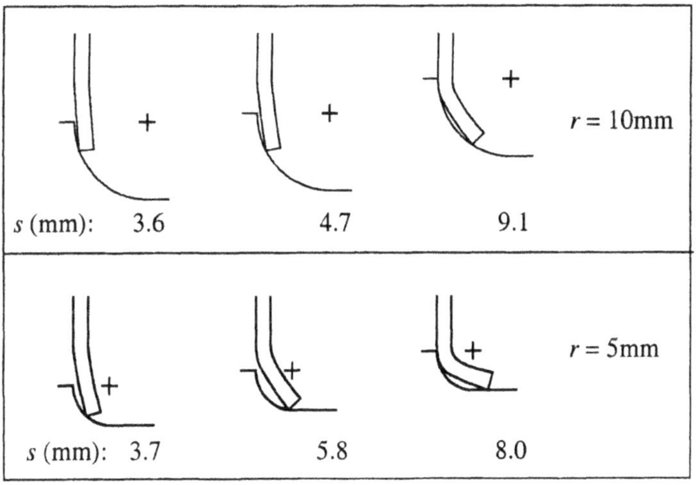

Figure 6. Stage 1 deformations predicted by rigid-plastic analysis.

References

1. Alexander, J.M. An approximate analysis of the collapse of thin cylindrical shells under axial loading, *Quart Journal of Mechanics and Applied Mathematics* **13** (1960) 10-15.
2. Al-Hassani, S.T.S., Johnson, W. and Lowe, W. T. Characteristics of inversion tubes under axial loading, *J Mech Engng Sci* **14** (1972) 370-382.
3. Al-Qureshi, H.A. and Demorais, G. A. Analysis of multi-inversion of tube ends, *Design Engng. Conf.*, ASME paper No. 77-DE-35 (1977).
4. Calladine, C.R. *Theory of Shell Structures*, Cambridge University Press, Cambridge, 1983.
5. Calladine, C.R. Some problems in propagating plasticity, in *Plasticity and Impact Mechanics*, edited by Gupta, N.K., Wiley Eastern, New Delhi, 1993, pp.71-97.
6. de Oliveira, J.G. and Wierzbicki, T. Crushing analysis of rotationally symmetric plastic shells, *Journal of Strain Analysis* **17** (1982) 229-236.
7. Guist, L.R. and Marble, D.P. Prediction of the inversion load of a circular tube, NASA TND 3622 (1966).
8. Harrigan, J. J. Internal inversion and nosing of laterally constrained metal tubes, *Ph.D. Thesis*, UMIST (U. K.) 1995.
9. Harrigan, J.J., Reid, S.R. and Peng, C. Inertia effects in impact energy absorbing materials and structures, *Int J Impact Engng* **22** (1999) 955-979.
10. Hodge, P.G. *Limit Analysis of Rotationally Symmetric Plates and Shells*, Prentice Hall, Englewood Cliffs, NJ, 1963.
11. Johnson, W. and Reid, S.R. Metallic energy dissipating systems, *Appl. Mech. Rev.* **31** (1978) 277-288; See also *Applied Mechanics Update* (Edited by C.R. Steele), 303-319 ASME, New York, 1986.
12. Onat, E.T. and Prager, W. Limit Analysis of Shells of Revolution, *Proc Roy Neth Acad Sci* **B57** (1954) 534-548.
13. Reddy, T.Y. Tube inversion - an experiment in plasticity, *Int J Mech Engng Educ* **17** (1989) 277-291.
14. Reid, S.R. Plastic deformation mechanisms in axially compressed metal tubes as impact energy absorbers, *Int J Mech Sci* **35** (1993) 1035-1052.
15. Reid, S.R. and Harrigan, J.J. Transient effects in the quasi-static and dynamic plastic internal inversion and nosing of metal tubes, *Int J Mech Sci* **40** (1997) 263-280.
16. Robinson, M. The teaching of the theory of yield surfaces for perfectly plastic thin shells, *Int J Mech Engng Educ* **19** (1991) 1-8.

FAILURE CRITERIA FOR SHELLS ON LOCAL BRACKETS

J.M.F.G. Holst, J.M. Rotter, M. Gillie and M. Münch
School of Civil and Environmental Engineering, University of Edinburgh
The King's Buildings, University of Edinburgh, Edinburgh, EH9 3JN, U.K.
m.holst@ed.ac.uk

Abstract Several criteria are given in design standards for determining the failure load of cylindrical shell structures. When the loading is nonuniform, a nonlinear finite element analysis is required to capture fully the deformation characteristics of the shell and to obtain a true estimate of the failure load. However, even if a fully nonlinear analysis is performed, the determination of the strength of the shell is not always easily achievable. Several different forms of nonlinear load-deformation behaviour are found. The adequacy of existing design guidelines for obtaining an estimate of the design strength is examined. For certain types of behaviour alternative methods are suggested that may provide a more definitive measure. The numerical results are based on silos supported locally on brackets. The results are, however, more widely applicable to shells under a general set of loads. The bracket-supported silo is used in the calculations as it provides a good example of a simple design, but with a relatively complex pattern of deformations leading to a highly nonuniform stress state.

Keywords: buckling, failure criteria, local brackets, plastic collapse, silos, Southwell plot

1. INTRODUCTION

Most granular solids storage containers such as silos are discharged by gravity and so need a hopper with access space beneath. Where the silo is discharged into vehicles or conveyors, the clear space must be extensive and the silos will therefore often be supported on columns. Large silos generally require a heavy ring beam, Fig. 1, to redistribute the local forces from the columns into the silo shell. This design is expensive to construct and smaller silos can be built successfully with a much cheaper arrangement in which the columns support brackets attached to the side of the silo, Figs 2 and 3. While such designs are often used, their

H.R. Drew and S. Pellegrino (eds.),
New Approaches to Structural Mechanics, Shells and Biological Structures, 315–327.

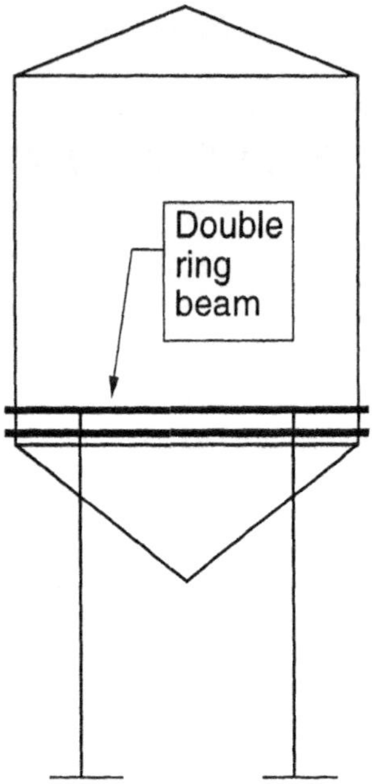

Figure 1. Ring-supported silo.

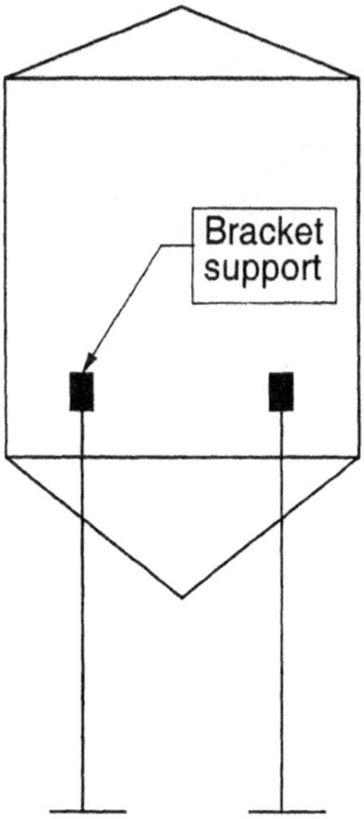

Figure 2. Bracket-supported silo.

strength has never been systematically investigated by researchers and limitations on their usage have only been discovered through in-service failures. A detailed study is currently underway which is attempting to address this shortcoming and to define appropriate design guidelines (Holst, Münch, and Rotter, 2001; Gillie and Holst, 2002; Gillie, Holst, Rotter and Münch, 2002).

In general, the first concern for the designer of a silo is to establish its ultimate load capacity. Several methods are given in the recent European Standard for Steel Shells (ENV 1993-1-6, 1999) for determining the failure load of a cylindrical shell structure. The standard attempts to define the strength of shells at several critical limit states and under various loading regimes. At the same time, it provides the designer with several design methods, the choice of which will depend on the complexity of the silo and the designer's ability to perform different computational analyses. Thus, where the conditions for the application of a particular method are satisfied, the designer may use either stress design or a global numerical analysis. In stress design, the stresses are obtained either by linear shell analysis or from membrane theory. In a global numerical analysis, the design strength is found from

- eigenvalue analysis based on linear shell analysis (LA),

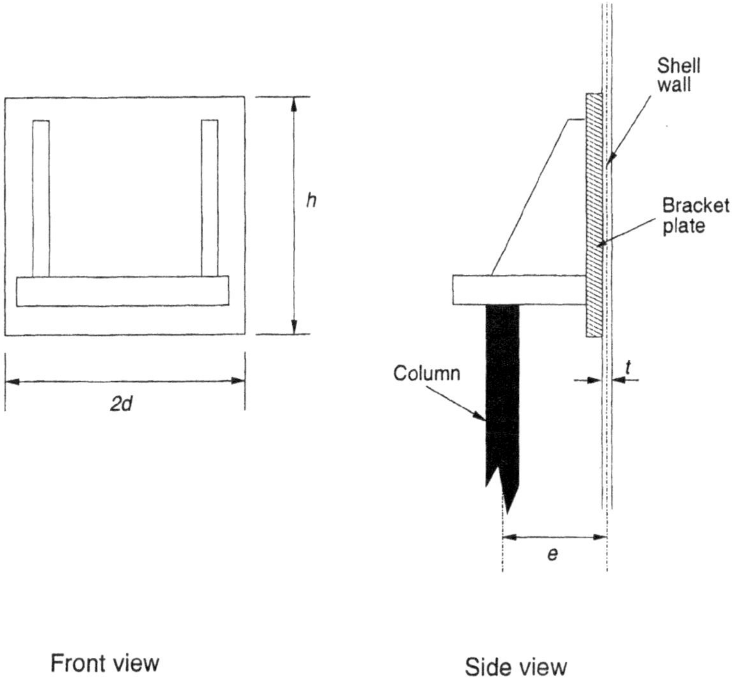

Figure 3. Detail of an eccentric bracket.

- geometrically nonlinear elastic analysis (GNA),
- materially nonlinear analysis (MNA),
- geometrically and materially nonlinear analysis (GMNA),
- geometrically and materially nonlinear analysis with imperfections included (GMNIA).

The design of most silos in general use is governed by the plastic and buckling limit states. For silos where the stress distribution is significantly nonuniform around the circumference, such as those supported on local brackets, the deformation behaviour tends to be highly nonlinear. In some circumstances, design by linear stress analysis can still provide a reasonable estimate of the true strength for such shells, but in most cases a nonlinear analysis is essential if the behaviour under load is to be captured reliably and a good estimate of the strength is to be obtained.

However, even if a fully nonlinear analysis is performed, it is not always easy to decide on a value for the strength of the shell. This paper explores several different types of nonlinear load-deformation behaviour and examines how the existing design guidelines can be applied to obtain an estimate of the design strength. For certain types of behaviour it suggests alternative methods that may provide a more definitive measure. Throughout the paper, the numerical results are based on silos supported locally on brackets. The results are, however, more widely applicable to shells under a general set of loads. The bracket-supported silo is used as it provides a good example of a simple design, but which undergoes a relatively complex pattern of deformations and leads to a highly nonuniform stress state involving both membrane and bending effects.

This paper is intended to stimulate more extensive discussion on the subject of failure criteria for shells under nonuniform stress states.

2. NUMERICAL MODEL

The geometric form of a typical silo is shown in Fig. 4. The cylindrical shell portion has a radius r, thickness t, and height H. For all the example structures presented in this paper, the shell has a thickness $t = 10$ mm and a height $H = 4r$. The silo is supported on four equally spaced brackets. Each of these consists of a rigid plate of dimensions $h \times 2d$ welded to the shell wall, Fig. 3. The plate is rigidly connected to a further radially projecting plate with attached gusset plates. The column supports are located at an eccentricity e from the shell wall and the load from the silo is transferred to the columns by means of the bracket arrangement. In the numerical analyses discussed in this paper the brackets are positioned midway up the cylindrical shell to ensure that there is little interference from the shell boundaries. The relative dimensions of the bracket are kept constant in all analyses ($h/H = 0.007$; $2d/R = 0.028$).

Silos are designed for the storage of various types of bulk material. Some of the total load due to the bulk material is transferred directly to the shell wall by means of friction. However, for a silo of average height, a large proportion of the total load, typically 80%, is carried by the base or hopper and then transferred indirectly to the cylindrical shell. For the analyses presented here, the loading was simplified so that a uniform axial load was applied only to the lower edge of the shell, corresponding to the dominant loading.

The cylindrical shell was modelled using the commercial finite element analysis package ABAQUS (HKS, 2000). To reduce the computational expense, the analyses modelled one eighth of the shell only, with symmetry boundary conditions applied along the meridional edges. One of these edges passes through the middle of a bracket and the other one

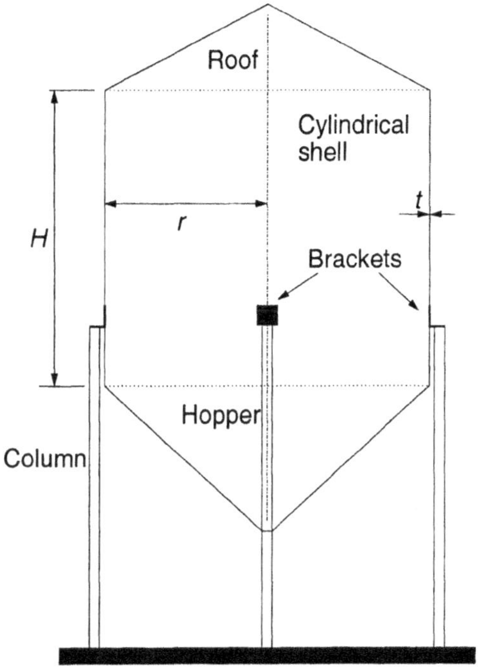

Figure 4. Details of silo geometry.

Figure 5. Displaced finite element mesh.

lies midway between two brackets. The finite element mesh is shown in Fig. 5.

The supporting columns were assumed to be rigid in the axial and circumferential directions. Hence the bracket was restrained against axial and circumferential displacements, but radial displacements were permitted, as were rotations about the bracket-column junction. To represent the restraining effect of the hopper and roof of the silo, the top and bottom of the cylindrical shell were restrained radially and circumferentially and against rotations about the radial and axial directions.

The material properties of the cylindrical shell were taken to be those of mild steel. Thus the material was assumed to be elastic perfectly plastic with a Young's modulus of elasticity $E = 2 \times 10^5$ N/mm^2, Poisson's ratio $\nu = 0.3$ and yield stress $\sigma_y = 250$ N/mm^2.

The nonlinear analyses were performed using the modified Riks arc-length technique (HKS, 2000).

3. PLASTIC LIMIT STATE

Steel structures failing by plastic collapse are characterised by a ductile response which allows a yield plateau to develop in the load-deflection curve in a small-displacement analysis. If a global numerical analysis is used to evaluate the plastic limit resistance then ENV 1993-1-6 (1999) permits two analysis types.

According to ENV 1993-1-6 (1999), if an MNA analysis is used, then the strength may be taken as the largest value attained in the analysis. In most cases, and in particular where strain-hardening is not an issue, this equates to the classical plastic limit load. However, in problems involving highly localised mechanisms, such a plateau may not be reached within a practically achievable number of increments of the analysis; and the analyst must attempt to identify the true plastic limit load from the available information, see the MNA curve in Fig. 6. The maximum load reached is often a very conservative measure of the strength, but extrapolation of the load-deflection curve to find the asymptotic limit is not a reliable process.

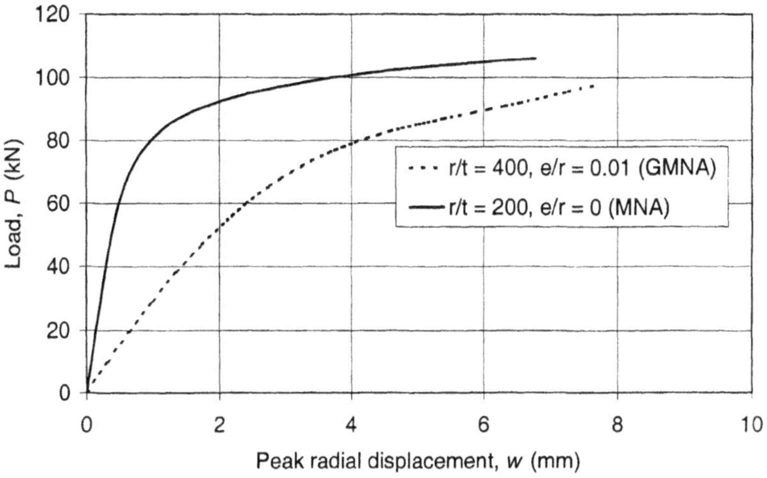

Figure 6. Load against peak radial displacement: plateau behaviour.

It is proposed here that a Southwell type plot can be used to provide a much better estimate of the asymptotic plateau load. The Southwell plot was originally conceived (Southwell, 1932) as a method for obtaining an accurate estimate of the elastic buckling load P_{cr} from experiments on columns. Thus, if the geometric imperfections are in the form of the buckled shape, the measured deflection w can be written as

$$w = \frac{w}{P} P_{cr} - w_0 \tag{1}$$

where P is the applied load and w_0 is the initial imperfection amplitude in the fundamental buckling mode. A plot of w against w/P gives a straight line with slope P_{cr} and intercept w_0. A slightly different form of the above equation is frequently known as the modified Southwell plot (see e.g. Horne and Merchant, 1965)

$$P = P_{cr} - w_0 \frac{P}{w} \tag{2}$$

In this case the asymptotic load is found as the intercept on one of the axes. By choosing this form, calculations that get closer and closer to the desired load level give points that are closer and closer to the target intercept. Much has been written since on the applicability of Southwell type plots to other structures and in other circumstances (e.g. Roorda, 1967; Spencer and Walker, 1975; Calladine, 1983).

The key feature of the Southwell plot is that it transforms a hyperbolic function into a linear one, so if a phenomenon is exactly hyperbolic, it should transform into an exactly linear one. In cylindrical shells, the elastic axisymmetric amplification of initial axisymmetric imperfections follows the hyperbolic path closely, so it is worth examining whether the Southwell plot may also give satisfactory results for the analyses detailed here. In the case of the MNA analysis shown in Fig. 6 where plastic behaviour dominates, the original theory of Southwell relating to elastic columns clearly no longer applies. Furthermore, the elastic-plastic behaviour is not precisely hyperbolic, even close to the asymptotic load. Nevertheless, a modified Southwell plot of this case provides a good means of obtaining the required reference load, Fig. 7.

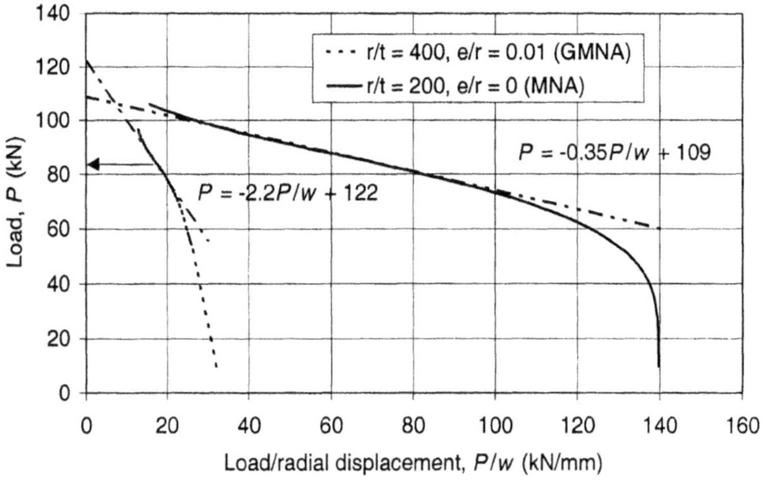

Figure 7. Modified Southwell plot: plateau behaviour.

The MNA curve for the bracket supported shell is initially vertical, indicating a linear load-deflection behaviour. At higher loads the slope of the modified Southwell plot is again linear over a large range before a slight increase in the slope is observed towards the end of the curve. If a tangent is drawn to the second linear portion of the curve then its intersection with the load axis gives an estimate of the plateau load.

The plastic limit resistance may also be obtained from a GMNA analysis. If a maximum occurs on the load-displacement path then this is taken as the resistance for design purposes. However, if the load-displacement relationship rises progressively, see the GMNA curve with $e/r = 0.01$ in Fig. 6, then ENV 1993-1-6 (1999) indicates that the resistance should be taken as the load at which a specified limiting plastic strain occurs anywhere in the shell. For cases such as bracket supported silos, where very high strains occur locally, such a criterion is not practical. In such circumstances the limiting plastic strain may be reached before transverse displacements become excessive or there is any danger of global collapse.

A modified Southwell plot again provides a useful means of defining the design resistance. The GMNA curve does not display an extended portion of uniform slope as in the previous case, Fig. 7. Nevertheless, a point of inflexion can easily be observed and this is bounded by a shorter portion of curve which again has a uniform slope. The intersection of the tangent to this part of the curve with the load (P) axis could again provide a limiting load. However, in this case it is clear that such an extrapolated load would greatly exceed any load that was calculated and might therefore be unsafe.

A safer approach is to select the point of inflexion itself as a bounding value. The curve below this point is nominally hyperbolic and represents the reduction in stiffness due to the shell's nonlinearity. Above the point of inflexion, the stiffness increases because of the change of geometry associated with the large displacements.

4. BUCKLING LIMIT STATE

The European Standard on Shells (ENV 1993-1-6, 1999) gives three criteria for determining the buckling load from a GMNA analysis. The first criterion limits the buckling resistance to the maximum load on the load-deformation curve. Curves of load against peak radial displacement are shown in Fig. 8 for two bracket-supported thin shells with different values of r/t.

In both cases the peak load can be determined easily. It is not clear whether this peak load constitutes failure (Holst and Rotter, 2001). The curve for $r/t = 1000$ rises again after a small postbuckling minimum and the supported load rises. The peak load corresponds to a local dimple

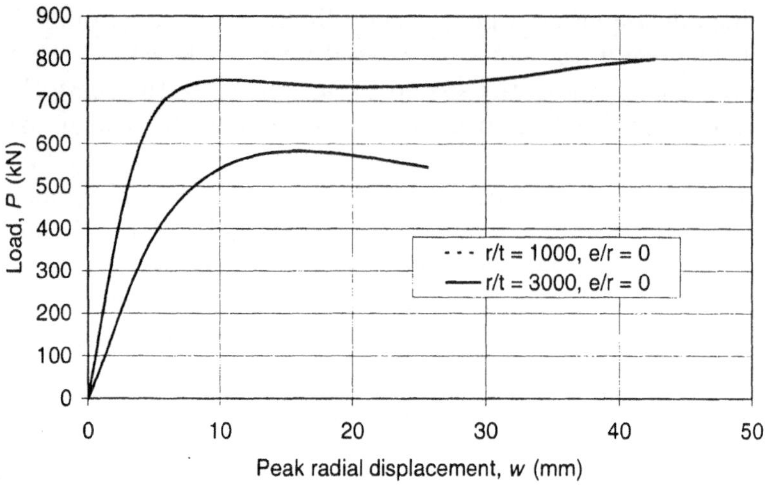

Figure 8. Load-displacement plot: peak behaviour.

forming just above the bracket. This may simply constitute a cosmetic issue for the silo owner.

The second criterion sets the buckling load equal to the bifurcation load if this occurs before a limit point is reached. Since there are no bifurcations in the analyses presented here, this criterion is not further examined.

Finally, if the deformation exceeds the largest tolerable value before either of the other limits is reached, the buckling load is taken as the load corresponding to this deformation. No value is given in the code for this limit. This may be because it is difficult to define such a limit for all circumstances. In practice, the largest tolerable deformation might be taken as a radial displacement $w = 2t$. Such a value cannot be justified with a mechanical argument in terms of loss of stiffness or stability. In the present example with $r/t = 3000$ and $e/r = 0.01$, Fig. 9, the load-displacement curve rises progressively, with a point of inflexion occurring on the load path. It would be useful to link the limit on the displacement to a physical phenomenon.

One solution is to plot the curve again as a modified Southwell plot, Fig. 10. This curve can then be split into two distinct regions, each of which can be approximated by a straight line. For the first portion of the curve, the slope is uniform right from the start, indicating that the initial load-displacement behaviour is close to hyperbolic and therefore asymptotic to a certain load. This load can be found by extrapolating the initial portion of the curve to the load axis. It could be used as the

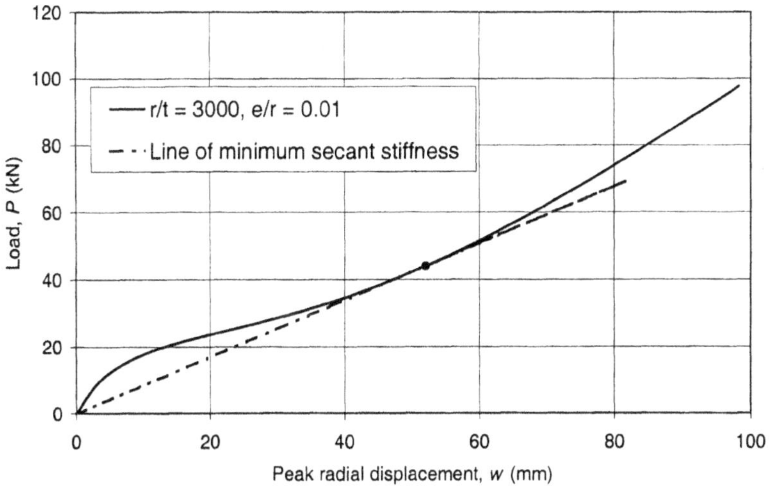

Figure 9. Load-displacement plot: point of inflexion.

buckling load as it constitutes a measure of the initial load capacity of the shell if subsequent stiffening is neglected.

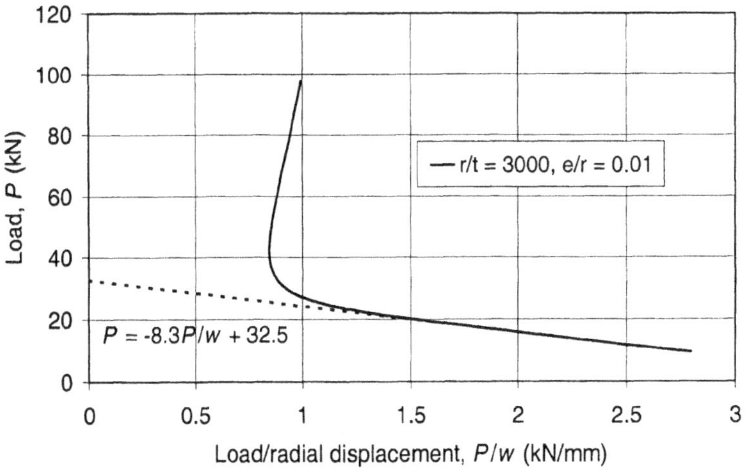

Figure 10. Modified Southwell plot: point of inflexion.

Another solution is to take the buckling load as the load defined by the minimum secant stiffness. This is the greatest flexibility that the structure ever displays. Its value is defined by the minimum value of P/w in the modified Southwell plot.

A further possibility is to interpret the rule about largest tolerable deformations in terms of rotations of the shell surface. This is a better measure since the human eye does not detect deflections, but it does observe rotations. It may then not be necessary to set limits on the displacements at all (Gillie, Holst, Münch and Rotter, 2002).

5. CONCLUSION

This paper has highlighted some of the difficulties that designers might find in implementing the failure criteria given in the recent European Standard for Shells (ENV 1993-1-6, 1999). There are two cases which cause particular difficulties. In the first of these, where a plastic plateau load can only be established through very lengthy analysis, a modified Southwell plot may provide an efficient and accurate means for obtaining this plateau.

In the second case, where the load-displacement curve rises progressively without the occurrence of a plateau, limit point, or bifurcation point, the modified Southwell plot may be used to estimate the asymptotic load from the initial deformation behaviour, discarding any subsequent strength gains due to geometric stiffening.

The modified Southwell plot also provides an easy means of obtaining the minimum secant stiffness, which corresponds to the most flexible deformation mode of the structure.

6. ACKNOWLEDGEMENTS

Financial support for the work described in this paper is gratefully acknowledged from the UK Engineering and Physical Sciences Research Council (EPSRC) under Grant GR/N39647.

References

Calladine, C.R. (1983) *Theory of Shell Structures*, Cambridge University Press, Cambridge.

ENV 1993-1-6 (1999) *Eurocode 3: Design of Steel Structures, Part 1-6: General Rules: Supplementary Rules for Shell Structures.*

ENV 1993-4-1 (1999) *Eurocode 3: Design of Steel Structures, Part 4-1: Silos.*

Gillie, M. and Holst, J.M.F.G. (2002) Structural behaviour of discretely supported silos, Report no. R02-01, School of Civil and Environmental Engineering, University of Edinburgh..

Gillie, M., Holst, J.M.F.G., Münch, M. and Rotter, J.M. (2002) Behaviour of silos supported on discrete brackets, *International Journal of Structural Stability and Dynamics*, In Press.

HKS (2000) *ABAQUS/Standard User's Manual*, Hibbitt, Karlsson and Sorenson, Inc., Pawtucket, Rhode Island.

Holst, J.M.F.G., Münch, M. and Rotter (2001) Failure criteria in thin shells, Report no. R01-09, School of Civil and Environmental Engineering, University of Edinburgh.

Holst, J.M.F.G. and Rotter (2001) Geometric imperfections, residual stresses and buckling in cylindrical shells with local support settlement, Presented at *Euromech 424, Rolduc-Kerkrade, September 2001.*

Horne, M.R. and Merchant, W. (1965) *The Stability of Frames*, Pergamon Press, Oxford.

Roorda, J. (1967) Some thoughts on the Southwell plot, *Journal of the Engineering Mechanics Division, ASCE* **93**, 37-48.

Southwell, R.V. (1932) On the analysis of experimental observations in problems of elastic stability, *Proceedings of the Royal Society of London, Series A* **135**, 601-616.

Spencer, H.H. and Walker, A.C. (1975) Critique of Southwell plots with proposals for alternative methods, *Experimental Mechanics* **15**, 303-310.

PCCP SHELLS

K. Miura
University of Tokyo (Retired)
3-9-7 Tsurukawa, Machida, Tokyo 195-0061, Japan
miurak@gakushikai.jp

Abstract: The concept of pseudo-cylindrical concave polyhedral (PCCP) shells is based on the hypothesis that the post-buckling cylindrical shape known as the Yoshimura pattern provides a structural form of a polyhedrally-stiffened cylindrical shell. It supplements the design domain of circular cylindrical shells, particularly for environments where the external pressure criterion is the major factor in design. The recent successful application of the concept to beverage cans demonstrates the potential of the concept for the first time. The potential of PCCP shells for undersea pressure hulls is briefly reviewed.

Key words: Shell structures, Folded plate structures, Circular cylinders

1. INTRODUCTION

The circular cylindrical form is, without doubt, one of the most widely used structural concepts in both natural and engineered structures. Its popularity may certainly be due to its intrinsic stiffness, strength, and stability. In addition, the simplicity of this form and the ease of mass production are other factors for its abundance today.

This brings up the question "Are there any other structural forms, which are 'globally cylindrical' in shape and with a continuous skin, that can be manufactured with present-day technology?" One answer to this question is the pseudo-cylindrical concave polyhedral (PCCP) shell shown in Fig. 1.

The concept of PCCP shell has been proposed thirty years ago (Miura, 1969, 1970). It is based on the hypothesis that the post-buckling cylindrical shape known as the Yoshimura pattern, Fig. 2, provides a structural form of a polyhedrally-stiffened cylindrical shell. Its shape is globally cylindrical and continuous. However, in the strict sense of the word it is not cylindrical.

H.R. Drew and S. Pellegrino (eds.),
New Approaches to Structural Mechanics, Shells and Biological Structures, 329–339.

Figure 1 PCCP shell

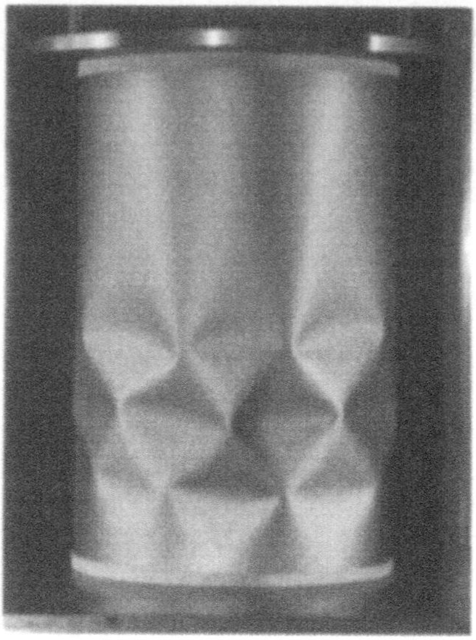

Figure 2 Yoshimura pattern

Moreover, due to its geometry its structural properties are different from those of cylindrical shells in many respects. These are the reasons why the PCCP shell can be considered as a new concept for the design of circular cylindrical shells.

The numerical and experimental studies by Knapp (1974, 1977) and Tanizawa and Miura (1975) have established a database of structural properties of PCCP shells. One drawback of the PCCP shell is its difficulty of mass production and, therefore, industrial applications have not been possible for many years. However, in recent years there have been some noticeable advances in two areas of application.

At a beverage can manufacturing company in Japan, some engineers have become interested in the application of the PCCP shell to beverage cans. The advantage of PCCP shells over conventional circular cylindrical shells stems from the fact that PCCP shells have high buckling strength, which is a very desirable property for depressurized cans. In other words, a PCCP-can could be made thinner than conventional cans. After some years of development, the company has finally succeeded in developing a manufacturing process, and beverage drinks in PCCP-type cans are now marketed in Japan.

In the USA an interesting study has been made on undersea pressure hulls to depths of about 1500 m (Knapp and Le, 1998). The PCCP shell, also known as PC (polyhedral cylinder) shell concept, offers a design alternative to the conventional ring-stiffened cylindrical shells for undersea pressure hulls. It has been confirmed by the previous investigation that an hydrostatically loaded PCCP shell may possess several times the buckling resistance of a comparably sized circular cylinder having the same wall thickness and overall dimensions. Extensive parametric studies covering a broad range of materials suitable for undersea pressure hulls have been carried out.

In this paper, first, some particular properties of the PCCP shell are reviewed especially in comparison to circular cylindrical shells. Second, the current applications to beverage cans and undersea pressure hulls are shown. The discussion in this paper is focused on thin, closed, PCCP shells.

2. GEOMETRIC AND STRUCTURAL PROPERTIES

2.1 Geometry

The geometric parameters of a PCCP shell are defined in the following. As shown in Fig. 1, the PCCP surface (the mid-plane of the shell) is a concave polyhedral surface consisting of identical diamond shaped repeating modules. Each module is formed by a pair of triangles. The two possible repeating modules are the diamond or the hexagon (Miura, 1969). Typical triangular faces are either the equilateral triangle or the isosceles right triangle, with base angles of 60° or 45°, respectively. The geometric parameters defining a PCCP shell are as follows:

λ	: Base length of a triangle in circumferential direction
α	: Base angle of a triangle; typical values are $45^\circ, 60^\circ$
N	: Number of circumferential triangles
γ	: Inclination of triangles relative to the cylindrical axis
R	: Radius of equivalent cylinder $(= N\lambda / 2\pi)$
T	: Thickness of skin

If one of these triangles is selected, the number of independent parameters are three, for instance N, R and T.

Therefore, the shape of a PCCP shell can be characterized with the sole parameter N in addition to the two standard parameters R and T defining a circular cylindrical shell. It is noted that the parameter N defines the shape of the shell not only in the circumferential direction, but also in the axial direction. If N is a small number, such as 6 to 12, the inclination γ of triangles to the cylindrical axis is substantial, that is, the concavity is significant. On the contrary, as N increases, for instance $N = 30$, it becomes closer to a smooth circular cylinder. A more detailed treatment of PCCP surface geometry can be found in Miura (1969).

From an intrinsic geometry viewpoint, it is important to note that the PCCP surface and the equivalent circular cylindrical surface are isometric with each other. In other words, the circular cylindrical surface can be transferred inextensionally (without stretching or contracting) into the equivalent PCCP surface. This property follows from the fact that the PCCP shell concept was originally induced from the post-buckling Yoshimura-pattern, that is characterized by the inextensional transformation of the original circular cylindrical shell. This property played an important role in the manufacturing process of PCCP shells from circular cylindrical shells, see Section 3.1.

2.2 Elastic Properties

A limited amount of information exists on the elastic behavior of PCCP shells and will be presented in the following. The data is presented as values relative to the equivalent circular cylinder having the same thickness and overall dimensions.

The stress distribution in a PCCP shell is complicated, due to its complex polyhedral geometry. Figure 3 shows experimental values of the membrane stress σ_m and bending stress σ_b distributions in the face elements of a PCCP shell subjected to an axial load P. It is shown that both bending and membrane stresses concentrate along the polyhedral edges.

The axial stiffness of a PCCP shell relative to the equivalent circular cylindrical shell is very small. For instance, the relative axial stiffness of the case shown in Fig. 3 is 0.153. This is due to the zigzag longitudinal section of the PCCP shell.

Figure 4 shows experimental axial buckling loads of PCCP shells relative to the critical buckling load of circular cylindrical shells, $P_{cc} = (1/2)\pi ET^2$. It is interesting to note that, although the PCCP shell is inferior in terms of axial stiffness, it is relatively strong in axial stability. These results on the axial stiffness and stability support the idea that the PCCP shell can be considered as a polyhedrally-stiffened cylindrical shell.

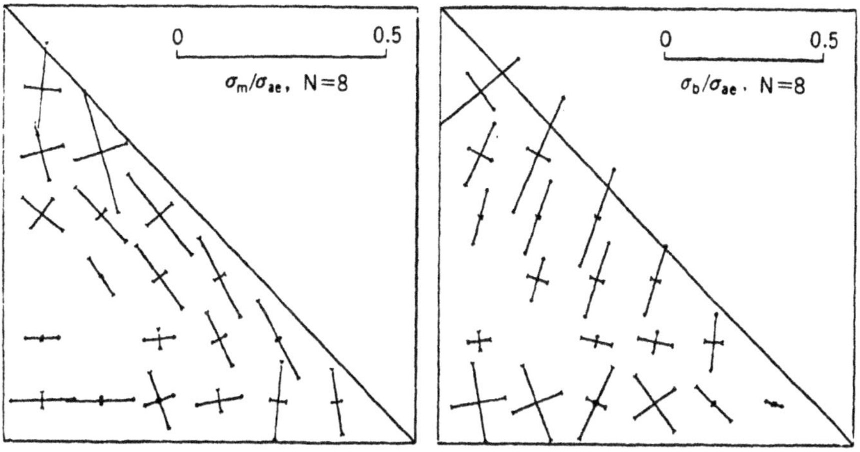

Figure 3. Experimental stress distribution under axial compression (in half of a triangular element) $R/T = 104, N = 8$, σ_m = membrane stress, σ_b = bending stress, $\sigma_{ae} = P/2\pi RT$

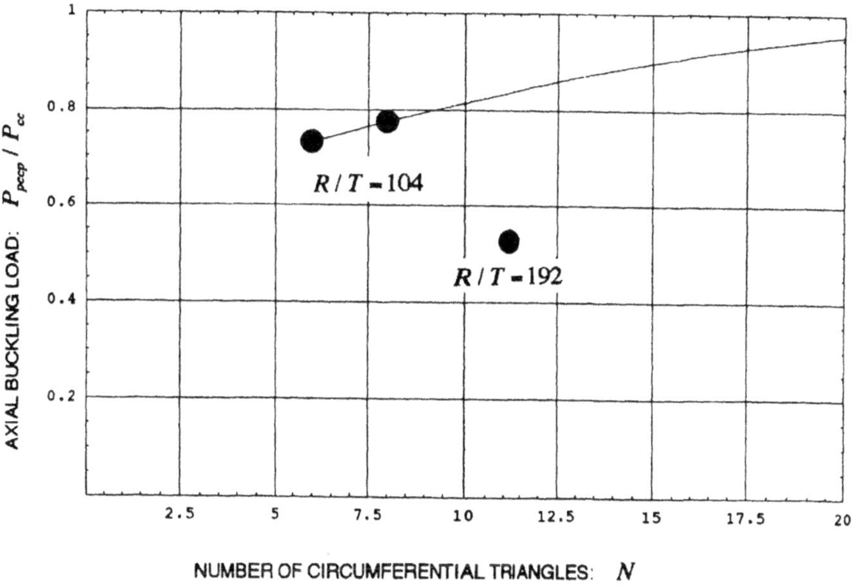

Figure 4. Buckling of PCCP shells under axial load; P_{cc} = buckling load of circular cylindrical shell, P_{pccp} = buckling load of PCCP shell

Figure 5 shows the experimental buckling pressure values of PCCP shells subject to hydrostatic pressure. The models tested and shown in this figure are made of plastic (Polystyrene) (N =6, 8), and steel (N =11), and have a base angle $\alpha = 45^\circ$. The length-to-radius ratios of these models are between 3.5 and 4.0. Both models were made by forming on a mould. Again, all data presented are relative to the equivalent buckling pressure of a perfect circular cylindrical shell. The figure shows that the buckling pressure of a PCCP shell is significantly higher for these models. For the same model cited in Fig. 3 (N =8), the buckling pressure is 5.92 times that of the equivalent circular cylindrical shell. As N increases the buckling load gradually decreases, but it remains high. For the model with N =11, which is used for the beverage can design in Section 3, this value is 3.14. Also, as N becomes very large, the buckling pressure ratio should approach a value of unity. A solid line shows the approximate trend of the buckling behavior.

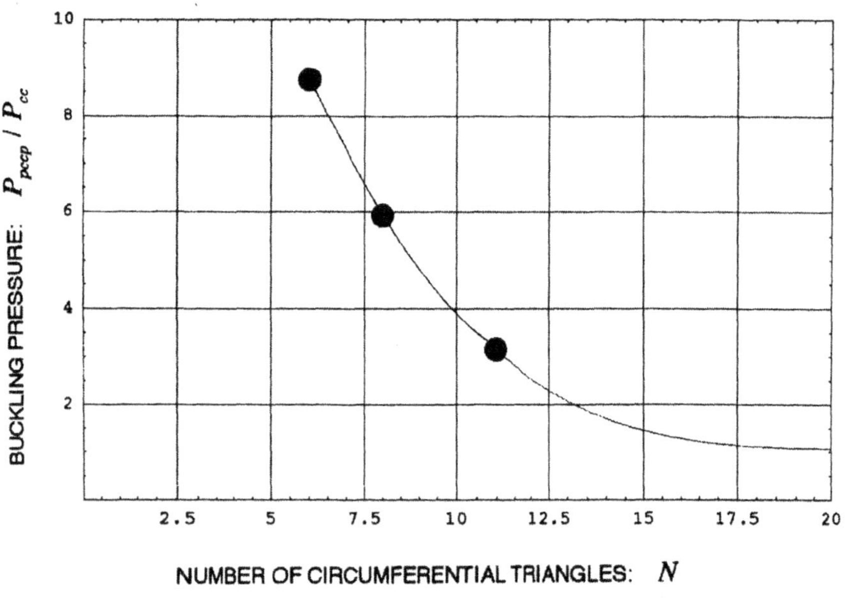

Figure 5. Buckling of PCCP shells under hydrostatic pressure

Table 1 is a summary of the elastic properties of PCCP shells relative to equivalent circular cylindrical shells. The data are based on the experiments on joint-less specimens, and the values are given to one significant figure, except for one case. From these data, it can be concluded that the most suitable design domain for PCCP shells is where the stability to hydrostatic pressure is a key requirement. In this respect, they definitely extend the design domain of circular cylindrical shells.

Table 1. Properties of PCCP shells relative to circular cylindrical shell

N (*R/T*)	6 (100)	8 (100)	11 (200)
Axial stiffness	0.1	0.15	0.1
Axial stability	0.7	0.8	0.6
Hydrostatic stability	9.0	6.0	3.0

3. PCCP -TYPE BEVERAGE CANS

3.1 Depressurized PCCP-cans

Probably the first practical application of the PCCP shell was the beverage can for soft drinks. In the 1990's, Toyo Seikan Co., the leading can manufacturer in Japan, succeeded in mass producing of the PCCP-type beverage cans. The product is called TULC diamond-cut can by the company, Fig. 6.

Figure 6. PCCP-type beverage can (Courtesy of Toyo-Seikan Co.)

The story of its development is very interesting. One of the structural design criteria for beverage cans for soft drinks, such as coffee and tea, is stability to hydrostatic pressure. This is because such cans are depressurized to prevent the decomposition of the contents, such as milk. For this reason, structures for beverage cans of this sort tend to have smaller diameter and thicker skin. In order to save material, it is desirable to reduce the skin thickness without losing stability. In addition, the shape has to be globally that of a circular cylinder to fit the existing distribution infrastructure, and has to be compatible with the high-speed manufacturing lines. Thus a structural concept that provides high hydrostatic stability, under these constraints, was needed.

In this respect, the PCCP shell concept seemed very attractive. After years of R & D efforts, the company finally developed a manufacturing line to produce PCCP shell-type cans. The core manufacturing process consists of forming circular cylindrical shells into PCCP shells. This is done by rotating internal moulds travelling inside circular cylindrical shells. As mentioned before, the circular cylindrical surface is transformed inextensionally into the equivalent PCCP surface. Therefore, in theory, this metal forming process can be done by bending only, without in-plane stretching of the metal sheets. Without doubt, this intrinsic property of the PCCP shells played an important role for the success of the manufacturing process.

The specification of the model in Fig. 6 are as follows, $N = 11$, $\theta = 45^\circ$ and $R/T = 192$. The critical hydrostatic pressure is 3.14 times of the equivalent circular cylindrical can. As a result, the skin thickness of the commercial product can be reduced as much as 30%.

3.2 Pressurized PCCP-cans

It is interesting to note that the opposite effect is now exploited in the design of the pressurized can shown in Fig. 7.

Figure 7(a) shows the pressurized state of a can that contains carbonated alcoholic drink. The skin of the can is rather thin, similar to current beer and coke cans. Although a faint diamond pattern can be observed, it is very close to a circular cylinder. If one opens the can by pulling the tab, then the pressure is released, and the form of the can suddenly changes to a PCCP shape, Fig. 7(b). The resulting, rigid PCCP shell provides firm holding for a drink. This trick takes advantage of the intrinsic property of the isometric transformation between a circular cylindrical shell and a PCCP shell. It provides a rare opportunity for watching a phenomenon of 'isometric transformation' between two shell surfaces.

(a) Before opening (b) After opening

Figure 7. Transformation of pressurized PCCP-can

4. PCCP SHELLS FOR UNDERSEA HULLS

As shown in Section 2, it has been confirmed by numerical analysis and experimental investigation that a hydrostatically loaded PCCP shell may possess several times the buckling resistance of a comparatively sized circular cylindrical shell having the same wall thickness and overall dimensions.

Applications of PCCP shells to undersea habitats and pressure hulls have been studied extensively at the University of Hawaii (Knapp, 1975, 1977, Knapp and Le, 1998). In the following, the author presents brief highlights of these studies.

Figure 8 shows an example of the configuration and geometric parameters for undersea pressure hulls that have been studied. It consists of a PCCP shell, transition sections and hemispherical end caps. The design study aimed to determine PCCP shell geometrical parameters that optimize structural performance for a specified material. Optimal structural performance requires that the PCCP hull design fails by buckling and the material yields at the same time, over a range of depths. The designs developed in this study were based on nearly coincident failure modes.

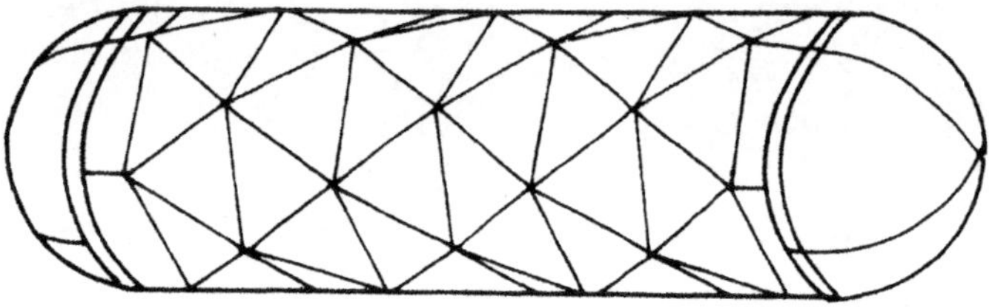

Figure 8. Undersea pressure hull (Knapp and Le, 1998)

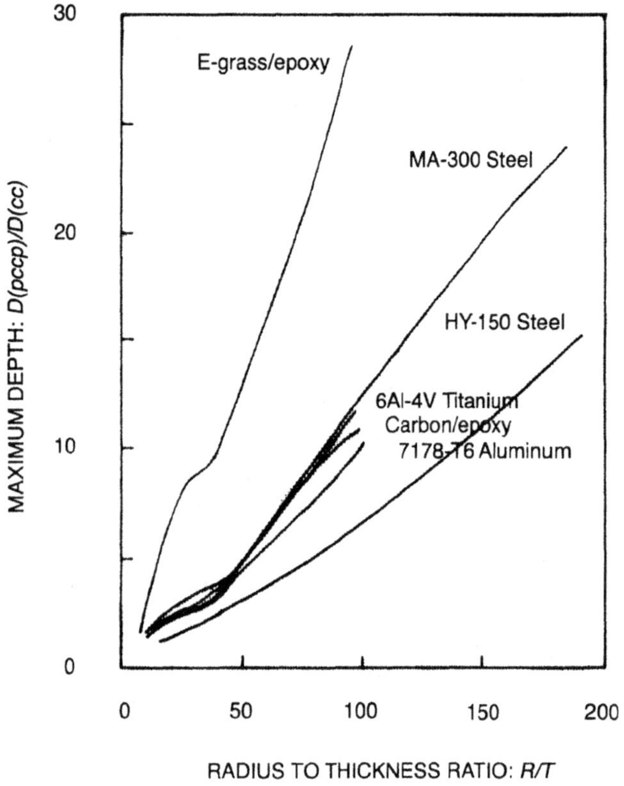

Figure 9. Maximum depth for undersea PCCP hulls (Knapp and Le, 1998); D_{cc} = depth of circular cylinder; D_{pccp} = depth of PCCP.

To evaluate the performance of the PCCP hull relative to the circular cylindrical hull, the PCCP hull failure curve is divided by the circular cylindrical hull buckling curve at particular values of R/T to obtain a single depth ratio curve, $Dpccp/Dcc$. Such curves are presented in Fig. 9 for all materials investigated. This figure reveals that the PCCP hull performs as a stiffened shell with depth ratios substantially greater than unity for R/T ratios greater than approximately 10.

Another interesting result is the sensitivity to fabrication errors. To determine the sensitivity of PCCP shell buckling resistance to fabrication errors, a finite element study was conducted on E-glass/epoxy. The resulting buckling pressure ratio increases with increasing geometric imperfection, between 1.5-1.8 for radial imperfections of 0 to 0.5%. The PCCP shell is out-of-round by design and its buckling strength is less sensitive to variations in its shape than a circular cylindrical shell. The fabrication tolerances might be relaxed to reduce manufacturing cost.

5. CONCLUDING REMARKS

The geometric and elastic properties of the PCCP cylindrical shell have been briefly reviewed. It has been shown that the concept of the PCCP shell extends the design domain of circular cylindrical shells, particularly for environments where resistance to external pressure is a major factor in design. The recent successful application of the concept to beverage cans has demonstrated the potential of the concept for the first time. The potential of the PCCP shell for undersea pressure hulls has also been briefly reviewed.

References

Knapp, R.H. (1977) Pseudo-cylindrical shells: A new concept for undersea structures, *Trans ASME J Eng. for Industry*, **99**, 485-492.

Knapp, R.H. and Le, T.T. (1998) Polyhedrally stiffened shells for undersea pressure hulls, *Int. J. of Offshore and Polar Engineering*, **8**, 207-212.

Knapp, R.H. (1975) Numerical and experimental analysis of a pseudo-cylindrical shell, *Bulletin of the IASS, **59**, 23-35.*

Miura, K. (1969) Proposition of pseudo-cylindrical concave polyhedral shells, *Institute of Space and Aeronautical Science, Report* **442**, University of Tokyo.

Miura, K. (1970) Proposition of pseudo-cylindrical concave polyhedral shells, *IASS Symposium on Folded Plates and Prismatic Structures*, Wien.

Tanizawa, K. and Miura, K. (1975) Stress analysis of a concave polyhedral shell, *Institute of Space and Aeronautical Science, Report,* **523**, University of Tokyo.

POGORELOV'S THEORY OF CREASES, AND POINT LOADS ON THIN CYLINDRICAL SHELLS

Andrew Palmer
Department of Engineering, University of Cambridge, Trumpington Street, Cambridge CB2 1PZ, U.K
acp24@eng.cam.ac.uk.

Abstract A.V. Pogorelov developed a little-known theory of thin elastic shells. He considered a class of piecewise inextensional deformations, in which relatively sharp creases separate areas of inextensional deformation. Within a crease the deformation is inextensional in the plane perpendicular to the crease. The local form of the crease minimises the strain energy, which has bending and stretching components. The paper outlines the theory, applies it to radially-loaded open-topped cylindrical shells, and compares it with measurements by Holst and Lukasiewicz.

Keywords: Shell structures

1. INTRODUCTION

Aleksei Vasil'evich Pogorelov (1919-) is a distinguished mathematician who has worked extensively on the geometry of surfaces and its application to shell theory. He was for many years in Kharkov, latterly at the Institute for Low Temperature Physics, and is now at the Steklov Mathematical Institute in Moscow. His research into shell theory began in the 1950s, and is described in a series of monographs, notably a book on the geometric theory of shells (Pogorelov, 1966). Two of his books have been translated (Pogorelov, 1973,1988), and he wrote an article for an early thin shell conference (1967), but his research is not well known in the English-speaking world. It seems also to be disconnected from mainstream shell theory in the former Soviet Union. Vol'mir's review of plate and shell theory (1956) includes one reference to an early paper by Pogorelov, but does not mention him otherwise.

Part of the objective of this paper is to bring Pogorelov's research to the attention of a wider audience, and to encourage interest in powerful ideas that

H.R. Drew and S. Pellegrino (eds.),
New Approaches to Structural Mechanics, Shells and Biological Structures, 341–354.

have much to offer. It seems peculiarly appropriate to consider him at this conference. Pogorelov is distinguished by his emphasis on the central role of geometry, by his readiness to make bold simplifications that reveal a deep feeling for the underlying mechanics, and by his disinclination to be much influenced by received opinion or by previous work. These are qualities that many of us associate with Chris Calladine's immense contribution to applied mechanics.

2. PIECEWISE INEXTENSIONAL DEFORMATIONS OF THIN CONVEX SHELLS

Purely inextensional deformations are not possible in most convex shells, because they are not consistent with the boundary conditions. Pogorelov considered a wider class of deformations, within which regions of pure inextensional deformations are separated by sharp creases (Russian: rebro). Deformations of this kind are frequently seen in thin shells. If you push a finger into a thin rubber ball (pierced so that the internal pressure does not alter), you observe the formation of a dimple bounded by a sharp crease. Inside the dimple, the curvature is reversed , and the deformed configuration inside the crease is the mirror image of the original spherical configuration, reflected in the plane of the crease.

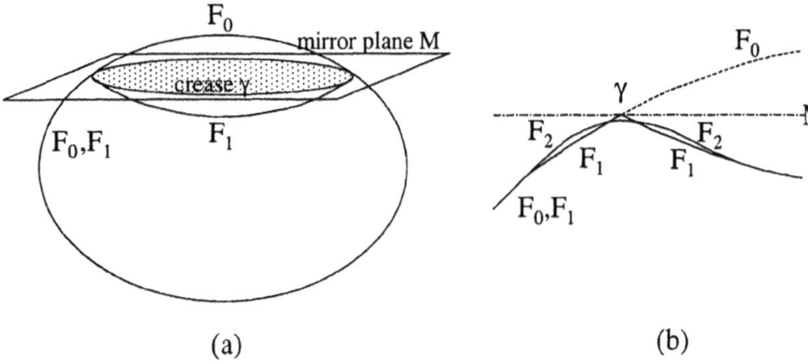

Figure 1. Deformations of a convex shell.

Figure 1(a) depicts this kind of deformation. F_0 is the original form, convex but not necessarily spherical. A reflection in the mirror plane M takes the surface to a deformed configuration F_1, in which the section of F_0 inside the intersection with M has been replaced by its mirror image. The crease γ lies in M. The deformation from F_0 to F_1 is inextensional, and that it can be continuous can be seen by imagining the mirror plane in Figure 1(a) to be translated downwards, from a starting point just in contact with F_0 to the final position shown in the diagram.

The sharp crease in F_1 implies a finite relative rotation at γ. In an elastic shell, the crease must be locally smoothed by the interaction of bending and extensional deformation. A second deformation takes F_1 to the final configuration F_2, identical to F_1 except in a comparatively narrow region close to γ. Figure 1(b) depicts this in a close-up.

Pogorelov determined the form of the deformation F_1 to F_2 that minimizes the strain energy of bending and extension. The notes below summarise the version of his argument in Pogorelov (1966). It is set out in greater detail in his books, though slightly differently in different versions.

The crease γ is a plane curve . Figure 2 is a section on the plane N normal to γ at an arbitrary point Q, and shows the coordinate system and notation used to determine the strain energy per unit length of crease in the neighbourhood of Q. The origin of coordinates r and z is taken at the centre of the osculating circle to γ at Q, whose radius is ρ. 2α is the angle between

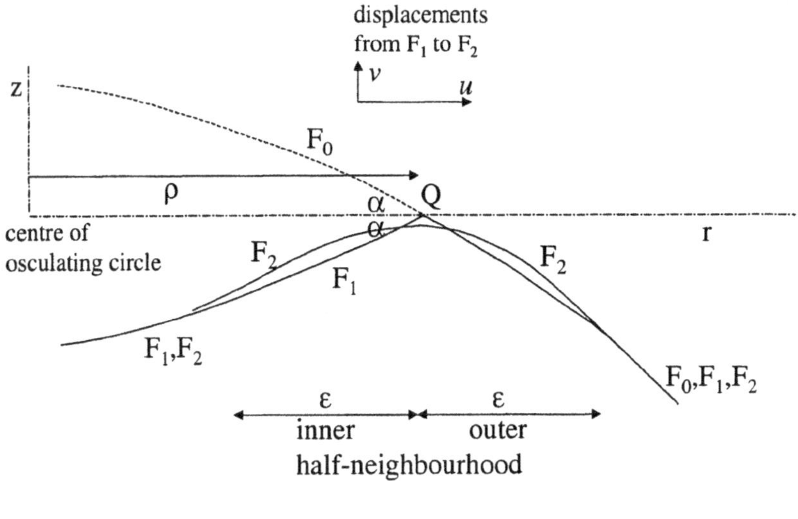

Figure 2. Section perpendicular to crease.

F_0 and F_1 measured in the plane N. The displacements from F_1 to F_2 lie in N, and are u and v in the r and z directions.

The strain in the direction perpendicular to N is u/ρ. The deformation is taken to be inextensional in the plane N. Two neighbouring points on the surface are at (r,z) and $(r+dr,z+dz)$ in configuration F_1 and at $(r+u,z+v)$ and $(r+dr+u+du,z+dz+v+dv)$ in configuration F_2. Since the distance between them does not alter

$$dr^2 + dz^2 = (dr + du)^2 + (dz + dv)^2 \tag{1}$$

and so

$$0 = \frac{du}{dr} + \frac{dz}{dr}\frac{dv}{dr} + (1/2)\left((\frac{du}{dr})^2 + (\frac{dv}{dr})^2 \right) \tag{2}$$

Pogorelov argues that the third term can be neglected because $du/dr<<1$. To the left of Q $dz/dr \cong \alpha$, and to the right of Q $dz/dr \cong -\alpha$. Accordingly, and denoting differentiation with respect to r by a prime,

$$0 = u' \pm \alpha v' + (1/2)v'^2 \tag{3}$$

where the + sign is taken on one side of Q and the – sign on the other.

Consider now bending in the plane N, and denote the change of curvature from F_0 to F_1 by Δk. The change of curvature from F_1 to F_2 is v'' and so the change of curvature in the deformation from F_0 to F_2 is $\Delta k+v''$. The plane N does not necessarily coincide with one of the principal directions of bending, and additional twisting and bending in the perpendicular direction are both ignored. The total strain energy per unit length of γ is

$$U' = (1/2)D' \int_{-\varepsilon}^{\varepsilon} (u/\rho)^2 ds + (1/2)D \int_{-\varepsilon}^{\varepsilon} (v''^2 + 2v''\Delta k + \Delta k^2) ds \tag{4}$$

where

D is the plate flexural stiffness $Et^3/12(1-\nu^2)$
D' is the membrane stiffness $Et/(1-\nu^2)$
t is the shell thickness
E is the elastic modulus
ν is Poisson's ratio
s is distance measured along the section

and the integrals are taken over a region of breadth 2ε where F_2 departs from F_1.

Pogorelov argues that the third term in the second integral of (4) is negligibly small. The second term in that integral can be evaluated by taking advantage of the fact that Δk is constant in each half-neighbourhood, and is

$$(1/2)D \int_{-\varepsilon}^{\varepsilon} 2v''\Delta k ds = D\alpha(-2k + k_e + k_i) \tag{5}$$

where

- k is the curvature of the intersection between F_0 and plane N
- k_e is the curvature of the intersection between F_1 and plane N, in the external half-neighbourhood, and
- k_i is the curvature of the intersection between F_1 and plane N, in the inner half-neighbourhood

The boundary conditions on u and v are that F_2 is smooth at Q, and that it approaches F_1 asymptotically at the outer boundaries of each half-neighbourhood.

The form taken up by the shell in the neighbourhood of γ is defined by u and v as functions of r. It minimises the local strain energy U', and so the next task is to find u and v that accomplish the minimisation. The minimising deformation is symmetric with respect to Q.

At this point it is helpful to introduce new nondimensional variables. Let

$$\begin{aligned} \varepsilon &= \sqrt{\frac{t}{\sqrt{12}\rho\alpha}} \\ \underline{u} &= \frac{u}{\varepsilon\rho\alpha^2} \\ \underline{v} &= \frac{v'}{\alpha} \\ \underline{s} &= \frac{s}{\rho\varepsilon} \end{aligned} \tag{6}$$

and note in particular that the new $\underline{v}$ is the derivative v' nondimensionalised, and not the old v.

In the new variables, taking account of (5), and now omitting the underbars for simplicity, the minimised strain energy per unit length of γ is

$$U' = \frac{J_0 E t^{5/2}}{12^{3/4}(1-\nu^2)} \int_\gamma \frac{\alpha^{5/2}}{\rho^{1/2}} ds' + \frac{Et^3}{12(1-\nu^2)} \int_\gamma \alpha(-2k + k_e + k_i) ds' \tag{7}$$

where J_0 is the minimum value of the integral

$$J = \int_0^\infty (v'^2 + u^2)\, ds \tag{8}$$

subject to the nonholonomic constraint

$$0 = u' + v + \frac{v^2}{2} \tag{9}$$

and to the boundary conditions

$$\begin{aligned} u(0) &= 0 \\ v(0) &= -1 \\ u(\infty) &= 0 \\ v(\infty) &= 0 \end{aligned} \tag{10}$$

In (10) the first condition comes from the symmetry of the deformation at the crease, the second from the condition that F_2 is smooth at Q, and the other two conditions from the coincidence of F_2 and F_1 at large distances from Q.

3. THE VARIATIONAL PROBLEM

The variational problem represented by equations (8) through (10) can be tackled in several ways.

Pogorelov (1966) determines the minimum by solving the Euler equation for a linearised version of (9) with the v^2 term omitted, expressing the solution for v as a power series of powers and products of the solutions to the linearised problem, and then finding the solution to the unconstrained problem with the full form of the constraint (9) incorporated with a Lagrangian multiplier. The resulting integral-differential equation leads to a solution for the coefficients of the power series through a series of successive approximations. The minimum J_0 is approximately 1.2, and the maximum values of $|v'|$ and $|u|$ are about 1 and 0.5 respectively.

Pogorelov also adopted an ingenious alternative approximation. The half-neighbourhood of the crease is divided into two regions, an inner region $[0,\sigma]$ in which the meridional curvature v' is taken as constant and an outer region $[\sigma,\infty)$ in which the $(1/2)v^2$ term in (9) is neglected; σ is a free parameter. Integrating and applying the boundary conditions, J is readily calculated as a function of σ. Minimising with respect to σ, the minimum value is 1.15. This is an approximate solution but not a bound. He goes on to mention a value of 1.1156 determined by a computer, but gives no details.

Another approximate solution, but one which is known to be an upper bound, can be determined by taking the relationship between v and s to be

$$v(s)=\begin{cases} -1+\dfrac{(1+\mu)s}{\sigma} & 0\le s\le\sigma \\ \mu\dfrac{(\lambda+1)\sigma-s}{\lambda\sigma} & \sigma\le s\le(1+\lambda)\sigma \\ 0 & s>(1+\lambda)\sigma \end{cases} \tag{11}$$

which has three free parameters μ,σ and λ. It corresponds to a constant convex outward curvature in an inner region, and a different constant concave outward curvature in a broader outer region. Conditions (10) are met if

$$\lambda = -\frac{\mu^2+2\mu-1}{\mu(\mu+1)} \tag{12}$$

u is determined by integrating (9). The integral J can then be evaluated and minimised with respect to the remaining free parameters μ and σ. The minimum value is 1.219.

Kerrigan (2001) solved the problem as a constrained optimisation, with the control variables the N values of v at a regular spacing $\lambda/N+1$ between 0 and an upper integration limit λ. u was determined by integrating (9) from 0. The cost function was determined by sampling u and v', and the one constraint was that $u(\lambda)=0$. The solution applied Matlab version 6.0.0.88 together with the Optimisation Toolbox version 2.1 function FMINCON. The minimum J_0 is found to be 1.156, and that value will be taken from now on.

4. RADIAL POINT LOAD ON A THIN-WALLED SPHERICAL SHELL

Pogorelov's theory would be only a curiosity if we could not use it to solve problems, but we can. He analysed a thin spherical shell loaded by a concentrated point load (Pogorelov, 1966), and found the relationship between load P and deflection w under the load to be

$$\frac{Pa}{Et^3} = 3\pi c\left(\frac{w}{t}\right)^{1/2} \tag{13}$$

where

a	is the shell radius
t	is the shell thickness
E	is the elastic modulus
ν	is Poisson's ratio
c	is $J_0/12^{3/4}(1-\nu^2)$

He tested his analysis by comparing it with experiments on thin spherical shells made from electrodeposited copper. Figure 3 compares equation (13) with his measurements on spherical shells 150 mm radius and wall thickness 0.056, 0.048 and 0.037 mm. E is not given explicitly in the original reference, but is back-calculated to be 85 kN/mm^2 from theoretical plots.

The same problem was tackled by Ashwell (1959), who took a completely different approach. He recognised that the sphere formed a reversed dimple, applied the Reissner small-deflection equations separately to the outer original sphere and to the inner reversed spherical cap, and found a solution that satisfies equilibrium and matches the deflections at the crease. He made measurements on relatively thick aluminium shells (a 518 mm, t 1.6 mm), but the deformations were partly inelastic, as he remarks and as his plots confirm.

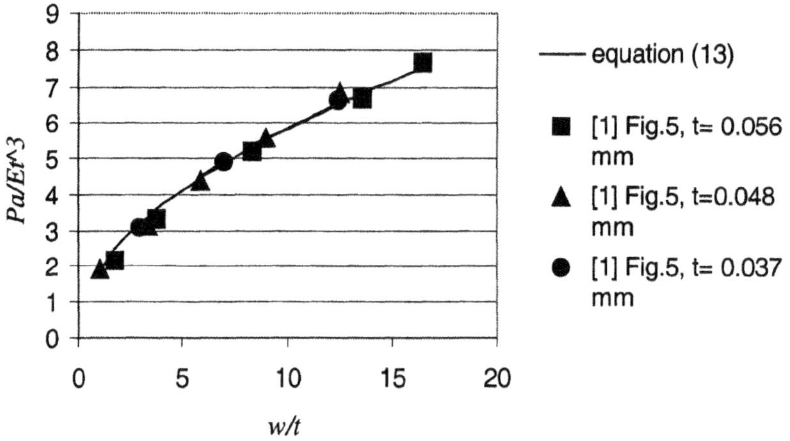

Figure 3. Spherical shells under radial point loads: comparison between Pogorelov theory and measurements.

The good agreement between the observations and Pogorelov's analysis encourages us to apply the analysis to a more difficult problem with less symmetry.

5. RADIAL POINT LOAD ON A THIN-WALLED CYLINDER

Figure 4(a) depicts a thin-walled open-topped elastic circular cylindrical shell F_0 with uniform radius a and wall thickness t, together with X Y and Z axes whose origin is the centre of the cylinder at the upper edge. The Z axis is the axis of the cylinder. The shell is loaded by an inward-directed radial force P, acting at the upper edge and parallel to the X axis.

Now consider a mirror plane M inclined at an angle ϕ to the YZ plane. The figure also shows a deformed surface F_1, in which the part of the cylinder to the right of the mirror plane has been replaced by its reflection in the plane. The surface F_1 is related to F_0 in the way described earlier. The lowest point on the intersection between M and the cylinder F_0 is at $(a,0,-H)$, where H is arbitrary. Figure 4(b) shows the section of the cylinder by the mirror plane, referred to new axes x,y,z whose origin is at $(a,0,-H)$. The complete section of the cylinder by the mirror plane is an ellipse with semi-axes a and $a\text{cosec}\phi$.

From here on ϕ will be taken as small. The relevant portion of the ellipse is the crease γ between F_0 and F_1, and can then be idealised as the parabola

$$y \;=\; x^2/2a\phi \tag{14}$$

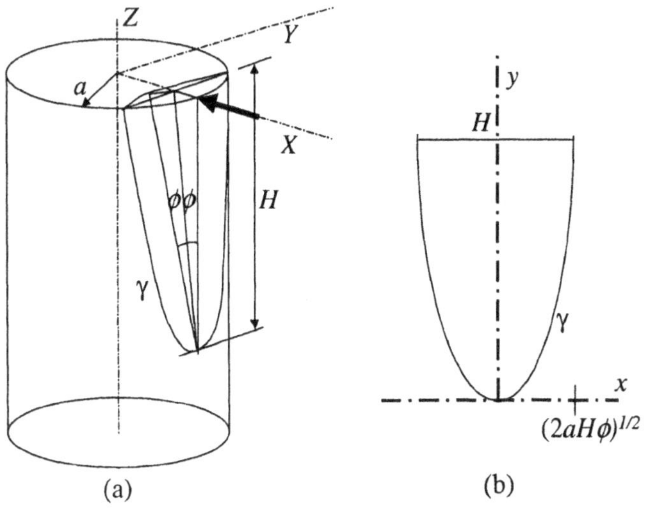

Figure 4. Radial load on cylindrical shell.

The cylinder F_0 is the surface

$$X^2 + Z^2 = a^2 \tag{15}$$

and the unit vector perpendicular to the cylinder at X, Y, Z is therefore

$$(X/a)\underline{i} + (Y/a)\underline{j} + 0\underline{k}$$

where $\underline{i}$ $\underline{j}$ and $\underline{k}$ are the unit vectors corresponding to the X Y and Z axes. The unit vector perpendicular to the mirror plane is

$$\underline{i}\cos\phi + 0\underline{j} + \underline{k}\sin\varphi$$

and the angle α between the mirror plane and the normal to the cylinder F' at (X,Y,Z) is

$$\alpha = \arccos\left(\frac{X}{a}\cos\phi\right) \tag{16}$$

2α is the angle between the surfaces F and F' at the crease. Since ϕ is small, and using the equation of the cylinder,

$$\alpha^2 = \phi^2\left(1+\left(\frac{x}{a\phi}\right)^2\right) \tag{17}$$

The curvature κ of the crease is

$$\kappa = \frac{\dfrac{d^2y}{dx^2}}{\left(1+\left(\dfrac{dy}{dx}\right)^2\right)^{3/2}} = \frac{\dfrac{1}{a\phi}}{\left(1+\left(\dfrac{x}{a\phi}\right)^2\right)^{3/2}} \tag{18}$$

and the length of element ds along γ is

$$ds = \left(1+\left(\frac{x}{a\phi}\right)^2\right)^{1/2} dx \tag{19}$$

and so

$$a^{5/2}\kappa^{1/2}ds = \frac{\phi^2}{a^{1/2}}\left(1+\left(\frac{x}{a\phi}\right)^2\right)dx \tag{20}$$

Using the earlier result (7), integrating over the whole crease, and applying the relation between ϕ, H and the displacement w at the point of application of the load

$$\phi = \frac{w}{2H} \tag{21}$$

the strain energy in the crease is

$$U' = \frac{Et^{5/2}J_0}{12^{3/4}(1-\nu^2)}\left(\frac{2}{3}w^{3/2}a^{-1} + \frac{1}{2}w^{5/2}H^{-2}\right) - \frac{8}{3}\frac{Et^3}{12(1-\nu^2)}w^{1/2}a^{-3/2}H \tag{22}$$

The other component of strain energy is the bending strain energy U'' of the area inside the crease, whose curvature changes from $1/a$ to $-1/a$. That strain energy is easily shown to be

$$U'' = \frac{8}{3}\frac{Et^3}{12(1-\nu^2)}w^{1/2}a^{-3/2}H \tag{23}$$

Adding (22) and (23), and noting that (23) and the last term in (22) are equal and opposite, the strain energy is

$$U = \frac{Et^{5/2}J_0}{12^{3/4}(1-\nu^2)}\left(\frac{2}{3}\frac{w^{3/2}}{a} + \frac{1}{2}\frac{w^{5/2}}{aH}\right) \tag{24}$$

If no other strain energy components are significant, the force is

$$P = \frac{\partial U}{\partial w} = \frac{Et^{5/2}J_0}{12^{3/4}(1-\nu^2)}\left(\frac{w^{1/2}}{a} + \frac{5}{4}\frac{w^{3/2}}{aH}\right) \tag{25}$$

The ratio of the second term to the first is $(5/4)w/H$, and therefore small if ϕ is small. Neglecting the second term, the force is given by the simple formula

$$P = \frac{Et^{5/2}J}{12^{3/4}(1-\nu^2)}\frac{w^{1/2}}{a} \tag{26}$$

Unexpectedly, the analysis does not determine H: (26) suggests that H will take the maximum value compatible with the length of the cylinder and the boundary conditions at the other end, though that is not what is observed experimentally (Holst, 1996).

It has been noted that the strain energy U'' of the area inside the crease is equal and opposite to the last term in the equation (22) of the crease strain energy U'. This is also the case for the corresponding expressions for a spherical shell. It is tempting to conjecture that it is generally true, but this has not been demonstrated.

6. COMPARISON WITH EXPERIMENT

The simple result (26) can now be compared with experiment.

Holst (1996) constructed thin open-end cylindrical shells with radius 88.7 mm and thickness 0.125, 0.25 and 0.35 mm out of Melinex (polyethylene teraphthalate) sheet, elastic modulus 5.13 GPa and Poisson's ratio 0.3. A possible objection to this material is that the sheets are naturally flat, so that bending them into a cylinder creates bending stress. Since the problem is nonlinear, the response to loading may then not be quite the same as for an initially stress-free cylinder.

Holst loaded the shells radially at the free end and at points below the free end. Figure 5 compares equation (26) with his measurements on a 0.25 mm thick cylinder loaded at its free edge. The agreement is encouraging. The calculated force is slightly higher than the observed force, as one would expect because the calculation imposes constraints on the allowed deformations. Figure 6 is a photograph of the cylinder under load. It can be seem that the crease is not particularly concentrated.

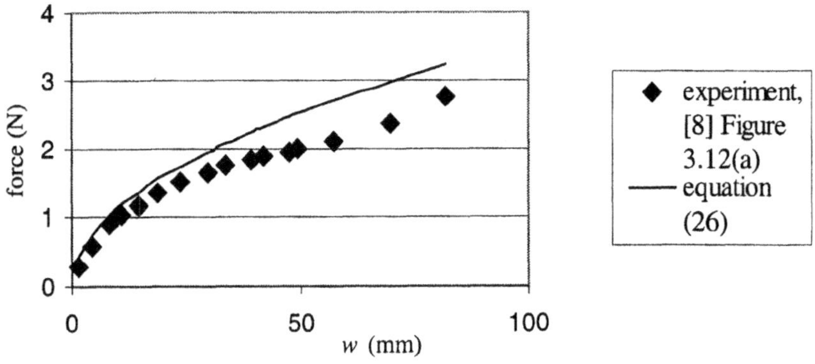

Figure 5. Comparison between equation (26) and Holst's experiment on a 0.25 mm shell.

A stronger comparison is between tests on shells with different wall thickness and made from different materials. In terms of nondimensional variables, (26) becomes

$$\frac{P(1-\nu^2)}{Et^2} = \frac{J_0}{12^{3/4}} \frac{\sqrt{wt}}{a} \tag{27}$$

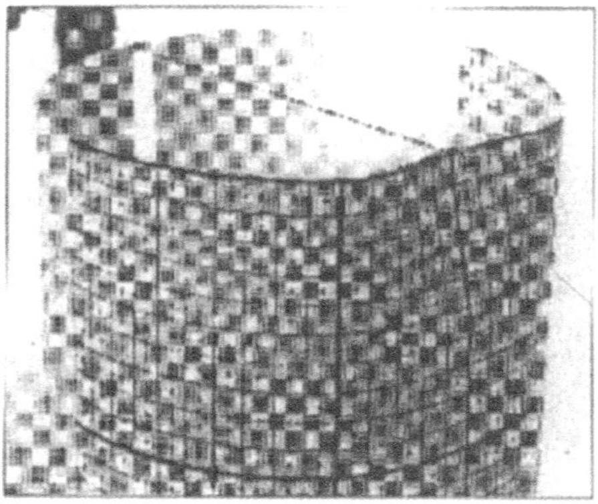

Figure 6. Photograph of loaded shell.

Figure 7 plots $P(1-\nu^2)/Et^2$ against $\sqrt{(wt)}/a$, and compares (27) with Holst's measurements on shells with all three thicknesses, 0.125, 0.25 and 0.35 mm. It also includes measurements by Lukasiewicz and Nowinka (1991) on a thin aluminium shell, radius 100 mm, thickness 0.3 mm, E 69 GPa and ν 0.33.

The agreement is again encouraging, surprisingly so considering the simplicity of the theory and the approximations involved.

7. CONCLUSION

The results suggest that analysis based on Pogorelov's approach agrees well with observation, despite its audacious approximations. It has much to offer, and a number of other problems suggest themselves.

Acknowledgement

The author thanks the Jafar Foundation for its continuing support. He thanks Eric Kerrigan for the numerical solution of the variational problem, Mark Holst for his generous permission to quote data from his dissertation, Gleb Makarov and Sergio Pellegrino for assistance in locating Pogorelov and interpreting his publications, and Sergio Pellegrino and Chris Calladine for comments on an earlier version.

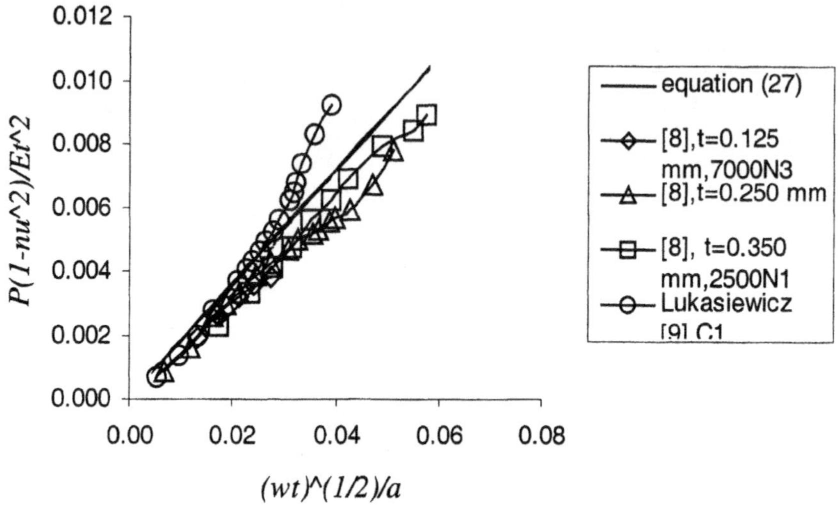

Figure 7. Comparison between equation (27) and experiments by Holst and Lukasiewicz.

References

Ashwell, D.G. (1959) On the large deflexion of a spherical shell with an inward point load. Proceedings, IUTAM Symposium on the theory of thin elastic shells, Delft, 43-63, North-Holland.

Holst, J.M.F.G. (1996) Large deflection phenomena in cylindrical shells. PhD dissertation, University of Cambridge.

Kerrigan, E.C. (2001) Private communication.

Lukasiewicz, S. and Nowinka, J. (1991) Geometric elements in the elastic-plastic analysis of a cylindrical shell. Applied Mechanics Reviews, **44** (11:2), 181-193.

Pogorelov, A.V. (1966) Geometriskaya teoriya ustoichivostei obolochek (Geometric theory of the stability of shells. Nauka, Moscow.

Pogorelov, A.V. (1967) Geometrical methods in the nonlinear theory of shells. Proceedings, IUTAM symposium on the theory of thin shells (ed. F.I. Niordson), Copenhagen.

Pogorelov, A.V. (1969) Vneshnaya geometriya vypukhlykh poverkhnostei, Izdatel'stvo Nauka, Moscow; translation, Extrinsic geometry of convex surfaces, American Mathematical Society, Providence, RI (1973).

Pogorelov, A.V. (1986) Izgibaniya poverkhnostei i ustoichivost' obolochek, Nauka, Moscow; translation, Bendings of surfaces and stability of shells, American Mathematical Society, Providence, RI (1988).

Vol'mir, A.S. (1956) Gibkie plastinki i obolochki, Gosurdarstvennoye Izdatel'stvo tekhniko-teoreticheskoi literatury, Moscow; translations, Biegsame Platten und Schalen, VEB Verlag fur Bauwesen, Berlin (1962); Flexible plates and shells, TR AFFLD-TR-66-216, Wright-Patterson Air Force Base, Dayton, OH (1967).

SHELL BUCKLING AND COLLAPSE ANALYSIS FOR STRUCTURAL DESIGN

J. Michael Rotter
University of Edinburgh, Edinburgh EH9 3JN, Scotland
michael@civ.ed.ac.uk

Abstract: A very large literature exists on the static analysis of thin shell structures, covering elastic stress states, elastic buckling, plastic collapse and elastic-plastic failures. Modern approaches to these problems usually involve the use of computer programs to perform a variety of different alternative types of calculation. However, the way in which different analyses should be used in the design process has received much less attention until recently. The new European Standard for the Strength and Stability of Shells sets out a philosophical framework that accommodates the results of both algebraic analysis and different levels of sophistication in computer analyses. This paper describes the philosophy on which the new standard is based, and outlines the way in which the various analyses should be used, when applied in design practice. It also identifies some of the remaining difficulties which pose challenges of definition and interpretation, and which need debate in the scientific community.

Key words: Analysis, Buckling, Collapse, Design, Finite elements, Nonlinear, Plastic, Shell structures.

1. INTRODUCTION

The final goal of engineering analysis is its application into the design process. The use and appropriate interpretation of the results of different kinds of analysis is therefore of paramount importance in engineering. The concepts used in structural design have largely been developed in relation to single structural members (beams and columns) and to assemblages of these members into structural frameworks, for both of which there is a wealth of experimental data. However, thin shell structures exhibit rather more

H.R. Drew and S. Pellegrino (eds.),
New Approaches to Structural Mechanics, Shells and Biological Structures, 355–378.

complex behaviour than structural members, and many shell geometries and loading conditions have never been the subject of experimental study. Shell structural design is inevitably based on the same philosophical framework as structural members, though some simplifications are no longer valid for shells. Moreover, modern shell structural design must rely heavily on numerical analyses for cases for which no test data exist.

The design of civil and structural engineering structures is regulated by a system of codes and standards, and the development of the Eurocodes represents a collaboration and harmonisation into common rules across Europe. The new Standard for the Strength and Stability of Shells (ENV 1993-1-6, 1999) offers a substantial advance in the scope, structure and permitted methods within such standards (Rotter, 1998a). Rules for all types of metal shells are covered within this generic standard.

All structural engineering design standards have a long history of assessment based on hand calculation. As a result, hand calculation concepts provide the framework for calculations on shells too.

In general, structural engineering standards have defined the assessed strength of individual members, but have paid less regard to the manner in which the forces acting on each member were calculated (known traditionally as the "analysis" of the structure). For thin shells, many different types of approximate analysis may be used, each with a different outcome. The shells standard defines a much more explicit relationship between the type of analysis and the criteria of failure, and this is evident in the steps required to achieve the same goal in each analysis. It classifies each analysis and regulates its usage to provide a consistent framework that gives approximately the same reliability, irrespective of the analysis choice. This appears to be the first time that such a task has been attempted. However, in the process of devising this consistent framework, many aspects of the relationship between different analyses have been brought into sharp focus. This paper attempts to provide a clear description of these relationships in a manner that will permit users and researchers to understand the value of each type of analysis and to exploit it as part of the design process.

A huge literature exists (e.g. Timoshenko and Gere, 1961; Brush and Almroth, 1976; Calladine, 1983; Bushnell, 1985; Teng, 1996) on the static analysis of thin shell structures, covering elastic stress states, elastic buckling, plastic collapse and elastic-plastic failures. Although a huge research effort has developed computer programs to perform these analyses, much less attention has been paid to the process of interpretation of the predictions into reliable design information. The new shells standard sets out a single philosophical framework that formally defines these interpretations, but it remains in need of further development.

2. THE CONCEPTUAL FRAMEWORK FOR STRUCTURAL ENGINEERING DESIGN

2.1 Structures composed of structural members

Structural engineering education, and most structural engineering research, has long focussed on the behaviour of line elements: beams and columns. These elements are the simplest to test in the laboratory, and the response found on testing an individual element is generally also relevant to statically determinate frameworks assembled from these elements. The serviceability requirements for civil engineering beams generally mean that the deformations do not affect the equilibrium conditions, so material failure is the dominant consideration. When the beam model is extended to a rigid-jointed frame, the redundant system can often be similarly treated as sensitive only to material failure, with small-deformation theory being used to predict plastic collapse when a full incremental deformation mechanism has developed (Fig. 1). The difference in load between the first hinge formation (first section full plasticity) and the complete collapse mechanism is usually not very great (Horne, 1971; Heyman, 1971).

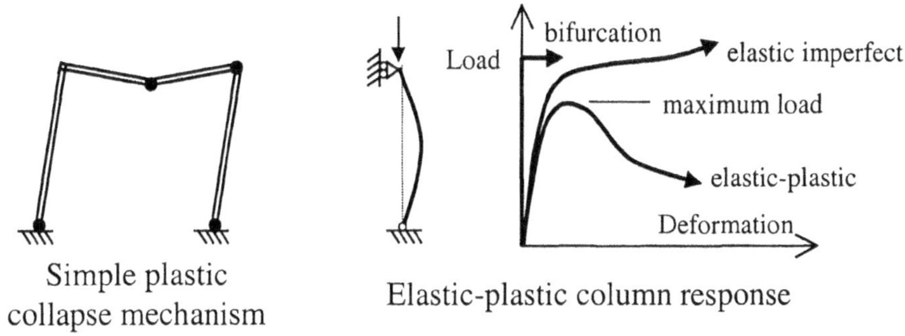

Figure 1. Structural behaviour of beams, columns and frames.

In columns, the potential for bifurcation into a different mode and the geometrically nonlinear response to axial load mean that stability phenomena are most often seen in relation to column behaviour (Fig. 1). In slender members, elastic buckling may be seen, but in shorter or geometrically imperfect members, the combination of imperfections, geometrical nonlinearity and material nonlinearity leads to a clearly defined maximum load (Fig. 1). This load is characterised by the "column curve" (Fig. 2), here expressed in terms of the relative slenderness $\lambda = \sqrt{\sigma_Y/\sigma_{cr}}$, where σ_Y is the yield stress and σ_{cr} is the elastic critical stress which depends on the elastic modulus E. This curve, which might better be termed a "capacity curve", henceforth plays a major role in defining the interaction between plasticity and stability effects. In very stocky columns, strain-hardening leads to increased strengths, and in extremely stocky columns, three-dimensional stress states may raise the strength even further. However, these short-column phenomena are almost universally ignored in design. It may also be noted that the large-deflection load-displacement response of a very slender imperfect column does not display a maximum load that can be used to characterise the strength, and this difficulty is also ignored. These features do not impinge on frame design because they do not appear often in typical civil engineering geometries, but the same is not true for shell structures. In addition to the above, the simplification to a two-parameter material model (E, σ_Y) for most calculations, so useful for mild steel, has led to a less developed methodology for application to other steels and other metals.

Such is the conceptual framework derived from frame structures and adopted into the domain of shell structures.

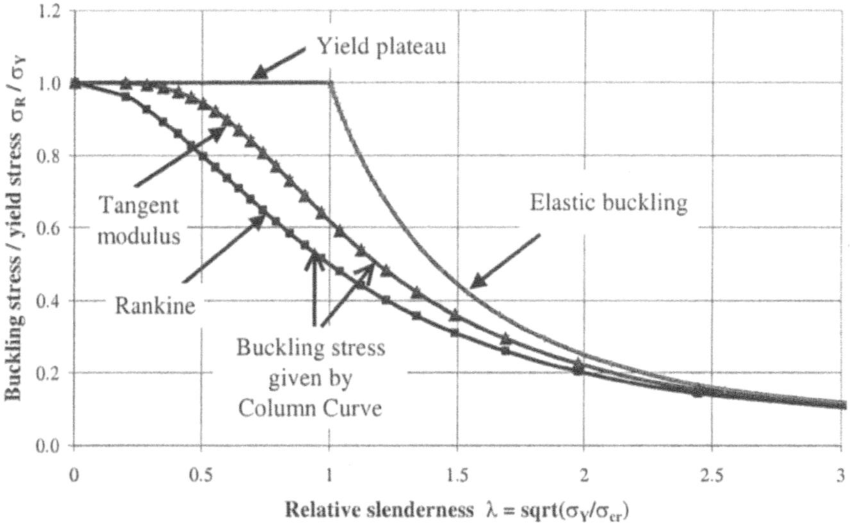

Figure 2. Elastic-plastic member failure defined by a "column curve" or "capacity curve".

It is useful to redraw capacity curves of the type shown in Fig. 2 in a different format, with the ratio of buckling stress σ_R to yield stress σ_Y plotted against the ratio of buckling stress to elastic critical stress σ_{cr}. The generalised Rankine (1866) capacity curves

$$\sigma_R = \frac{\sigma_{cr}\,\sigma_Y}{\sqrt[n]{\sigma_{cr}^{\ n} + \sigma_Y^{\ n}}} \tag{1}$$

are shown in this form in Fig. 3, where $n=1$ gives Rankine's original formula, which was illustrated in Fig. 2, and $n \to \infty$ leads to no interaction between yielding and buckling, as shown by the yield plateau and elastic buckling curves in Fig. 2.

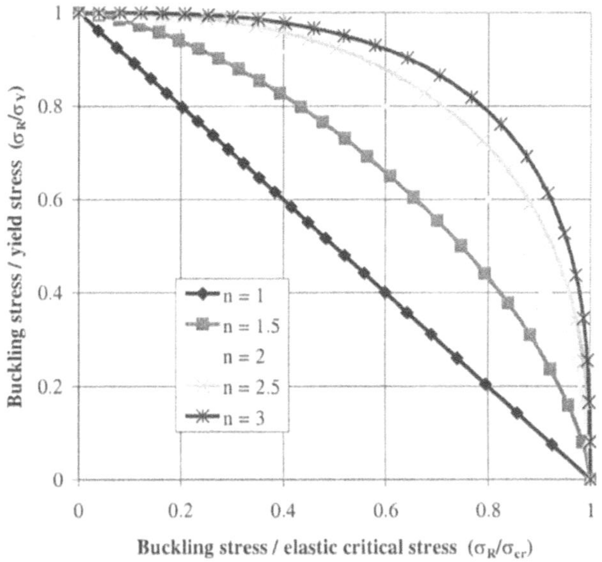

Figure 3. Dimensionless buckling interaction curves for columns.

2.2 Shell structures

Shell structures are formed from a two-dimensional surface curved in three-dimensional space. This form is highly efficient in supporting distributed loads, but its optimised form potentially leads to a high susceptibility to minor deviations from perfection. It is internally highly redundant, permitting the load to be carried by more than one stress system, and this leads to considerable complexity of behaviour. The simple design concepts for frame structures are carried over, first to plates, then to shells; but additional features appear at every turn, making simplified statements more difficult to formulate.

The equivalent of a small-displacement plastic collapse analysis is considerably more complex in a shell structure. In a flat plate, the discrete plastic hinges of the frame structure become yield lines whose location must be optimised to find the lowest load, and the yield criterion involves bending in more than one direction: nevertheless, these factors are relatively straightforward and are described by yield line analysis. However, in shell structures, the collapse mechanism involves stretching as well as bending in two directions (Massonnet and Save, 1972), so that the yield criterion is more difficult to define, and the kinematics of the mechanism become simultaneously more obscure for all but the simplest problems. As a result, it is natural for the designer to adopt a first-yield, or a first full-plasticity-at-a-section measure to provide a lower bound estimate of the collapse load; but often these simplifications lead to very low lower bounds.

Most shell design problems involve multiple shell segments connected together, with applied loading that varies throughout the structure and is unsymmetrical to any simple axis. By contrast, most tests of the strength of shells have been conducted using simple geometries of single shell segments under perfectly uniform or almost uniform loads. The interpretation of analyses of real shell structures in terms of the knowledge base from testing is not easy, because there are so many uncertainties. Moreover, the mathematics of formal algebraic solution of the governing equations is so unmanageable that numerical analyses must dominate the field.

As a result, the majority of progress in the understanding and design of shell structures now comes from numerical analysis. These numerical analyses can take many different forms, depending on the assumptions made, and all have some difficulty in matching the results of the full range of experiments. The new shells standard has attempted to define each type of analysis and regulate the manner in which its predictions should be used to ensure that all proper safety margins are included in design.

3. SCOPE AND COVERAGE OF THE STANDARD

3.1 Introduction

The enormous literature on the analysis of shell structures covers small and large deformations, elastic, elastic-plastic, hardening, bifurcation, snap-through, imperfection-sensitivity, stability and collapse analyses. These analyses are mostly concerned to evaluate either a load-displacement path, or a stress state, or a load at which “failure” may be deemed to have occurred. There is a lack of clarity in the shells literature on the relationship between a given analysis and the appropriate criterion of failure. There are also relatively few clear descriptions identifying which analysis should be used for which purpose, and what margins of safety are necessary in the use of each criterion of failure. This paper outlines the manner in which these matters have been addressed within the new European standard.

Although the legal framework requires that civil engineering design satisfies an appropriate standard, there is often a large gap between the design rules in a standard on shell structures (e.g. ACI 334, 1992) and the sophistication of the analyses presented in the research literature. It is often difficult for the practising engineer to see how a more sophisticated analysis may be exploited in the design process. For example, many designers are accustomed to performing only membrane-theory analyses of their shells, and if they move to a bending theory analysis, much higher local stresses are calculated (for which they may not realise that different failure criteria should apply), and this often discourages them from moving towards greater realism in their design calculations.

3.2 Overall aims and scope

The overall aim of the drafters of the new standard was to produce a forward-looking document that could accommodate analyses of any sophistication and produce results of comparable safety and reliability. A more thorough analysis should, in general, produce a more precise evaluation of the actual strength of the structure, and by eliminating some uncertainty should open opportunities for cheaper, safer structures. In addition, the standard is intended to provide researchers, who may be developing new understandings and new design rules, with clear guidance on the methodology that they should adopt to ensure an easy accommodation into the wider framework of design rules. Thus, the standard sets out a hierarchy of analyses and draws appropriate failure criteria for each.

This standard is notionally for steel shells, so its compass is effectively restricted to metal shells, though some of its provisions may be useful for wider application. Its scope is limited to shells about which much is known. This is why it is restricted to multiple-segment shells formed from axisymmetric segments or panels, and to static loading or quasi-static cyclic loads. The loading pattern on the shell may take any form, but the hand analysis of shells under unsymmetrical and local loads is not simple, and the criteria of failure for these analyses are almost always very conservative for complex load cases. The scope is currently limited to unstiffened shells, but this is a shortcoming that will be addressed when time and resources permit.

3.3 Alternative analyses of the structure

As indicated above, the standard attempts to codify the use of a wide range of analyses. To this end, each is defined with care, using the abbreviations and meaning shown in Table 1.

Membrane theory uses only static equilibrium, predicts only membrane stresses and is a poor representation near discontinuities or for loading that is severely unsymmetrical, especially if this acts tangential to the plane of the shell surface. However, it is widely used in design, and hand analysis of the shell is normally predicated on a membrane theory analysis.

Linear shell theory (LA) defines the normal operating condition of the shell under small-displacement elastic conditions, and produces a complete membrane and bending distribution with six stress resultants at every point. Interpretation of the results into values relevant to limit states is not always easy, since very high local bending stresses can occur that do not affect the strength much (associated with satisfying boundary conditions), but under other conditions similar bending stresses may be needed to satisfy equilibrium (Rotter, 1987). A linear analysis does not distinguish between these two cases. Also within the framework of linear theory is a linear bifurcation or eigenvalue analysis (here termed LEA), performed on a perfect shell structure, yielding the elastic critical load for the system. It should be remembered that the critical load depends only on the membrane stresses and is unaffected by bending stresses.

Table 1. Structural analysis types

Type of analysis	Term	Shell theory	Material law	Shell geometry
Membrane theory of shells		membrane equilibrium	not applicable	perfect
Linear elastic shell analysis	LA	linear bending and stretching	linear	perfect
Geometrically non-linear elastic analysis	GNA	non-linear	linear	perfect
Materially non-linear analysis	MNA	linear	non-linear	perfect
Geometrically and materially non-linear analysis	GMNA	non-linear	non-linear	perfect
Geometrically non-linear elastic analysis with imperfections	GNIA	non-linear	linear	imperfect
Geometrically and materially non-linear analysis with imperfections	GMNIA	non-linear	non-linear	imperfect

The nonlinear analyses are identified according to the additional phenomena included. The materially nonlinear analysis (MNA) uses small-displacement theory with plasticity and its principal use is to determine plastic limit loads. The geometrically nonlinear elastic analysis (GNA) uses large-displacement elastic theory and can trace nonlinear load paths. It is needed to detect elastic snap-though buckling events (elastic limit loads). It should also be able to follow the load path through a bifurcation onto the post-buckling path, but often this can only be achieved if an appropriate small perturbation is introduced, which the numerical analysis can amplify into the new mode.

Geometrically and materially nonlinear analysis (GMNA) combines the above analyses, and provides good estimates of the strengths of perfect structures that are not imperfection-sensitive. However, when imperfection-sensitivity is present, this must be expanded to the GMNIA analysis, explicitly including geometric imperfections of appropriate amplitudes in appropriate modes. The latter task makes the use of GMNIA analysis far from straightforward.

3.4 Obtaining the characteristic strength for design

Researchers in applied mechanics often appear to think that the GMNA or GMNIA analyses are the only ones that are required, since they give the "right answer" to the problem in question. It should therefore be very clearly stated that these analyses do not solve problems in a general manner, because the significance of the result cannot be recognised unless it is placed in the context of other reference loads. Thus the buckling stress of a column is only useful when its yield stress and slenderness are known. If this concept is transferred across to a general shell structure, the GMNIA result is only useful if both an MNA and an LEA or GNA analysis is performed to identify the relevant reference loads.

The target of all the analyses described here is a definition of the characteristic strength of the structure R_k, which is notionally the 95 percentile value in the distribution of possible strengths. The distribution arises from variability in material strengths, actual structural geometry, construction defects and idealisations made in the analysis. This 95 percentile strength is deemed to be calculated by adopting appropriately low material properties and structural section thicknesses. In addition, the 95 percentile strength must be separated from the consequences of the "worst" loading event, so the partial factor γ_M is used to reduce the characteristic strength to a design value R_d.

3.5 Limit States

The standard defines four separate limit states. These are:

a) Limit State 1: Plastic limit state
b) Limit State 2: Cyclic plasticity
c) Limit State 3: Buckling
d) Limit State 4: Fatigue

3.5.1 LS1: Plastic limit state

The plastic limit state was originally termed "plastic collapse" in the standard before anxious committee members denied the usage of the word "collapse". It represents the state of material rupture or unacceptably large plastic deformations, when stability phenomena do not intervene. Its simplest and clearest expression is in the classical plastic limit load, which is derived from a small-displacement theory analysis. Lower-bound approximations to this load are used when simpler analyses are involved, but the standard permits the use of true limit loads in "direct design" where a formal limit load is known algebraically.

Because the framework described below relies on this limit load to define slenderness, and the slenderness is needed to locate a structure correctly in the elastic-plastic interaction space, research studies that evaluate formal limit loads continue to have a vital role in the development of design rules.

The inclusion of geometric nonlinearity in an evaluation of the plastic limit state requires caution for both geometrically hardening and softening structures. For softening structures, the distinction between the plastic limit and buckling is lost, and the peak load is the same as that for plastic buckling, unless bifurcation occurs. In hardening structures, the plastic limit depends on the maximum permitted deformation, as discussed below.

3.5.2 LS2: Cyclic plasticity

The cyclic plasticity limit state is concerned with a few cycles of loading that induce large strains, generally reversing in sign, and producing plastic damage in both directions. It limits the amount of plastic straining to avoid cracking. It is usually required in large shell structures where local membrane deformation incompatibilities lead to severe local bending. It has often been ignored in the past, but it remains a distinct limit state from that of fatigue, and one that should not be forgotten in a limit state assessment.

3.5.3 LS3: Buckling

The section on buckling is by far the largest in the standard. It is based on the concepts used in hand analysis, but it defines how these should be extended when numerical analyses are used.

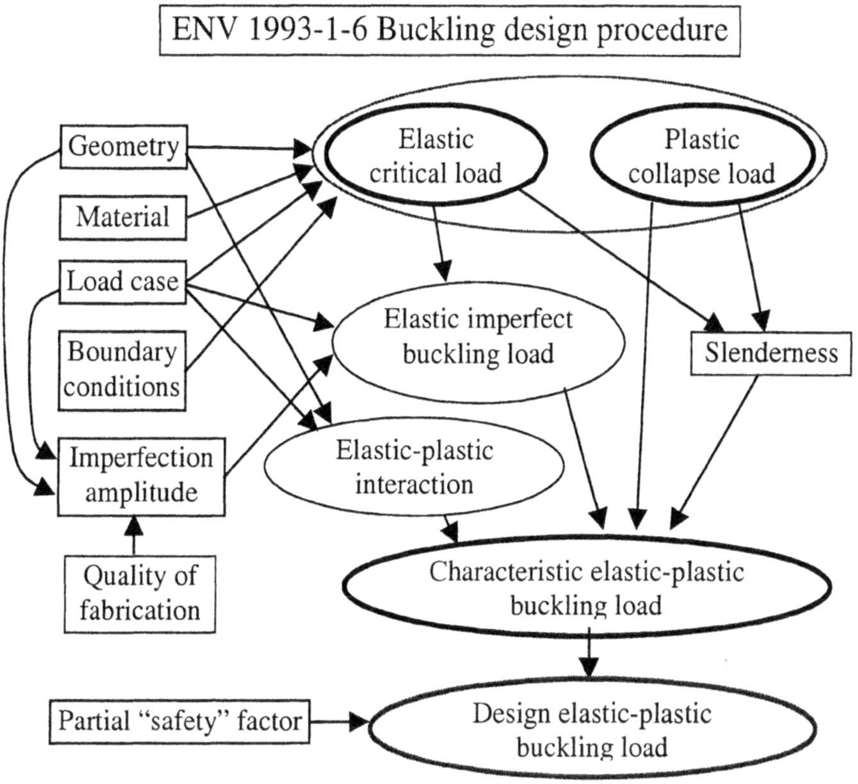

Figure 4. Calculation concepts and the route to design strength.

The process of buckling design-strength assessment is illustrated in Fig. 4. The geometry of the structure, material properties, load case and boundary conditions are used to define the two key reference loads: the plastic limit load and the elastic critical load. These may be either accurately or approximately assessed (e.g. first yield gives a lower bound on the plastic limit load), but both are needed for the strength assessment to proceed. The relation between the two defines the slenderness of the structure.

The geometry, quality of fabrication and load case determine the amplitude of the imperfection that must be assumed to exist in the structure. Through a strength-imperfection sensitivity relationship (knock-down factor), the imperfect shell elastic buckling strength is next found. Finally, the slenderness is used to define the relevant point on the elastic-plastic interaction between the imperfect shell elastic buckling strength and the plastic limit load. The parameters controlling this interaction curve depend on the shell geometry and load case, so that new configurations with different characteristics may be added at any time. Finally the characteristic buckling strength found in the above process must be reduced by an appropriate partial factor to give a design strength; but the appropriate statistical calibration is not discussed further here.

The above process must appear to the reader to be very complicated. However, there are good reasons for this. The individual parts of the strength assessment are separated into distinct processes, so that the imperfection sensitivity, the elastic-plastic interaction, the limit load evaluation etc. may be individually upgraded by the use of improved information, and each new piece of knowledge can make a useful improvement to the design. Moreover, it is easy to extend the model to new problems for which only partial information exists.

An improved analysis can produce a more direct and accurate evaluation of individual parts of this structure. Thus an MNA analysis may give a better plastic collapse load, and a GMNA analysis may give a better perfect shell strength, but the imperfection sensitivity must be separately evaluated.

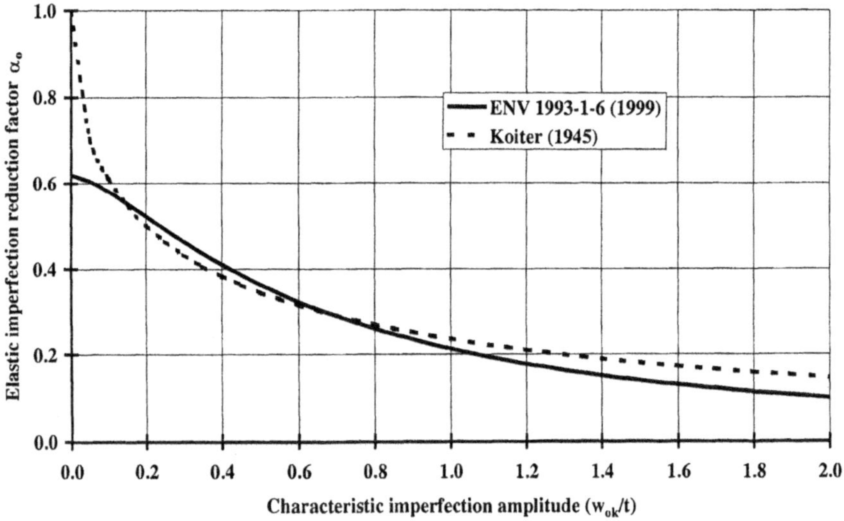

Figure 5. Strength-imperfection relationship for elastic buckling.

The quality of fabrication is explicitly acknowledged as part of the design process for the first time, bringing the large literature on imperfection sensitivity into the design process, and making an explicit connection between the tolerances required of the completed construction and the assumptions in the design. Three qualities of fabrication are defined, with levels of imperfection to be adopted accordingly. The standard includes extensive material on tolerance-assessment to ensure that the assumed quality translates into the final construction.

The adopted imperfection-sensitivity relationship (elastic imperfection reduction factor α) for axially compressed cylinders is shown in Fig. 5, and compared with Koiter's (1945) original analysis for reference: its derivation from the earlier empirical ECCS (1988) recommendations was described by Rotter (1998b).

3.5.4 LS4: Fatigue

The limit state of fatigue is concerned only with high cycle failures in the elastic stress regime. Shells provide additional problems because multi-directional stresses are the norm and very high stresses can occur near boundary conditions. Because these high stresses are quite sensitive to the rigidity of the restraint at the boundary, they can be difficult to assess properly. Moreover, shell structures often have construction details whose fatigue strength may differ considerably from the corresponding values for straight elements. This is a consequence of the additional stresses that arise from the local curvature of the shell, which is often forgotten or ignored.

Fatigue need not be considered further here because different analyses do not lead to very different outcomes for this limit state.

3.6 The capacity curve for the shells standard

To accommodate all forms of static structural analysis within the framework of the same design philosophy, it is necessary to address questions of plasticity, stability and imperfection sensitivity within the framework of a reliability assessment. The interaction between plasticity and stability is traditionally defined by the capacity curve, which gives a transition between a) fully plastic failure unaffected by geometric nonlinearity, and b) elastic stability failure unaffected by material nonlinearity. In recent decades, the "slenderness" parameter, which defines where a particular structure lies in this interaction, has been more usefully defined in generalised dimensionless terms, applicable to any system:

$$\bar{\lambda} = \sqrt{R_{pl} / R_{cr}} \qquad (2)$$

where $\bar{\lambda}$ is now the overall relative slenderness of the whole structure, R_{pl} is its plastic limit resistance and R_{cr} is its elastic critical resistance. This definition permits all structures to be treated in a uniform manner (but see additional comments below).

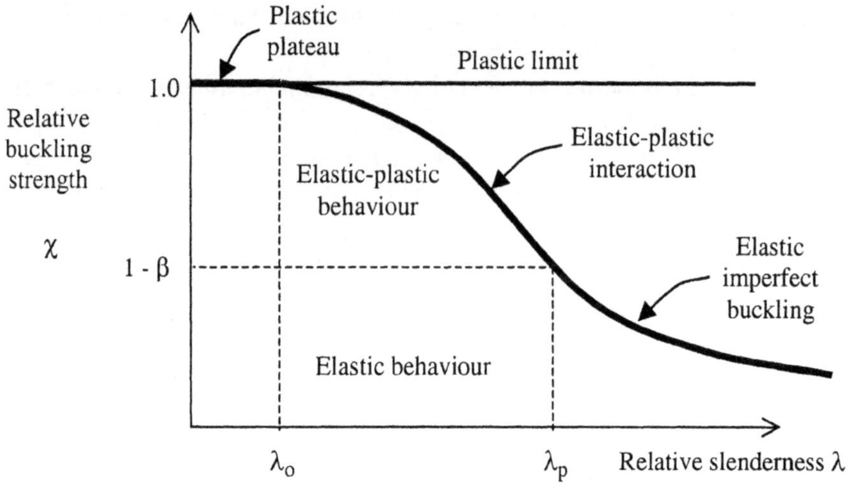

Figure 6. Shell buckling capacity curve: generic form.

The form of the capacity curve for shell structures must accommodate a number of different features because of the great variety of behaviours seen in different geometries and different loading conditions. Imperfection sensitivity in slender structures (Calladine, 1995), buckling strengths affected by yielding at either high or low stress levels, stocky structures whose strength well exceeds the plastic limit load, and a variety of shapes for different severities of plasticity effects at intermediate slendernesses.

These aspects were all included as variable features of the capacity curve (Rotter, 1999), shown in Fig. 6, which is defined by $\chi = R/R_{pl}$ as

$$\chi = 1 \qquad \text{when} \qquad \bar{\lambda} \leq \bar{\lambda}_0 \tag{3}$$

$$\chi = 1 - \beta\left(\frac{\bar{\lambda} - \bar{\lambda}_0}{\bar{\lambda}_p - \bar{\lambda}_0}\right)^{\eta} \qquad \text{when} \qquad \bar{\lambda}_0 < \bar{\lambda} < \bar{\lambda}_p \tag{4}$$

$$\chi = \frac{\alpha}{\bar{\lambda}^2} \qquad \text{when} \qquad \bar{\lambda}_p \leq \bar{\lambda} \tag{5}$$

in which $\bar{\lambda}$ is the relative slenderness, α is the elastic imperfection reduction factor, β is the plastic range factor, η is the interaction exponent and $\bar{\lambda}_0$ is the squash limit relative slenderness, all of which are separately defined for each geometry and loading case.

The plastic limit relative slenderness $\bar{\lambda}_p$ is given by:

$$\bar{\lambda}_p = \sqrt{\frac{\alpha}{1 - \beta}} \tag{6}$$

These relations permit a very wide range of elastic-plastic interactions to be accommodated, and the chief challenge for code writers and researchers is to find the different parameters α, β, $\bar{\lambda}_0$ and η from the results of GMNIA analyses. A few of the resulting shapes of these curves are shown in Fig. 7.

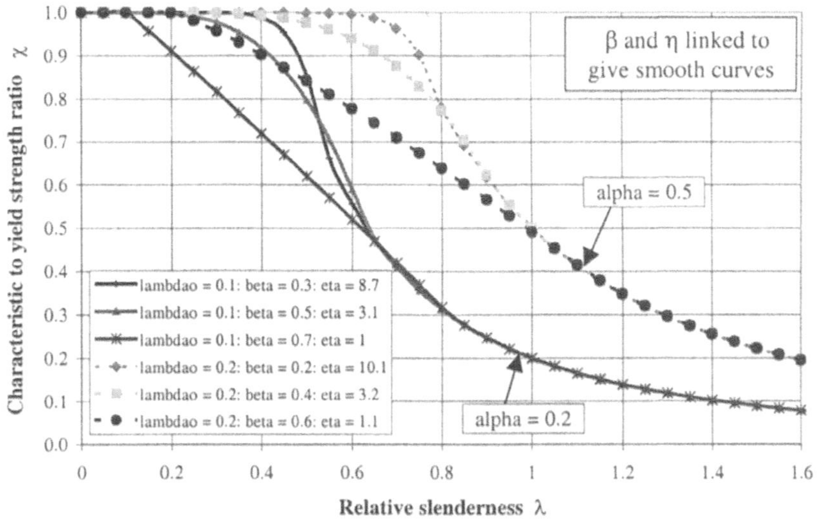

Figure 7. Different capacity curves arising from different parameter values.

It should be noted that the relative slenderness $\bar{\lambda}$ is defined in a very general manner (Eq. 2), so that it should be possible to apply the methodology to all shell structures, with a great variety of geometries, loadings and boundary conditions. However, a certain amount of difficulty arises here: so long as the loading is perfectly axisymmetric and in simple patterns (in Yamaki's (1984) terms "fundamental loads"), there is no ambiguity about the definition of the reference loads R_{pl} and R_{cr}. But as soon as realistic load conditions are applied, both R_{pl} and R_{cr} begin to change, so the relative slenderness appears to be sensitive to small changes in the design load pattern. An example of such a change in load is the introduction of internal pressure in an axially compressed cylinder. This problem is currently ignored in all known formulations, and the relative slenderness is treated as being associated only with the fundamental load case. However, if this interpretation is retained, the relative slenderness immediately loses its generality, and care must be exercised when identifying the relative slenderness for each load case in a complicated real structure. This problem requires much further careful thought before a general treatment can be found.

4. EXPLOITATION OF DIFFERENT ANALYSES

4.1 The reference design framework

The conceptual framework of Fig. 4 is interpreted in terms of a generalised capacity curve, shown in Fig. 6, to define the complete strength. The following briefly outlines the manner in which each limit state is addressed by each analysis.

4.2 Hand analysis: Membrane Theory

Membrane theory is widely used by designers, at least for the initial proportioning of their structures. Figure 4 applies directly, with the plastic limit load conservatively evaluated as first membrane yield at any point in the structure, and the elastic critical load assessed by simple determination from classical buckling formulae. The characteristic load is found by using an elastic imperfection reduction factor α ("knock-down") on the critical load and the capacity curve to cover elastic-plastic responses. This approach becomes ineffective when the shell is subject to unsymmetrical or local loads, for which membrane theory is either inaccurate or inapplicable.

4.3 Linear elastic shell analysis

When a linear elastic (LA) analysis is performed ("shell bending theory"), a much better description is provided of the stress state in the structure, and unsymmetrical or local loads and discontinuities are all properly accommodated. The framework shown in Fig. 4 is again used.

The plastic limit load can still only be estimated. The presence of high bending stresses at discontinuities, junctions and near boundaries means that some methods of evaluating the plastic limit load will give a lower estimate than that of membrane theory, which is unfortunate. The drafting committee chose to make only a small move towards addressing this problem, because of their lack of confidence about the generality of any less conservative approach. They defined the plastic limit load R_{pl} as first yield at any point based on the Ilyushin criterion (Massonnet and Save, 1972; Calladine, 1985), which corresponds to an ideal elastic-plastic cross-section response with full plasticity permitted at the limits of pure membrane forces and pure bending. This criterion is rather conservative, chiefly because it only considers yield at a single point, and not an incremental mechanism.

Despite the conservatism of the single point yield criterion, it should be also noted that high local surface stresses are often found in an LA analysis, which can cause the cyclic plasticity limit state to be attained after a small number of load cycles. Cyclic plasticity may then become the controlling design condition, though further work is needed to identify the relative severities of the two limit states in a probabilistic context.

Where a linear-eigenvalue analysis is not undertaken, the buckling evaluation must accommodate buckling under a combination of different stress components. The standard uses an interaction rule between the maximum calculated values of each of the three components of membrane stress (meridional, circumferential and membrane shear). The interaction strength criterion is based on buckling conditions for shells under completely uniform stresses where there are few restrictions on buckle size. This criterion is very conservative for problems with unsymmetrical and local loads because no account is taken of the finite size of buckles and the consequent need for a highly stressed zone of size comparable with the critical buckle size.

Moreover, the maximum value of each membrane stress component may occur at a different place in the structure, and the current rule is even more conservative because it defines the same interaction between maximum values of different component stresses even though the maxima are widely separated in space. Values of stress at widely separated locations that are deemed to interact are indicated in Fig. 8.

There are currently two possible approaches to this problem:

a) The stresses at any point may be combined into the interaction formula, and the buckling check applied at all points in the structure. This should always be conservative, because the finite size of buckles is ignored, but it has not yet been shown to be safe in all cases.

b) The highest compressive stress of each kind, irrespective of where it occurs, should be combined together (with limitations against the use of stresses near boundary restraints (Fig. 8)) into the interaction formula. This approach is certainly conservative, but for many cases it must be very conservative, since the stresses inserted into the formula must often not interact. An improved rule is certainly needed in the standard.

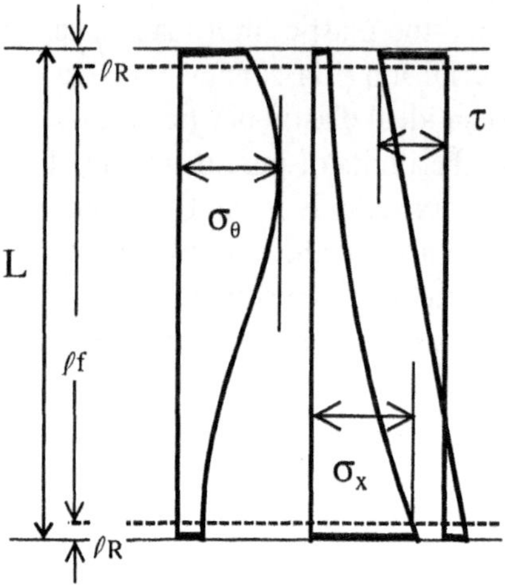

Figure 8. Peak values of different stress components that are deemed to interact according to the interaction formula.

The linear-elastic eigenvalue analysis (LEA), based on LA stresses, is classed as part of the LA analysis, though it constitutes a considerable additional step. It accurately determines the elastic critical buckling load R_{cr}, as indicated in Fig. 9. This is an important reference load for all analyses, since it is required in the formal definition of slenderness needed to interpret the results of other analyses, as shown in Fig. 4. This load is unique, always calculable when compressive stresses occur in the structure, and can usually be represented by a relatively simple equation following a parametric numerical study. Linear eigenvalue analysis cannot detect snap-through buckling, so cautionary notes are given concerning problems where snap-through phenomena are known to occur.

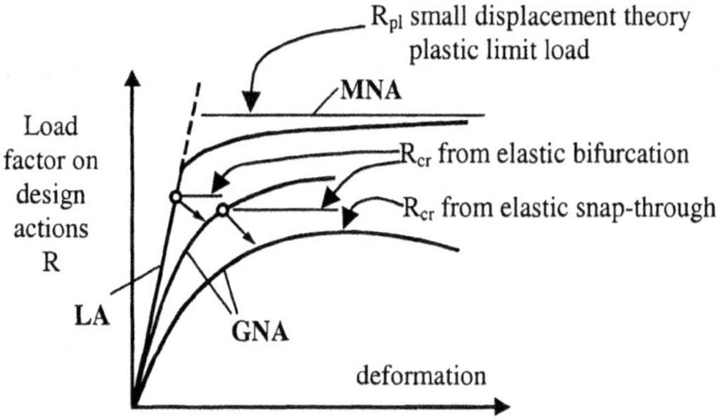

Figure 9. Calculated load-displacement paths for LA, LEA, MNA and GNA analyses.

4.4 Materially nonlinear analysis (MNA)

This analysis uses small-displacement theory with material yielding and can so determine the classical plastic limit load of the system (Massonnet and Save, 1972; Calladine, 1985), as indicated in Fig. 9. Strictly, an ideal elastic-plastic material model is needed. The calculated load may be identified as the correct value for the reference load R_{pl}, which is needed for the definition of slenderness as shown in Fig. 4 and the plateau in Fig. 6. This is its primary role.

This load is unique, always calculable, and can usually be represented by a relatively simple equation following a parametric numerical study. It is sometimes necessary to transform the data to obtain a precise estimate of the limit load (Holst et al, 2002).

The second role of an MNA analysis relates to the cyclic plasticity limit state. The MNA analysis can often be used to show that the structure can achieve shake-down into stress cycles that lie within the yield boundary, eliminating the need to consider cyclic plasticity further.

4.5 Geometrically nonlinear analysis (GNA)

This analysis captures the large-displacement behaviour of the perfect structure, but models the material response as elastic. It is the minimum analysis required to capture snap-through buckling, and for structures susceptible to snap-through (arches, shallow spherical caps, shallow conical roofs), it is needed to define the reference critical load R_{cr}, as shown in Fig. 9.

It may be used to explore bifurcation into non-symmetric modes, but the values so obtained are not very useful. The reason is that these values do not include imperfections, so they are only reference values for the perfect structure. Moreover, the reference buckling strength R_{cr} is based on the linear eigenvalue critical load, so the GNA analysis, which includes non-linear prebuckling stress redistributions, actually does not give the required result. Thus, on this occasion, the more "accurate" analysis is of less value. In addition, in geometrically hardening structures, it may be difficult to extract an appropriate maximum load from this analysis.

4.6 Geometrically and materially nonlinear analysis (GMNA)

For structures that are not imperfection-sensitive, a GMNA analysis provides a rather accurate evaluation of the true strength of the structure. It is therefore the reference precise analysis for frames and similar structures.

It is an appealing analysis, because the difficulties of choosing appropriate imperfections are avoided, and there is only a single calculation in place of a whole series with different imperfection amplitudes needed for GMNIA.

However, since this analysis does not explicitly include geometric imperfections, and shell structures are generally imperfection sensitive, the standard requires that the maximum load determined from a GMNA analysis should be reduced according to the imperfection sensitivity factor $\alpha_{ov,GMNA}$ assessed using hand methods for similar structures, the similarity to take account of buckling modes, imperfection sensitivity, yielding sensitivity and postbuckling behaviour, together with the fabrication tolerance quality class.

Thus the standard requires that the characteristic load R_k be taken as

$$R_k = \alpha_{ov,GMNA} \; R_{GMNA} \tag{7}$$

where R_{GMNA} is the calculated load. The reduction $\alpha_{ov,GMNA}$ is satisfactory for relatively slender structures, but it becomes very conservative for stocky shells. The reason is that it does not account for the reduced imperfection sensitivity of yielding structures (note the different imperfection sensitivity at large slenderness in Fig. 7 but the close proximity of the curves at small slenderness). For a very stocky shell, it returns the very conservative value of αR_{MNA}, and certainly needs improvement.

However, an improvement of this rule is not simple. Unless the relative slenderness of the structure can be identified, the extent to which the imperfection sensitivity α should influence the calculation outcome cannot be determined. Thus, even a GMNA analysis is of very limited value for imperfection-sensitive structures unless it is accompanied by another means of determining the slenderness (either using MNA and LEA analyses, or an approximate evaluation from an LA analysis). Thus the two key reference loads R_{pl} and R_{cr} are almost always needed, even for this high quality analysis.

One last problem with the GMNA analysis is common with the GMNIA: in geometrically and strain hardening structures: it may be difficult to identify the maximum load. This is discussed below.

4.7 Geometrically and materially nonlinear analysis including geometric imperfections

This very sophisticated GMNIA analysis should, ideally, be able to identify the true characteristic strength of the structure, and obtain calculated values that are rather close to the requirements of the design. However, several difficulties arise in seeking a description applicable to all shells. The different forms of load-deformation plot are shown in Fig. 10.

Three criteria are given for identifying the failure load:

a) the maximum load,

b) the lowest bifurcation load, and

c) the largest tolerable deformation.

The latter, needed for geometrically and/or materially hardening structures, is unfortunately difficult to define well, because no plateau load is achieved in a hardening structure, and the "maximum load" is rather arbitrarily defined, as indicated in Fig. 10. Moreover, tolerable deformations tend to be very problem-specific. It is important that the criterion is interpreted flexibly in terms of either displacements or rotations of the surface of the structure, and it should be noted that the latter are often more detectable by the naked eye. It may be that this restriction should be applied as a serviceability limitation at the operational load level.

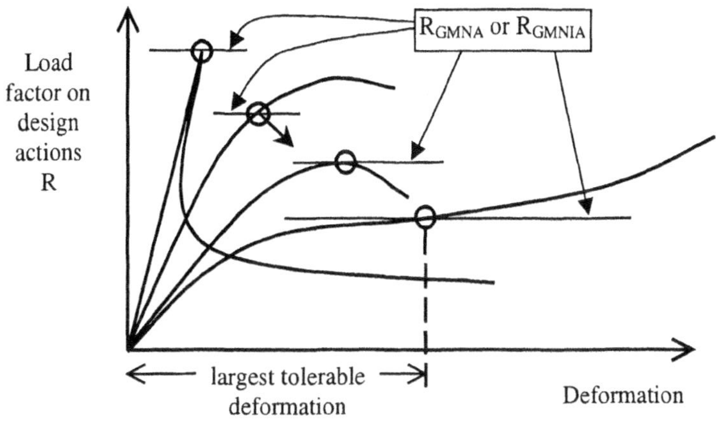

Figure 10. Criteria of failure for GMNA and GMNIA analyses.

The challenge of choosing an appropriate imperfection form and amplitude to capture an appropriately low strength in GMNIA analysis is not simple. The standard requires that the mode be chosen to give the most unfavourable effect on the buckling behaviour, and suggests both eigenmode imperfections (in the form of the critical mode from LEA analysis) and forms related to the fabrication method (e.g. weld depressions).

The amplitude of imperfection to be adopted is again related to the assumed fabrication tolerances through a quality class, and a warning is given that sometimes an imperfection of smaller amplitude can produce a lower buckling load (this is typically true for large amplitude imperfections on short shells).

Other forms of imperfection are always present: the load is not applied as perfectly as is commonly assumed, the boundary conditions are imperfect, and residual stresses always arise as a result of the fabrication process.

Whilst the latter may not be very damaging (Holst et al, 1996), non-uniformities of load often affect the strength considerably. Thus, it must be recognised that the GMNIA analysis is a deterministic evaluation seeking to establish a characteristic value and it must be calibrated against similarly calculated known results for similar shells to ensure that an adequate margin is included for the scatter of strengths associated with unknown factors.

5. CHALLENGES IN EXPLOITATION OF THE NEW FRAMEWORK

Unlike other structural engineering standards, the shells standard clearly defines the four different ultimate limit states against which the structure must be designed, and has a chapter on each. For each limit state, the method by which an analysis of any given complexity or simplicity must be used, and the criteria against which its predictions should be judged, are all defined. The committee that drafted this standard found that the task of defining these criteria was sometimes very difficult, because it is not yet clear what should be classed as failure in the most general case.

The load-deflection curves can take a wide variety of forms, as indicated in Figs 9 and 10, so a general prescription of what constitutes an unacceptable or ultimate limit state is not straightforward.

Key questions that need answers are:

a) How should be maximum acceptable deformation be defined?
b) The effects of material strain-hardening are normally ignored: should the effects of geometric hardening and strain hardening be permitted to be exploited to give permitted strengths in excess of the plastic limit load?
c) When should a bifurcation be treated as a limitation on the structural resistance (does a little dent matter if it is followed by hardening?)
d) The relative slenderness is currently defined in an apparently general manner, but it is really defined in terms of fundamental (axisymmetric) loads. How can the description be kept general without introducing too much complexity where realistic load cases are used?
e) The buckling resistance depends on the amplitude of geometric imperfections. How should these be characterised and controlled by tolerance measurements?
f) The hand-calculation stress-combination rule for interactions between different stresses needs further exploration and development, as illustrated in Fig. 8.

Assistance from the academic community in finding better solutions to these challenging problems would be most valuable during the trial usage period of the next few years, before the new standard is made mandatory.

6. CONCLUSIONS

This paper has outlined the philosophical framework used in the new European standard on shells (ENV 1993-1-6, 1999). It has shown how a great variety of different structural behaviours can be accommodated within a single design framework, including bifurcation buckling, imperfection sensitivity, snap-through, nonlinear geometric effects, plasticity, elastic-plastic buckling, plastic limit loads, strain hardening and geometric hardening. It has outlined the way in which different types of analysis may be exploited. Although Fig. 4 is based on hand analysis concepts, each type of numerical analysis may be used to upgrade the evaluation of one or other part of this system.

The overall strength evaluation must almost always be based on the two key parameters of the elastic critical load R_{cr} and the plastic limit load R_{pl}. Even geometrically and materially nonlinear analyses cannot be properly exploited unless these two key values are determined: this problem has not been recognised previously. Also, the design calculation process currently assumes that neither R_{cr} nor R_{pl} can be exceeded: there is clearly room for this restriction to be lifted.

The paper has also outlined how new results from geometrically and materially nonlinear analyses including imperfections may be interpreted into consistent design rules that fit within the design framework.

This standard currently has the status of a draft for trial application (ENV), but will become mandatory (EN) in about 2003. It is therefore most desirable that experts in shell analysis and design should make trial usage of it and give feedback to the project team or their standards body on helpful improvements.

Acknowledgements

The author, Convenor of the Project Team CEN 250/SC3/PT4 who drafted the European standard ENV 1993-1-6, would like to express his sincere thanks to all members of the project team, together with the members of ECCS TWG8.4 Buckling of Shells, who also contributed greatly to the development. In particular, special thanks are expressed for the major contributions of Professors Herbert Schmidt and Richard Greiner, and for the most valuable comments of Professor Greiner on the original manuscript.

References

ACI-334 (1992) Concrete Thin Shell Design and Construction, ACI 334-92, American Concrete Institute, Detroit.

Brush, D.O. and Almroth, B.O. (1976) *Buckling of Bars, Plates and Shells*, McGraw-Hill, New York.

Bushnell, D. (1985) *Computerized buckling analysis of shells*, Dordrecht.

Calladine, C.R. (1983) *Theory of Shell Structures*, Cambridge University Press, Cambridge.

Calladine, C.R. (1985) *Plasticity for Engineers*, Ellis Horwood, Chichester.

Calladine, C.R. (1995) Understanding imperfection-sensitivity in the buckling of thin-walled shells, *Thin Walled Structures*: Special Issue on Imperfection-Sensitive Shells, ed. G.D. Galletly, **23**, 215-235.

ECCS (1988) *European Recommendations for Steel Construction: Buckling of Shells*, 4th edition, European Convention for Constructional Steelwork, Brussels.

ENV 1993-1-6 (1999) *Eurocode 3: Design of steel structures, Part 1.6: General rules - Supplementary rules for the strength and stability of shell structures*, Eurocode 3 Part 1.6, CEN, Brussels.

Heyman, J. (1971) *Plastic design of frames: Volume II Applications*, Cambridge University Press, Cambridge, England.

Holst, J.M.F.G., Rotter, J.M. and Calladine, C.R. (1996) Geometric imperfections and consistent residual stress fields in elastic cylinder buckling under axial compression, *Proc. Int. Workshop: Imperfections in Metal Silos: Measurement, Characterisation and Strength Analysis*, CA-Silo, Lyon, France, 19 April, pp 199-216.

Holst, J.M.F.G., Rotter, J.M., Gillie, M. and Münch, M. (2002) Failure criteria for shells on local bracket supports, *New Approaches to Structural Mechanics, Shell and Biological Structures*, edited by H. Drew and S. Pellegrino, Kluwer Academic Publishers, London.

Horne, M.R. (1971) *Plastic Theory of Structures*, Nelson, London.

Koiter, W.T. (1945) *On the Stability of Elastic Equilibrium*, (in Dutch) PhD Thesis, Delft University, (see also Translation AFFDL-TR-70-25 Wright Patterson Air Force Base, 1970).

Massonnet, C.E. and Save, M.A. (1972) *Plastic Analysis and the Design of Plates Shells and Disks*, North-Holland.

Rankine, W.J.M. (1866) *Useful Rules and Tables*, London.

Rotter, J.M. (1987) Bending Theory of Shells for Bins and Silos, *Transactions of Mechanical Engineering*, Institution of Engineers, Australia, Vol. ME12 No.3 September, pp 147-159.

Rotter, J.M. (1997) Pressurised axially compressed cylinders, *Proc., Int. Conf. on Carrying Capacity of Steel Shell Structures*, Brno, 1-3 October, pp 354-360.

Rotter, J.M. (1998a) Shell structures: the new European standard and current research needs, *Thin-Walled Structures*, Vol. 31, pp 3-23.

Rotter, J.M. (1998b) Development of proposed European design rules for buckling of axially compressed cylinders, *Advances in Structural Engineering*, Vol. 1, No. 4, July, pp 273-286.

Rotter, J.M. (1999) Proposal for generalisation of the elastic-plastic buckling interaction rule from Eurocode 3 Part 1.6, submission to CEN TC250/SC3/PT4 and ECCS TWG8.4 Buckling of Shells.

Teng, J.G. (1996) Buckling of thin shells: recent advances and trends, *Applied Mechanics Reviews*, Vol. 49, No. 4 April, pp 263-274.

Timoshenko, S.P. and Gere, J.M. (1961) *Theory of Elastic Stability*, 2nd edn, McGraw-Hill, New York.

BIFURCATION BUCKLING OF SPHERICAL SANDWICH SHELLS UNDER EXTERNAL PRESSURE IN PLASTIC RANGE

S. C. Shrivastava
Department of Civil Engineering and Applied Mechanics
McGill University
Montreal, Quebec, Canada H3A 2K6
suresh.shrivastava@mcgill.ca

Abstract: An exact analysis of bifurcation buckling of spherical sandwich shells, stressed by uniform external pressure beyond the elastic limits of the component materials is presented. The analysis accounts for transverse shear effects and considers buckling in axisymmetric and nonaxisymmetric modes. The material behaviour in the inelastic range is assumed according to both the J_2-incremental and J_2-deformation theories of plasticity. Results for elastic buckling follow as special cases.

Keywords: Shell buckling, Sandwich structures, Plasticity

1 INTRODUCTION

Bifurcation buckling of a thin homogeneous elastic complete spherical shell under uniform external pressure was dealt with by Zoelly (1915), Schwerin (1922), and Van der Neut (1932); see Timoshenko and Gere (1963). The present paper is a generalization of this problem in two respects: (1) the shell is of sandwich type with high-modulus inner and outer face shells of equal thickness bonded to a low-modulus core shell, and (2) the component materials may undergo plastic straining before buckling. This requires consideration of the transverse shear effects and the constitutive relations of a strain-hardening plasticity theory.

The first theoretical work dealing with the plastic buckling of plates and shells was that of Bijlaard (1949). He used the J_2-deformation theory of plasticity and considered buckling of sandwich plates, but not of sandwich shells. The present work employs the constitutive relations of both the competing theories of plasticity, namely the J_2-deformation and J_2-incremental theories. The approach followed is similar to the author's previous works (1995a, b).

H.R. Drew and S. Pellegrino (eds.),
New Approaches to Structural Mechanics, Shells and Biological Structures, 379–391.

The constitutive relations for both theories are based on the Mises yield condition ($J_2 = c$) and on Shanley's concept of bifurcation under continued loading ($dJ_2 > 0$). Although the incremental theory of plasticity is considered to be the correct phenomenological theory, it is well known that when applied to bifurcation buckling of plates, it predicts loads that are unrealistically higher than experimental data. In contrast, the load results from the deformation theory for plates are in safe-side agreement with experiments. However, for the plastic buckling of cylindrical shells in axial compression, the differences in buckling load predictions of the two theories are not as significant (Batterman,1965). The present paper provides this comparison for spherical shell buckling. Additionally, although the classical bifurcation results for shells are rendered useless by imperfection sensitivity in the elastic range, they might become meaningful in the plastic range, as was demonstrated experimentally by Batterman (1965) for cylindrical shells.

The presented analysis is exact and leads to formulas generalizing the classical results. The analytical results are illustrated by plotting the buckling stress for sandwich shells of 24S-T3 aluminum alloy faces and a low-modulus elastic core. The wrinkling instability of individual shell faces is precluded.

2 CONSTITUTIVE RELATIONS

Figure 1 shows a spherical sandwich shell and the coordinate system used for the buckling analysis. The middle-surface radius is denoted by R, the core thickness by h, and the face thickness by t. The total thickness $(2t + h)$ is assumed much smaller than the mid-surface radius R so that the shell can be considered thin. The prebuckling states of stress and strain are assumed to be those induced by a uniform all-round pressure q at the outer surface. For a complete sphere, an isotropic material behaviour implies a prebuckling state of uniform strain, $\epsilon^{\circ}{}_{\phi\phi} = \epsilon^{\circ}{}_{\theta\theta} =$ constant in the faces as well as the core, independent of the ϕ, θ, and r coordinates, in the elastic as well as plastic ranges. However, owing to the differences in material moduli, the stress values are different constants for the faces and the core. When the sphere buckles, the states of strain and stress are changed from $\epsilon^{\circ}{}_{ij}$, $\sigma^{\circ}{}_{ij}$ to $\epsilon^{\circ}{}_{ij} + \epsilon_{ij}$, $\sigma^{\circ}{}_{ij} + \sigma_{ij}$. Thus, in the present notation, ϵ_{ij} and σ_{ij} denote the increments in strain and stress components due to buckling.

The common assumptions of the two plasticity theories are taken to be: (1) that the strains are small, with the plastic part being volume-preserving, and (2) that the material is loaded from a virgin state, remains isotropic, and obeys the Mises yield condition, namely $J_2 = \sigma^2/3$, where σ is the (true) stress recorded in the uniaxial compression test of the virgin material in plastic range. Invoking the usual engineering approximation that $\sigma_{zz} = 0$, the relationship between the increments σ_{ij} of the stress and the increments ϵ_{ij} of the strain, for the case of loading $(dJ_2 > 0)$ from an equi-biaxial state of stress, $\sigma^{\circ}_{\phi\phi} = \sigma^{\circ}_{\theta\theta}$ are taken (Bijlaard, 1949) as

$$\sigma_{rr} = B'\epsilon_{rr} + C'\epsilon_{\theta\theta},\ \sigma_{\theta\theta} = C'\epsilon_{rr} + B'\epsilon_{\theta\theta},\ \sigma_{ij} = 2F'\epsilon_{ij}\,(i \neq j) \tag{1}$$

where $B' = E(\lambda + 3 + 3e)/\{(1 + \lambda - 2\nu)(2 + 2\nu + 3e)\}$,

$C' = E(1 - \lambda + 4\nu + 3e)/\{(1 + \lambda - 2\nu)(2 + 2\nu + 3e)\}$, $F' = E/(2 + 3e + 2\nu)$,

$\lambda = E/E_t$, and $e = (E/E_s) - 1$.

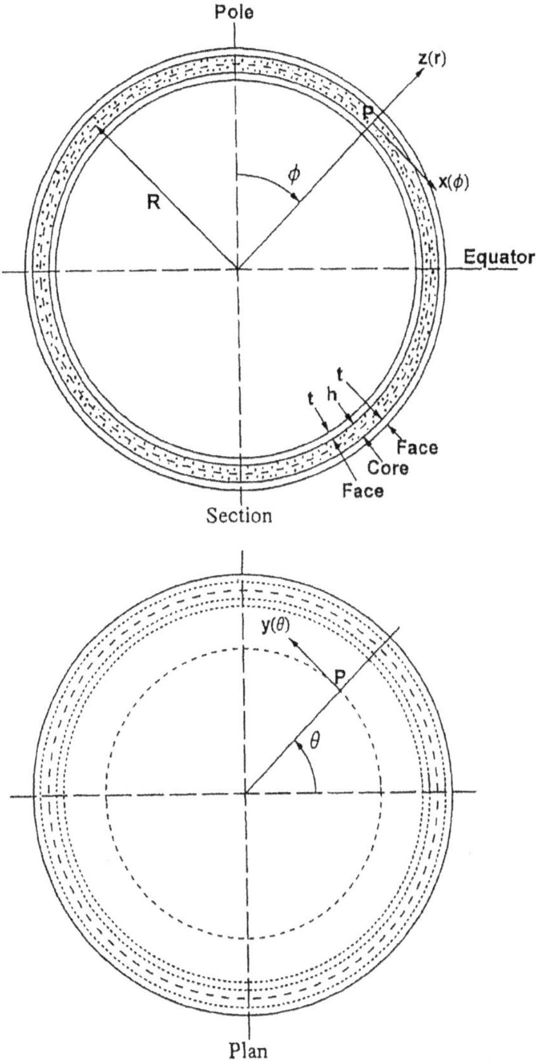

Figure 1. Spherical sandwich shell and coordinate system.

In Equation (1) E is Young's modulus of elasticity, and ν is Poisson's ratio. E_t and E_s are respectively the tangent and secant moduli obtained from a uniaxial compression $(\sigma - \epsilon)$ curve of the virgin material at a stress level $\sigma = \sigma_{eff}$. For the Mises yield criterion, and the pre-buckling stress state $(\overset{\circ}{\sigma}_{\phi\phi} = \overset{\circ}{\sigma}_{\phi\phi} = \overset{\circ}{\sigma}, \overset{\circ}{\sigma}_{zz} \approx 0)$, $\sigma_{eff} = \sqrt{3J_2} = \overset{\circ}{\sigma}$ where J_2 is the second invariant of the pre-buckling deviatoric stress tensor. As written, Equation (1) holds for the J_2-deformation theory of plasticity. However, it applies to the J_2-incremental theory for $e = 0$, and also to the linear isotropic elastic case for $e = 0$ and $\lambda = 1$. Additionally, $C' = B' - 2F'$ in all three cases.

3 GOVERNING EQUATIONS AND THEIR SOLUTION

The governing equations are derived by a modification of the standard Love-Kirchhoff kinematic hypothesis: a normal to the undeflected middle surface remains straight but not necessarily perpendicular to the deflected middle surface. The ϕ, θ, r displacement components arising due to buckling are therefore taken as

$$\overline{u} = u(\phi,\theta) - z\alpha(\phi,\theta),\ \overline{v} = v(\phi,\theta) - z\beta(\phi,\theta), \overline{w} = w(\phi,\theta) \tag{2}$$

where $u \equiv u(\phi,\theta)$, $v \equiv v(\phi,\theta)$ are the in-plane (membrane) displacements, $\alpha \equiv \alpha(\phi,\theta)$, $\beta \equiv \beta(\phi,\theta)$ are the rotations of the normal, and $w \equiv w(\alpha,\theta)$ is the radial displacement. The mid-surface radius R is defined by $z = 0$. Neglecting z/R terms in comparison to unity ($r = R + z \approx R$), the matrix of the buckling displacement gradients is $[L] =$

$$\begin{bmatrix} l_{\phi\phi} & l_{\phi\theta} & l_{\phi r} \\ l_{\theta\phi} & l_{\theta\theta} & l_{\theta r} \\ l_{r\phi} & l_{r\theta} & l_{rr} \end{bmatrix} = \begin{bmatrix} \frac{\partial u}{R\partial\phi} - \frac{z}{R}\frac{\partial\alpha}{\partial\phi} + \frac{w}{R} & \frac{1}{R\sin\phi}(\frac{\partial u}{\partial\theta} - z\frac{\partial\alpha}{\partial\theta}) - \frac{\cot\phi}{R}(v - z\beta) & -\alpha \\ \frac{\partial v}{R\partial\phi} - z\frac{\partial\beta}{R\partial\phi} & \frac{1}{R\sin\phi}(\frac{\partial v}{\partial\theta} - z\frac{\partial\beta}{\partial\theta}) + \frac{\cot\phi}{R}(u - z\,\alpha) + \frac{w}{R} & -\beta \\ \frac{\partial w}{R\partial\phi} - \frac{u}{R} & \frac{\partial w}{R\sin\phi\partial\theta} - \frac{v}{R} & 0 \end{bmatrix} \tag{3}$$

The components of the corresponding strain field, $[\epsilon] = \frac{1}{2}\{[L]^T + [L]\}$ are

$$\epsilon_{\phi\phi} = \frac{\partial u}{R\partial\phi} - \frac{z}{R}\frac{\partial\alpha}{\partial\phi} + \frac{w}{R},\ \epsilon_{\theta\theta} = \frac{1}{R\sin\phi}(\frac{\partial v}{\partial\theta} - z\frac{\partial\beta}{\partial\theta}) + \frac{\cot\phi}{R}(u - z\,\alpha) + \frac{w}{R},\ \epsilon_{rr} = 0,$$

$$\epsilon_{\phi\theta} = \frac{1}{2}\{(\frac{\partial v}{R\partial\phi} - z\frac{\partial\beta}{R\partial\phi}) + \frac{1}{R\sin\phi}(\frac{\partial u}{\partial\theta} - z\frac{\partial\alpha}{\partial\theta}) - \frac{\cot\phi}{R}(v - z\beta)\}, \tag{4}$$

$$\epsilon_{\phi r} = \frac{1}{2}(\frac{1}{R}\frac{\partial w}{\partial\phi} - \frac{u}{R} - \alpha),\ \epsilon_{\theta r} = \frac{1}{2}(\frac{1}{R\sin\phi}\frac{\partial w}{\partial\theta} - \frac{v}{R} - \beta)$$

The equilibrium equations and the boundary conditions, appropriate to the above kinematic assumptions can be obtained by applying the principle of virtual work, Washizu (1975), suitable for initially stressed bodies:

$$\int_V (\sigma_{ij}\delta L_{ij} + \overset{\circ}{\sigma}_{ij} L_{mi}\delta L_{mj})dV \equiv \int_V \{\mathrm{tr}([\sigma][\delta L]) + \mathrm{tr}([\overset{\circ}{\sigma}][L]^T[\delta L])\}dV = 0 \tag{5}$$

where $\overset{\circ}{\sigma}_{ij}$ stands for the initial (prebuckling) stress field (with $\overset{\circ}{\sigma}_{\phi\phi} = \overset{\circ}{\sigma}_{\theta\theta}$, and other components are zero), σ_{ij} denotes the stresses arising due to buckling (where only σ_{rr} is assumed zero), and L_{mi} and δL_{mj} are respectively the buckling displacement gradients and their virtual variations. The equilibrium equations (independent of the material behaviour) are obtained as

$$-\frac{\partial N_{\phi\phi}}{\partial\phi} - \cot\phi(N_{\phi\phi} - N_{\theta\theta}) - \frac{\partial N_{\phi\theta}}{\sin\phi\partial\theta} - Q_\phi + P_{cr}\{(\nabla^2 - \mathrm{cosec}^2\phi)\frac{u}{R} - \frac{2\cot\phi\partial v}{R\sin\phi\partial\theta} + \frac{2\partial w}{R\partial\phi}\} = 0$$

$$-\frac{\partial N_{\theta\theta}}{\sin\phi\partial\theta} - \frac{\partial N_{\phi\theta}}{\partial\phi} - 2\cot\phi N_{\phi\theta} - Q_\theta + P_{cr}\{(\nabla^2 - \mathrm{cosec}^2\phi)\frac{v}{R} + \frac{2\cot\phi\partial u}{R\sin\phi\partial\theta} + \frac{2\partial w}{R\sin\phi\partial\theta}\} = 0$$

$$-\frac{\partial Q_\phi \sin\phi}{\sin\phi\partial\phi} - \frac{\partial Q_\theta}{\sin\phi\partial\theta} + N_{\phi\phi} + N_{\theta\theta} + P_{cr}\{(\nabla^2 - 2)w - \frac{2\partial u}{R\partial\phi} - 2\cot\phi\frac{u}{R} - \frac{2\partial v}{R\sin\phi\partial\theta}\} = 0 \tag{6}$$

$$\frac{\partial M_{\phi\phi}}{R\,\partial\phi} + \frac{\partial M_{\phi\theta}}{R\sin\phi\,\partial\theta} + (M_{\phi\phi} - M_{\theta\theta})\frac{\cot\phi}{R} - Q_\phi = 0$$

$$\frac{\partial M_{\theta\theta}}{R\sin\phi\,\partial\theta} + \frac{\partial M_{\phi\theta}}{R\,\partial\phi} + \frac{2\cot\phi}{R}M_{\phi\theta} - Q_\theta = 0$$

$$\text{where } \nabla^2 \equiv \frac{\partial^2}{\partial\phi^2} + \cot\phi\frac{\partial}{\partial\phi} + \mathrm{cosec}^2\phi\frac{\partial^2}{\partial\theta^2}. \tag{7}$$

P_{cr} is the buckling load parameter

$$P_{cr} = 2\overset{\circ}{\sigma}^f t + \overset{\circ}{\sigma}^c h = \frac{q_{cr}R}{2} \tag{8}$$

in which q_{cr} is the critical external pressure at which bifurcation occurs, and $\overset{\circ}{\sigma}^f$ and $\overset{\circ}{\sigma}^c$ denote the corresponding uniform prebuckling compressive stresses in the faces and core respectively.

We adopt Shanley's concept of plastic buckling under increasing load, so that there is no unloading from plasticity when the perfect spherical shell experiences bifurcation into a buckled shape. Therefore, the loading Equation (1), applies throughout the thickness of the plate, and the buckling stress resultants, obtained by integration through the thickness, can be expressed as

$$N_{\phi\phi} = (\frac{B_1}{R}\frac{\partial u}{\partial \phi} + \frac{C_1}{R\sin\phi}\frac{\partial v}{\partial \theta} + \frac{C_1\cot\phi}{R}u) + \frac{w}{R}(B_1 + C_1)$$

$$N_{\theta\theta} = (\frac{C_1}{R}\frac{\partial u}{\partial \phi} + \frac{B_1}{R\sin\phi}\frac{\partial v}{\partial \theta} + \frac{B_1\cot\phi}{R}u) + \frac{w}{R}(B_1 + C_1) \tag{9}$$

$$N_{\phi\theta} = \frac{F_1}{R}(\frac{\partial v}{\partial \phi} - v\cot\phi + \frac{1}{\sin\phi}\frac{\partial u}{\partial \theta})$$

$$M_{\phi\phi} = -h^2(\frac{B_2}{R}\frac{\partial \alpha}{\partial \phi} + \frac{C_2}{R\sin\phi}\frac{\partial \beta}{\partial \theta} + \frac{C_2\cot\phi}{R}\alpha)$$

$$M_{\theta\theta} = -h^2(\frac{C_2}{R}\frac{\partial \alpha}{\partial \phi} + \frac{B_2}{R\sin\phi}\frac{\partial \beta}{\partial \theta} + \frac{B_2\cot\phi}{R}\alpha) \tag{10}$$

$$M_{\phi\theta} = -h^2\frac{F_2}{R}(\frac{\partial \beta}{\partial \phi} - \beta\cot\phi + \frac{1}{\sin\phi}\frac{\partial \alpha}{\partial \theta})$$

$$Q_\phi = G(\frac{1}{R}\frac{\partial w}{\partial \phi} - \frac{u}{R} - \alpha), \quad Q_\theta = G(\frac{1}{R\sin\phi}\frac{\partial w}{\partial \theta} - \frac{v}{R} - \beta)$$

where: $B_1 = B_c'h\,\{1 + \frac{tB_f'}{hB_c'}\}$, $C_1 = C_c'h\{1 + \frac{tC_f'}{hC_c'}\}$, $F_1 = F_c'h\,\{1 + \frac{tB_f'}{hB_c'}\}$,

$B_2 = \frac{B_c'h}{12}\{1 + \frac{B_f'}{B_c'}f(t)\}$, $C_2 = \frac{C_c'h}{12}\{1 + \frac{C_f'}{C_c'}f(t)\}$, $F_2 = \frac{F_c'h}{12}\{1 + \frac{F_f'}{F_c'}f(t)\}$,

$G = kF_c'h$, and $f(t) = 6(t/h) + 12(t/h)^2 + 8(t/h)^3$

in which k is a correction factor to account for the (actual) non-constant variations of shear strain and shear stress, contrary to the kinematic assumption of the theory. As is well known $k = 5/6$ for homogeneous plates and shells, and it can be shown that $k \approx 1$ for sandwich cases of moderate face to core thickness ratios, say $t/h < 0.2$. Substitution of these integrated relations in the equilibrium equations yields the following governing equations for the kinematic variables u, v, w, α, and β:

$$-B_1(\nabla^2 - \mathrm{cosec}^2\phi)\frac{u}{R} - (2F_1 - G)\frac{u}{R} + \frac{B_1 - F_1}{R\sin^2\phi}\frac{\partial^2 u}{\partial \theta^2} + \{\frac{B_1 + F_1}{R}\frac{\cot\phi}{\sin\phi}\frac{\partial v}{\partial \theta} - \frac{B_1 - F_1}{R\sin\phi}\frac{\partial^2 v}{\partial \phi \partial \theta}\}$$

$$-(2B_1 - 2F_1 + G)\frac{\partial w}{R\partial \phi} + G\alpha + P_{cr}\{(\nabla^2 - \mathrm{cosec}^2\phi)\frac{u}{R} + \frac{2\partial w}{R\partial \phi} - \frac{2\cot\phi\partial v}{R\sin\phi\partial \theta}\} = 0 \tag{11}$$

$$-\frac{B_1-F_1}{R\sin\phi}\frac{\partial^2 u}{\partial\theta\partial\phi}-\frac{B_1+F_1}{R}\cot\phi\frac{\partial u}{\partial\theta}-F_1(\nabla^2-\mathrm{cosec}^2\phi)\frac{v}{R}-\frac{B_1-F_1}{R\sin^2\phi}\frac{\partial^2 v}{\partial\theta^2}-\frac{2F_1-G}{R}v$$

$$-(2B_1-2F_1+G)\frac{\partial w}{R\sin\phi\partial\theta}+G\beta+P_{cr}\{(\nabla^2-\mathrm{cosec}^2\phi)\frac{v}{R}+\frac{2}{R\sin\phi}\frac{\partial w}{\partial\theta}+\frac{2\cot\phi\partial u}{R\sin\phi\partial\theta}\}=0 \quad (12)$$

$$(2B_1-2F_1+G)(\frac{\partial u\sin\phi}{R\sin\phi\partial\phi}+\frac{\partial v}{R\sin\phi\partial\theta})+G(\frac{\partial\alpha\sin\phi}{\sin\phi\partial\phi}+\frac{\partial\beta}{\sin\phi\partial\theta})-G\nabla^2\frac{w}{R}+4(B_1-F_1)\frac{w}{R}$$

$$+P_{cr}\{(\nabla^2-2)\frac{w}{R}-\frac{2\partial\sin\phi u}{R\sin\phi\partial\phi}-\frac{2\partial v}{R\sin\phi\partial\theta}\}=0 \quad (13)$$

$$G\frac{u}{R}-G\frac{\partial w}{R\partial\phi}-\frac{h^2B_2}{R^2}(\nabla^2-\mathrm{cosec}^2\phi)\alpha+\frac{h^2(B_2-F_2)}{R^2\sin^2\phi}\frac{\partial^2\alpha}{\partial\theta^2}+(G-\frac{2h^2F_2}{R^2})\alpha$$

$$-\frac{h^2(B_2-F_2)}{R^2\sin\phi}\frac{\partial^2\beta}{\partial\phi\partial\theta}+\frac{h^2(B_2+F_2)\cot\phi}{R^2\sin\phi}\frac{\partial\beta}{\partial\theta}=0 \quad (14)$$

$$G\tfrac{v}{R}-G\frac{\partial w}{R\sin\phi\partial\theta}-\frac{h^2(B_2-F_2)}{R^2\sin\phi}\frac{\partial^2\alpha}{\partial\theta\partial\phi}-\frac{h^2(B_2+F_2)\cot\phi}{R^2\sin\phi}\frac{\partial\alpha}{\partial\theta}-\frac{h^2F_2}{R^2}(\nabla^2-\mathrm{cosec}^2\phi)\beta$$

$$-\frac{h^2(B_2-F_2)}{R^2\sin^2\phi}\frac{\partial^2\beta}{\partial\theta^2}+(G-\frac{2h^2F_2}{R^2})\beta=0 \quad (15)$$

where we have used the relations $C_1=B_1-2F_1$ and $C_2=B_2-2F_2$. To solve these simultaneous partial differential equations, we effect a separation of variables by assuming:

$$u(\theta,\phi)=u'(\phi)\cos m\theta,\quad v(\theta,\phi)=v'(\phi)\sin m\theta,\quad w(\theta,\phi)=w'(\phi)\cos m\theta \quad (16)$$

$$\alpha(\theta,\phi)=\alpha'(\phi)\cos m\theta,\quad \beta(\theta,\phi)=\beta'(\phi)\sin m\theta$$

in which m is not necessarily an integer. The substitution yields the following simultaneous ordinary differential equations in the independent variable ϕ:

$$-B_1(\nabla_1^2-\mathrm{cosec}^2\phi)\frac{u'}{R}-(2F_1-G)\frac{u'}{R}-m^2\frac{B_1-F_1}{\sin^2\phi}\frac{u'}{R}+m(B_1+F_1)\frac{\cot\phi}{\sin\phi}\frac{v'}{R}-m\frac{B_1-F_1}{R\sin\phi}\frac{dv'}{d\phi}$$

$$-(2B_1-2F_1+G)\frac{dw'}{Rd\phi}+G\alpha'+P_{cr}\{(\nabla_1^2-\mathrm{cosec}^2\phi)\frac{u'}{R}+\frac{2dw'}{Rd\phi}-\frac{2m\cot\phi\, v'}{R\sin\phi}\}=0 \quad (17)$$

$$m\frac{B_1-F_1}{R\sin\phi}\frac{du'}{d\phi}+m(B_1+F_1)\cot\phi\frac{u'}{R}-F_1(\nabla_1^2-\mathrm{cosec}^2\phi)\frac{v'}{R}+m^2\frac{B_1-F_1}{\sin^2\phi}\frac{v'}{R}-(2F_1-G)\frac{v'}{R}$$

$$+m(2B_1-2F_1+G)\frac{w'}{R\sin\phi}+G\beta'+P_{cr}\{(\nabla_1^2-\mathrm{cosec}^2\phi)\frac{v'}{R}-\frac{2m}{\sin\phi}\frac{w'}{R}-\frac{2m\cot\phi}{\sin\phi}\frac{u'}{R}\}=0 \quad (18)$$

$$(2B_1 - 2F_1 + G)(\frac{du'\sin\phi}{R\sin\phi d\phi} + \frac{m}{\sin\phi}\frac{v'}{R}) + G(\frac{d\alpha'\sin\phi}{\sin\phi d\phi} + \frac{m}{\sin\phi}\beta') - G\nabla_1^2\frac{w'}{R} + 4(B_1 - F_1)\frac{w'}{R}$$

$$+ P_{cr}\{(\nabla_1^2 - 2)\frac{w'}{R} - \frac{2du'\sin\phi}{R\sin\phi d\phi} - \frac{2m}{\sin\phi}\frac{v'}{R}\} = 0 \tag{19}$$

$$G\frac{u'}{R} - G\frac{dw'}{Rd\phi} - \frac{h^2B_2}{R^2}(\nabla_1^2 - \mathrm{cosec}^2\phi)\alpha' - m^2\frac{h^2(B_2 - F_2)}{R^2\sin^2\phi}\alpha' + (G - \frac{2h^2F_2}{R^2})\alpha'$$

$$- m\frac{h^2(B_2 - F_2)}{R^2\sin\phi}\frac{d\beta'}{d\phi} + m\frac{h^2(B_2 + F_2)\cot\phi}{R^2\sin\phi}\beta' = 0 \tag{20}$$

$$G\frac{v'}{R} + G\frac{m}{\sin\phi}\frac{w'}{R} + m\frac{h^2(B_2 - F_2)}{R^2\sin\phi}\frac{d\alpha'}{d\phi} + m\frac{h^2(B_2 + F_2)\cot\phi}{R^2\sin\phi}\alpha' - \frac{h^2F_2}{R^2}(\nabla_1^2 - \mathrm{cosec}^2\phi)\beta'$$

$$+ m^2\frac{h^2(B_2 - F_2)}{R^2\sin^2\phi}\beta' + (G - \frac{2h^2F_2}{R^2})\beta' = 0 \tag{21}$$

wherein $\nabla_1^2 \equiv \frac{d^2}{d\phi^2} + \cot\phi\frac{d}{d\phi} - m^2\mathrm{cosec}^2\phi$ is the Legendrian operator.

The above equations can be simplified (1) by eliminating w' from equations (17) and (18) and also from equations (20) and (21), and (2) by forming $\nabla_1^2 w'$ by combining equations (17) and (18) and also equations (20) and (21). These combinations, replacing equations (17), (18), (20), (21), plus equation (19) can be written as

$$\nabla_{m,\gamma_1}^2(\frac{mu'}{R\sin\phi} + \frac{d\sin\phi v'}{R\sin\phi d\phi}) - \frac{G}{F_1 - P_{cr}}(\frac{m\alpha'}{\sin\phi} + \frac{d\sin\phi\beta'}{\sin\phi d\phi}) = 0 \tag{22}$$

$$\nabla_{m,\gamma_2}^2(\frac{m\alpha'}{\sin\phi} + \frac{d\sin\phi\beta'}{\sin\phi d\phi}) - \frac{GR^2}{h^2F_2}(\frac{mu'}{R\sin\phi} + \frac{d\sin\phi v'}{R\sin\phi d\phi}) = 0 \tag{23}$$

$$- (B_1 - P_{cr})\nabla_{m,\delta_1}^2(\frac{d\sin\phi u'}{R\sin\phi d\phi} + \frac{mv'}{R\sin\phi}) + G(\frac{d\sin\phi\alpha'}{\sin\phi d\phi} + \frac{m\beta'}{\sin\phi})$$

$$- (2B_1 - 2F_1 + G - 2P_{cr})\nabla_1^2\frac{w'}{R} = 0 \tag{24}$$

$$- \frac{h^2B_2}{R^2}\nabla_{m,\delta}^2(\frac{m\beta'}{\sin\phi} + \frac{d\sin\phi\alpha'}{\sin\phi d\phi}) + G(\frac{mv'}{R\sin\phi} + \frac{d\sin\phi u'}{R\sin\phi d\phi}) - G(\nabla_1^2\frac{w'}{R}) = 0 \tag{25}$$

$$(2B_1 - 2F_1 + G)(\frac{du'\sin\phi}{R\sin\phi d\phi} + \frac{m}{\sin\phi}\frac{v'}{R}) + G(\frac{d\alpha'\sin\phi}{\sin\phi d\phi} + \frac{m}{\sin\phi}\beta') - G\nabla_1^2\frac{w'}{R} + 4(B_1 - F_1)\frac{w'}{R}$$

$$+ P_{cr}\{(\nabla_1^2 - 2)\frac{w'}{R} - \frac{2du'\sin\phi}{R\sin\phi d\phi} - \frac{2m}{\sin\phi}\frac{v'}{R}\} = 0 \tag{26}$$

where the new symbols are $\nabla_{m,\gamma}^2 \equiv \frac{d^2}{d\phi^2} + \cot\phi\frac{d}{d\phi} - m^2\mathrm{cosec}^2\phi + \gamma^2$ and

$$\gamma_1^2 = \frac{(2F_1 - G)}{(F_1 - P_{cr})},\ \gamma_2^2 = (2 - \frac{GR^2}{F_2 h^2}),\ \delta_1^2 = \frac{(2F_1 - G)}{(B_1 - P_{cr})},\ \delta_2^2 = (\frac{2F_2}{B_2} - \frac{GR^2}{B_2 h^2}) \tag{27}$$

We now introduce two "strain functions" χ_1 and χ_2 such that

$$\frac{u'}{R} = \frac{d\chi_1}{d\phi},\ \frac{v'}{R} = -\frac{m\chi_1}{\sin\phi},\ \text{and}\ \alpha' = \frac{d\chi_2}{d\phi},\ \text{and}\ \beta' = -\frac{m\chi_2}{\sin\phi} \tag{28}$$

Then, by this device, equations (22) and (23) are satisfied identically, and the three remaining equations in χ_1, χ_2, and w'/R can be written as

$$\begin{bmatrix} -(B_1 - P_{cr})\nabla^2_{m,\delta_1}\nabla^2_1 & G\nabla^2_1 & -(2B_1 - 2F_1 + G - 2P_{cr})\nabla^2_1 \\ G\nabla^2_1 & -\frac{h^2 B_2}{R^2}\nabla^2_{m,\delta_2}\nabla^2_1 & -G\nabla^2_1 \\ (2B_1 - 2F_1 + G - 2P_{cr})\nabla^2_1 & G\nabla^2_1 & (P_{cr} - G)\nabla^2_1 + 2(2B_1 - 2F_1 - P_{cr}) \end{bmatrix} \begin{bmatrix} \chi_1 \\ \chi_2 \\ \frac{w'}{R} \end{bmatrix} = \begin{bmatrix} 0 \\ 0 \\ 0 \end{bmatrix} \tag{29}$$

These equations, with appropriate boundary conditions (not needed and not listed in this paper), can be used to investigate the bifurcation of a sector of a spherical sandwich shell under external pressure, for example, a spherical cap. The solutions, by virtue of the operator ∇^2_1, will involve Legendre functions.

Here, only the bifurcation of a complete shell is considered. For such a shell, firstly, m must be an integer to have the required periodicity in θ. Secondly, only the first kind of associated Legendre functions $P_n^m(\cos\phi)$, Murnaghan (1963), which remain finite at the poles ($\phi = 0, \pi$), can be considered as mode shape functions. The mode shapes are assumed as:

$$\chi_1 = A_1 P_n^m(\cos\phi), \chi_2 = A_2 P_n^m(\cos\phi), \frac{w}{R} = A_3 P_n^m(\cos\phi) \tag{30}$$

where A_1, A_2, A_3 are constants. Now, for non-zero values of $P_n^m(\cos\phi)$ over a complete sphere, n must be a non-negative integer, and m may be any integer not exceeding n, i.e., $0 \le m \le n$. Physically, $2m$ and $n - m + 1$ are the number of "half-waves" in the equatorial and azimuthal directions respectively. Recalling the properties of the associated Legendre functions:

$$\nabla^2_1 P_n^m(\cos\phi) = (\frac{d^2}{d\phi^2} + \cot\phi\frac{d}{d\phi} - m^2\text{cosec}^2\phi)P_n^m(\cos\phi) = -\mu P_n^m(\cos\phi) \tag{31}$$

$$\nabla^2_{m,\delta}P_n^m(\cos\phi) = (\frac{d^2}{d\phi^2} + \cot\phi\frac{d}{d\phi} - m^2\text{cosec}^2\phi + \delta^2)P_n^m(\cos\phi) = -(\mu - \delta^2)P_n^m(\cos\phi) \tag{32}$$

where $\mu = n(n+1)$, the condition for a nontrivial solution (i.e. the bifurcation condition) is found to be ($\mu \neq 0$)

$$\begin{vmatrix} \mu - \dfrac{2F_1 - G}{B_1 - P_{cr}} & \dfrac{G}{(B_1 - P_{cr})} & (2 - \dfrac{2F_1 - G}{B_1 - P_{cr}}) \\ \dfrac{GR^2/h^2}{B_2} & \mu - \dfrac{2F_2 - GR^2/h^2}{B_2} & \dfrac{GR^2/h^2}{B_2} \\ (2 - \dfrac{2F_1 - G}{B_1 - P_{cr}})\mu & \dfrac{G\mu}{(B_1 - P_{cr})} & \dfrac{(G - P_{cr})\mu + 2(2B_1 - 2F_1 - P_{cr})}{B_1 - P_{cr}} \end{vmatrix} = 0 \qquad (33)$$

This implicit equation in P_{cr} has terms dependent on n but not on m. Therefore, P_{cr} remains unchanged for any allowable number of circumferential half-waves $(0 \leq 2m \leq 2n)$. In other words, for one eigenvalue P_{cr} correspoding to a chosen n, there are $n+1$ eigenmodes. An apparently similar result was obtained by Van der Neut (1932) for the simpler case of bifurcation of homogeneous elastic spherical shells under external pressure. A logical reason for the multiplicity of modes is given in Flügge (1967). The mode shapes resulting from the radial displacement w for $n = 12$, and $m = 0$, 4, 6 and 12 (obtained by using *Mathematica*) are shown in Fig. 2.

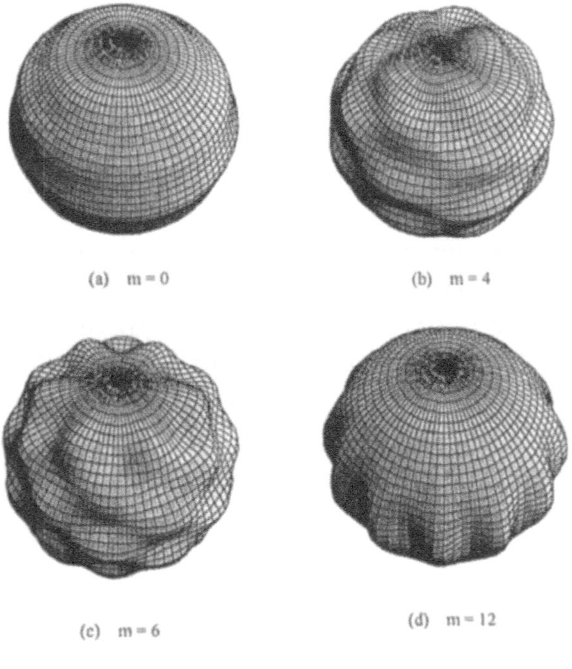

Figure 2. Mode shapes due to radial displacement w for $n = 12$ and $m = 0$, 4, 6 and 12

In view of the above discussion, Equation (33) turns out to be identical with the one derived (in a different way) in Shrivastava (2001), dealing with only the axisymmetric ($m = 0$) bifurcation. This equation can be used to determine in any particular case the exact numerical value of the minimum buckling load and also the corresponding number of the azimuthal half-waves. However, an approximate closed form expression may be obtained by assuming that $P_{cr} \ll B_1$. Then $B_1 - P_{cr} \approx B_1$ and

$$P_{cr} = \frac{(B_1 - F_1)\{4(R/h)^2 F_1 G_1 - 2(G_1 - 2F_1)(\mu B_2 - 2F_2)\} + \mu B_1 G_1(\mu B_2 - 2F_2)}{G_1(R/h)^2(\mu B_1 - 2F_1) + (\mu B_2 - 2F_2)(\mu B_1 + G_1 - 2F_1)} \tag{34}$$

Compact expressions for the minimum P_{cr} and the corresponding μ may now be obtained by using Mathematica. This value of μ is derived by treating it as a continuous variable and requiring $\frac{dP_{cr}}{d\mu} = 0$. Noting that $R/h \gg 1$, we obtain, Shrivastava (2001),

$$\mu = \frac{2F_1}{B_1} + (R/h)\sqrt{\frac{4F_1(B_1 - F_1)}{B_1 B_2}}\{1 + (\frac{h}{R})\sqrt{\frac{4F_1(B_1 - F_1)}{B_1 B_2}} \times \frac{B_2}{G}\} \tag{35}$$

and the minimum buckling load as

$$P_{cr}(\min) = \frac{2B_2}{(R/h)}\sqrt{\frac{4F_1(B_1 - F_1)}{B_1 B_2}}\{1 - (\frac{h}{R})\sqrt{\frac{4F_1(B_1 - F_1)}{B_1 B_2}} \times \frac{B_2}{2G}\} \tag{36}$$

where in the expression for $P_{cr}(\min)$ we have omitted some additional higher-order terms not containing the core shear stiffness G. These formulas generalize the classical results of the elastic thin shell buckling to the plastic case (J_2-incremental or J_2- deformation) by virtue of the moduli used and to the sandwich case by addition of the transverse shear terms, i.e. the terms containing G. These terms, both for μ and P_{cr}, are rendered negligible when G is large or h/R is small.

The results pertaining to plastic buckling of one-material (homogeneous) shells of thickness h are obtained by taking $t/h = 0$ and by considering the core as an elastic/plastic material. The substitution of $\lambda = 1$, $e = 0$, $t/h = 0$, $G \to \infty$ reduces the above $P_{cr}(\min)$ to the exactly equivalent expression obtained by Zoelly (1915) and Schwerin (1922) for thin shell elastic buckling.

4 NUMERICAL RESULTS

Numerical results are presented to illustrate the differences in predictions of the buckling loads according to the two plasticity theories. For this purpose, the face shells are considered to consist of 24S-T3 aluminum alloy with moduli as $E_f = 11,100\,\text{ksi}$, $\nu_f = 1/3$ and a stress-strain $(\sigma - \epsilon)$ curve:

$$\epsilon = \frac{\sigma}{11,\,100} + 0.002(\frac{\sigma}{43.5})^7 \tag{37}$$

(σ in ksi units) valid up to a stress of 45 ksi (310 MPa), with plasticity setting around 25 ksi (170 MPa). The core shell is assumed to remain elastic with moduli $E_c = 53.2$ ksi (367 MPa) and $\nu_c = 0.4$. Figure 3 shows plots of the critical nominal buckling stress σ_{cr} defined by

$$\sigma_{cr} = \frac{P_{cr}}{(2t+h)} = \frac{R\,q_{cr}}{2(2t+h)} \tag{38}$$

against R/h for $t/h = 0.1,\ 0.15$. This stress is close to the buckling stress in the aluminum faces. The results are similar to those obtained for the axially compressed cylindrical sandwich shells, Shrivastava (1995b).

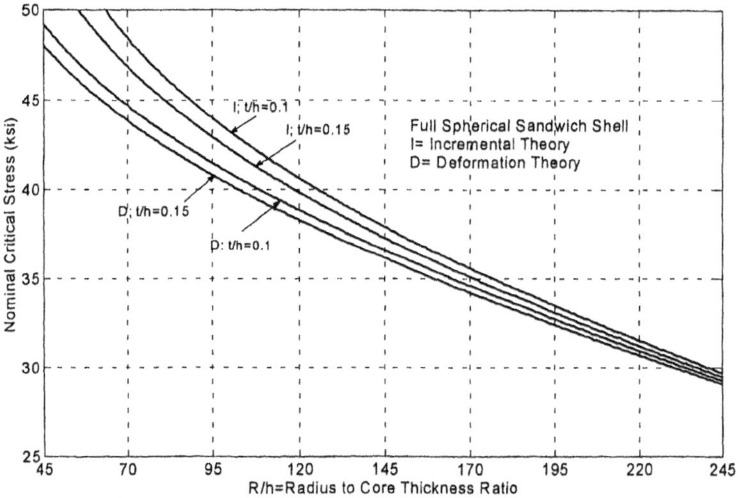

Figure 3. Nominal critical stress as a function of radius to corehickness ratio

Evidently, the incremental theory results are always higher than those for the deformation theory; the differences increase from nearly negligible at the beginning of plasticity to roughly 10 % at the upper end of the $\sigma - \epsilon$ curve. It may be noted that, for either theory, buckling occurs somewhat earlier, i.e. at a lower face stress, for shells with thicker faces and therefore relatively thinner cores $(t/h = 0.15$ versus $t/h = 0.1)$. Experiments to test the validity of the above theoretical results remain to be done.

References

Batterman, S.C. (1965). Plastic Buckling of Axially Compressed Cylindrical Shells. *AIAA Journal*, 3, 316-325.

Bijlaard, P.P. (1949). Theory and Tests on the Plastic Stability of Plates and Shells. *J. Aero.Sc.*, 16:9, 529-541.

Flügge, W. (1967). *Stresses in Shells*, Springer Verlag Inc., New York.

Murnaghan, F.D. (1963). *Introduction to Applied Mathematics*, Dover Publications, New York.

Schwerin, E. (1922). Zur Stabilität der dünnwandigen Hohlkugel unter gleichmäßigem Außendruck. *Z. angew. Math. u. Mech.*, 2:2, 81-91.

Shrivastava, S.C. (1995a). Inelastic Buckling of Rectangular Sandwich Plates. *Int. J. Solids Structures,* 32, 1099-1120.

Shrivastava, S.C. (1995b). Plastic Buckling of Cylindrical Sandwich Shells in Axial Compression. *Abstract, Society of Engineering Science 32nd Annual Meeting*, October 29-November 2, New Orleans, 547-548.

Shrivastava, S.C. (2001). Plastic Buckling of Spherical Sandwich Shells Under External Pressure. *Proceedings of the International Conference on Structural Engineering, Mechanics and Computation*, April 2 – 4, Cape Town, South Africa, Ed. A. Zingoni, Elsevier, vol. 1, 557-564.

Timoshenko S. and Gere J. M. (1963). *Theory of Elastic Stability*, McGraw Hill, New York.

Van der Neut, A. (1932). *De Elastische Stabiliteit van den Dunwandigen Bol*, H. J. Paris, Amsterdam.

Washizu, K. (1975). *Variational Methods in Elasticity and Plasticity*, Pergamon Press, Oxford.

Zoelly, R. (1915). *Über ein Knickungsproblem am der Kugelschale*, Dissertation, Zürich.

THE PREREQUISITES FOR AN ADVANCED DESIGN METHODOLOGY IN SHELLS PRONE TO BUCKLING*

Josef Singer, Haim Abramovich and Tanchum Weller
Faculty of Aerospace Engineering
Technion - Israel Institute of Technology
Haifa 32000, Israel.

Abstract The performance goals of future aerospace vehicles will require an unprecedented degree of structural efficiency, which can be obtained by an advanced design methodology, often called "high-fidelity analysis". Such a methodology combines accumulated geometrical, material and loading imperfections and boundary- condition data with the available computer codes. The prerequisite for a successful application of such a high fidelity analysis to shell type structures is the existence of data banks of measured imperfections and data on the effective boundary conditions. These data banks have to be not only extensive, but also well correlated with the relevant manufacturing processes, loading conditions and practical end fixtures. The experimental work on stiffened metal shells that produced the available data banks is briefly reviewed and their present limitations are emphasized. The urgent need for coordinated extensive international efforts by industry and research laboratories to measure imperfections and effective boundary conditions, and correlate them with experimental buckling loads as well as fabrication processes, is emphasized. Nondestructive vibration correlation techniques are also briefly discussed as a means for determining the effective boundary conditions. Finally, assuming that the necessary prerequisites will become available, the consecutive stages of the advanced design methodology are outlined.

Keywords: Shell buckling, High-fidelity analysis, Imperfections, Boundary-conditions.

---

* Sponsored in part by the Jordan and Irene Tark Aerospace Structures Fund. Extended version originally presented at the 42nd AIAA/ASME/ASCE/AHS Structures, Structural Dynamics and Materials Conference, Seattle WA, April 16-19, 2001, and published there as AIAA Paper No. 2001-1396.

H.R. Drew and S. Pellegrino (eds.),
New Approaches to Structural Mechanics, Shells and Biological Structures, 393–411.

1. INTRODUCTION

In recent years several authors (see for example Arbocz (1982), Arbocz & Hol (1995), Arbocz et al. (1998), Arbocz (2000), Singer (1983), Singer & Abramovich (1995) and Bushnell (1985)) have pointed out that, in order to achieve the required performance goals of future aerospace vehicles, such as the next generation of reusable launch vehicles, an unprecedented degree of structural efficiency is necessary. This efficiency can be obtained only by an advanced design methodology, often called "high fidelity analysis". This methodology combines the accumulated, measured geometrical, material and loading imperfections and effective boundary- condition data that has been correlated with manufacturing processes, with the high capacity computational power available.

It should be remembered, as already emphasized by Singer & Abramovich (1995) that "...Today we are at a crossroads. If the initial imperfection shapes and types are known, or can be reliably predicted, and the boundary conditions (including the load eccentricity) are well defined, adequate analytical tools and computer codes are available for calculation of the buckling load of the shell structure. *But, if we do not know the imperfections and the boundary conditions, we cannot improve our predictions of the buckling loads, no matter how sophisticated our codes are and how large and fast our computers become!*"

For reliable prediction of the likely initial imperfections and accurate definition of the effective boundary conditions, there are three essential prerequisites: (1) the extensive collection and reduction of imperfection data, (2) the assessment of actual boundary conditions and (3) a comparison with benchmark tests of the various available, large analytical and finite element codes.

For clarity, the discussion in this paper will focus on cylindrical shells, the most commonly used shell configuration. Before surveying the influence of imperfections and boundary conditions on the buckling behavior of shells, it may be useful to outline this behavior for typical thin cylindrical shells.

2. BUCKLING AND POSTBUCKLING BEHAVIOR OF ISOTROPIC, STIFFENED AND COMPOSITE SHELLS

The buckling behavior of shells in general and cylindrical shells in particular depends on the type of construction (primarily isotropic, stiffened or composite) and on the type of loading, but can always be divided into prebuckling, buckling and postbuckling. The buckling behavior of a thin cylindrical shell is exhibited most clearly when the shell is subjected to axial compression; see for example Singer et al. (2001). For other types of loading, such as external pressure, torsion or bending, the buckling behavior is essentially similar, though snap-through is less pronounced.

It should, however, be pointed out that the term "buckling" actually embodies two types of instability failure: limit load and bifurcation, as discussed in detail in Chapter 1 of Bushnell (1985) or in Chapters 2 and 16 of Singer et al. (1998 & 2002). Figure 1 (following Bushnell (1985)) summarizes the two types of failure. Usually, bifurcation precedes the limit load, but for relatively thick-walled shells axisymmetric plastic collapse may occur at a lower load and then this mode of failure predominates.

Buckling and collapse behavior of stiffened cylindrical shells is more complex, and they can fail in one of three modes: *local shell instability* between stiffeners (elastic or inelastic), *global (general) instability* (elastic or inelastic) or *axisymmetric plastic collapse*. Often, a stiffened shell will first buckle locally and, after considerable deformation in that mode, eventually fail by global instability (Singer 2002). Since optimization usually results in shells with closely-spaced stiffening, which generates global instability failure, that is the dominant mode; see for example Singer (1969) and (1972) or Singer & Rosen (1976). Closely stiffened shells have been satisfactorily analyzed by *"smeared"-stiffener theory*, taking into account the eccentricity of stiffeners, whether they are on the outside or inside of the shell, (Baruch & Singer, 1963), Chapter 13 of Singer et al. (1998 & 2002) and the numerous references cited there.

The eccentricity effects for stringer- and ring-stiffened cylindrical shells have been extensively studied and experimentally verified; see for example Singer (2002),Thielemann & Esslinger (1965), Chapter 13 of Singer et al. (1998 & 2002) and Card & Jones (1966). Outside stiffeners have usually been found to stiffen the shell more than inside ones, depending on the loading and shell geometry.In a certain range of shell geometries an inversion of the eccentricity effects occurs, which is most pronounced in ring- stiffened shells under external pressure or torsion, see for example Singer et al. (1966).

Contrary to isotropic shells, closely stiffened cylindrical shells buckle in well defined unique global buckling modes, hardly affected by initial imperfections. On the other hand, the influence of boundary conditions (discussed later), in particular of the axial restraint on stringer-stiffened shells, is predominant, as are the eccentricity effects. Their influence can, however, be estimated if the boundary conditions are well defined; and experiments will usually verify these predictions with relatively little scatter.

Recall that the length of the shell affects the relative effectiveness of closely spaced stringers or rings in stiffening of a cylindrical shell. For instance, under axial compression, stringers are more efficient than rings in short shells, whereas for long shells rings are as effective as stringers (see for example Figs. 3 or 7 of Seggelke & Geier (1967)). The stiffening effectiveness depends, however, also on the boundary conditions, as pointed out in Section 4 below.

The buckling behavior of *laminated composite shells* generally follows that of isotropic shells, except that it significantly depends not only on the basic lamina material properties as in metal shells, but also on the laminate lay-up. Various secondary parameters, such as material anisotropy, radial inhomogeneity, and differences in tension and compression elastic moduli also strongly influence the buckling behavior of composite shells and curved panels.

Furthermore, *laminated composite cylinders* are fabricated by either filament winding or prepreg lay-up, each resulting in a skin construction that possesses inherent material characteristics as well as geometrical ones, which influence the behavior and buckling sensitivity of the shell to a different degree.

Filament winding is today the most common manufacturing method for advanced composite cylindrical shells, since it provides an automatic, cost effective process. However, filament-winding introduces complex fiber geometries in regions where material is inter-woven during winding, which may affect the buckling behavior of the shell; see for example Jensen & Hipp (1991), Hahn et al. (1994) or Chapter 14 of Singer et al. (1998 & 2002). Various studies have indicated that the filament winding process has to be carefully controlled to ensure shells of satisfactory quality with corresponding buckling capacity.

Prepreg-lay-ups are usually employed for structural elements of non-cylindrical configurations, but often also for circular cylindrical shells; see for example Tennyson (1975), Leissa (1985) or Chapter 14 of Singer et al. (1998 & 2002). The common manual lay-up does not always ensure uniformity and therefore automated lay-up machines have been developed in several aerospace industries.

Present-day advanced composites have low transverse shear moduli, and therefore in most laminated composite shells, except for very thin-walled ones, irrespective of their method of fabrication, transverse shear effects may be significant. This implies that shear correction factors have to be introduced, see for example Simitses (1966) or Chapter 14 of Singer et al. (1998 & 2002).

One further characteristic of the buckling of composite shells is their significant postbuckling load-carrying capacity, resulting from their stable postbuckling behavior, that occurs at loads somewhat smaller than the initial buckling loads. This characteristic is reflected by the imperfection sensitivity of laminated composite shells, as pointed out in the next sections.

Extensive studies of this behaviour have been carried out in the last decades and appropriate computational tools have been developed, see for example Tennyson (1975), Leissa (1985), Simitses (1966) or Chapter 14 in Singer et al. (1998 & 2002). It was realized in these investigations that due to the many parameters involved, experimental verification of the theory and the predicted buckling behavior is here even more important than in unstiffened or stiffened metal shells. However, such experiments are still very scarce in the open literature, though they are essential for the development of reliable computational tools for buckling of laminated composite shells. One of the few careful and well documented experiments is the series of axial compression tests on 10 laid-up graphite-epoxy cylindrical shells carried out at NASA Langley in the mid eighties, see Knight et al. (1986). These tests became benchmarks for verification of numerical computations.

Another example of extensive benchmark tests on the buckling behavior of composite, unstiffened cylindrical shells under axial compression, torsion and their combinations is the program carried out at DLR Braunschweig Germany within the framework of the 4th European Union Initiative (see Geier et al. (1997)). Correlations with theoretical results were found to be very good, and thus the benchmark tests provided verification for the predictions of imperfect composite shells. It should be emphasized, however, that many more such benchmark tests are still needed as a base, or prerequisite, for reliable predictions.

3. INFLUENCE OF IMPERFECTIONS ON BUCKLING BEHAVIOR

In the past three decades it has been firmly established that initial imperfections are a prime factor in the buckling behavior of shells, see for example for axial compression in Fig. 1, as stated in Babcock's (1974) review "Experiments in Shell Buckling": "The most important problem confronting the experimentalist is the influence of initial imperfections on the buckling loads obtained experimentally". Though there has been considerable progress in both analysis and measurement techniques, see for example Singer & Abramovich (1995), Babcock's statement is still true today, especially if "the experimentalist" is replaced by "the designer".

One can differentiate between three types of initial imperfections: geometrical shape imperfections, thickness variations and material nonhomogeneities (including delaminations). In metal shells, shape imperfections exert the dominant influence and have therefore been most extensively studied, see for example Singer & Abramovich (1995) or Chapter 10 Singer et al. (1998 & 2002), while thickness variations and material irregularities represent only secondary effects.

Among the geometrical imperfections, those that resemble the pattern of the inherent buckling modes of the perfect shell, contribute most significantly to the degradation of the buckling loads, see for example Singer et al. (1971) or (1979).

This characteristic is most prominent in anisotropic and stiffened shells, whose inherent or basic buckling modes are in general uniquely defined and therefore identification of the dominant geometric imperfection shapes becomes feasible.

It may be recalled here that in the early fifties a concept of "maximum out-of-roundness" was introduced as the criterion for the imperfection of a shell, see for example, Holt (1952). The "out of roundness" extracted only the amplitude of the geometric imperfection and disregarded its shape (which is actually the prime factor that determines the influence of the imperfection on the buckling load and behavior), and therefore it could not provide the desired correlation between prediction and experiment.

Only in the late sixties and early seventies were complete and automated imperfection surveys introduced, the first being those developed simultaneously at GALCIT (Caltech) and at Stanford University in 1968, see for example Singer & Abramovich (1995). Both these systems and many later ones used as a reference in the data reduction a perfect cylinder, determined by a least-squares fit of the measured imperfection data.

In most imperfection-scanning systems today, the records of the measured radial displacements at the chosen points (along the circumference and at different heights) are the deviations of the inner surface of the shell from the cylindrical surface generated by the measuring probe during its scanning path. These raw scans are usually plotted on X-Y recorders, or the like, during the scan, and have to be "adjusted" for data reduction.

The initial imperfections are therefore represented by the deviations from an imaginary cylindrical reference surface defined as the best-fit cylinder to the measured data, computed by a least-squares method. Therefore, the recorded measured displacements are recalculated with respect to this best-fit reference surface. In this manner any rigid body displacements of the shell with respect to the scanning system are removed, yielding the "adjusted" imperfection scans, like Fig. 2 (taken from Singer & Abramovich, 1979). These "adjusted" scans are then used to calculate the harmonic components of two double Fourier series, the standard representation proposed by Arbocz & Williams (1977) and then adopted in many imperfection data. See for example Arbocz (1982), Singer et al. (1971) &

(1979), Singer & Abramovich (1979) or Chryssanthopoulos et al. (1991) for details. This representation has also been adopted in the International Imperfection Data Bank (Singer & Abramovich,1995, Singer et al, 1979, Chryssanthopoulos et al, 1991, Arbocz & Abramovich, 1979 or Abramovich et al, 1981).

Once the initial imperfections have been determined, the buckling load of the imperfect shell can be calculated by a multimode analysis MIUTAM (Arbocz & Babcock, 1976) or other appropriate codes like STAGS C1. In the actual calculation of the buckling load, employing MIUTAM, one faces the problem of choosing the modes, or identifying the "important" modes. As already mentioned, it has been found that the imperfection modes, which are at or close to those corresponding to the lowest eigenvalue, dominate, even if their amplitudes are small. Other modes of initial imperfection have a noticeable influence only if their equivalent amplitude is very large. Hence, by mapping out of the linear theory buckling loads one obtains some guidelines for the mode selection. For details see for example Singer et al. (1979), Abramovich et al. (1981) or Arbocz (1974).

In metal shells, thickness variations were usually rather small and therefore the geometrical imperfections were represented solely by the shape imperfections. But in composite shells, the thickness variations may become significant and require consideration. Early studies on the effect of thickness variations, carried out at the University of Toronto Institute for Aerospace Studies, see for example Tennyson et al. (1971), dealt only with intentionally introduced axisymmetric imperfections. More recent investigations, however, have measured the shape imperfections of their specimens on both the inside and outside cylindrical surface, see for example Fig. 3 and Chryssanthopoulos et al. (1991), Giavotto et al. (1991) or Bisagni (1996). Figure 3 (from Chryssanthopoulos et al. (1991)) shows the adjusted scans (after reference to a best-fit cylinder) for a typical angle-ply cylindrical shell, with symmetrical lay-up $(45/-45)_S$.

One notes in the figure that the geometric imperfection patterns of the internal and external surfaces differ considerably, a difference that implies some thickness variations, in particular local thickness variations due to overlapping layers. The local sharp peaks (short wave-length modes), that occur on the external surfaces are rather dissimilar to the long wave shapes that correspond to the inherent lowest eigenvalues, and hence their influence on the buckling behavior will be minimal. Thus it was concluded with the support of a statistical analysis, see Chryssanthopoulos et al. (1991), that measurement of the internal surfaces suffices and this strategy was adopted in later investigations; see for example Bisagni (1996).

In general, however, the effect of the significant thickness variations common in laminated composite shells has to be investigated. As there are hardly any comprehensive studies of this effect reported in the literature, such experimental studies become a pre-requisite for reliable analysis of the buckling of laminated composite shells.

The third type of initial imperfections, *material inhomogeneities*, are not of prime importance in metallic shells, except for residual stresses, that in the case of welded construction can be very significant. These initial or *residual stresses*, which may remain in a shell as a result of its fabrication process, can lead to a considerable degradation in the buckling load. The magnitude, location and influence of the residual stresses should therefore be ascertained prior to detailed calculations of the buckling strength.

It may be reiterated here that the residual stresses are usually determined experimentally by one of two methods: the *Direct Measurement Method*, in which strain changes are measured before and after welding (or other fabrication); or the *Sectioning Method*, in which sections of the shell are cut up into short narrow pieces, whose strain changes are measured as the initial stresses are relieved; see for example Somerville et al. (1977), Wohlfahrt (1986) or Pineault & Brauss (1995).

In laminated composite shells, delaminations may represent another significant form of imperfections, which might influence the buckling and failure behavior. It is common practice today to reject composite components that contain delaminations beyond a predetermined size. However, the relation of the typical size of delaminations with a possible buckling failure is not well established yet. The magnitude and locations even of small delaminations should therefore be carefully measured. The typical delaminations of these shells provide some kind of imperfection data for their buckling analysis, that yields a corresponding degradation in buckling behavior. Comparison of the predicted values with experimental ones may then provide realistic knock-down factors for the analyses of relevant composite shells with similar delaminations. In the assessment of the influence of typical delaminations, one must bear in mind that delaminations often grow under load and this growth, in particular in regions of stress gradients, has to be included in the delamination-imperfection data. The influence of delaminations and their growth on buckling behavior is a field that warrants much additional work.

4. INFLUENCE OF BOUNDARY CONDITIONS ON INSTABILITY OF CYLINDRICAL SHELLS

The emphasis in the studies of the buckling of thin shells has in recent decades been directed towards the effect of geometrical imperfections, and has thus overshadowed the equally important problem of the influence of boundary conditions. In the 1960's the effect of in-plane boundary conditions was extensively investigated for isotropic cylindrical shells, see for example Hoff (1966). But, thereafter, the boundary conditions were relegated to second place both for unstiffened and composite shells by most researchers, and indeed their influence has been found to be less severe than that of geometric imperfections.

For stiffened shells, however, and in particular closely stiffened shells, the effect of geometric imperfection is less pronounced, whereas the influence of the boundary conditions dominates the buckling behavior. Their salient features can be briefly recapitulated as follows.

In ring-stiffened shells under axial compression and external pressure the boundary conditions affect buckling in a manner similar to isotropic shells, see for example Weller et al. (1971). In stringer-stiffened cylindrical shells, however, the influence of boundary conditions differs appreciably from that of unstiffened ones and the in-plane boundary conditions become predominant; see for example Singer & Rosen (1974), Weller (1978) or Chapter 11 of Singer et al. (1998 & 2002). Under axial compression, for example, axial restraint ($u = 0$, SS2 or SS4 boundary conditions) may raise the buckling load by 50% or more if the shell is fairly long. These effects depend on shell and stiffener geometry. A better definition of the actual boundary conditions, in particular with respect to axial and rotational restraint, will therefore lead to appreciably more accurate predictions of buckling loads. One should note here that the influence of boundary conditions on the lower natural frequencies of vibrations of ring- and stringer-stiffened shells closely resembles that on their buckling loads, see for example Singer & Rosen (1974), Rosen & Singer (1974) & (1976).

Correlation between vibration and buckling was therefore studied extensively at the Technion, see for example Singer & Abramovich (1979), Singer & Rosen (1974), Rosen & Singer (1976), Singer (1979) or Chapter 15 of Singer et al. (1998 & 2002) and yielded a nondestructive experimental tool for the definition of the boundary conditions, which led to a more accurate prediction of the buckling loads. This nondestructive method, known as vibration correlation technique (VCT), for determination of boundary conditions, essentially consists of an experimental determination of the lower natural frequencies for a loaded shell, and evaluation of equivalent elastic restraint representing the actual boundary conditions. It is based on the similarity between the strong influence of axial and rotational restraints on free vibrations of stiffened shells, in particular for the lower natural frequencies whose mode shapes resemble the buckling modes, and that observed for buckling loads. Figure 4 shows the variation of frequency squared with axial load for a typical "simply supported" stringer-stiffened cylindrical shell RO-41 tested at the Technion and that for a nominally identical shell RO-42, with nominally "clamped" boundary conditions.

The technique is discussed in detail in Singer (1979) & (1983) or Chapter 15 of Singer et al. (1998 & 2002). It correlates the behavior of actual imperfect shells, measured in the vibration tests, with a theoretical model of a perfect shell. For the method to be valid one has therefore to show that initial imperfections have similar effects on buckling and vibrations. This motivated theoretical studies on the influence of imperfections on vibrations of cylindrical shells at the Technion, which confirmed the similarity of the effect of imperfections on buckling and vibrations implicitly assumed in VCT.

The vibration correlation technique for the definition of boundary conditions was applied to shells of different Technion test series for axial compression loading, as well as external pressure and combined loading. The details are presented in Singer (1979) & (1983) or Chapter 15 of Singer et al. (1998 & 2002). As seen in Fig. 4, the nondestructive experimental determination of the boundary conditions resulted in a significant reduction in scatter. It should be noted that the 35 shells in Fig. 5 include also shells with practical boundary conditions, as shown in Fig. 7 of Singer et al. (2001). This vibration correlation technique (VCT) has been extensively tested and proven as a reliable tool on a laboratory scale. The technique has been extended to different types of practical boundary conditions, and modified to detect the presence of significant load eccentricity and to account for it, once it has been identified. It has also been successfully applied to larger shells of different constructions, see Singer (1979) & (1983), Weller et al. (1986) or Chapter 15 of Singer et al. (1998 & 2002). Hence VCT can be considered a practical tool for improved assessment of boundary conditions and therefore for improved buckling predictions of stiffened shells. Since the importance of well-defined boundary conditions for buckling and collapse calculations of shells are universally accepted nowadays, VCT or similar nondestructive methods represent a definite prerequisite to modern high fidelity analyses. Many additional vibration correlation tests should therefore be run to facilitate the establishment of a *boundary condition data bank* for the designer.

One should mention here three phenomena that can actually be considered as additional "effective" boundary conditions: (1) prebuckling deformations, (2) load eccentricities and (3) load nonuniformities. The influence of these phenomena, which are discussed in detail for example in Bushnell (1985), Singer (1969), (1972) & (2002), Singer & Rosen (1976) or Chapters 9 and 18 of Singer et al. (1998 & 2002), should be considered in the calculations of the buckling and postbuckling of unstiffened, stiffened and laminated composite shells. Again, these effects should be further investigated, correlated with experiments and collected in an appropriate data bank for the designer.

5. MEASUREMENT OF IMPERFECTIONS AND ASSESSMENT OF BOUNDARY CONDITIONS

Though sporadic imperfection measurements date back to the early 1950's, the early complete imperfection surveys appeared only in the late sixties and early seventies. The concept of the GALCIT (Caltech) system, developed at that time, was soon transferred to the Technion as a result of close research cooperation between the two institutes, which also consolidated the "imperfection measurement lobby", that led to the International Imperfection Data Bank.

Until the late seventies, the leaders of that "lobby", Babcock, Arbocz and Singer had been rather lonely, but then a most important change occurred: the interest in geometric imperfection measurements greatly increased. Some type of geometric imperfection measurement became an integral part of a properly carried out shell buckling test, be it on a laboratory scale or on a large scale!

With the realization of the importance of the shape and amplitude of initial imperfections, various measurement systems were developed in different laboratories, see for example Arbocz (1982), Singer & Abramovich (1995), Singer (1983), Bisagni (1996) or Weller et al. (1996).

As mentioned earlier, only recent systems measure shape imperfections scanned on both the inside and outside of the specimens, see for example Fig. 3, Chryssanthopoulos et al. (1991), Giavotto (1991) or Bisagni (1996); this is essential for laminated composite shells.

As pointed out in the previous section, *well defined boundary conditions* are of prime importance for accurate prediction of buckling behavior and therefore, in addition to an imperfection data bank, a *boundary conditions data bank* is a prerequisite to reliable high fidelity analyses. The establishment of such a data bank requires extensive tests, that correlate nominal assumed boundary conditions with the actual measured ones, by VCT or force/stiffness techniques, see for example Chapter 15 of Singer et al. (1998 & 2002), and associate the resulting ratios with typical assembly processes. Considerable experimental and numerical work is still needed to make such a data bank a practical design tool that will enhance the value of the proposed high fidelity analyses.

6. ANALYSES AND COMPUTATIONAL TOOLS

There are a considerable number of computer codes commercially available, that have buckling analysis capabilities and could therefore constitute the core of the proposed high-fidelity analysis. Programs like NASTRAN, ADINA, MARC, ANSYS, ABAQUS, BOSOR 4, BOSOR 5 and STAGS C are probably the most frequently used. The majority of these codes are finite-element codes, except the finite difference based BOSOR 4. However, no comprehensive comparison of the suitability of these codes for calculation of buckling behavior of shells, with reliable results, has yet been reported in the literature.

Some of the above mentioned codes, in particular BOSOR 4, BOSOR 5, STAGS, ABAQUS or MARC, have been extensively compared with experimental data and have produced reliable results. They can therefore be considered appropriate high fidelity computational tools, provided they can handle all prerequisite data. But great care is advised if they are to be employed for types of shells and loadings, whose buckling behavior has not yet been investigated experimentally.

To improve the reliability of these codes for buckling calculations, a series of benchmarks should be chosen, with which all the relevant computer codes should be compared and assessed. Such a benchmark comparison is urgently needed and represents another prerequisite to the successful application of the proposed high-fidelity analysis.

7. DESIGN METHODOLOGY

When the necessary prerequisites, discussed above, have been complied with, an advanced buckling design methodology for imperfect shells, that employs high fidelity analysis, will hopefully be adopted. This methodology was originally proposed in 1993, see Singer & Abramovich (1995), but appears now nearer to realization. This design method can be outlined as follows:

(1) In the preliminary design stage, various shell construction methods will be evaluated with respect to their availability, cost and imperfection-degradation characteristics, taken from the International Imperfection Data Bank. A shell type and fabrication process will thus be selected, probably with the aid of some optimization method or a stochastic approach.

(2) The expected imperfection shape and magnitude of this designed shell, obtained from the imperfection data bank, see for example Singer & Abramovich (1995), Singer et al. (1979), Arbocz & Abramovich (1979),

Abramovich et al. (1981) or Singer et al. (1998 & 2002), will be input into one of the nonlinear codes for imperfect shell analysis, which by then will also have been further improved. The required dimension of the shell will thus be determined.

(3) A similarity based small-scale model of the shell will then be made by a fabrication process as close as possible to that to be used in the full- scale production. Its imperfection will be measured and compared with that used in the design. The relevant scale factors for this comparison will be obtained from the various data banks (imperfections, boundary conditions, etc.). The actual effective boundary conditions will also be determined by a nondestructive technique, like VCT, see for example Singer (1979) & (1983), Singer & Rosen (1976), Weller et al. (1986) or Singer et al. (1998 & 2002) and compared with computed or assumed ones. Finally, the model will be tested under the scaled design loads.

(4) The model test results will serve as a second iteration to the shell design, determined in (2) and, if necessary, the dimensions will be adjusted to bridge the gap between the two iterations. In extreme cases where this gap is large, a second model test could be carried out to serve as a third iteration.

(5) The full-scale shell fabrication can then proceed. An imperfection scan will then be made of the first full-scale shell, and the results of this scan will be compared with those of the model scan. The effective boundary conditions will also be verified on the full-scale shell. The scan and the boundary conditions will also be employed to recalculate the buckling load of the shell.

(6) The full-scale imperfection scanning setup could eventually become a quality control device for the production run, and an acceptable imperfection pattern will be one of the key quality control parameters.

(7) Eventually, if the production run justifies it, a full-scale buckling test will be performed and possibly utilized to relax the acceptance requirements in the quality control.

This design approach is not simple, but it leads to much more accurate and less conservative buckling load predictions. Therefore it yields more efficient shell structures, which cannot be obtained without comprehensive imperfection and boundary conditions data banks, as well as a series of benchmark comparisons for the computer codes employed.

8. CONCLUSIONS

The prerequisites for reliable application of the proposed high-fidelity analyses for the buckling behavior of shells (primarily measured imperfection data and assessments of effective boundary conditions) have been enumerated. It is indicated that preparation of these prerequisites will involve thorough experimental investigations, to compensate for their scarcity in the literature. In parallel a series of benchmark tests for comparison of the suitability of the available computer codes is also needed. Once these prerequisites have been accomplished, the high-fidelity analyses will provide the designer with the necessary approach and tools to create more efficient shell structures.

References

Abramovich, H., Singer, J. and Yaffe, R. (1981) Imperfection characteristics of stiffened shells - Group I, TAE Report 406, Faculty of Aerospace Engineering, Technion - Israel Institute of Technology, Haifa, Israel.

Arbocz, J. (1974) The effect of initial imperfections on shell stability, in *Thin Shell Structures, Theory Experiment and Design*, Y.C. Fung and E.E. Sechler, Eds., Prentice-Hall, Englewood Cliffs, N.J., 205-245.

Arbocz, J. and Babcock, C.D. (1976) Prediction of buckling loads based on experimentally measured imperfections, in *Buckling of Structures*, Proc. of IUTAM Symp., Harvard University, Cambridge, MA, June 1974, B. Budiansky, Ed., Springer-Verlag, Berlin, 291-311.

Arbocz, J. and Williams, J.G. (1977) Imperfection survey on a 10-ft diameter shell structure, *AIAA J.* **15**, 949-956.

Arbocz, J. and Abramovich, H. (1979) The initial imperfection data bank at the Delft Univ. of Techn. - Part I", Report LR-290, Dept. of Aerospace Engineering, Delft, The Netherlands.

Arbocz, J. (1982) Imperfection data bank, a means to obtain realistic buckling loads, in *Buckling of Shells*, Proc. of the State-of-the-Art Colloquium, Universität Stuttgart, Germany, 1982, E. Ramm, Ed., Springer-Verlag, Berlin, Heidelberg, 535-567.

Arbocz, J. and Hol, J.M.A. (1995) Collapse of axially compressed cylindrical shells with random imperfections, *Thin-Walled Structures*, **23**(1-4) 131-158.

Arbocz, J., Starnes, J. and Nemeth, M.P. (1998) Towards a probabilistic criterion for preliminary shell design, in Proc. 39th AIAA/ASME/ASCE/AHS/ASC Structures, Structural Dynamics and Material Conf., 20-23 April, 1998, Long Beach, CA, AIAA Paper No. 98-2051, 2941-2955.

Arbocz, J. (2000) The effect of imperfect boundary conditions on the collapse behavior of anisotropic shells, *Int. J. Solids & Structures* **37**, 6891-6915.

Babcock, C.D. (1974) Experiments in shell buckling, in *Thin Shell Structures, Theory Experiment and Design*, Y.C. Fung and E.E. Sechler, Eds., Prentice-Hall, Englewood Cliffs, N.J.,345-369.

Baruch, M. and Singer, J. (1963) The effect of eccentricity of stiffeners on the general instability of stiffened cylindrical shells under hydrostatic pressure, *J. Mech. Eng. Sci.* (England), **5**(1), 23-27.

Bisagni, C. (1996) Buckling and postbuckling behavior of composite cylindrical shells, Proc., ICAS, 20th Congress on the Intern. Council of the Aeronautical Sciences, Sorrento, Napoli, Italy, **1**, 1148-1155.

Bushnell, D. (1985) *Computerized Buckling Analysis of Shells*, Martinus Nijhoff Publishers, Dordrecht/ Boston/Lancaster.

Card, M.F. and Jones, R.M. (1966) Experimental and theoretical results for buckling of eccentrically stiffened cylinders, NASA TN D-3639.

Chryssanthopoulos, M.K., Giavotto, V. and Poggi, C. (1991) statistical imperfection models for buckling analysis of composite shells, in *Buckling of Shell Structures, on Land in the Sea and in the Air*, J.F. Jullien, Ed., Elsevier Applied Science, London and New York, 43-52.

Geier, B., Klein, H. and Zimmermann, R. (1997) Buckling tests at DLR, BRITE-EURAM Project BE-7550 "DEVILS", DLR Report WPO4.OK/DLR (1,2,3), DLR, Braunschweig, Germany.

Giavotto, V., Poggi, C., Chryssanthopoulos, M. and Dowling, P. (1991) Buckling behaviour of composite shells under combined loading, in *Buckling of Shell Structures, on Land, in the Sea and in the Air*, J.F. Jullien, Ed., Elsevier Applied Science, London and New York, 53-60.

Hahn, H.T., Jensen, D.W., Claus, S.J. and Hipp, P.A. (1994) Structural design criteria for filament wound composite shells, NASA CR-195125.

Hoff, N.J. (1966) The perplexing behavior of thin cylindrical shells in axial compression, *Israel J. of Technology* **4**(1), 1-28.

Holt, M.A. (1952) A procedure for determining the allowable out-of-roundness for vessels under external pressure, *Trans. ASME* **74**, 1225-1230.

Jensen, D.W. and Hipp, P.A.(1991) Compressive testing of filament-wound cylinders, *Composites*, Proc. of the 8th Intern. Conf. on Composite Materials (ICCM/8), Honolulu, Hawaii, July 15-19, 35-F-1 to 35-F-9.

Knight, N.F., Starnes, J.H., Jr. and Waters, W.A., Jr. (1986) Postbuckling behavior of selected graphite-epoxy cylindrical panels loaded in axial compression, Proc. AIAA/ASME/ASCE/AHS 27th Structures, Structural Dynamics, and Materials Conf., San Antonio, TX, May 19-21, Paper No. 86-0881-CP.

Leissa, A.W. (1985) Buckling of laminated composite plates and shell panels, Report AFWAL-TR- 85-3069, AF Wright Aeronautical Laboratories.

Pineault, J. and Brauss, M. (1995) Stress mapping: A new way of tackling the characterization of residual stresses, *Experimental Techniques* **19**(2), 17-19.

Rosen, A. and Singer, J. (1974) Vibrations of axially loaded stiffened cylindrical shells, *J. Sound & Vibration* **34**(3), 357-378.

Rosen, A. and Singer, J. (1976) Vibrations and buckling of axially loaded stiffened cylindrical shells with elastic restraints, *Int. J. Solids & Structures* **12**(8), 577-588.

Seggelke, P. and Geier, B. (1967) Das Beulverhalten versteifter Zylinderschalen", Zeitschrift für Flug- wissenschaften, **15**(12), 477-490.

Simitses, G.J. (1966) Buckling of moderately thick laminated cylindrical shells: a review, *Composites*, Part B, 27B, 1996, pp. 581-587.

Singer, J., Baruch, M. and Harari, O. (1966) Inversion of the eccentricity effect in stiffened cylindrical shells buckling under external pressure, *J. of Mechanical Eng. Sci.* (England), **8**(4), 363-373.

Singer, J., Baruch, M. and Harari, O. (1967) On the stability of eccentrically stiffened cylindrical shells under axial compression, *Int. J. Solids & Structures*, **3**(4), 445-470.

Singer, J. (1969) The influence of stiffener geometry and spacing on the buckling of axially compressed cylindrical and conical shells, in Proc. of the 2nd IUTAM Symp., *Theory of Thin Shells*, Copenhagen, September 1967, F.I.N. Niordson, Ed., Springer-Verlag, 239-263.

Singer, J., Arbocz, J. and Babcock, C.D. (1971) Buckling of imperfect stiffened cylindrical shells under axial compression, *AIAA J.*, **9**(1), 68-75.

Singer, J. (1972) Buckling of integrally stiffened cylindrical shells - a review of experiment and theory, in *Contributions to the Theory of Aircraft Structures*, Delft University Press, 325-357.

Singer, J. and Rosen, A. (1974) Design criteria for buckling and vibration of imperfect stiffened cylindrical shells, *ICAS Proceedings 1974*, Proc. of the 9th Congress of the International Council of the Aeronautical Sciences, Haifa, August, The Weizmann Science Press of Israel, Jerusalem, 495-517.

Singer, J. and Rosen, A. (1976) Influence of boundary conditions on the buckling of stiffened cylindrical shells, in *Buckling of Structures*, Proc. of IUTAM Symposium of Structures, Harvard University, Cambridge, USA, June 17-21, 1974, B. Budiansky, Ed., Springer-Verlag, Berlin, 227-250.

Singer, J. (1979) Recent studies on the correlation between vibration and buckling of stiffened cylindrical shells, *Zeitschrift für Flugwissenschaften und Weltraumforschung* **3**(6), 333-343.

Singer, J. and Abramovich, H. (1979) Vibration techniques for definition of practical boundary conditions in stiffened shells, *AIAA J.* **17**(7), 762-769.

Singer, J., Abramovich, H. and Yaffe, R. (1979) Initial imperfection measurements of stiffened shells and buckling predictions, Proc., 21st Israel Annual Conf. on Aviation and Astronautics, *Israel J. of Technology*, 324-338.

Singer, J. (1983) Vibrations and buckling of imperfect stiffened shells - recent developments, in *Collapse: The Buckling of Structures in Theory and Practice*, J.M.T. Thompson and G.W. Hunt, Eds., Cambridge University Press, Cambridge, 443-481.

Singer, J. and Abramovich, H. (1995) The development of shell imperfection measurement techniques, *Thin-Walled Structures* **23**(1-4) 379-398.

Singer, J., Arbocz, J. and Weller, T. (1998 and 2002) *Buckling Experiments - Experimental Methods in Buckling of Thin-Walled Structures*, John Wiley & Sons, Chichester, UK and New York, Vol. 1 and Vol. 2.

Singer, J. (2002) Stiffened Cylindrical Shells, Chapter 14 in *Buckling of Thin Metal Shells*, J.-G. Teng and J.M. Rotter, Eds., E&FN Spon, London.

Singer, J., Abramovich, H. and Weller, T. (2001) The prerequisites for an advanced design methodology in shells prone to buckling, Proc. AIAA/ASME/ASCE/AHS 42nd Structures, Structural Dynamics, and Materials Conf., Seattle, WA, April 16-19, Paper No. 2001-1396-CP.

Somerville, W.L., Swan, J.W. and Clarke, J.D. (1977) Measurement of residual stresses and distortions in stiffened plates, *J. Strain Analysis* **12**(2), 107-116.

Tennyson, R.C., Chan, K.H. and Muggeridge, D.B. (1971) The effect of axisymmetric shape imperfections on the buckling of laminated anisotropic circular cylinders, *Trans. of the Canadian Aeronautics and Space Institute* **4**(2).

Tennyson, R.C. (1975) Buckling of laminated composite cylinders: a review, *Composites* 17-24.14.

Thielemann, W. and Esslinger, M. (1965) Über den Einfluss der Exzentrizität von Längssteifen auf die axiale Beullast dünnwandiger Kreiszylinder- schalen, *Stahlbau* **34**, 332-333.47.

Weller, T., Baruch, M. and Singer, J. (1971) Influence of in-plane boundary conditions on buckling under axial compression of ring-stiffened cylindrical shells, Proc. of the 5th Annual Conference of Mechanical Engineering, *Israel J. of Technology* **9**(4), 397-410.

Weller, T. (1978) Combined stiffening and in-plane boundary condition effects on the buckling of circular cylindrical stiffened-shells, *J. Computers & Structures* **9**, 1-16.

Weller, T., Abramovich, H. and Singer, J. (1986) Application of nondestructive vibration correlation techniques for buckling of spot welded and riveted stringer stiffened

cylindrical shells, *Zeitschrift für Flugwissenschaften und Weltraumforschung* **10**(3), 183-189.
Wohlfahrt, H. (1986) Residual stresses due to welding: their origin, calculation and evaluation, in *Residual Stresses*, Macherauch, E. and Hauk, V., Eds., DGM Informationsgesellschaft Verlag, Oberursel, Germany.

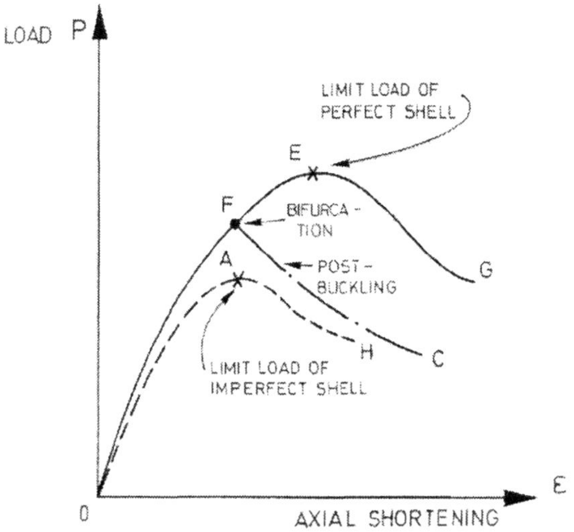

Figure 1. Two types of buckling failure in axially compressed, moderately thick cylindrical shells: limit load and bifurcation, as predicted by a general nonlinear analysis.

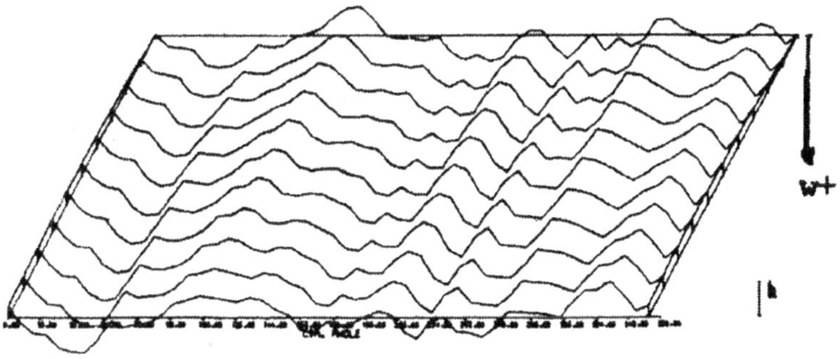

Figure 2. Typical adjusted scan for initial shape imperfections for an axially compressed stringer-stiffened cylindrical shell AB-6 tested at Technion (from Singer et al. (1979)).

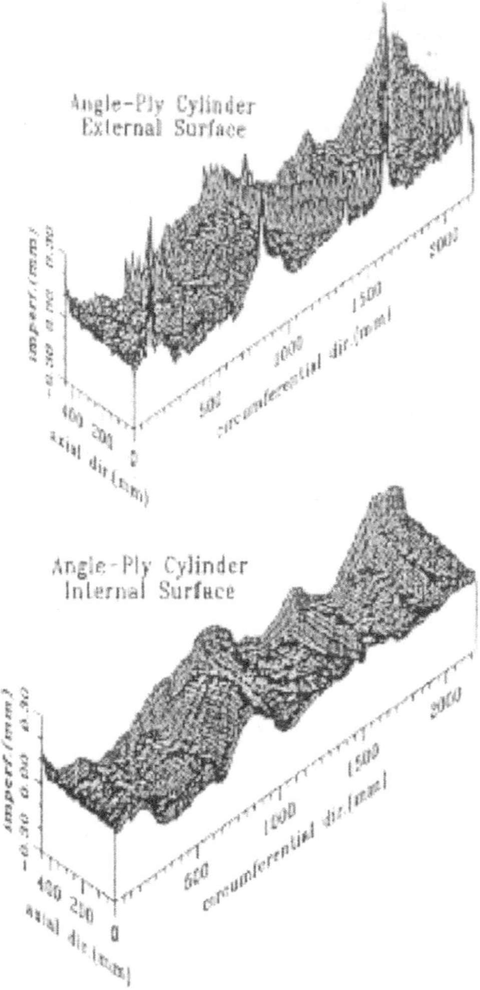

Figure 3. Typical imperfection surfaces of an angle-ply composite cylindrical shell tested at Politecnico di Milano (from Chryssanthopoulos et al. (1991)).

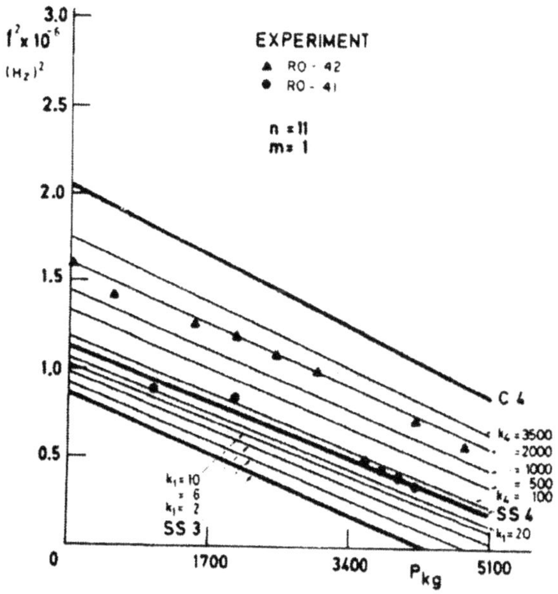

Figure 4. Typical plot of natural frequency squared versus axial load for axially compressed stringer-stiffened shells RO-41 and Ro-42 tested at Technion (from Singer (1983)).

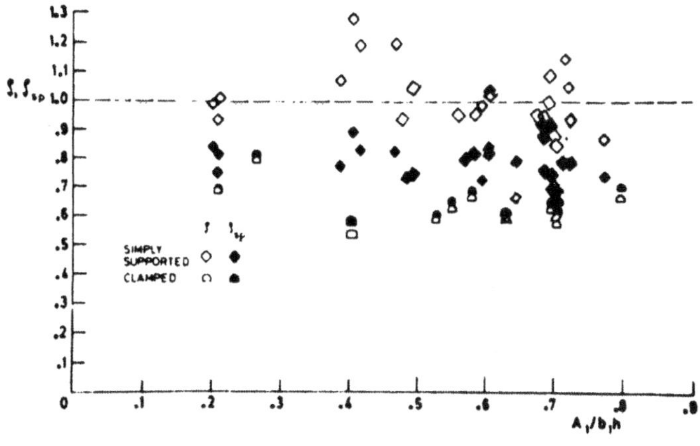

Figure 5. "Knockdown factor" of 35 axially compressed, simply-supported and clamped stringer-stiffened cylindrical shells of RO, AB, KR, SN, RS, DUD and DK series, corrected for experimentally (nondestructively) VCT determined boundary conditions, ρ_{sp}. The VCT corrected "knockdown factor" is compared to the original ρ_{SS3} and ρ_{C4} values (from Singer (1983)).

CHRIS CALLADINE AND BIOLOGICAL STRUCTURES: A PERSONAL ACCOUNT

Aaron Klug
MRC Laboratory of Molecular Biology
Hills Road, Cambridge CB2 2QH, UK

Abstract I give an account of my interactions with Calladine over certain structural problems in molecular biology.

Keywords: Virus shells, Geodesic domes, Flagella, DNA, Supercoiling

1. VIRUS SHELLS AND GEODESIC DOMES

I first met Chris Calladine in 1962 at Peterhouse, when I became a teaching Fellow there in Natural Sciences. I soon found him an interesting man to talk to with a natural curiosity extending well beyond structural engineering. Our first scientific conversations were about the structure of spherical viruses on which I was working at the time and the relations of their protein shells to geodesic domes. My colleague John Finch and I, and Donald Caspar in the USA, had found that these shells which form the outer coats of the virus particles, enclosing the infective nucleic acid within, were built out of many small, chemically identical protein subunits and that these were arranged with the symmetry of the icosahedral point group [1,2,3]. This implied that the number of subunits was either 60, or a multiple of it. Now we knew that in several cases there were a hundred or more subunits, and since no perfectly regular arrangement of more than 60 is possible, it meant that all these units could not be all arranged equivalently with respect to each other. Yet this would be required if they were to "self-assemble" themselves into a specific shell without an external scaffold, as was known experimentally to be the case.

In 1962, after much puzzling Caspar and I were able to reconcile these apparently divergent results. We asked how could one build a spherical shell out of large numbers of units, abandoning the assumption that the symmetry had to be perfect. Molecular structures are, after all, not built to conform to exact mathematical concepts but rather to satisfy the condition that the system be in a minimum energy condition. Dropping the requirement of

H.R. Drew and S. Pellegrino (eds.),
New Approaches to Structural Mechanics, Shells and Biological Structures, 413–419.

mathematical equivalence, but allowing identical units to be quasi-equivalently related, either through slight adjustments of the bonding contacts between units or through a measure of internal flexibility, we showed that the optimum design for such a shell required icosahedral symmetry, since the distortions in the specific bonds are then minimized. Certain, but not all, multiples of 60 subunits could be accommodated in such designs, which Caspar and I enumerated in terms of triangulated surfaces with icosahedral symmetry [4]. Many of the hemi-spherical, "geodesic" domes designed by the architect Buckminster Fuller are built according to a similar geometry, but whereas these have to be assembled by following a fairly elaborate code, the virus shell, because of the flexibility of its units, can, so to speak, build itself. All virus shells so far known conform to one of the predicted icoshaedral designs.

The subunits have automatically to build a shell of the right size. To explain how this could be done, I built various models to explain the principle of our theory of shell construction. This associated with each subunit a "built-in" Gaussian curvature so that only a fixed number were used to close a particular shell. In a triangulated surface lattice, in which each vertex represented a subunit, the spherical defect at each vertex, that is the difference between the sum of the angles meeting at the vertex and 360°, is a discrete measure of the Gaussian curvature at the point. The sum of all such contributions would add up to 4π, the total curvature of a sphere.

Calladine used the idea of Gaussian curvature to good effect in his major book on shell structures published in 1983. Here he developed a novel approach, by representing the bending and stretching of a shell under load by two separate imaginary surfaces, but coupled together by the requirement that the changes in Gaussian curvature must be identical in each. This relatively simple scheme enabled him to make powerful analyses of complex shell structures.

Our discussions on shells and Buckminster Fuller's domes also led Calladine to study the "tensegrity" structures devised by Fuller. They violated "Maxwell's rule" for the construction of rigid structures, but Calladine and his then student Pellegrino found ways of explaining the behaviour of these kinematically indeterminate structures.

2. BACTERIAL FLAGELLA

The first biological problem on which Calladine worked, and to which I introduced him, was concerned with the construction of bacterial flagella. These slender, rigid corkscrew-like shaped filaments are used for propelling bacteria through water. A flagellum is made out of a helical array of a single type of protein subunit – flagellin – thus constituting a cylindrical surface lattice. If all subunits were in identical environments with respect to their neighbours, the filament would have to be straight. But the flagellum is actually curved into a helical shape, so the identical units cannot be strictly equivalently related. Indeed, it was the existence of this feature of bacterial flagella that I had used to adduce evidence for the flexibility of proteins, or for small variations in the contacts between them, in constructing the protein shells of viruses. This was at a time when some protein chemists thought that our proposal for quasi-equivalent relationships in such shells was "far-fetched".

Having heard of this problem in understanding the construction of flagella, Calladine took up the challenge. He suggested that the solution might be in the idea of some sort of bistable switch between two states of the subunit, most simply in the form of alternative connection points between neighbouring subunits. Now it was known that flagella could adopt different helical shapes, either because of mutations in the flagellin or, artificially, when the flagellins from different strains of the same bacterial species were mixed. Calladine noticed that the distinct polymorphic helical waveforms could be grouped into a family of discrete forms, and later realised that they could be correlated with the arrangement of the flagellin molecules in the surface-lattices of the various forms. The latter were determined by a method introduced by DeRosier and myself of optical filtering of electron micrographs of filamentous structures [5].

Finch and I had shown that the basic surface lattice of a flagellum consisted of 11 near-longitudinal helices (i.e. nearly parallel to the flagellar axis) and two families of 6-start and 5-start helices running obliquely across the surface, respectively right handed and left handed [6]. Calladine realised that the angle of inclination of the 11 start longitudinal helices in the different helical wave forms was a measure of the twist of the surface lattice relative to the axis of the filament. To him, as a structural engineer, "it seemed obvious that one should map the available data on the radius and pitch of the various polymorphic helices into a space of (uniform) curvature κ and twist τ". When he did this he found that the experimental points corresponding to six members of the family of discrete forms lay on a single sinusoidal curve, and moreover that they lay close to an array of 12 points equally spaced on the curve: the two extreme points lay at points of zero curvature, but were respectively of left handed and right handed twist. It did not escape his notice that 12=11+1, where 11 is the number of longitudinal helices in the surface lattice. He reasoned that the extra state arose because there were *two* straight forms, but with opposite inclinations of the longitudinal rows of subunits, and hence of opposite twist. This remarkable prediction was later confirmed.

To fit the data on the twist-curvature curve, Calladine built a simple model with a bistable connections, which could account for the different helical forms, according to how many of the eleven near-longitudinal rows adopt the second state of the subunit. An analysis of the elastic strain showed that rows in the same state would cluster together, and this produced a curved shape in the filament. The curvature would change with the number of such rows. This simple mechanical model reproduced all the main aspects of the forms of bacterial flagella.

An extension of the model by Namba [7] which uses two coupled sets of interconnections does somewhat better, but all such models are, of course, far from the structure of the real flagellin molecule. The work of Namba on the X-ray structure of the flagellum has already shown that only the outer part and not the inner part of the flagella, carry the alternative intersubunit connections. Nevertheless Calladine's simplifications show remarkable physical insight into the problem and grasp the essence of the matter (see Namba, this Symposium).

3. THE DNA DOUBLE HELIX

A second major biological system for which Calladine has used applied mechanics to further our understanding is that of the physical properties of the DNA double helix, its transformation between different forms, how it bends and how its detailed structure depends on the particular sequence of bases of which it is composed. Horace Drew will describe these in detail at this Symposium, but I would like to tell a little of the background history.

The DNA in a chromosome is a single molecule, several centimetres long, but it must be highly folded, by a factor of several thousand. It is largely in the form of chromatin, where it is tightly associated with an equal weight of a set of proteins, called histones. In 1974 my postdoc Roger Kornberg showed the existence of a structural unit in chromatin, called the nucleosome. This consists of about 200 base pairs of the DNA double helix wrapped in two superhelical turns around a spool (or core formed by a complex of eight histone molecules [8,9]). Later we showed that this DNA is sealed at the point where it enters and leaves the spool by a fifth type of histone, called H1. The H1 also mediates the coiling of the nucleosome filament into a helical or 'solenoidal' structure to form the next level of structure [10]. Indeed, the wrapping of DNA on a protein core is only the first step in a hierarchical series of foldings which eventually results in a 10,000-fold linear condensation of the DNA into the compact chromosomes seen at the metaphase state of cell division. Through our frequent meetings at Peterhouse, Calladine became aware of this work and, with its wrappings and foldings, saw it as a rich field for applying mechanics.

Now it is clear from the above that, in the cell, the DNA double helix is supercoiled in various ways, and, although it is not possible to extract pure supercoiled DNA from cells, this can be done from certain DNA viruses or single cell organisms, where the DNA has the form of a covalently closed circle. This circular supercoiled DNA can exist in different configurations, either toroidal or interwound, or even made into a simple circular form by intercalating dye molecules between adjacent base pairs to untwist the double helix. Calladine's first paper discussed these configurations, their range of stability and the changes between them in terms of an elastic model. To do this he had to use the geometrical/topological concepts of twisting Tw, linkage Lk (between the two strands of the DNA) and writhing Wr, which are related by the Brock Fuller relation, Lk = Tw + Wr. This paper of Calladine's is perhaps a minor one, in terms of his later major output on the structures of DNA, but I, and others, learned much from it. I benefitted personally from seeing how he went about tackling the topological problem. The subject is still being studied with more sophisticated mathematics (J.M.T. Thompson, this Symposium).

By this time I had myself become interested in the structure of the DNA helix, and had shown by physical chemical and enzymatic methods, that, while the overall or average solution structure was the well known B-form of X-ray crystallography, in fact the local configuration at the base-pair steps of the double helix depended on the precise sequence of bases there (11). Despite initial scepticism this view was confirmed from further data by others, and by the X-ray crystal analyses of defined sequence DNA by Drew and Dickerson [12].

This led to Calladine's first paper (in 1982) on the structural mechanics of the DNA double helix. Calladine explained the local variations in helix twist angle in terms of one simple cause: steric hindrance or clash between purine bases (the large bases) on opposite strands of the helix at adjacent base pairs. Calladine formulated his theory in quantitative form by including two disposable parameters to be fitted to the experimental data. Subsequently other sequences of DNA were analysed in structure and they all fitted "Calladine's rules" – at least, if they were of the B-form type.

Encouraged by this success, Calladine went on to explain the mechanics of the well known transition (discovered by Rosalind Franklin) from the (wet) B-form to the (dry) A-form. It had already been realised (particularly from the work of Levitt) that there was greater overlap between adjacent base pairs in the B-form, driven by hydrophobic or stacking forces. Calladine's contribution was to provide a conceptual reference frame for following the manoeuvres required for the transition. He did this literally in this case, since he effectively used a local system of coordinates to describe the base pair steps, in contrast to crystallographers who used "laboratory" coordinates, based on the overall helix axis. Calladine introduced the local motions "roll" and "slide", to add to "helical twist", to describe the relative arrangement of adjacent base pairs. These are easily visualised and provide the most important three of the six degrees of freedom required to relate one base pair to the next, because they carry most of the variations in changing DNA structure. The paper by Calladine and Drew, in which the B-to-A transition was described, thus became a minor classic, in transforming the way that the structuralists thought about DNA structure, and led to a new standard nomenclature for nucleic acid parameters.

The introduction of the new parameter "roll" had important consequences for analysing the bending of DNA around a protein complex, as for example around the histone spool of the nucleosome, or in the case of intrinsically curved (naked) DNA, as introduced by Crothers. In considering such matters Calladine arrived at the major result that the curvature κ of a length of DNA double helix is the Fourier transform of the roll angles ρ_n, at a period equal to the sequence repeats:

$$k = \frac{1}{s}\sum_{n=1}^{s} \rho_n \exp(2\pi i\, n / s)$$

I will not describe all the applications of this powerful result, only how it helped me in settling a controversy. This was about the origins of intrinsic curvature which occurs when a piece of DNA contains tracts of about 5 A-T base pairs (4-6 in protein), phased regularly at intervals of about 10 base pairs, the periodicity of the helix. Various explanations were offered, for example that there was a sharp discontinuity been a homogeneous A-T tract and the "heterogeneous" DNA sequence following (the "junction model"). Others argued that it was the A-T tracts themselves which were curved and hence the source of the curvature (the so-called "wedge model") – it was the "obvious" reason.

To settle the question we grew crystals of a dodecamer of DNA containing 6 A-T base pairs in its centre, and in 1987 my post doc, Hillary Nelson (13) solved the structure. The dodecamer was only gently curved, and the A-T tract was absolutely straight! The key to understanding came, however, from following Calladine's approach and analysing the roll angles between successive A-T steps. These were all close to zero, a result quite untypical of mixed base pair steps. These show roll angles which are both positive and negative, but on the average more positive. How then is B-DNA normally straight? The answer comes from Calladine's equation above. If we consider a long piece of DNA of "random" sequence then, whatever value the accumulated roll angle has after 5 base pairs, it will be cancelled (because of the helical twist) by the remaining 5 base pairs in the 10-fold period. This means that in the phased A-T tract structure, roll will accumulate in the 5 bases of the non- A-T pair region, but this will not be compensated by the total roll of the 5 bases of the A-T pair region, which is zero. According to Calladine's formula, the net roll over the 10 basepair period will lead to the molecule being curved.

4. CONCLUDING REMARKS

Calladine has gone on to tackle other problems in biological structure, more recently the packing of protein helices and liposomes. He has the knack of picking out problems which are tractable, or which he makes tractable, by using physical insight to simplify them to their essentials and produce imaginative solutions. Coupling these with simple but powerful mathematics, he explains the phenomena quantitatively, but at the same time provides clear pictures which all can understand. We are in his debt.

References (to papers other than by Calladine)

1. Caspar, D.L.D. Structure of tomato bushy stunt virus, *Nature* **177** (1956) 928-930.

2. Klug, A., Finch, J.T. and Franklin, R.E. Structure of turnip yellow mosaic virus, *Nature* **179** (1957) 683-684.

3. Finch, J.T. and Klug, A. Structure of poliomyelitis virus, *Nature* **183** (1959) 1709-1714.

4. Caspar, D.L.D. and Klug, A. Physical principles in the construction of regular viruses, *Cold Spring Harb. Symp. Quant. Biol.* (1962) **27** 1-24.

5 Klug, A. and DeRosier, D.J. Optical filtering of electron micrographs: reconstruction of one-sided images, *Nature* (1966) **212** 29-32.

6. Finch, J.T. and Klug, A. The helical surface lattice of bacterial flagella, *in 'The Generation of Subcellular Structures' (Eds. R. Markham and J. Bancroft). First John Innes Symp., Norwich,* (1972).

7. Namba, K. *Nature Struc. Biol.* **5** (1998) 125-132.

8. Kornberg, R.D. *Ann. Rev. Biochem.* **44** (1977) 161-182.

9. Finch, J.T., Lutter, L.C., Rhodes, D., Brown, R.S. Rushton, B., Levitt, M. and Klug, A. Structure of nucleosome core particles of chromatin', *Nature* (1977) **269,** 29-36.

10. Finch, J.T. and Klug, A. A solenoidal model for superstructure in chromatin, *Proc. Nat. Acad. Sci. USA*, (1976) **73** 1897-1901.

11. Klug, A., Jack, A., Viswamitra, M.A., Kennard, O., Shakked, Z. and Steitz, T.A. A hypothesis on a specific sequence-dependent conformation of DNA and its relation to the binding of the lac-repressor protein, *J. Mol. Biol.* (1979) **131** 669-680.

12. Dickerson, R.E. and Drew, H.R. Structure of a B-DNA dodecamer: influence of base sequence on helix structure, *J. Mol. Biol.* **149** (1981) 761-786.

13. Nelson, H.C.M., Finch, J.T., Luisi, B.F. and Klug, A. 'The structure of an oligo (dA)-oligo (dT) tract and its biological implications, *Nature* (1987) **330,** 221-226.

CALLADINE'S ENTRY TO THE WORLD OF DNA

Horace R. Drew
CSIRO Molecular Science, P.O. Box 184, North Ryde 1670, New South Wales, Australia
Horace.Drew@csiro.au

Abstract Chris Calladine's work in theoretical biology is summarized with special emphasis on bacteria flagella and double-stranded DNA.

Keywords: Bacterial flagella, DNA, Steric clash, Bistability, Curvature and twist

"No amount of method and industry can make up for a lack of mother-wit, either in science or in practical life." -- T.H. Huxley, 1870.

1. INTRODUCTION

In this era of scientific specialization, it seems worthwhile to enquire how any scientist, no matter how bright, can achieve substantial results in more than one field such as engineering, physics, chemistry or biology? Furthermore, how can work by an outsider ever be accepted by vested interests within another speciality? Thus, a well-known Cambridge nucleic-acid specialist once said about Chris Calladine, "What would an engineer know about DNA?"

However, the cross-fertilization of scientific disciplines is often essential for progress to be made, as was recognized by T.H. Huxley when he became involved with setting up the University of London in the 19th century. Calladine is clearly a contemporary exponent of that notion, since he has succeeded in exerting a large influence on both modern engineering and biology. How did such a rare and exceptional effort come about?

H.R. Drew and S. Pellegrino (eds.),
New Approaches to Structural Mechanics, Shells and Biological Structures, 421–433.

2. EARLY STUDIES ON BACTERIAL FLAGELLA

In the 1970's, Chris Calladine was able to reconcile a plethora of phenomenological data concerning the structure and dynamics of bacterial flagella (which produce bacterial motility), in terms of a mechanical model involving ten supercoiled forms of the same multi-stranded protein filament [4-7]. It had been proposed previously that any small protein subunit or building block within the large flagellar filament might require a bistable feature, in order to generate the various filament conformations as observed. Calladine was able to construct a simple, three-dimensional model which reproduced in detail all of those filament conformations, by using just a single bistable feature: either within the protein molecule or at a protein-protein interface in an otherwise elastic subunit. A key feature of his approach was to analyse different supercoiled forms in terms of their "curvature and twist", which are local quantities relevant to the process of assembly.

By 1998, a high-resolution x-ray study of bacterial flagella [41] was shown to be compatible with the Calladine model, and concurred that "bistable switching" of protein-to-protein contacts was responsible for various supercoiled forms (although details of the switching at an atomic level remain controversial).

This work was accomplished before my time, so I can only report on its acceptance by flagella researchers afterwards. One famous biologist at Yale commented, "He did it just by thinking!" Another at Columbia said: "Perhaps rare people such as Calladine or Crick can make progress by thinking, but the rest of us should stick to experiments."

Clearly those flagella specialists were quite expert in the experimental and genetic aspects of their work, but had never been exposed to certain concepts from mathematics or engineering. They were collecting raw data and uncovering fascinating phenomena simply in the hope that "a pattern would emerge". Indeed a unifying model eventually was developed, but by someone in the field of engineering where the appropriate intellectual concepts were available.

3. "STERIC CLASH" AS APPLIED TO DNA

Following one small paper on DNA supercoiling in terms of an elastic model [8], Calladine next applied his flagellar ideas to DNA and its many polymorphic forms. It was well known by 1980 that double-stranded DNA could adopt a variety of different helical shapes, known alphabetically as "A", "B", "C" or "Z", but no one understood how those different conformations might relate to one another, nor the mechanics of any transition between forms.

Prior work on the x-ray structures of simple mononucleotides, as well as NMR spectra of the same, had led specialists in DNA structure to formulate what was known in 1970 as a "rigid nucleotide" model. It was hypothesized that the five-membered deoxyribose sugar ring within any mononucleotide could assume just two rigid shapes known as C2'-*endo* or C3'-*endo*. Those local shapes could then supposedly determine the conformation of a long double-helical polynucleotide as either "B" or "A" respectively, by a difference in "sugar pucker".

This "rigid nucleotide" model was well accepted for 10 years, yet by 1980 some cracks in its edifice were emerging. First, it was never clear how different sugar rings in a long polynucleotide could communicate with one another at a distance: why did they all have to adopt the same shape? Secondly, detailed calculations by M. Levitt cast doubt on whether the DNA sugar (as opposed to the RNA sugar) might be sufficiently rigid to mediate such effects [28] Thirdly, A. Klug argued that the conformation of double-helical DNA might depend on its precise sequence of bases A (adenine), G (guanine), C (cytosine) or T (thymine), in which case the sugar would not be a determining factor [24]. Fourthly, D. Rhodes [35] and L. Peck [34] showed that the helical twist repeat of DNA in solution was 10.6 base pairs per 360° turn, versus an integral 10.0 or 11.0 in fiber models [1, 21, 27], and also that such a twist repeat could vary with base sequence. How then could the sugar be rigid?

Two anecdotal aspects of this problem should be mentioned. First, very few biologists learn at school about the difference between "geometry" and "mechanics", so not many of them in 1980 were concerned with the mechanical forces that hold DNA together. Also, in private discussion, F. Crick told me that he had never taken the rigid-nucleotide model seriously, and was still at a loss to understand what held DNA together, once the base pairs had formed!

It was in this confused climate that Calladine proposed his "steric clash" model of 1982. As shown in Fig. 1, he argued that a lengthwise "propeller twist" between any two bases in the same Y-R base pair (where Y = C or T, and R = A or G), would create a steric clash between the large R bases on one side (as indicated by a "star" in the Figure). Then by use of elastic beam mechanics, he found that such a clash could deform the helical structure of B-form DNA in various ways, for example by altering the twist between two adjacent base pairs in a predictable fashion [9].

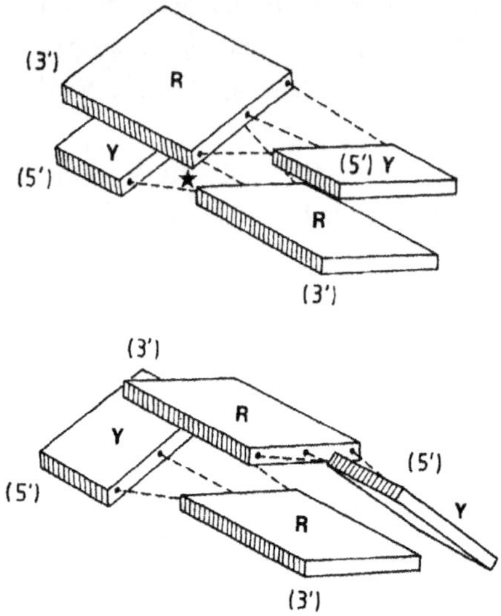

Figure 1. Bistable stacking of DNA base pairs in the "B" form (upper) or "A" form (lower) of a pyrimidine-purine Y-R step.

I learned about this steric-clash model while attending a Cold Spring Harbor Symposium in 1982, where A. Klug provided a preprint of Calladine's work. Since the new model was based on details seen in our dodecamer x-ray structure [17] of 1981, we were perturbed not to have thought of it ourselves; but we agreed that I should talk to Calladine once I arrived in Cambridge after the meeting.

4. "BISTABILITY" AS APPLIED TO DNA

By 1984, several other x-ray structures of small fragments of defined-sequence DNA had been analyzed in single crystals [e.g. 31, 37], and these showed a variety of sequence-dependent effects that could not be explained by Calladine's 1982 model. The problem was that DNA could adopt a wide range of double-helical forms in aqueous crystallizing solution: some right-handed, some left-handed; some with steric clash, some without; so that a simple perturbation model based on B-form DNA would not apply in a general case.

After careful study, Calladine and I concluded [10] that the energies of base-to-base stacking might themselves be bistable, so as to select for two general helical forms known as "B" and "A" without much influence of the sugar. Thus the upper drawing in Fig. 1 represents "B" form DNA, while the lower drawing represents "A" form DNA, with a significant energy barrier between the two states (this has subsequently been confirmed, see refs. 20, 23, 33, 38).

We also concluded that bistable stacking of bases at YR steps could be transmitted over a distance through the double helix, by means of a sideways "slide" motion that would vary cooperatively along both polynucleotide chains.

Finally, we introduced the use of Euler parameters (three translations and three rotations) to describe the geometry of DNA in a local sense, rather than use an imaginary "helix axis" as others had done previously. This subtle point deserves further explanation: prior to 1984, all workers had described the polymorphism of DNA in terms of base pair "tilt" from an imaginary vertical helix axis as shown in Fig. 2. Yet such a time-honored description became unsuitable with the advent of sequence-dependent DNA structure, as well as with the investigation of curved DNA (see below). Hence we introduced a novel description of DNA structure in terms of the six Euler parameters at any base-pair step, the three most flexible of which are shown in Fig. 3 as τ (twist), ρ (roll), and σ (slide).

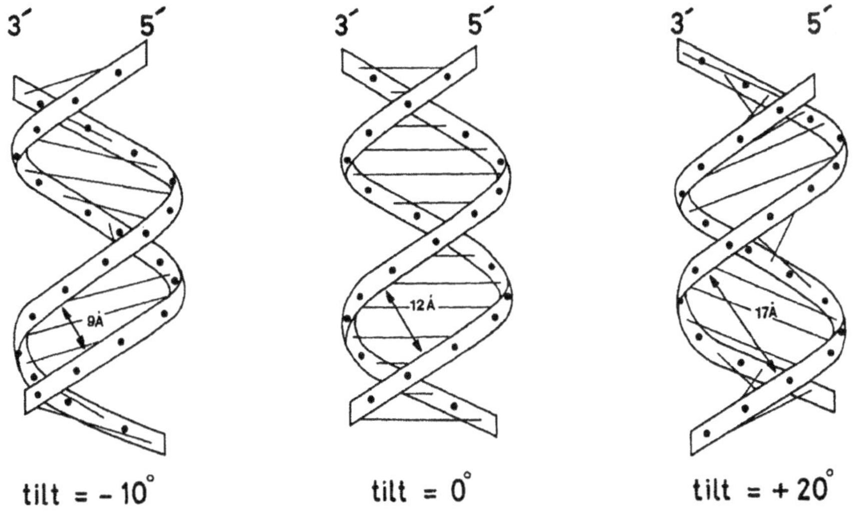

Figure 2. Traditional description of DNA in terms of "tilt" from an imaginary vertical helix axis.

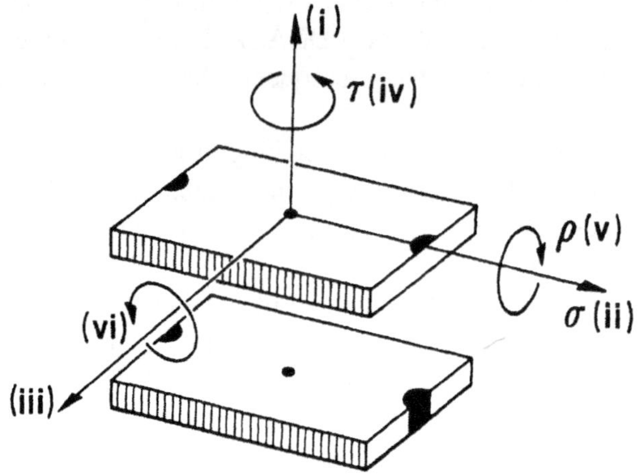

Figure 3. Six degrees of freedom may be defined at any DNA base-pair step, the three most important of which are (iv) τ(twist), (v) ρ (roll), and (ii) σ (slide).

None of those innovations seem radical in retrospect, but they were opposed by DNA specialists at the time. One told me at a conference that we were "re-inventing Newton"! The original paper [10] was rejected from *Nature* in 1983, and only gained acceptance once chemists confirmed the validity of our base-stacking models, and once DNA crystallographers needed Euler angles for their own work, having found the "helix axis" unsuitable. Recent x-ray studies of the B-to-A transition at high resolution would seem to confirm our model as originally stated [18, 32, 40].

5. CURVATURE OF DNA AROUND PROTEINS: FOURIER VARIATION OF ROLL

While we were working on these important problems, which involve DNA free in solution, T. Richmond and A. Klug were working on the x-ray structure of a eukaryotic nucleosome, which represents a complex of eight chromosomal histone proteins with 145 base pairs of DNA [29]. There a long DNA molecule of (145 / 10.2) = 14 turns wraps tightly about eight proteins into 1.75 turns of left-handed supercoil, with a tight curvature of 45° per double-helical turn (where 10.2 is the local helical repeat). The long DNA also shows sequence-dependent effects in its positioning about the protein, where AT-rich minor grooves point in, while GC-rich minor grooves point out [19].

There was no understanding prior to 1985 how this complex form of behavior might come about. Yet Calladine suggested to me from model-building studies, that the roll angles between base pairs in nucleosomal DNA (i.e., angles of separation down their long axes) should show a periodic Fourier distribution when the DNA bends.

Thus, base-pair steps which open to one side (the "minor groove") might favor a positive roll angle of +9°, while steps which open to the other side (the "major groove") might favor a negative roll angle of –9°. The overall periodicity of variation in roll would then equal 10 base pairs to match one double-helical turn; and the cosine-weighted sum of variation in roll would equal 45° per turn as a Fourier integral (approximately). By this means, the smooth bending of DNA about any protein could be accomplished, without creating an energetically unfavorable kink of 22°-45° as others had proposed.

He therefore encouraged me to analyze the sequences of nucleosomal DNA in a statistical fashion, to see whether any Fourier periodicities might be detectable? S. Satchwell, A. Travers and I soon found [36] that lengthwise distributions of certain dinucleotide steps (mainly AA/TT and GC/GC) in chicken nucleosomal DNA were highly periodic, with the expected double-helical repeat of 10.2 base pairs. Steps such as AA/TT which prefer low roll were found where the minor groove points in towards the protein, whereas steps such as GC/GC which prefer high roll were found where the minor groove points out.

MATRIX OF ROTATIONAL PREFERENCES, BASED ON ALIGNED CHICKEN SEQUENCES

PHASE ANGLE IN DEGREES

0	30	60	90	120	150	180	
1.48	1.21	0.96	1.03	1.05	0.68	0.65	CG
1.23	1.06	0.95	1.07	1.02	0.84	0.88	TG/CA
0.84	0.89	0.83	0.96	1.14	1.23	1.05	TA
1.32	1.16	0.98	1.05	0.91	0.83	0.84	GG/CC
1.04	1.07	1.15	0.94	0.92	0.96	0.88	AG/CT
1.04	1.02	1.01	0.98	0.99	1.02	0.91	GA/TC
0.66	0.69	0.86	1.03	1.15	1.29	1.30	AA/TT
0.95	0.93	0.92	0.96	0.97	1.09	1.34	AT
0.91	1.10	1.14	0.96	0.96	0.90	0.98	GT/AC
1.07	1.07	1.22	0.75	0.96	1.08	0.76	(Y)GC
0.96	1.33	1.16	1.02	0.81	0.82	0.77	(A)GC
1.28	1.48	1.28	1.07	0.66	0.50	0.73	(G)GC

210	240	270	300	330	360	
0.68	1.05	1.03	0.96	1.21	1.48	CG
0.84	1.02	1.07	0.95	1.06	1.23	TG/CA
1.23	1.14	0.96	0.83	0.89	0.84	TA
0.83	0.91	1.05	0.98	1.16	1.32	GG/CC
0.96	0.92	0.94	1.15	1.07	1.04	AG/CT
1.02	0.99	0.98	1.01	1.02	1.04	GA/TC
1.29	1.15	1.03	0.86	0.69	0.66	AA/TT
1.09	0.97	0.96	0.92	0.93	0.95	AT
0.90	0.96	0.96	1.14	1.10	0.91	GT/AC
1.08	0.96	0.75	1.22	1.07	1.07	(Y)GC
0.82	0.81	1.02	1.16	1.33	0.96	(A)GC
0.50	0.66	1.07	1.28	1.48	1.28	(G)GC

Figure 4. Experimental data on rotational preferences for DNA curvature in chicken nucleosome core particles. A phase angle of 0° or 360° tells where the minor groove faces out, while a phase of 180° tells where the minor groove faces in, towards a central protein.

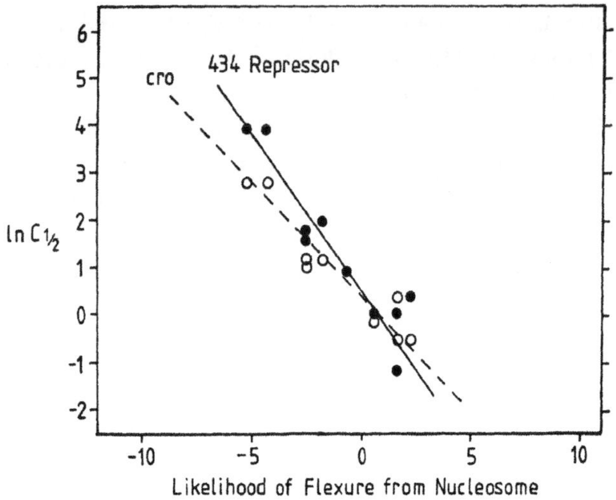

Figure 5. Chicken core rotational preferences also explain the sequence-dependent curvature of DNA about other proteins, for example the *434* or *cro* repressors. The symbol "*ln C 1/2*" on left represents binding free-energy. Each of those proteins binds most tightly where the DNA sequence bends appropriately.

A summary of the original experimental data as averaged over 25,000 base pairs is shown in Fig. 4. The same scheme also explains many other protein-DNA contacts, for example the curvature of DNA about *cro* or *434* repressors: see Fig. 5. S. Nedospasov confirmed that these same relations hold true for human nucleosomal DNA [2]; while J. Reeve confirmed that they hold true even for archaeal histone-bound DNA where the protein is very different [3].

Calladine and I wrote down the theory of this important effect in simple explanatory terms [11], which should be accessible to engineers as well as biologists. It is therefore the Fourier transform of roll angles which depends on base sequence, that provides for the local mechanical stability and definition of DNA in chromosomes.

6. INTRINSIC CURVATURE OF DNA IN SOLUTION: ROLL COMBINED WITH TWIST

While these studies of chromosomal DNA were proceeding, a number of workers noted that free DNA moves more slowly through electrophoretic gels, when it contains AA/TT sequences located periodically at a spacing of might be intrinsically curved due to its base sequence, although the theory of this effect, as well as structural details of the curved DNA, remained uncertain.

Formulation of a definite theory was slow and confusing for three reasons: (i) most biochemists did not yet understand how to describe DNA curvature in terms of a Fourier transform of roll angles; (ii) DNA curvature in solution involves variation in both roll and twist, making it a more difficult problem than for roll alone; (iii) the mechanisms of normal DNA gel-running were not yet understood.

In a preliminary paper, Calladine successfully described the reduced velocity of curved DNA through gels, in terms of an increased "hole size" that it would require to move forward [13]. As shown in Fig. 6, intrinsically curved DNA should require a larger space between gel fibers in order to move forward under the influence of an external electric field, than would straight DNA of the same length.

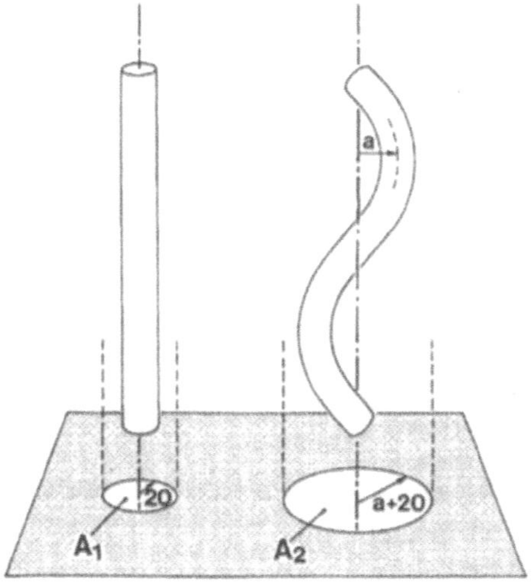

Figure 6. Intrinsically curved DNA in solution may form a stable superhelix of 100-300 base pairs per turn with superhelical radius *a*, that requires a larger cross-sectional area of *A2* rather than *A1* to pass through any two-dimensional gel slice.

Next as shown in Fig. 7, the ratio of speeds R^* for curved versus straight DNA may be related to increased area for forward motion *A2-A1*, by a simple empirical formula for many curved sequences. A model using *A2-A1* as defined in Fig. 6, and only four roll and twist parameters for "AA/TT" sequences versus "others", can generate the data as observed.

Those slight variations of roll and twist then create a three-dimensional superhelix for curved DNA of increased radius *a*, which is of large cross-section and cannot pass statistically through as many "holes" (by analogy with protein electrophoresis).

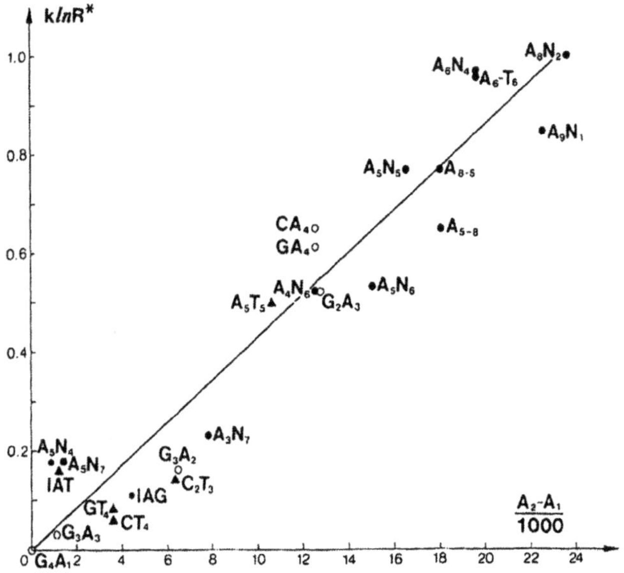

Figure 7. The ratio R^* of gel speeds for curved versus straight DNA varies in proportion to the increased superhelical area $A2 - A1$ of any curved repetitive sequence.

This semi-empirical theory was subsequently extended to the general case of DNA motion through three-dimensional gels, when additional data became available [14]. Thus, a three-dimensional statistical distribution may be used to describe the gel, which "sieves" small DNA by size at low voltage, or else "deforms" to accept large DNA at high voltage (cf. laminar versus turbulent flow in hydrodynamics). An electron microscopic study confirmed the shape and size of curved DNA, and thereby provided an absolute calibration for gel-running models [14].

A similar "sieving-deformation" model was described at the same time by B. Kozulic [26]. Many other models for DNA gel-running do not include a plausible mechanism for gel motion as in refs. 14 and 26; and hence their conclusions about DNA structure, often at variance with x-ray results, may be questioned.

7. DNA STRUCTURE AND THE HUMAN GENOME

Some engineers and physicists may also wish to work in biology but fear that everything has already been done! Well, there is no reason to despair. Since 1995, an enormous amount of new information has emerged from genomic sequencing efforts—human, mouse, fly and micro-organisms—yet hardly any of it has been interpreted in a theoretical sense: see <www.ncbi.nlm.nih.gov>.

Nor have today's biologists become any less specialized than in the past. They are trying to interpret these data in purely abstract terms or "bio-informatics", without much notion of the structures or engineering principles involved [39]. For example, A. Clark writes [16]: "the noncoding regions of DNA play a vital role in regulating gene expression, but what sort of processes might produce such sequence distributions?" Might the fundamental properties of DNA contribute to these poorly-understood biological functions?

Thus in this new era where many different disciplines are trying to merge with biology (e.g., physicists with astrobiology, or chemists with nanotechnology), a good engineer can still emulate the success of Chris Calladine and have a major influence on widely disparate fields.

The key factors for success in Calladine's case were perhaps the following: (i) reliance on mother-wit rather than method or industry, to develop new ways of studying problems: (ii) building simple physical models as predictors of complex behavior; and (iii) collaboration with a range of expert biologists to become familiar with the literature, and also to learn of any new data which might need to be explained [15, 30]. An introduction to DNA structure may be obtained from our text on the subject [12].

To conclude, I will recount a brief anecdote from a scientific meeting where Calladine went to explain his DNA curvature theory, to an audience of technically-minded experimental biologists. Some fellow objected that Calladine's simple first-order theory would not explain a tiny feature of the gel-running data; hence his theory must be wrong! Chris then explained that even the best theory would often explain only 90% of the facts; and that to explain the remaining 10%, one might have to make the theory much more complicated so that it would no longer be useful. This was a revelation to many of those present; while others went away shaking their heads, wondering what it might all mean.

References

1. Arnott, S. and Hukins, D. Optimized parameters for A-DNA and B-DNA, *Biochem. Biophys. Res. Comm.* **47** (1972) 1504-1509.
2. Azizov, M.M. and Nedospasov, S.A. Characterization of a library of mononucleosomal DNA from the chromatin of human cells, *Doklady Biochemistry* **322** (1992) 415-420.
3. Bailey, K.A. *et al.* Archaeal histone selection of nucleosome positioning sequences, *J. Molecular Biology* **303** (2000) 25-34.
4. Calladine, C.R. Bacteria can swim without rotating flagellar filaments, *Nature* **249** (1974) 385.
5. Calladine, C.R. Construction of bacterial flagella, *Nature* **255** (1975) 121-124.
6. Calladine, C.R. Design requirements for the construction of bacterial flagella, *J. Theoretical Biology* **57** (1976) 469-489.
7. Calladine, C.R. Change of waveform in bacterial flagella: the role of mechanics at the molecular level, *J. Molecular Biology* **118** (1978) 457-479.
8. Calladine, C.R. Toroidal elastic supercoiling of DNA, *Biopolymers* **19** (1980) 1705-1713.
9. Calladine, C.R. Mechanics of sequence-dependent stacking of bases in B-DNA, *J. Molecular Biology* **161** (1982) 343-352.
10. Calladine, C.R. and Drew, H.R. A base-centered explanation of the B-to-A transition in DNA, *J. Molecular Biology* **178** (1984) 773-782.
11. Calladine, C.R. and Drew, H.R. Principles of sequence-dependent flexure of DNA, *J. Molecular Biology* **192** (1986) 907-918.
12. Calladine, C.R. and Drew, H.R. *Understanding DNA: the molecule and how it works,* Academic Press, London, 1997.
13. Calladine, C.R. *et al.* The intrinsic curvature of DNA in solution, *J. Molecular Biology* **201** (1988) 127-137.
14. Calladine, C.R. *et al.* A study of electrophoretic mobility of DNA in agarose and polyacrylamide gels, *J. Molecular Biology* **221** (1991) 981-1005.
15. Calladine, C.R. *et al.* How to untwist an alpha helix: structural principles of an alpha-helical barrel, *J. Molecular Biology* **305** (2001) 603-618.
16. Clark, A. The search for meaning in noncoding DNA, *Genome Research* **11** (2001) 1319-1320.
17. Dickerson, R.E. and Drew, H.R. Structure of a B-DNA dodecamer: influence of base sequence on helix structure, *J. Molecular Biology* **149** (1981) 761-786.
18. Dickerson, R.E. and Ng, H.-L. DNA structure from A to B, *Proc. Natl. Acad. Sci. USA* **98** (2001) 6986-6988.
19. Drew, H.R. and Travers, A.A. DNA bending and its relation to nucleosome positioning, *J. Molecular Biology* **186** (1985) 773-790.
20. El Hassan, M.A. and Calladine, C.R. Propellor-twisting of base pairs and the conformational mobility of dinucleotide steps in DNA, *J. Molecular Biology* **259** (1996) 95-103.
21. Franklin, R.E. and Gosling, R.G. Studies on sodium thymonucleate fibers, *Acta Crystallographica* **6** (1953) 673-677.
22. Hagerman, P.J. Sequence dependence of the curvature of DNA: a test of the phasing hypothesis, *Biochemistry* **24** (1985) 7033-7037.
23. Hunter, C.A. Sequence-dependent DNA structure: the role of base stacking interactions, *J. Molecular Biology* **230** (1993) 1025-1054.

24. Klug, A. *et al.* A hypothesis on a specific sequence-dependent conformation of DNA, *J. Molecular Biology* **131** (1979) 669-680.
25. Koo, H.S. *et al.* DNA bending at adenine-thymine tracts, *Nature* **320** (1986) 501-506.
26. Kozulic, B. A model of gel electrophoresis, *Applied and Theoretical Electrophoresis* **4** (1994) 117-159.
27. Langridge, R. *et al.* The molecular configuration of deoxyribose nucleic acid, *J. Molecular Biology* **2** (1960) 19-37.
28. Levitt, M. How many base pairs per turn does DNA have in solution and in chromatin? *Proc. Natl. Acad. Sci. USA* **75** (1978) 640-644.
29. Luger, K. *et al.* Crystal structure of the nucleosome core particle at 2.8 A resolution, *Nature* **389** (1997) 251-260.
30. Lutter, L.C. *et al.* Topological measurement of protein-induced DNA bend angles, *J. Molecular Biology* **261** (1996) 620-633.
31. McCall, M.J. et al. The crystal structure of d(GGGGCCCC), *J. Molecular Biology* **183** (1985) 385-396.
32. Ng, H.-L. *et al.* The structure of a stable intermediate in the A-B DNA helix transition, *Proc. Natl. Acad. Sci. USA* **97** (2000) 2035-2039.
33. Packer, M.J. *et al.* Sequence-dependent DNA structure: tetranucleotide conformational maps, *J. Molecular Biology* **295** (2000) 85-103.
34. Peck, L.J. and Wang, J.C. Sequence-dependence of the helical repeat of DNA in solution, *Nature* **292** (1981) 375-378.
35. Rhodes, D. and Klug, A. Sequence-dependent helical periodicity of DNA, *Nature* **292** (1981) 378-380.
36. Satchwell, S.C., Drew, H.R. and Travers, A.A. Sequence periodicities in chicken nucleosome core DNA, *J. Molecular Biology* **191** (1986) 659-675.
37. Shakked, Z. *et al.* Sequence-dependent conformation of an A-DNA double helix: the crystal structure of d(GGTATACC), *J. Molecular Biology* **166** (1983) 183-201.
38. Suzuki, M. *et al.* Use of a 3-D structure database for understanding sequence-dependent conformational aspects of DNA, *J. Molecular Biology* **274** (1997) 421-435.
39. Tobias, I. *et al.* Elastic stability of DNA configurations, *Physical Review* **E61** (2000) 747-758.
40. Vargason, J.M. *et al.* A crystallographic map of the transition from B-DNA to A-DNA, *Proc. Natl. Acad. Sci. USA* **98** (2001) 7265-7270.
41. Yamashita, I. *et al.* Structure and switching of bacterial flagellar filaments studied by x-ray fiber diffraction, *Nature Structural Biology* **5** (1998) 129-132.

MORPHOLOGICAL AND TOPOLOGICAL TRANSFORMATION OF LIPOSOMES

H. Hotani, F. Nomura, S. Takeda, T. Inaba, K. Takiguchi, T.J. Itoh and A. Ishijima
Department of Molecular Biology, Graduate School of Science, Nagoya University, Nagoya 464-8602, Japan
hotani@nucc.cc.nagoya-u.ac.jp

T. Umeda
Kobe University of Mercantile Marine, Higashinada, Kobe 658-0022, Japan.

Abstract Liposomes are micro-compartments made of lipid bilayer membranes with characteristics quite similar to those of biological membranes. To form artificial cell-like structures, we generated liposomes that contained subunit proteins of cytoskeletons: tubulin or actin. Spherical liposomes were transformed into bipolar or cell-like shapes by mechanical forces generated by the polymerization of encapsulated subunits of microtubules. Disk- or dumbbell-shaped liposomes were developed by the polymerization of encapsulated actin. Dynamic processes of morphological transformations of liposomes were visualized by high intensity dark-field light microscopy.

Topological changes, such as fusion and division of membrane vesicles, play an essential role in cellular activities. To investigate the mechanism of these processes, we visualized in real time the liposomes undergoing topological transformation. A variety of novel topological transformations were found, including the opening-up of liposomes and the direct expulsion of inner vesicles.

Keywords: Liposome, Cytoskeleton, Actin, Microtubule, Topology

1. INTRODUCTION

Phospholipids assemble into a bilayer membrane spontaneously in aqueous solution, and necessarily form liposomes that are closed membrane vesicles. Liposomes have been well studied as simplified models of biological membranes, and are now used as such in a number

H.R. Drew and S. Pellegrino (eds.),
New Approaches to Structural Mechanics, Shells and Biological Structures, 435–446.

of applications from pharmacology to bioengineering, for example as carriers of DNA vectors or anti-cancer drugs for internal delivery.

Cellular organelles, such as the nucleus, the Golgi apparatus and the endoplasmic reticulum, are all compartmentalized by lipid bilayer membranes and show characteristic shapes depending on their specific functions. In order to study the mechanism that determines vesicular shape, we investigated the transformation processes of liposomes. Electron microscopy is not suitable for visualizing the three-dimensional morphology of liposomes, because the liposomes often become deformed during the preparation of dried specimens. Thus, the dynamic characteristics of liposomes were visualized using dark-field light microscopy and their images were recorded using a highly sensitive SIT video camera.

Living cells themselves also have specific shapes depending on their physiological roles. That morphology is thought to be determined and maintained by cytoskeletal networks of microtubules or actin filaments. Therefore, we generated liposomes containing subunit proteins of those cytoskeletons to study their roles in the morphogenesis of the cells.

The morphologies of cellular organelles are quite changeable in response to physiological conditions. In particular, topological changes such as the fusion and the division of membranes play essential roles in cellular activities, for example, in endocytosis or in exocytosis that occurs during vesicle trafficking within cells. It has been thought that such topological transformations are made possible through interactions with proteins. However, direct observation of liposomes reveals that the lipid bilayer itself possesses the ability to undergo topological transformation.

2. MORPHOLOGICAL TRANSFORMATION OF LIPOSOMES

2.1 Transformation Pathways

Many years ago, we reported a systematic study on liposomes undergoing transformation [2]. Liposomes were found to transform sequentially in a well-defined manner through one of several transformation pathways. A circular biconcave form was the initial shape in all pathways and it then transformed into a stable thin flexible filament or small spheres via a variety of regularly shaped vesicles which possessed geometrical symmetry, Fig. 1. The driving force for those transformations was found to be osmotic pressure that decreases the liposome volume thereby altering the surface to volume ratio.

Based on the mechanical principle of the least bending energy of membranes, we investigated the stability and shape transformation of

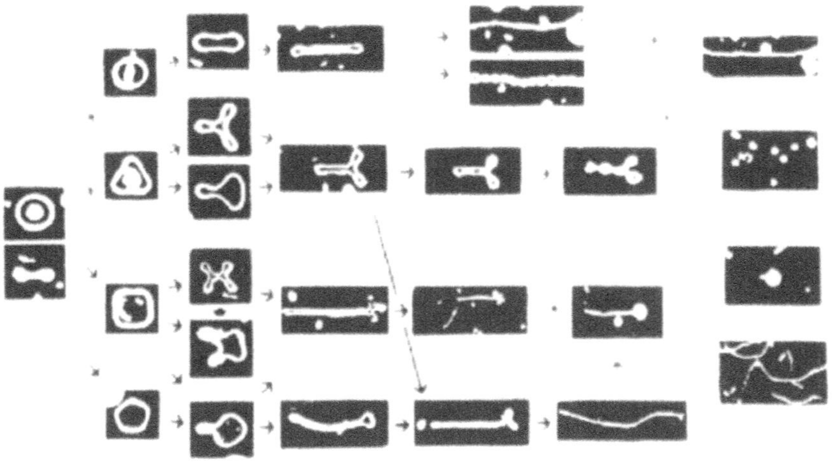

Figure 1. Transformation pathways of liposomes. The biconcave form transformed *via* one of the possible pathway as shown by the arrows.

liposomes. We evaluated the bending energy within constraints of the constant surface area and volume. We also assumed that any shape transformations due to volume change would occur without altering the total surface area of the liposome [2]. The bending energy stored in the lipid bilayer was calculated numerically for each liposome shape formulated by the modified Cassini equation. By computer simulations, we showed that there are energetically stable liposome shapes that possess intrinsic geometrical symmetry [11]. We found that by reducing the volume, the stable shape of a liposome changed from a circular biconcave shape to elliptical, triangular, square or other polygonal shapes.

Based on the theory of mechanics, Pamplona and Calladine analyzed these transformations in more detail [8, 9]. They used a finite-deformation theory of axisymmetric, quasi-static thin shells to theoretically analyze bifurcations and changes of liposome shape under decreasing volume. The main structural response in a lipid bilayer is basically agreed to be its elastic resistance to bending of a two-dimensional liquid.

2. 2 Transformation Caused by Microtubule Growth

Cell morphology is thought to be determined and maintained by cytoskeletal networks of microtubules (MTs) and actin filaments (F-actins). To study the role of MTs on the morphogenesis of cells, we made model liposomes encapsulating tubulin [3, 5]. The encapsulated tubulin was polymerized into MTs by raising the temperature. By monitoring individual liposomes, it was found that the elongation of MTs

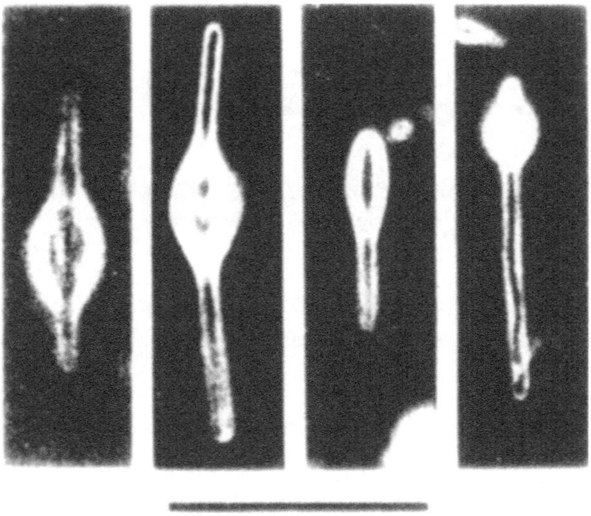

Figure 2. Bipolar liposomes formed by microtubule growth viewed by dark-field microscopy. Bar = 10 μm.

by polymerization of tubulin caused shape changes of the liposomes, indicating that the self-assembly process itself can generate a mechanical force strong enough to deform membranes. When tubulin polymerized into MTs, spherical liposomes were transformed into rugby-ball shapes due to the mechanical force generated by the MT polymerization. Tubular projections of the membranes then grew from both ends of the rugby-ball shaped liposomes, and the liposomes were eventually transformed into characteristic shapes consisting of a central ellipsoid and two straight tubes, Fig. 2. This morphology was designated as a bipolar shape.

2.3 Theoretical Analysis of Transformations Caused by Microtubule Assembly

We investigated the mechanical aspects of the shape transformation of liposomes caused by MT growth. We calculated the liposome shape using a mathematical model based on the notion of the minimum bending energy of the liposome membrane [13]. The force generated by the MT assembly is incorporated in the model by considering the local force balance to the membrane. Numerical analysis yields a series of shapes that are similar to those observed experimentally, Fig. 3. To develop the bipolar shape, the inside positive pressure was found to be indispensable in this theoretical model. The estimated pressure required to develop the bipolar shape can be easily generated from osmotic pressure due to an increase in the concentration of encapsulated tubulin because transformation causes a decrease in the liposome volume. The force-transformation relationship thus obtained predicts the existence of a critical force required for the formation of tubular projections, Fig. 3. Recently, this critical force has been measured experimentally by applying tension to polystyrene beads inside liposomes using laser tweezers (in preparation).

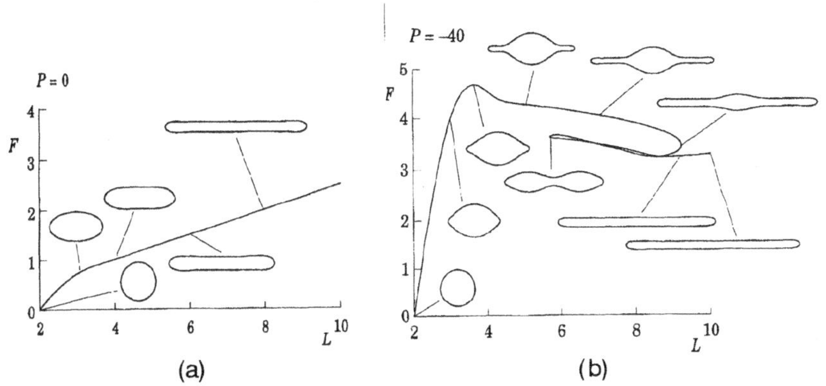

Figure 3. Length-force relations obtained by computer simulation and corresponding liposome shapes according to the pressure difference p = 0 (a) or p = -40 (b).

2. 4 Formation of Cell-Like Shaped Liposomes

We have developed a method to make multi-polar liposomes whose morphologies are similar to those of living cells. The key steps to obtain such multi-polar liposomes are as follows: (i) make giant-sized liposomes that are prepared slowly at low temperature, (ii) encapsulate large amounts of tubulin into the liposomes using high concentrations of tubulin in the preparation, (iii) assemble MTs in liposomes attached to the surface of a glass slide, and (iv) warm the liposome specimen to room temperature on the microscope stage. Many MTs grow in each liposome so that almost all liposomes that are attached to the glass surface began to transform. Unlike the case of bipolar liposomes, several membrane projections were formed on each of these liposomes [3].

The projections are different from each other in shape and direction of growth, and they are distributed randomly on each liposome. As a result, giant liposomes are transformed into a variety of multipolar liposomes similar to cultured living cells, Fig. 4.

Why do free-floating liposomes transform into bipolar liposomes, and why do liposomes attached to a glass surface transform into multipolar liposomes? In the former case, elongating MTs inside liposome can change their directions freely so that they can easily align along the long axis of the liposome. In the latter case however, MTs may interact with the flat surface of the liposome membrane attached to the glass surface. As a result, individual MTs can not alter their direction so that each MT elongates in its own direction.

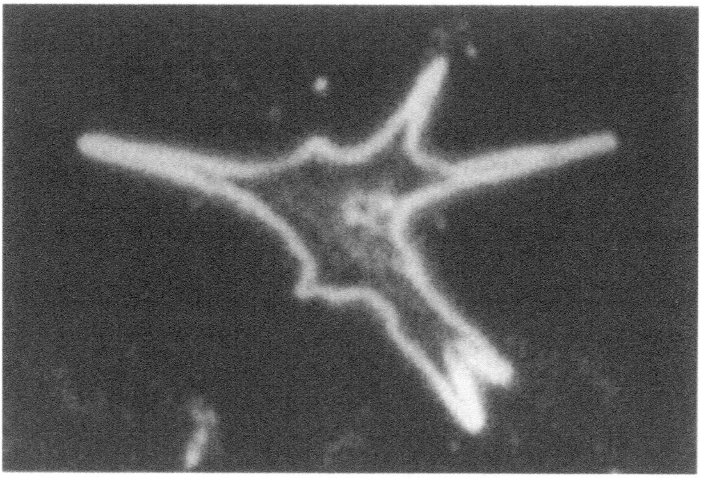

Figure 4. A multipolar liposome similar to living cells viewed by dark-field microscopy.

2. 5 Transformation Caused by Growth of F-Actin

To study the role of F-actin (another cytoskeletal filament) in the morphogenesis of cells, an experimental strategy similar to that for MTs was employed. Spherical liposomes that encapsulated G-actin (a subunit of F-actin) were prepared by swelling a dried lipid film in a solution of G-actin at 5 °C. Polymerization of the encapsulated G-actin into F-actin was achieved by raising the temperature, and subsequent shape changes were observed [6]. After actin polymerization, two shapes of liposome developed: dumbbell and disk shapes. Polarization microscopy showed that F-actins formed thick bundles in the liposomes and that these filaments were contiguous to the periphery of the liposomes. Localization of F-actin in the liposomes was confirmed by fluorescent labeling. Dumbbell- and disk-shaped liposomes were rigid and kept their shapes as long as the F-actins were stabilized, indicating that the actin bundles serve as a mechanical support for the liposome shapes.

The development of protrusions in living cells is indispensable to the process of cell motility. Membrane protrusions have been suggested to occur as a result of actin polymerization. In our work however, liposomes transformed only into dumbbell or disk shapes, and never formed protrusions.

It is well established that many kinds of actin binding proteins

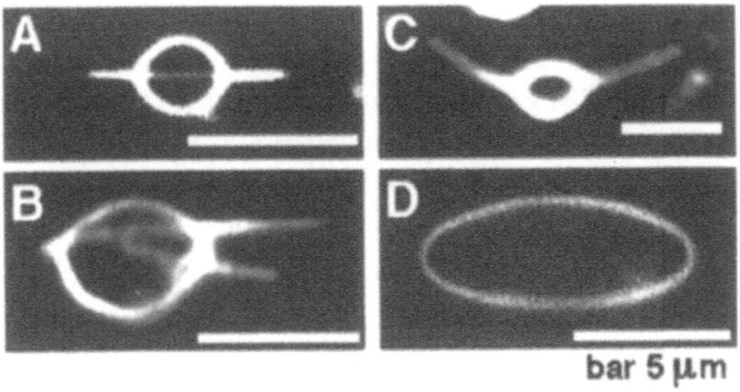

Figure 5. Liposomes caused by actin assembly with actin-crosslinking protein. Transformed liposomes with fascin (A), fascin in the presence of 5 mM MgCl2 (B), α-actinin (C) or filamin (D).

regulatory roles in the organization of F-actin networks in living cells. To study such organization roles, we characterized the transformation of liposomes encapsulating G-actin with its crosslinking proteins, fascin, α-actinin or filamin [1]. As the temperature increased, the encapsulated G-actin polymerized into F-actins and formed bundles or gels, depending on the type of crosslinking protein that was co-encapsulated, which caused various morphological changes of liposomes, Fig. 5. The differences in morphology seen in these transformed liposomes indicate that the actin-crosslinking proteins actually determine liposome shape by organizing their specific actin networks. Morphological analysis reveals that the manner of crosslinking, i.e. the distance and the angular flexibility between adjacent crosslinked F-actins, is essential for morphogenesis.

3. TOPOLOGICAL TRANSFORMATION OF LIPOSOMES

3. 1 Direct Observation of the Membrane Fusion Process

Membrane fusion plays an essential role in cellular activities such as exocytosis, vesicle transport among cell organelles and release of neurotransmitters at synapses. To investigate the mechanism of membrane fusion, we used liposomes as a model of biological vesicles, and monitored the fusion process in real time using optical dark-field microscopy [4]. To induce membrane fusion, we used the following peptides, i) HA (influenza hemagglutinin) peptide, a 20 residue peptide derived from the amino-terminal end of the HA2 subunit; ii) synthesized analogue peptides of HA, a negatively charged analogue (E5) or a positively charged analogue (K5); and iii) bee venom peptide: melittin.

We succeeded in visualizing the membrane fusion process induced by melittin or by a mixture of E5 and K5 at neutral pH. In addition to the fusion phenomenon, various behaviors of membrane vesicles, including aggregation, shrinkage and opening-up of liposomes, were observed depending on the buffer conditions. However, HA or E5 alone did not induce membrane fusion at acidic pH, which had been previously reported to be the optimal condition for membrane fusion. Such fusion occurred only in a narrow range of conditions involving the peptide concentrations, the ionic strength and the lipid composition of the membrane.

When unilamellar liposomes composed of phosphatidylcholine and phosphatidyl glycerol were mixed with E5 and K5 peptides, they transformed into multilamellar liposomes prior to the fusion. Their average diameters decreased by 50% and the light scattering intensity of each liposome increased remarkably. Only the multilamellar liposomes fused with each other, and the fusion was completed within several seconds, just as occurred with unilamellar liposomes. This result gives an insight into the mechanism of formation of multi-lamellar cell organelles such as lysosomes.

3.2 Opening-up of Liposomes

Recently, we found that band 4.1 family proteins make extremely stable holes in liposomes. When talin, a member of the band 4.1 family, was added to a liposome solution, liposomes opened stable holes and were subsequently transformed into cup-shaped liposomes [10]. The holes became larger with increasing talin concentration, and finally, at sufficiently high concentrations of talin, the cup-shaped liposomes were transformed into a lipid bilayer sheet. These morphological changes could be reversed by protein dilution, i.e. the sheets could be transformed back into closed spherical liposomes, Fig. 6. We demonstrated that talin is localized mainly along the membrane verges, presumably to avoid exposure of its hydrophobic portion at the edge of the lipid bilayer. This was the first demonstration that a lipid bilayer can stably maintain a free verge in aqueous solution, and refutes the established dogma that all lipid bilayer membranes inevitably form closed vesicles. This reversible opening may be used as a technique to encapsulate huge structures into liposomes.

This opening-up topological change was analyzed by a statistical theory [12], and was found to be due to the adsorption of talin to verges of cup-shaped liposomes. The adsorption isotherm of talin between an aqueous solution and a liposome membrane was analyzed by taking into account the bending energy of the membrane. The equilibrium was determined by the balance of the energy gain obtained by the adsorption of talin to the periphery of the liposome and the change of the bending energy of the membrane due to the shape change.

It is well known that cellular organelles, including the nucleus, the Golgi apparatus, and the endoplasmic reticulum, convert into small membrane vesicles when cells enter metaphase during mitosis. Further, those fragmented vesicles reverse into native organelles after mitosis. Such fragmentation and reassembling processes could be regulated by unknown proteins possessing activities similar to those of talin.

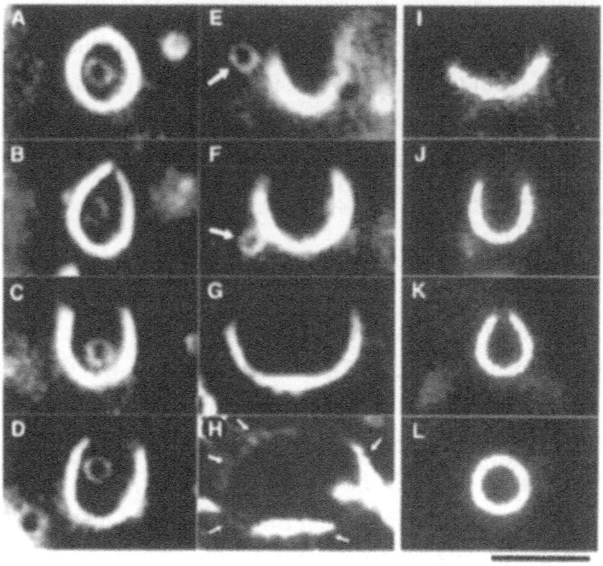

Figure 6. Opening-up of a liposome caused by an increase in talin concentration (A-H) and reverse transformation caused by dilution of talin (I-L). Bar = 5 μm.

3.3 Cyclic Shrinkage of Liposomes, Direct Expulsion of Inner Vesicles and Inside-out Inversion

Membrane fusion and lysis would seem to be opposite phenomena. Actually however, they are closely related behaviors, because the membranes must be lysed once in the vicinity of the fusion point to undergo the membrane fusion. Hence, we studied the effect of membrane-destabilizing reagents, such as surfactants, on liposome topology.

When Triton X-100, an electrically neutral detergent, was added to a liposome solution, the spherical liposomes became smaller in size while

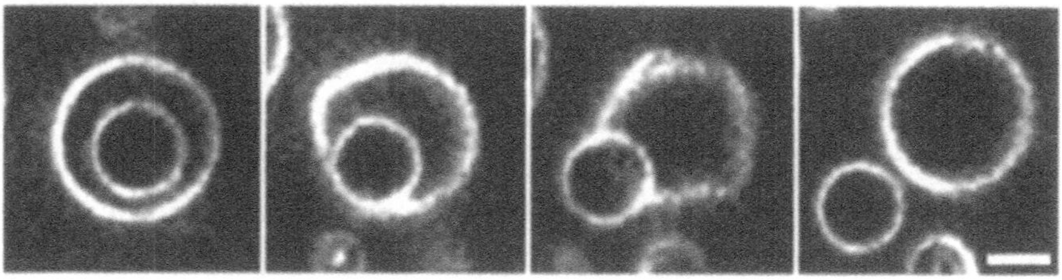

Figure 7. Expulsion of inner liposomes caused by detergent. Bar = 5 μm.

maintaining their spherical shape and membrane thickness. During this process, the shrinking liposomes exhibited two states alternatively, i.e. quaking and tense states. Liposomal quakes represent transitions from tense state to the transient state and are characterized by vigorous fluctuations in their spherical shape. Liposomes quaked individually and intermittently, and decreased their size step by step [7].

Some liposomes could be observed to encapsulate a few smaller liposomes within them. When Triton X-100 was added, the outer liposomes shrank, and surprisingly, their encapsulated liposomes escaped when the outer liposomes vigorously fluctuated in shape, Fig. 7. Since the outer liposomes prepared by our method are unilamellar, this is conclusive evidence that Triton X-100 transiently generates holes in liposomes and thereby excludes inner vesicles directly.

Liposomes were able to discharge water from their interiors, as well as encapsulated vesicles, only during the intermittent quaking by perforation of their lipid bilayers. Liposomes vigorously fluctuate their membranes within each quaking state due to membrane perforation and also due to the sudden decrease in internal pressure which results from the discharge of the water within. Recently, these processes have been investigated using 2-D Fourier analysis of video images.

In addition, we found a surprising inside-out topological inversion that might be due to asymmetry between the two leaflets in a bilayer membrane, Fig. 8. It seems likely that liposome membranes continuously lose their component lipids from only their outer leaflets by detergent solubilization. If there is little or no flip-flop movement of lipids between the two leaflets, the density of lipid molecules in the outer leaflet will continuously decrease. As a result, the surface area of the outer leaflet will tends to shrink while that of the inner leaflet is maintained. This imbalance in surface areas will cause a potential force to invert the curvature of the bilayer membrane, and once a liposome membrane is damaged at some point, an inside-out transformation will take place.

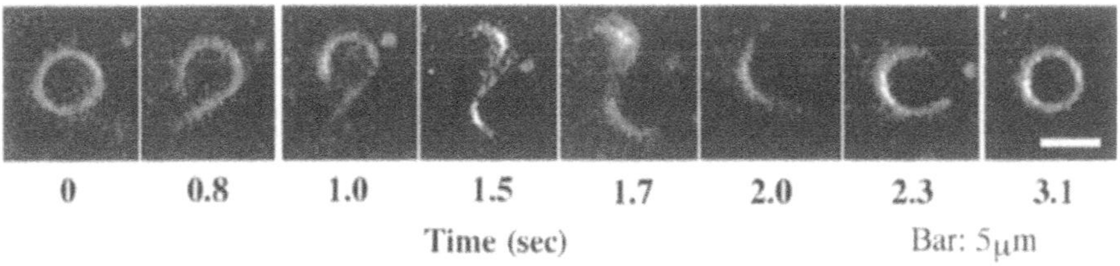

Figure 8. Inside-out topological inversion of a liposome.

References

1. Honda, M., Takiguchi, K. and Hotani, H. Morphogenesis of liposomes encapsulating actin depends on the type of actin-crosslinking, *J. Mol. Biol.* **287** (1999) 293-300.
2. Hotani, H. Transformation pathways of liposomes, *J. Mol. Biol.* **178** (1984) 113-120.
3. Hotani, H. and Miyamoto, H. Dynamic features of microtubules as visualized by dark-field microscopy, *Adv. Biophys.* **26** (1990) 135-156.
4. Ishikawa, S., Takahashi, S., Eto, K. and Hotani. H. Real time visualization of liposome usion induced by amphiphilic peptides, *Prog. Biophys. Mol. Biol. Suppl* **65** (1996) 129.
5. Kaneko, T., Itoh, T.J. and Hotani, H. Morphological transformation of liposomes caused by assembly of encapsulated tubulin and determination of shape by microtubule associated proteins (MAPs), *J. Mol. Biol.* **284** (1998) 1671-1681.
6. Miyata, H. and Hotani, H. Morphological changes of liposomes caused by polymerization of encapsulated actin and spontaneous formation of actin bundles, *Proc. Natl. Acad. Sci. USA* **89** (1992) 11547-11551.
7. Nomura, F., Nagata, M., Inaba, T., Hiramatsu, H., Hotani, H. and Takiguchi, K. Capabilities of liposomes for topological transformation, *Proc. Natl. Acad. Sci. USA* **98** (2001) 2340-2345.
8. Pamplona, D.C. and Calladine, C.R. The mechanics of axially symmetric liposomes, *J. Biochem. Eng.* **115** (1993) 149-159.
9. Pamplona, D.C. and Calladine, C.R. Aspects of the mechanics of liposomes, *J. Biochem. Eng.* **118** (1996) 482-488.
10. Saitoh, A., Takiguchi, K., Tanaka, Y. and Hotani, H. Opening-up of liposomal membranes by talin, *Proc. Natl. Acad. Sci. USA* **95** (1998) 1026-1031.
11. Sekimura, T. and Hotani, H. The morphogenesis of liposomes viewed from the aspect of bending energy, *J. Theor. Biol.* **149** (1990) 325-337.
12. Suezaki, Y., Ichinose, H., Takiguchi, K. and Hotani, H. A statistical mechanical theory for the adsorption of protein to liposomal membranes, *Biophys. Chem.* **80** (1999) 119-128.
13. Umeda, T., Nakajima, H. and Hotani, H. Theoretical analysis of shape transformations of liposomes caused by microtubule assembly, *J. Phys. Soc. Japan* **67** (1998) 682-688.

COMPUTATIONAL APPROACHES TO PREDICTING SEQUENCE-STRUCTURE RELATIONSHIPS IN DNA

C. A. Hunter and M. J. Packer
Centre for Chemical Biology, Krebs Institute for Biomolecular Science, Department of Chemistry, University of Sheffield, Sheffield S3 7HF (UK)
C.Hunter@shef.ac.uk

Abstract: We describe the development of a computational method for predicting the sequence-dependent properties of double-helical DNA. The approach relies on X-ray crystal structures of DNA oligomers to focus on the determinants of the major degrees of freedom observed experimentally. The key elements of the potential energy function are explained, and the utility of the method is illustrated with the successful structure prediction of short DNA oligomers.

Keywords: Sequence-dependent DNA structure, Base stacking, Backbone, X-ray crystal structure, Molecular modelling

1. INTRODUCTION

The sequence of DNA not only carries the information required for the synthesis of proteins, it also encodes a range of structural properties which are utilised by proteins in the packaging of DNA in the nucleus and regulation of gene expression. For example, Figure 1(c) illustrates the first step in the process that compacts the DNA macromolecule into the nucleus of a cell.[1] X-ray crystal structures of DNA oligomers tend to have rather linear structures (Figure 1(b)), so compaction requires distortion of the DNA,[2] and this is achieved by wrapping the polymer around a central protein core. However, the ability of the DNA to accommodate these distortions differs dramatically for different sequences, and so the stability of the complex and the precise positioning of the DNA are determined by the

H.R. Drew and S. Pellegrino (eds.),
New Approaches to Structural Mechanics, Shells and Biological Structures, 447–456.

base sequence.[3,4] This in turn determines the accessibility of the DNA to the proteins involved in DNA processing and gene regulation inside the cell.

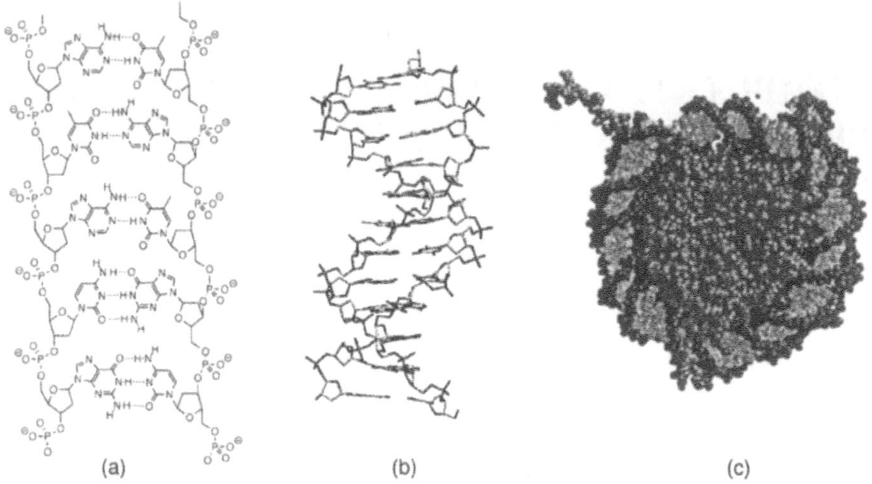

Figure 1. (a) The chemical structure of DNA (b) an X-ray crystal structure of a DNA oligomer (c) the X-ray crystal structure of the nucleosome complex.

Thus the development of an understanding of the relationship between DNA sequence and structural properties has important implications in many biological processes.[3-5] Chris Calladine is one of the pioneers of this field.[6-9] The structural differences discussed above are actually rather small at the single base level, but summed over hundreds of bases, the effects become very significant indeed, leading to differences in the stability of DNA-protein complexes of many orders of magnitude.[10,11] For the geometrical differences of 1 Å and 10° that are observed at the single base level, the only really useful experimental technique for obtaining reliable measurements is X-ray crystallography, and so the Calladine approach has relied on data from crystal structures of DNA oligomers.[12] His mechanical analysis of the structural properties of DNA has established the following important conceptual approaches now in common use to describe sequence-dependent DNA structure.

First, a base-centred frame of reference is key to the analysis of sequence-dependent structure,[9,13] since the differences between different DNA sequences lies in the chemical structure of the bases – the sugar-phosphate backbone is the same for all sequences, Figure 1(a). The six parameters used to describe the conformation of a dinucleotide step are illustrated in Figure 2 (parameters like "propeller" that describe variations in the relative orientation of the bases within a base pair, will be ignored for the purposes of this article).

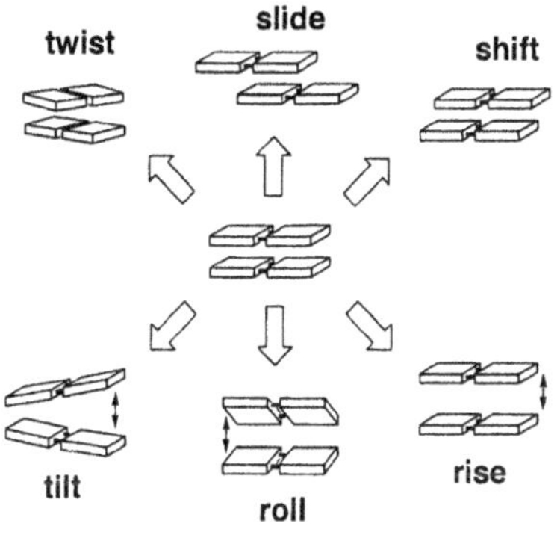

Figure 2. The six parameters used to describe the base pair stacking geometry in DNA.

Second, although a large number of parameters are required to describe the three-dimensional structure of a piece of DNA completely, many of these parameters are strongly coupled, so only a limited number of degrees of freedom are required to describe the structural variations observed experimentally. The six base-step parameters reduce to two degrees of freedom: slide-roll-twist and shift-tilt (rise is always about 3.4 Å, the thickness of a base). The conformation of the backbone is similarly described by two degrees of freedom that allow it to adjust to accommodate different the base stacking geometries. The torsion angles χ, P and δ are coupled to the slide-roll-twist degree of freedom, and the ε and ζ torsion angles are coupled to the shift-tilt degree of freedom, Figure 3.[14]

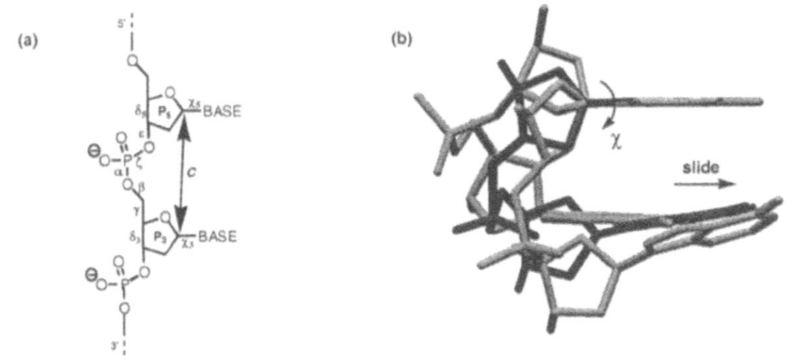

Figure 3. (a) Torsion angles describe the conformation of the DNA backbone. (b) Data from three X-ray crystal structures illustrate how changes in base stacking geometry are coupled to changes in the backbone torsion angles.

Third, the structural variations can be rationalised using simple physical principles, such as steric interactions associated with differences in the geometries of the base pairs.[6]

Fourth, the sequence-dependent variations found in X-ray crystal structures of short oligomers provide a basis for understanding a wide range of sequence-property relationships including interactions with proteins.[4,15]

2. DEVELOPMENT OF A PREDICTIVE MODEL

Efforts in our research group have been directed towards an atomic-level structural model for DNA within the Calladine framework. The aim is to use computational chemistry to extrapolate from the database of experimental X-ray crystal structures that contain a limited number of sequences to a semi-empirical theoretical model that will allow us to predict the structural properties of any sequence. The approach is illustrated in Figure 4. There are four different bases that lead to ten different base steps. If we know the conformational properties of these ten step types and the rules for connecting them together to make longer sequences, it will be possible to build the structure of any DNA oligomer.

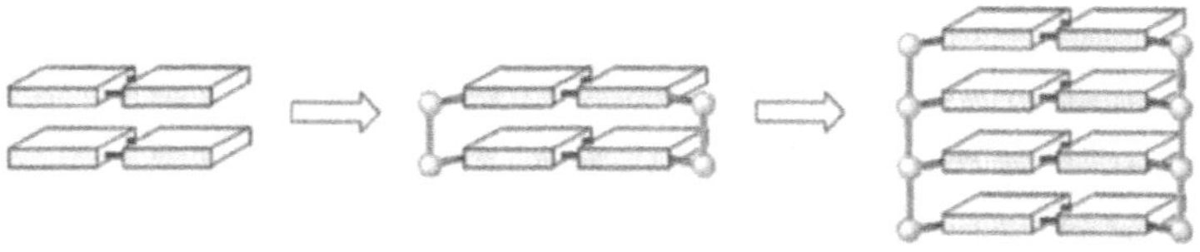

Figure 4. A DNA oligomer is built using potential energy functions for the base pair stacking interactions, constraints representing the backbone, and coupling terms that correlate the conformations of neighbouring base steps in a sequence.

One possible limitation of constructing molecular models of such complex systems is that a large number of empirical parameters are used, hence any small errors in these parameters can add up to huge errors for the macromolecule as a whole. In addition, some aspects of DNA structure such as electric charge, counterions and the solvent are still major challenges for computational chemists. We have therefore focussed on finding the factors that govern the major degrees of freedom as elucidated previously by Calladine. Thus we concentrated on base-stacking interactions as the primary driving force for sequence-dependent differences in DNA structure - the second two steps in Figure 4 relate to the restrictions imposed by the sugar-phosphate backbone and are therefore expected to be sequence-independent.

2.1 The Base Stacking Interactions

Calladine and El Hassan compiled an extensive database of the geometries of 400 dinucleotide base steps taken from X-ray crystal structures of DNA oligomers, and we have used this data extensively to benchmark our calculations.[12] Figure 5 shows how the base step geometry, optimised using a potential energy function for base stacking, compares with the corresponding experimental data from the Calladine-El Hassan database.[16] The parameters roll, tilt and rise are accurately predicted using this approach, because they all involve destacking of the bases (see Figure 2). The other three parameters, slide, shift and twist, can change without significantly disrupting stacking.

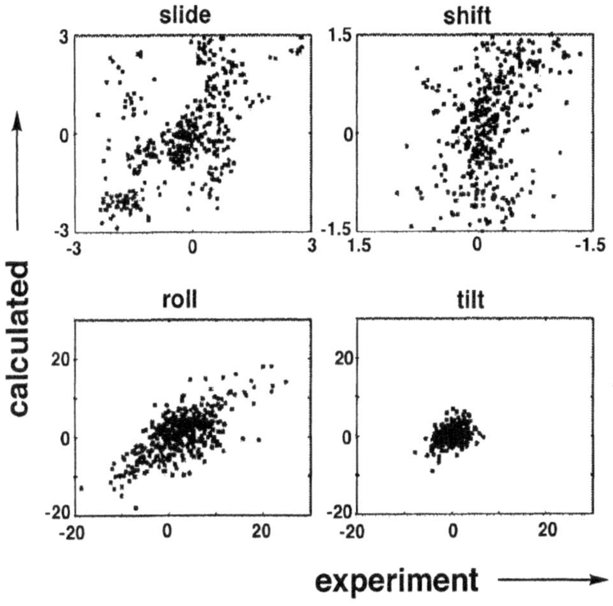

Figure 5. For the 400 base steps in the Calladine-El Hassan database, each of the six base step parameters were optimised using potential energy for base stacking. The resulting values are compared with the experimental values.

2.2 The Backbone

We next used the Calladine-El Hassan database to parameterise an empirical constraint to represent the backbone (step 2 in Figure 2).[14] A semi-flexible rod model gave the best results. The experimental data shows that the backbone is rigid with respect to changes in roll, tilt, rise and twist, but flexible with respect to slide and shift, Figure 6. When this constraint was added to the base stacking potential energy, we were able to predict the value of twist accurately.

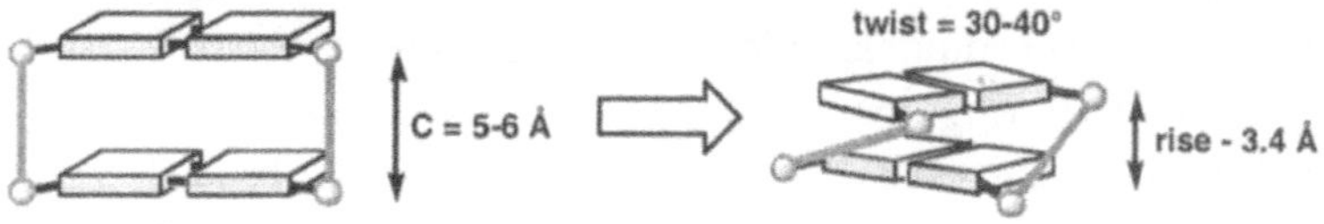

Figure 6. The backbone behaves as a rigid rod 5-6 Å long. Since the bases are 3.4 Å thick, base stacking requires a twist of 30°-40°.

At this point, the conformation of a base step can be completely described using just two parameters, slide and shift. Given the values of slide and shift, the other four step parameters can be calculated using the base stacking energy and the backbone constraint. In addition, the experimental correlations between the backbone torsion angles and the base step parameters allow construction of an all-atom model of the backbone compatible with this base stacking geometry. The parameter correlations observed previously follow naturally from this model.[17] For example, the coupling of roll and twist with slide is illustrated in Figure 7.

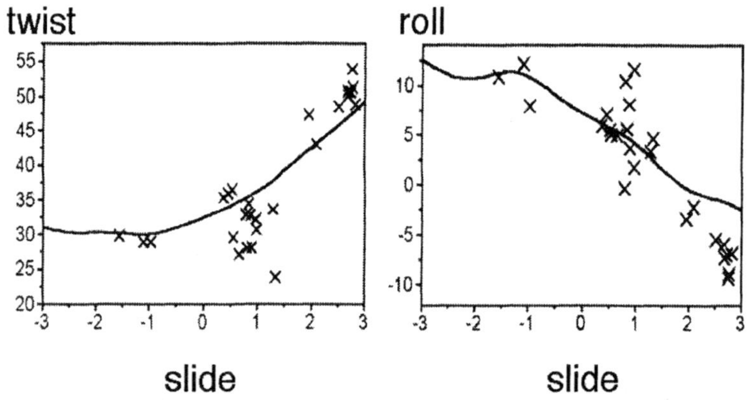

Figure 7. Correlations between the values of twist and roll with slide in CA/TG base steps. The line represents the optimised values calculated using our potential energy function, and the crosses are the experimental values from the Calladine-El Hassan database.

2.3 Coupling of Neighbouring Steps

The final stage in building a complete model of DNA structure is to investigate the effects of sequence context on the conformational properties of dinucleotide steps (step 3 in Figure 2).[18] An analysis of correlations between the conformations of neighbouring steps, taken from trinucleotide and tetranucleotide fragments in the Calladine-El Hassan database showed that roll, tilt, rise and twist are all context-independent, i.e. the values of these parameters can be defined at the base step level, and they are not strongly influenced by the conformation of any neighbouring step. In contrast, slide is strongly correlated with the values of slide at neighbouring steps, while shift is strongly anti-correlated with the values of shift at neighbouring steps. In other words, slide tends to be the same locally along a sequence, whereas shift tends to alternate.

The origin of these effects can be found in the coupling of the backbone torsion angles with the parameters describing the base stacking geometry. The sugar is common to two dinucleotide steps in any sequence, and its conformation is therefore coupled to the base-stacking geometry within either step. For example, slide is strongly coupled to the backbone torsion angle χ that links the base to the sugar, Figure 3. Since χ by definition is the same in both steps, the values of slide are constrained to be similar. A potential energy function that constrains the values of slide to be the same in neighbouring steps and constrains the values of shift to alternate was parameterised using the Calladine-El Hassan database. This allows accurate prediction of the values of slide and shift in tetranucleotides (Figure 8).

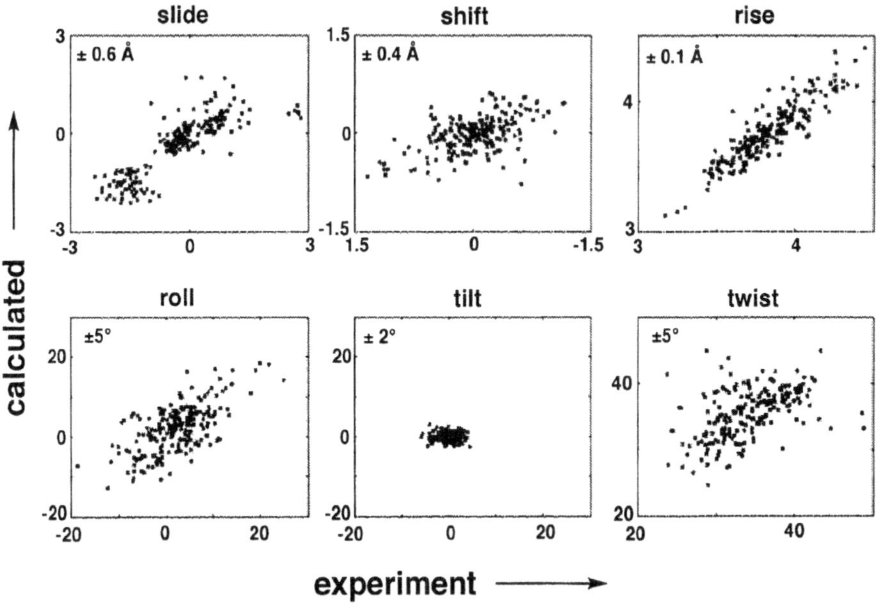

Figure 8. For the 200 tetranucleotides in the Calladine-El Hassan database, each of the six base-step parameters was optimised using potential energy terms for base stacking, the backbone constraint and slide-shift coupling. The resulting values are compared with the experimental values, and the rms differences are shown for each parameter.

2.4 Construction of DNA Oligomers

This model in principle allows us to construct DNA oligomers of any sequence and any length. For tetramers, exhaustive energy calculations are possible, i.e. the energy of every possible combination of slide and shift in each step can be evaluated in order to identify the optimum structure. However, for longer sequences this is impossible, and an alternative searching method is required to identify low energy conformations. We have implemented a genetic algorithm, in conjunction with a pre-calculated database containing the energies of all possible dinucleotide conformations, which allows fast optimisation of the structures of DNA oligomers.[19] For example, a full conformational search on the DNA octamer shown in Figure 9 takes less than a minute on a desktop PC.

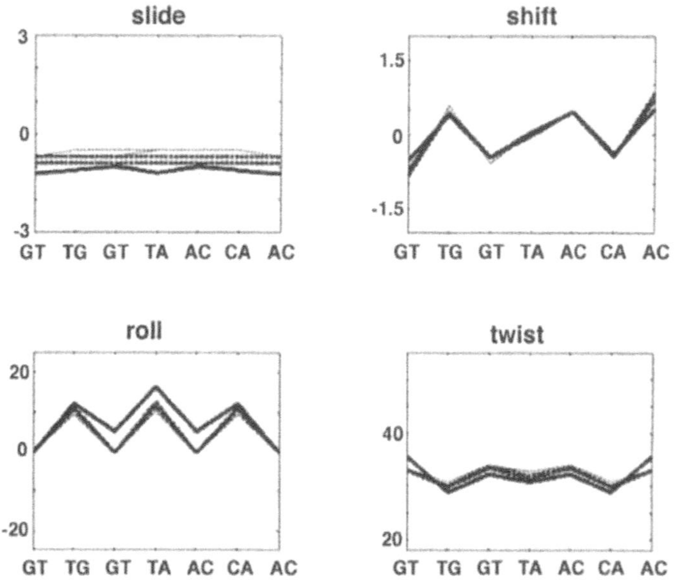

Figure 9. The key step parameters for the DNA octamer GTGTACAC. The solid line represents the experimental X-ray crystal structure, and the dotted lines are local minima calculated using a conformational search and our energy function.

Figure 10 shows how the calculated structures of 33 DNA oligomers compare with the experimental X-ray crystal structures. Half of the structures show excellent agreement (16 structures have an rms difference in slide of less than 0.5 Å). Figure 9 illustrates the quality of the agreement for the key step parameters, slide, shift, twist and roll, for one of these structures GTGTACAC. Most of the remaining structures show some discrepancies in the details, but are accurately predicted to be in the A-DNA or B-DNA family (11 structures have an rms difference in slide between 0.5 and 1.0 Å). The top 6 structures in Figure 10 show very poor agreement, but examination of the experimental data for these structures suggests that the discrepancies may be due to unusual crystal packing interactions. For example, three of the structures feature highly distorted CA/TG steps that are associated with an adverse intermolecular phosphate-phosphate interaction in the crystal.

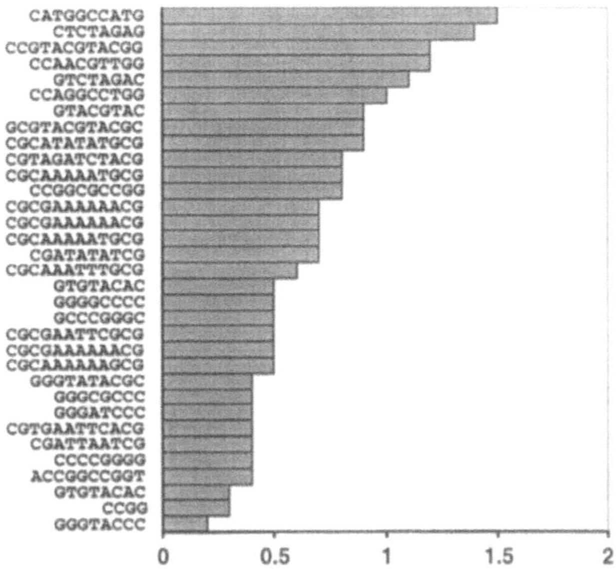

Figure 10. Comparison of the structures of DNA oligomers optimised using our energy calculation with experimental data from X-ray crystal structures. The horizontal axis is the rms difference in slide for the oligomer (Å) and is a measure of the quality of the calculated structure. A value of less than 0.5 Å is excellent agreement. A value of 0.5-1Å means that the structure is correctly predicted as A-form or B-form DNA, but the details are poor. Structures with values greater than 1 Å are poorly predicted.

3. CONCLUSIONS

Thus we have developed a novel computational method that allows us to predict the sequence-dependent structural properties of DNA with a reasonable degree of accuracy. The next challenge is to apply this approach to protein-DNA interactions, where the DNA may be substantially distorted away from its lowest energy conformation (Figure 1(c)). Calladine has analysed X-ray crystal structures of protein-DNA complexes and showed that there are two different categories of interaction: in one class of structure, the DNA utilises the same degrees of freedom found in 'naked' uncomplexed

DNA, but in the other class of structure, the DNA makes use of a new degree of freedom to accommodate the large structural changes required.[15] This is clearly something that we will have to implement in the future, if our calculations are going to handle very large structures such as the nucleosome (Figure 1(c)).

References

1. Luger, K., Mader, A. W., Richmond, R. K., Sargent, D. F., and Richmond, T. J. Crystal structure of the nucleosome core particle at 2.8 Å resolution, *Nature* **389** (1997) 251-260.
2. Lutter, L. C., Halvorson, H. R., and Calladine, C. R. Topological measurement of protein-induced DNA bend angles, *J. Mol. Biol.* **261** (1996) 620-633.
3. Anderson, J. D. and Widom, J. Sequence and position-dependence of the equilibrium accessibility of nucleosomal DNA target sites, *J. Mol. Biol.* **296** (2000) 979-987.
4. Drew, H. R. and Calladine, C. R. Sequence-specific positioning of core histones on an 860 base- pair DNA - experiment and theory, *J. Mol. Biol.* **195** (1987) 143-173.
5. Calladine, C. R., Collis, C. M., Drew, H. R., and Mott, M. R. A study of electrophoretic mobility of DNA in agarose and polyacrylamide gels, *J. Mol. Biol.* **221** (1991) 981-1005.
6. Calladine, C. R. Mechanics of sequence-dependent stacking of bases in B-DNA, *J. Mol. Biol.* **161** (1982) 343-352.
7. Drew, H. R., McCall, M. J., and Calladine, C. R. Recent studies of DNA in the crystal, *Ann. Rev. Cell Biol.* **4** (1988) 1-20.
8. Calladine, C. R. and Drew, H. R. Principles of sequence-dependent flexure of DNA, *J. Mol. Biol.* **192** (1986) 907-918.
9. Calladine, C. R. and Drew, H. R. A base-centered explanation of the B-to-A transition in DNA, *J. Mol. Biol.* **178** (1984) 773-781.
10. El Hassan, M. A. and Calladine, C. R. Curvature and torsion of infinitely repeating DNA sequences: a geometric investigation, *Proc. R. Soc. Lond. A* **453** (1997) 365-386.
11. Thastrom, A., Lowary, P. T., Widlund, H. R., Cao, H., Kubista, M., and Widom, J. Sequence motifs and free energies of selected natural and non- natural nucleosome positioning DNA sequences, *J. Mol. Biol.* **288** (1999) 213-229.
12. El Hassan, M. A. and Calladine, C. R. Conformational characteristics of DNA: Empirical classifications and a hypothesis for the conformational behaviour of dinucleotide steps, *Philos. Trans. A* **355** (1997) 43-100.
13. El Hassan, M. A. and Calladine, C. R. The assessment of the geometry of dinucleotide steps in double-helical DNA - a new local calculation scheme, *J. Mol. Biol.* **251** (1995) 648-664.
14. Packer, M. J. and Hunter, C. A. Sequence-dependent DNA structure: the role of the sugar-phosphate backbone, *J. Mol. Biol.* **280** (1998) 407-420.
15. El Hassan, M. A. and Calladine, C. R. Two distinct modes of protein-induced bending in DNA, *J. Mol. Biol.* **282** (1998) 331-343.
16. Hunter, C. A. and Lu, X. J. DNA base stacking interactions - a comparison of theoretical calculations with oligonucleotide crystal structures, *J. Mol. Biol.* **265** (1997) 603-619.
17. Packer, M. J., Dauncey, M., and Hunter, C. A. Sequence-dependent DNA structure - dinucleotide conformational maps, *J. Mol. Biol.* **295** (2000) 71-83.
18. Packer, M. J., Dauncey, M., and Hunter, C. A. Sequence-dependent DNA structure - tetranucleotide conformational maps, *J. Mol. Biol.* **295** (2000) 85-103.
19. Packer, M. J. and Hunter, C. A. Sequence-structure relationships in DNA oligomers - a computational approach, *J. Am. Chem. Soc.* **123** (2001) 7399-7406.

UNDERSTANDING BIOLOGICAL MACHINES USING HOUSEHOLD ITEMS: SOME INSIGHTFUL CONSTRUCTIONS FROM THE CALLADINE WORKSHOP

B.F. Luisi
Department of Biochemistry, University of Cambridge
80 Tennis Court Road, Cambridge, CB2 1GA, U.K.
ben@cryst.bioc.cam.ac.uk

Abstract: Recent crystallographic studies have revealed a novel protein architecture in which twelve α-helices form a cylindrical bundle. The geometrical principles of this architecture, referred to as the α-helical barrel, have been described by Chris Calladine. The helices in the barrel untwist to maintain their path on the surface of the cylinder, and they also translate co-axially to maintain interfacial contacts. These features, together with other geometrical requirements, have consequences for the choice of amino-acid side-chains at specified spatial locations. The sequence pattern may be useful in the engineering of designed peptide tubes. The Calladine model is described here together with other models developed by Chris for helical polymers (DNA) and macromolecular assemblies.

Keywords: α-helical barrel, DNA structure, Allostery, Nanotubes, Molecular assemblies, Coiled-coil

1. INTRODUCTION

During a conversation at Peterhouse more than thirty years ago, Aaron Klug brought to Chris Calladine's attention the interesting geometrical features of protein assemblies. The constructions discussed included the shells that package the genetic material of viruses, and the helical filaments that propel bacteria towards nutrients (Figures 1,2). These can self-assemble with apparently ideal symmetry, suggesting that the assemblies are constructed with a simple repetitive mechanism. However, despite their apparent symmetry, the viral shells, bacterial filaments and other biological assemblies are often built upon geometrical imperfection. For instance, the subunits in large icosohedral viral shells need to be structurally inequivalent in order to pack into the polyhedra (Caspar and Klug, 1962). Further, in cases of corkscrewing filaments, the spiralling trajectory of the polymer requires different path lengths on the inner and outer-most surfaces, so that the chemically identical subunits become physically distinct through non-equivalent contacts with their neighbours.

H.R. Drew and S. Pellegrino (eds.),
New Approaches to Structural Mechanics, Shells and Biological Structures, 457–474.

Figure 1. Symmetry can be deceptive, as illustrated by an icosohedral virus shell. The centers of symmetry shown are the five-folds (white faced pentagons) and six-folds (open hexagons). Perfect icosohedral symmetry requires 120 chiral subunits, but often multiples are found in nature. The subunits pack by sub-triangulation of the 20 triangular faces, resulting in quasi-equivalent contacts (Caspar and Klug, 1962). The model was constructed by Chris Calladine using cardboard.

Another intriguing property of the corkscrewing filament is its ability to reverse helical handedness under torsional strain. What rules govern the organization and structural changes of these assemblies?

The conversation three decades ago initiated a series of studies investigating the application of mechanics and geometry in the description of macromolecular assemblies. This paper will describe some of the ideas developed by Chris and his colleagues to explain the structure and mechanisms of a variety of biological molecules. An emphasis will be placed on Chris's most recent studies on naturally occurring tubes that are constructed of α-helices. In each system he studied, Chris constructed physical models to illustrate the mechanical and geometrical principles of the molecular assemblies, and some of these models will be described here.

2. DNA STRUCTURE

Twenty years ago, the DNA molecule was commonly depicted as a geometrically idealised object that principally exists in one of two right-handed forms, A or B, depending on environmental conditions. The A-form, studied by Rosalind Franklin and her colleagues using the fiber-diffraction method, is

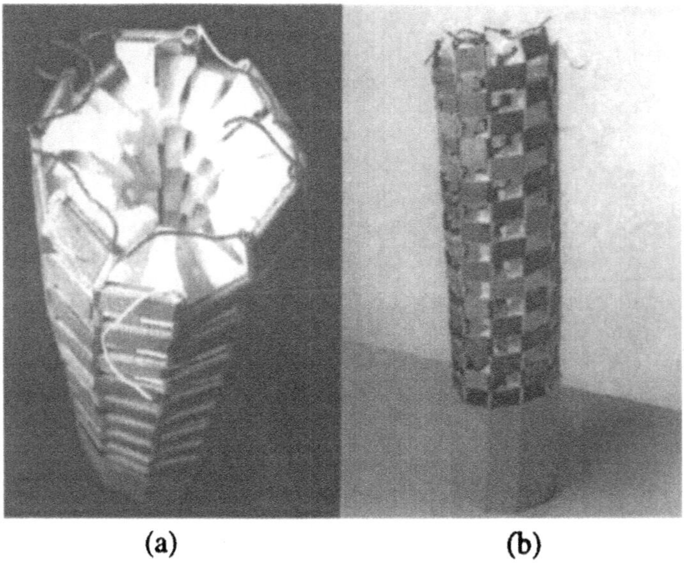

Figure 2. An early model of bacterial flagella constructed by Chris Calladine. The 11-sided tube is comprised of contacting subunits. This model shows subunits making four contacts, but in fact more recent fiber diffraction data show that six contacts are required to account for the subunit interactions. Two views are shown. Recent crystallographic studies by Keichi Namba and colleagues have elucidated the stereochemical details of the structure of the subunits (flagellin) and suggested how the protomers interact in the fiber (Samatey et al., 2001; Yamashita et al., 1998).

favoured under dehydrating conditions and has 11 base-pairs in a turn. In this form, the helix itself spirals in space and the base-pairs are inclined and displaced with respect the central axis (Figure 3(a)). The displacement of the base-pairs result in a central channel when viewed along the central axis (Figure 3(a)). The B-form, studied by Crick and Watson, is also a regular helix, but with 10 bases per turn in the idealized structure; the base-pairs stack so that the planes of the bases are roughly normal to the helical axis (Figure 3(a)). The helix presents two chemically distinct grooves. In the B form, the major groove is a more elaborate, wide canyon that is well presented for molecular recognition, whereas the minor groove is narrow but deep with respect to the phosphate-backbone rim. In the A-form the grooves are closer to equal width; the minor groove is shallow while the major groove is deep and narrow.

Roughly two decades ago, experimental evidence was just beginning to accumulate indicating that the DNA duplex is not conformationally uniform. These emerging data were obtained from solution studies and X-ray crystallography (Klug et al., 1979; Drew et al., 1981) and suggested that twist, groove width, and many other geometrical parameters could vary along the

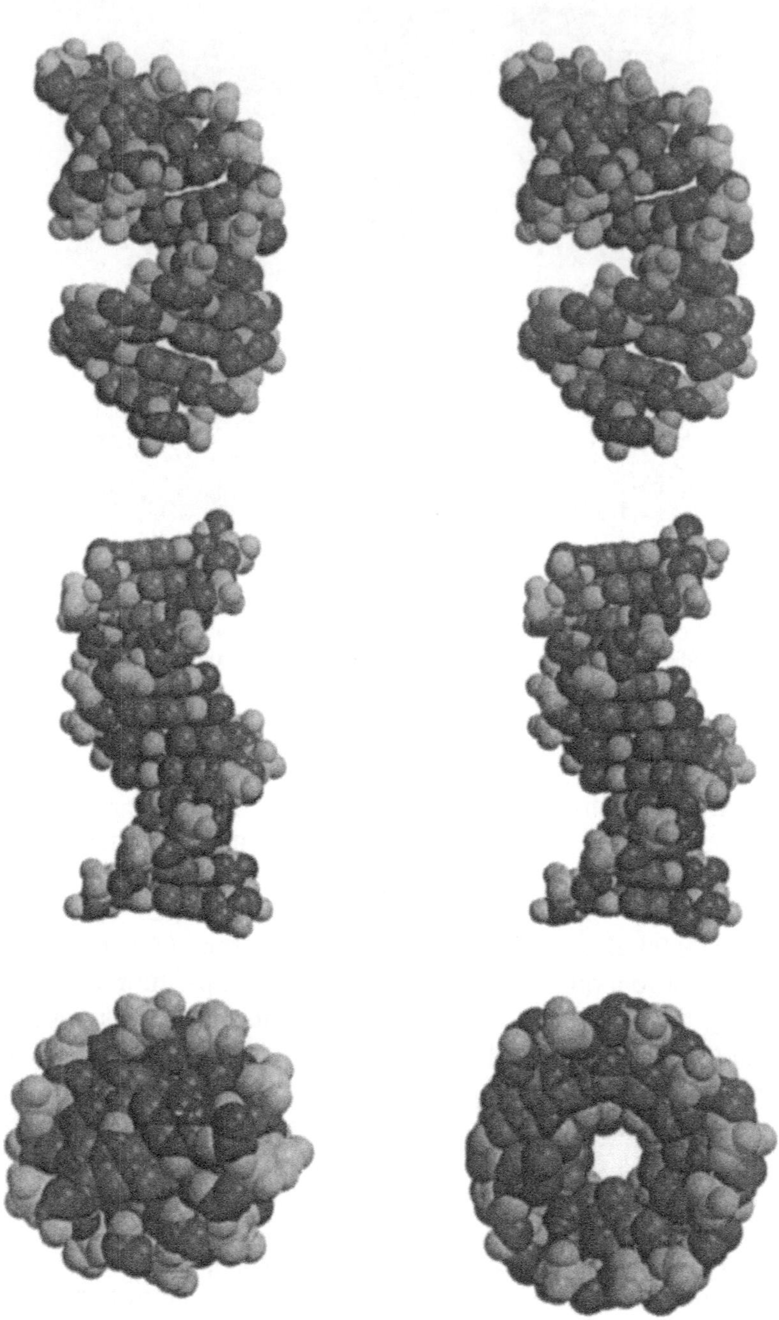

Figure 3(a)

(b)

Figure 3. (a) Space-filling representations of two structures from X-ray crystallography. A-DNA is shown in the top panel and B-DNA is in the middle panel. These two panels are stereoscopic, and the view is perpendicular to the helical axes, into the major groove. The view in the lower panel is along the helical axis (A-DNA on the right, B-DNA on the left). For clarity, none of the water molecules are included in the figures. The nucleic acid database entry codes for the A-DNA and B-DNA structures are adl025 and bdl084, respectively. (b) A stereoscopic view of the TATA-binding protein in complex with DNA. The structure database code is 1ais. (c) Simplified models of DNA, created by Chris Calladine, that illustrate how base-step geometry may influence the trajectory of the duplex. The base-pairs are represented by the rectangular plates, and the blocks between show the geometry of the base step. The darker blocks indicate where the consecutive base steps have a high degree of roll and are untwisted. The upper model in the figure on the left shows the path of the DNA in the TATA element described in part (b). When localized, an abrupt change occurs in the trajectory of the duplex to nearly 90 degrees, when projected onto a plane. In the model which is lower down in the illustration on the left, the center-to-center separation of regions of compression is 10 base-pairs, corresponding roughly to a helical repeat. The deformations thus add in phase and result in an in-plane curvature.

molecule in a manner that was somehow correlated with the base-sequence. Questions thus arose as to whether the sequence effects could be expressed as a simple code, and how important any such code would be for the recognition of the DNA by proteins and drugs. DNA, however is a flexible molecule, and the macroscopic properties such as axial path and torsion vary with thermal fluctuations. Thus, any pattern of microscopic variation would be expected to represent low-energy distributions, and would not necessarily be due to a one-state, static structure that might be associated with a deep-well energy potential.

In the late 70's, Horace Drew, Richard Dickerson and their colleagues at the University of California at Los Angeles elucidated the early crystal structures of synthetic DNA (Drew et al., 1981; Dickerson et al., 1982). These were among the first experimental studies of comparatively long pieces of DNA, 10 to 12 bases in length. They established the stereo-chemical details of the molecule, and demonstrated directly the potential for the sequence itself to affect conformational variations of the DNA. The aromatic bases appeared to interact in distinctive ways. Because they are non-polar, the bases will shield themselves from water by stacking, but their delocalised electrons confer electrostatic properties that influence the details of how they stack. This local effect of stacking will in turn affect the helical trajectory and groove widths. The crystal structures also indicated the subtle but important effects of hydration on DNA structure: a string of water molecules was observed to track along the minor groove side of the duplex, interacting with each other and with the exposed groups of the bases, and linking to the phosphate backbone (Drew and Dickerson, 1981; Dickerson et al., 1983). The patterns of hydration gave a first experimental clue about the delicate balances of thermodynamic forces that affect the equilibrium between the A-type and B-type conformations. The observations from crystallography alerted Horace to the potential rules for sequence-dependent effects, and these ideas soon found resonance in ideas developed by Chris on the mechanics of DNA.

Based on the available crystal structures, Chris proposed a model to explain how certain base-steps could affect the structural properties of duplex DNA (Calladine, 1982). The mechanical model was based on the observation that paired bases are not co-planar, but have a degree of rotation along their common long axis. This rotation, known as propeller-twist, brings one edge of the base pair down, and the other up. Certain steps with large bases (the purines, A and G) on opposite strands will tend to clash in either the major or minor groove, depending on the class of the step; i.e., purine-pyrimidine, pyrimidine-purine or purine-purine (which is equivalent to pyrimidine-pyrimidine). Chris proposed that the clash is relieved by rolling and sliding motions of the base pairs. The mechanics-based proposal was very useful for understanding the relationship between sequence and local helical geometry, and these came to be known as the Calladine rules. In retrospect, the rules may have been too simple to explain fully the complexity of the different base steps, but they proved to be a very effective paradigm, focusing attention on the importance of base-stacking interactions when defining the conformation of duplex DNA.

When Horace Drew came to Cambridge in the early 80's, he soon met Chris and they began an extremely productive collaboration that led to a series of studies on the nature of sequence-dependent effects. One of their first efforts was to define a reference frame for describing and comparing DNA structures. To define the full range of possible movements, they chose the base-step as the reference, with three *en block* rotations and translations of one base-pair, with respect to its partner pair of the step. From the available crystal structures, they focused on the base-step roll, the slide, and the twist as important parameters. Twist is the rotation of one pair with respect to its neighbour about the helical axis; slide is the relative displacement of adjacent base-pairs along their long axes; and roll is the angular opening of the step. The base-step reference frame was used to illustrate the structural differences between the A and B families of duplex DNA, and the mechanics of the transition between these states (Calladine and Drew, 1984). This reference frame showed, in a natural way, how the base-stacking changes during the transition between the two conformations. The principal parameters in this transition were proposed to be slide and roll, which are required together to account for the salient conformational features of the states. The model also explained how the mechanics of the movements would change the conformational preferences for the ribose sugar, which assumes distinctive conformations in the A and B states. Recent crystallographic studies of a series of DNA molecules with the same sequence, but under different environments, are consistent with the proposal that slide is an important parameter in the A/B transition (Vargason et al., 2001).

Chris and Horace went on to examine how local, base-step centered geometry affects properties such as anisotropic flexibility (Calladine and Drew, 1984). Anisotropic flexibility explains the propensity of DNA with particular sequence patterns to wrap around the surfaces of proteins, and also statistical properties such as the persistence length in solution. The models shown in Figure 3(c) illustrate how changes in base-step geometry can locally alter the trajectory of the DNA. If the changes in trajectory were to repeat in phase with the helical periodicity of the DNA, they would add together and make a macroscopic shape. These ideas were applied to investigate the propensities of DNA with certain base-steps to curve tightly, as seen for instance in the DNA that supercoils around the nucleosome core, an assembly of histone proteins that is the fundamental unit of chromatin. They were also applied to describe the rate at which DNA with periodic sequences migrate through a poly-acrylamide gel under the force of an electric field (Calladine et al., 1991). The mechanics based geometric properties of DNA were summarised in a lucid book written by Chris and Horace (1992, 1997).

One experimental limitation of crystallographic studies of DNA structures is that the details of the conformation will be affected by the packing of the neighbouring molecules in the lattice, because of DNA's intrinsic flexibility. Thus, the same sequence can be found in different forms in the same crystal, so that the DNA structure is highly dependent on the lattice contacts. If the duplex is deformable, how important are sequence-dependent conformational effects? The behaviour of the different base-steps were investigated by a detailed analysis of the structural database by Mustafa El-Hasan and Chris, which illustrated the plasticity of certain base-steps (El-Hassan and Calladine, 1996; 1997). Their analysis showed that different steps tended to show distinctive patterns of variation of the geometrical aspects, and that the distribution of certain properties could be strikingly biphasic for particular steps. This suggested that some steps might be more conformationally restricted, while others would be more plastic and might be more easily deformed by their dominating neighbours. Thus, the precise behaviour of a length of DNA might be expected to depend on the contextural rules of the steps.

In specific complexes, the protein-DNA interface is well matched and the molecular surfaces have shape-complementarity. The plasticity of the base-steps will affect the energy required to mould the DNA into a complementary shape, in order to match the recognition surface of a protein. This feature is referred to as conformability. The extension of the analysis of base-step geometry towards duplexes in crystal structures of DNA-protein complexes indicates that sequence-specific deformations are important in target site recognition (Dickerson, 1998; Olson et al., 1998). Thus the energetics of the conformability of DNA are crucial for specificity; *i.e.*, the amount of energy required to form the DNA into a defined shape that is complementary to a recognition surface. These data indicate qualitatively the contribution of sequence effects to the energetics of moulding the DNA interface.

Mustafa and Chris also undertook an analysis of the highly deformed DNA seen in the crystal structure of the TATA-binding protein complexed with its recognition sequence, alternating TA (El-Hassan and Calladine, 1998). The TATA-binding protein is a highly conserved protein that binds to an element upstream of promoters in eukaryotes (cells with nuclei) and archae (a separate phyla of anuclear organisms). The distinctive path will be recognized by other transcription factors, such as TFIIB (known as TFB in archae) to recruit and align the polymerase to transcribe the gene. Most DNA-binding proteins bind in the major groove and induce comparatively modest structural adjustments in the DNA. TBP, in contrast, binds in the minor groove, which is distinctive in itself, but what is even more remarkable is the distortion of the duplex (Figure 3(b)). The minor groove opens up to accommodate the protein, and the major groove becomes highly compressed. The path of the distorted DNA is seen in Chris's model shown in Figure 3(c). The highly distorted trajectory of the TATA element in the complex is highly conserved in evolution, and its importance might be related to the pre-organization of the binding site for TFIIB (TFB).

The distortions of the TATA element take the DNA into a completely different class of structural adjustments compared to the A and B states. Chris and Mustafa found that the A/B transitions seen in experimental crystal structures lie along a restricted region of conformational space of roll versus slide, but the TBP-bound TATA element lies distinctively off this trajectory. The transition is similar to the type of mechanical changes observed in sudden transitions in materials, such as fractures in glasses. The exaggerating rolling movements in the TATA element may originate from a type of pivoting effect. From the stereochemical perspective, the exocyclic groups in the major groove look like suitable candidates for the pivot, and they tend to make close contacts in the crystal structures of the TBP/TATA complexes.

It is now widely accepted that DNA sequence affects its structure, and that this in turn modulates protein-DNA interactions in an important aspect of recognition. This was considered rather controversial when Horace and Chris first started to work on these ideas. An important recent development in detailed prediction of base-stacking effects has included the contribution of electrostatics, alluded to earlier. Insightful ideas about electrostatic effects on base stacking have been elaborated by Chris Hunter and his colleagues at Sheffield University. The theory was refined with experimental data, and is now very effective for predicting sequence effects in duplex DNA.

3. THE α-HELICAL BARREL

My colleagues and I in the Biochemistry Department have been working closely with Chris over the past few years to study the packing of α-helices in proteins. We started with the analysis of the distinctive helices in the bacterial protein TolC (Figure 4). This protein is membrane-embedded and is part of a machine that transports proteins and drugs from certain classes of bacteria, known as the Gram-negative group, which includes members such as *Escherichia coli*. TolC engages two other proteins to form an energy-driven transport pump. TolC serves as a channel in the pumping mechanism and is shaped roughly like a tube in which one end is tapered to a close. This tapered end joins with two other proteins to form a transient assembly that functions as a pump. During the transport process, the closed end is probably opened, like a camera iris, through its interaction with the other protein components of the pump.

The membrane-embedded portion of TolC is comprised of a β-sheet (the flat ribbons in the diagram) that forms a barrel-like structure, and is embedded in a lipid bi-layer on the outer membrane of the bacterium. The long axis of the

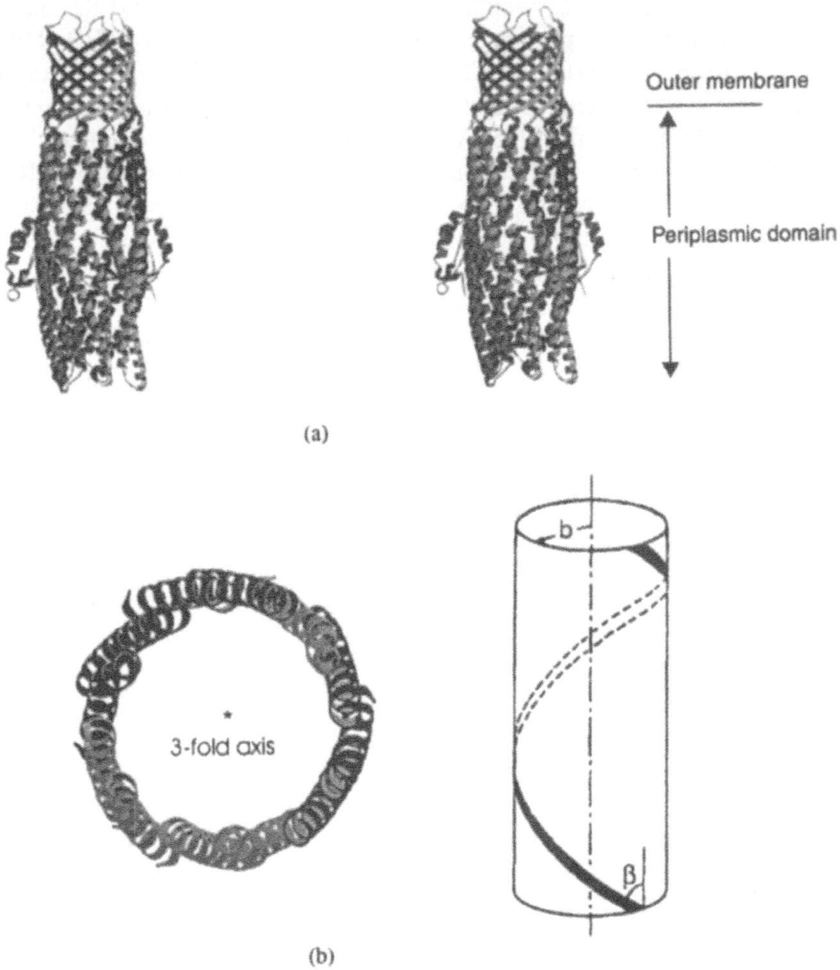

Figure 4. Properties of the helices of the α-helical barrel. (a) A stereoscopic view of the *Escherichia coli* outer membrane protein TolC (protein database entry 1EK9). For clarity, the side chains of the amino acids are not shown, and the peptide backbone is represented by ribbons. (b)A view of the α-helices forming a helical barrel in the top portion of the TolC periplasmic domain. The asterisk indicates the approximate position of the molecular 3-fold axis. The path of the helices in the α-helical barrel follows the surface of a cylinder. To remain on this surface, the helices must untwist and curve, which has implications for the packing of side chains at the interhelical interfaces, as shown in Figure 5.

TolC molecule is perpendicular to the plane of the membrane. The portion of the molecule outside the membrane is mostly α-helical. At the far end, the helices come together to seal the tube, but near the β-portion, the α-helices also form a type of barrel, which is referred to as the α-helical barrel.

The helices of the α-helical barrel lie on the surface of a cylinder, but they are inclined with respect to this long cylindrical axis. To remain on the surface of the cylinder, the inclined helices must distort from a straight geometry. Two deformations are required for the helix to maintain this idealised geometry: an in-plane bending, along with a twisting of the helix. The two parameters, twist (τ) and curvature (κ) are given by the formulae:

$$\kappa = (\sin^2\beta)/b$$
$$\tau = (\sin\beta\cos\beta)/b,$$

where β is the inclination angle of the α-helix on the surface of the cylinder (this is the angle with respect to the long molecular axis of TolC), and b is the radius to the center of the helix. Using these relationships, Chris deduced that the helices physically untwist so that their helical repeat changes from 3.6 residues in a period, characteristic of a relaxed α-helix, to almost 3.5 residues. This means that every seventh residue, corresponding to two turns, will lie on a stripe that follows the axis of the helix. Two of these helices coming together would have a matching stripe that maintains register along the entire lengths of the helices. Side chains of the amino acids would protrude along this matching stripe. Chris constructed a model of the TolC α-helical barrel from heptagonal columns, which was very useful for visualizing the packing contacts in the molecule (shown in Figure 5).

An α-helical barrel can be demonstrated to precisely match the periodic protrusions with periodic pockets, thus making a snug fit to stabilize the α-helical assembly. First, let us start with a straight, relaxed α-helix. This can be represented as a uniform cylinder in which the path of the peptide is a line on the surface of the cylinder and the residues are circles. Figure 6 illustrates an operation in which this cylinder, representing a helix with a repeat of 3.6 residues in a turn, is cut along a co-axial stripe and laid flat. This cylindrical projection results in a series of parallel lines, representing the path of the peptide, that are inclined with respect to the long axis. The slope of the lines is related to the helical repeat. Now, imagine that two similar helices are packed together. Their contacting face can be represented by overlaying the cylindrical projections of the two helices, taking care to ensure the sides corresponding to the outside surfaces of the α- helices meet (Figure 6(c)). If their helical axes were exactly parallel, they would make a contact along a face in which every 18th residue would be in register on the same co-axial stripe. Nearly fifty years ago, Crick proposed that if the helical nets were inclined with respect to each other by roughly 18^0, then the pattern of residues would make a perfect match, somewhat like a checkerboard. The intermeshed side chains represent knobs that fit into complementary holes. The two axial lines would correspond to an inter-helical contact. In three dimensions, the lines can be maintained with a gentle left-handed super-coil, which would have a periodic face every seventh residue. This is the basis of the heptad repeat in which a protruding non-polar residue of compatible size, such as leucine, repeats at every fourth and seventh position to lie on the two axial contact lines.

Analogous knobs-into-holes packing occurs for the helices of the TolC α-helical barrel. Representing the helices as helical nets, the set of parallel lines will have a different slope compared with the case of relaxed α-helices, because of their change in twist. The checkerboard pattern becomes slightly oblique, and the knobs-into-holes packing no longer has the same well-fitting intermeshing seen for the case of the relaxed α-helices. This requires a pattern of small residues at certain positions of the knobs and holes to prevent collision.

Since their periodicity is 7 residues, the helices of the TolC α-helical barrel can be represented as heptagons, and fourteen of such heptagons can be fitted into a barrel without distortion. However, only twelve are packed in TolC (four α-helices contributed from each of three protomers). Because there are only twelve helices rather than fourteen of the ideal cylinder, the adjacent helices of TolC must pack such that they are compressed a little on the innermost face and expanded a bit on the outside face. There will be a tendency for small amino acids to occur at the innermost face and larger, bulkier side chains to lie on the outer face, as shown in Figure 7. The required pattern of residues suggests a template, and this was used to identify a number of candidate structural homologues of TolC from the database sequence (Sharff et al., 2001).

The coiled-coil packing of the helices in the TolC α-helical barrel also explains a striking feature of the TolC molecule: with respect to the molecule axis, the strands of the β-barrel and the helices of the α-barrel have opposite inclination angles (Figure 4 (a)). The crossing angles of the strands are similar to those seen in β-sheets in globular proteins, while the crossing angles of adjacent helices of the α-helical barrel have the opposite hand. The helix crossing angle is in fact the same hand as seen in coiled-coils. Hence, the change in the trajectory at the junction of the α-barrel and the β-barrels is due to the twist of sheets on the one hand and the packing of helices in coiled-coils in the α-barrel on the other.

Chris's analysis of the packing of the helices in the TolC α-helical barrel also revealed that the helices undergo a relative axial translation. Chris derived a formulation for the relative shift of the two helices based on the differences between adjacent residues along lines of the co-axial stripes:

$$(\mathbf{l}_4\text{-}\mathbf{l}_3)_i - (\mathbf{l}_4\text{-}\mathbf{l}_3)_o = 4h,$$

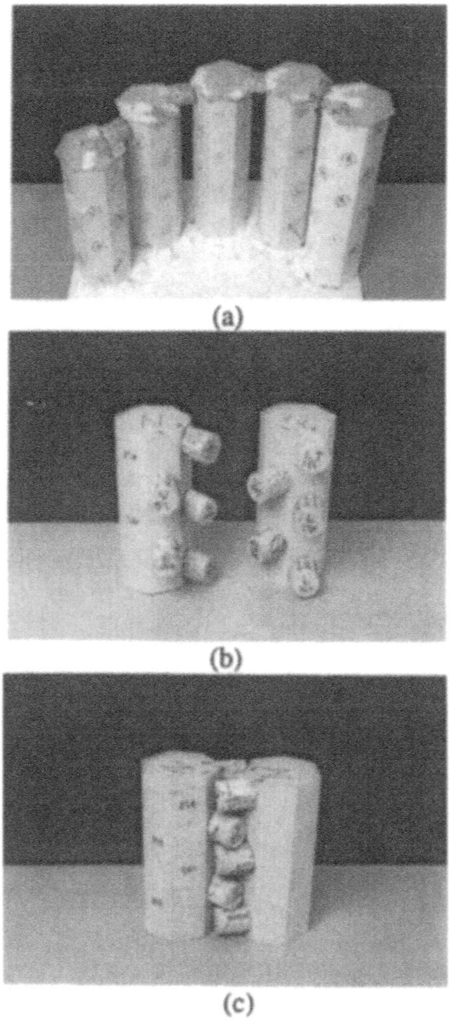

Figure 5. (a) Chris's model for the interaction of helices in the TolC α-helical barrel. There are four helices in each of the TolC subunits (indicated by heptagonal tubes). Five helices were included in the model to define all the possible helix-helix interactions in the barrel (the left-most tube is from the adjacent subunit). The helices are made from 7-sided cylinders, to reflect the 3.5 residue period, and the amino acids letters have been written on their corresponding position on each. Wires connecting residues indicate hydrogen-bonding interactions. (b) Knobs into holes packing of helices. The helices are represented by seven-sided cylinders. The corks represent side chains, and they are inclined with respect to the normal vector to the attached face. In view (c), the helices are packed together; each knob on one helix fits into holes made by a pattern of knobs on its partner helix. Side chains were not included in part (a) for clarity.

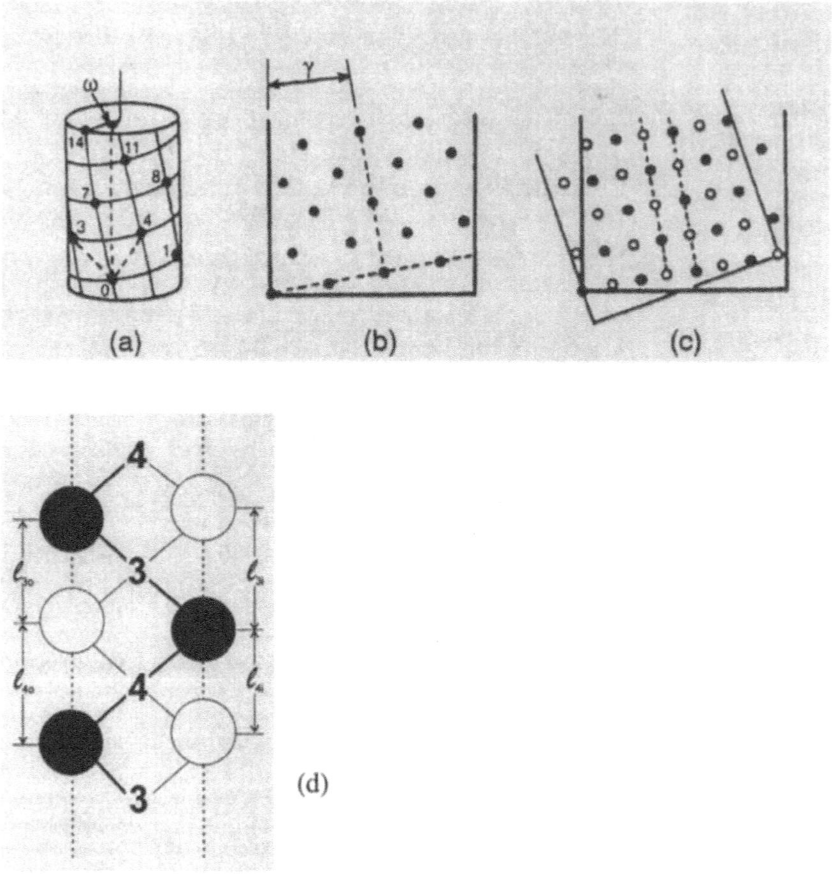

Figure 6. A schematic representation of an α-helix as a cylinder, with the successive amino acids numbered. If the cylinder is cut along the dotted line and then laid flat, as in (b), then the amino acids will lie on a lattice. The lattice is viewed from outside the cylinder. If a second helix is cut in a similar manner, the two lattices can be overlaid with a suitable tilt, as shown in (c). The second helix is viewed from the interior of its cylinder. The intermeshing of the side chains represents the "knobs-into-holes" interactions depicted in Figure 5(b,c). One set of side chains is shown as dark circles, and the set from the second helix as open circles. The lines indicate the connecting interface between the packing helices. The same lines are seen in part (d), which also defines the distances used in the text to calculate the relative axial shift of the coiled coils. This figure is adopted from Calladine et al. (2001).

where h is the relative axial displacement, and $\mathbf{l}_4$ and $\mathbf{l}_3$ are the C_β-C_β distances of the 3rd and 4th residues of the heptadic repeat from the knob of the neighbouring helix on the inside line, *i*, and the outside line, *o* (Figure 6(d)). This analysis was applied to conventional coiled-coils, which suggests that they can undergo relative axial translation, even in cases where the coiled-coil is made of two identical peptides (Calladine et al., 2001). This means that the structure becomes asymmetric, because the two peptides are no longer in identical environments with the co-axial shift. Our colleague Charlie Laughton at Nottingham University investigated this by calculating a molecular dynamics stimulation of the homo-dimeric coiled-coil from the dimerization domain for the DNA-binding protein GCN4. With energy minimization, the symmetrical structure relaxed into one with an axial shift. This shift could be propagated along the length of coiled-coil, although it became stuck at a point where asparagines interact to contact the peptide backbone on one of the two helices. The propagation continued when this contact was broken. Recent crystallographic studies of the muscle protein tropomyosin show that such translational effects of coiled-coils may be used in the macromolecular architecture of the muscle bundles (Brown et al., 2001), and we speculate that they may also serve allosteric roles in other protein machines.

In collaboration with Charlie Laughton and Venkatesh Pratap (Biochemistry Department, Cambridge), we are continuing the analysis of axial translations in coiled-coils, and we are now examining the structural database for examples in which this type of co-axial, symmetry-breaking movement within coiled-coils may be used by proteins to propagate allosteric signals. We have also been exploring the design of nanotubes based on the geometrical principles elaborated for the α-helical barrel.

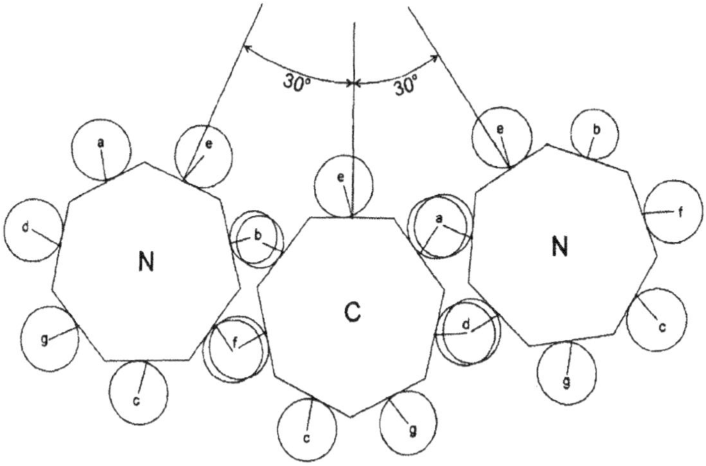

Figure 7. Three adjacent anti-parallel α-helices, with residues interlocked at the interfaces, are represented as heptagons, to represent the heptadic repeat. Residues have been labelled in sequence ***a,b,c,d,e,f,g***. The interlocking residues are labelled as ***a-d*** at one of the interfaces and ***b-f*** at the other, with residues ***a,b*** lying on the inside of the 12-helix barrel.

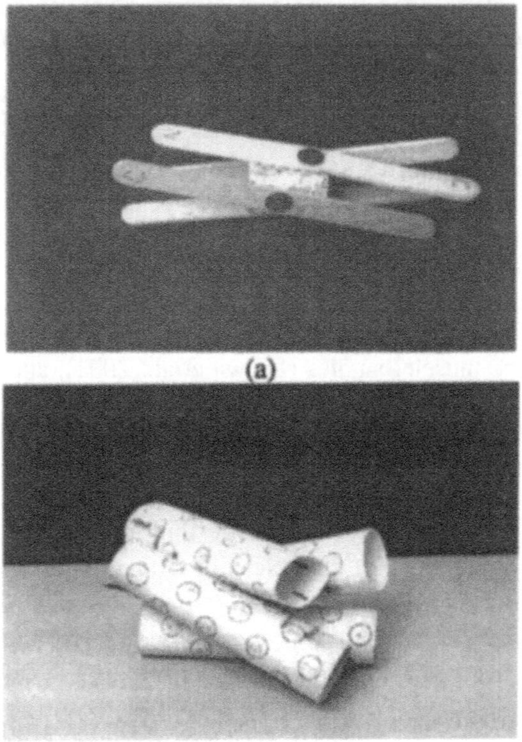

(a)

(b)

Figure 8. Extreme non-equivalence in a protein structure. Two models are shown of a four-helix bundle formed by a homo-tetrameric peptide from the tetramerization domain of a transcription factor. Experimental studies of the peptide show that it associates in non-equivalent ways. (a) A cork and popsicle stick representation of the four-helix bundle. The model was designed to allow rotation of the helices. N and C indicate the amino- and carboxy-termini, respectively. (b) The cylinders represent the α-helices, which are connected by yarn. The model can be moved into the two non-equivalent associations.

Looking back at the original model that Chris constructed to explain the packing of helices in the α-helical barrel and other helical assemblies (Figures 5,8), the utility of these inventive models was remarkable, and provided unexpected leads into new and stimulating areas of investigation. Somewhat like the best childhood toys, these models and others from Chris have inspired a lot of insight, fun, and imagination that have somehow transcended far beyond their deceptively simple construction.

References

Brown, J.H., Kim, K.H., Jun, G., Greenfield, N.J., Dominguez, R., Volkmann, N., Hitchcock-DeGregori, S.E. and Cohen, C. (2001) Deciphering the design of the tropomyosin molecule. *Proc. Natl. Acad. Sci. U.S.A.* **15**, 8165-8166.

Calladine, C.R. (1975) Construction of bacterial flagella. *Nature* **225**, 121-124.

Calladine, C.R. (1976) Design requirements for the construction of bacterial flagella. *J. Theor. Biol.* **57**, 469-489.

Calladine, C.R. (1978) Change of waveform in bacterial flagella: The role of mechanics at the molecular level. *J. Mol. Biol.* **118**, 457-479.

Calladine, C.R. and Drew, H.R. (1984) A base-centred explanation of the B-to-A transition in DNA. *J. Mol. Biol.* **178**, 773-782.

Calladine, C.R. and Drew, H.R. (1997) Understanding DNA: The molecule and how its works. Academic Press. 2nd edition.

Calladine, C.R., Collis, C.M., Drew, H.R. & Mott, M.R. (1991) A study of electrophoretic mobility of DNA in agarose and polyacrylamide gels. *J. Mol. Biol.* **221**, 981-1005.

Calladine, C.R., Sharff, A. & Luisi, B. (2001) How to untwist an α-helix: the principles of the α-helical barrel. *J. Mol. Biol.* **305**, 603-618.

Caspar, D.L.D. & Klug, A. (1962) *Cold Spring Harb. Symp. Quant. Biol.* **27**, 1-24.

Crick, F.H.C. (1953) The packing of α-helices: simple coiled coils. *Acta Crystallogr.* **6**, 689-697.

Dickerson, R.E. (1998) *Nucleic Acids Res.* **26**, 1906-1926.

Dickerson, R.E., Drew, H.R., Conner, B.N., Wing, R.M., Fratini, A.V., & Kopka, M.L. (1982) The anatomy of A-,B-, and Z-DNA. *Science* **216**, 475-485.

Dickerson, R.E., Drew, H.R., Conner, B.N., Kopka, M.L. & Pjura, P.E. (1983) Helix geometry and hydration in A-DNA, B-DNA, and Z-DNA. *Cold Spring. Harb. Symp. Quant. Biol.* **47**, 13-24.

Drew, H.R. and Dickerson, R.E. (1981) Structure of a B-DNA dodecamer. III. Geometry of hydration. *J. Mol. Biol.* **151**, 535-556.

Drew, H.R., Wing, R.M., Takano, T., Broka, C., Tanaka, S., Itakura, K. & Dickerson, R.E. (1981). Structure of a B-DNA dodecamer: conformation and dynamics. *Proc. Natl. Acad. Sci. U.S.A.* **78**, 179-183.

El-Hassan, M.A. and Calladine, C.R. (1996) Propeller-twisting of base-pairs and the conformational mobility of dinucleotide steps in DNA. *J. Mol. Biol.* **259**, 95-103.

El-Hassan, M.A. and Calladine, C.R. (1996) Two distinct modes of protein-induced bending in DNA *J. Mol. Biol.* **282**, 331-343.

El-Hassan, M.A. and Calladine, C.R. (1997) Conformational characteristics of DNA: empirical classifications and a hypothesis for the conformational behaviour of dinucledotide steps. *Phil. Trans. R. Soc. Lond A.* **335**, 43-100.

El-Hassan, M.A. and Calladine, C.R. (1997) Curvature and torsion of infinitely repeating DNA sequences: a geometric investigation. *Proc. R. Soc. Lond A.* **453**, 365-386.

Gestwicki, J.E. and Kiessling, L.L. (2002) Inter-receptor communication through arrays of bacterial chemoreceptors. *Nature* **415**, 81-84.

Klug, A., Jack, A., Viswamitra, M.A., Kennard, O., Shakked, Z. and Steitz, T.A. (1979). A hypothesis on a specific sequence-dependent conformation of DNA and its relation to the binding of the *lac*-repressor protein. *J. Mol. Biol.* **131**, 669-680.

Olson, W.K., Gorin, A.A., Lu, X.J., Hock, L.M., and Zhurkin, V.B. (1998) *Proc. Natl. Acad. Sci. U.S.A.* **95**, 11163-11168.

Pamplona, D.C. and Calladine, C.R. (1993) The mechanics of axially symmetric liposomes. *J. Biomech. Eng.* **115**, 149-159.

Pamplona, D.C. and Calladine, C.R. (1996) Aspects of the mechanics of lobed liposomes. *J. Biomech. Eng.* **118**, 482-488.

Samatey, F.A., Imada, K., Nagashima, S., Vonderviszt, F., Kumasaka, T., Yamamoto, M. and Namba, K. (2001) Structure of the bacterial flagellar protofilament and implications for a switch for supercoiling. *Nature* **410**, 331-337.

Sharff, A., Fanutti, C., Shi, J., Calladine, C.R. and Luisi, B. (2001). The TolC family in protein transport and multidrug efflux: From stereochemical certainty to mechanistic hypothesis. *Eur. J. Biochem.* **268**, 5011-5026.

Vargason, J.M., Henderson, K. and Ho, P.S. (2001) A crystallographic map of the transition from B-DNA to A-DNA. *Proc. Natl. Acad. Sci. U.S.A.* **98**, 7265-7270.

Yamashita, I., Hasegawa, K., Suzuki, H., Voderviszt, F., Mimori-Kiyosue, Y. and Namba, K. (1998) Structure and switching of bacterial flagellar filaments studied by X-ray fiber diffraction. *Nature Struct. Biol.* **5**, 125-132.

MEASUREMENT OF DNA BEND ANGLES USING DNA TOPOLOGY

L. C. Lutter, V. Tchernaenko, M. Radlinska, C. E. Drabik, J. Bujnicki, and H. R. Halvorson
Molecular Biology Research Program, Henry Ford Hospital, One Ford Place 5D, Detroit, Michigan 48202, USA
LLUTTER@cmb.biosci.wayne.edu

Abstract: A large number of studies have shown that an A-tract imparts curvature to the DNA double helix. Constructs of such A-tracts have been used as bend standards in a large number of both structural and functional studies, and A-tracts confer significant activation in transcription. An accurate value for the bend angle induced by an A-tract is of central importance to all these studies, but estimates for the extent of the bend angle of an A-tract vary significantly. To address this problem, we have used the rotational variant method [Lutter et al. J. Mol. Biol. **261** 620] to measure the angle of DNA curvature in an A6 tract. The original version of the method measured a protein-induced bend angle independent of external standards by comparing the effect of bent and straight forms of the sequence on the topology of a DNA plasmid in which the sequence is cloned as a series of tandem repeats. To adapt the approach to the measurement of an intrinsic bend, the straight form of the A-tract was generated by performing the relaxation at high temperature using a hyperthermophile topoisomerase. Fourteen different plasmids containing tandem repeats of A-tracts were constructed and topologically analysed. When relaxation was performed at 14° C, the bend value measured was 24° ± 2°.

Key words: DNA, bend, DNA topology

1. INTRODUCTION

Tracts of 4-8 adenines ("A-tracts") have been shown to cause the duplex axis of DNA to curve [reviewed in (Harvey *et al.*, 1995; Harrington, 1993; Hagerman, 1992; Hagerman, 1990; Crothers and Shakked, 1999)]. The exact structural details of how this curvature is accomplished are still uncertain, but A-tracts have found considerable utility in a wide number of

H.R. Drew and S. Pellegrino (eds.),
New Approaches to Structural Mechanics, Shells and Biological Structures, 475–484.

studies. For example, helically phased A-tracts have been used as electrophoretic mobility standards in studies to determine the bend angle induced when a protein binds to its site in DNA [see e.g. (Zinkel and Crothers, 1990; Thompson and Landy, 1988)]. In related studies, the protein-induced bend angle is estimated by comparing the alteration of electrophoretic mobility when an A-tract block is helically phased against a protein-induced bend (Zinkel and Crothers, 1990; Sitlani and Crothers, 1996; Gartenberg *et al.*, 1990; Kerppola and Curran, 1991b; Kerppola and Curran, 1991a; Kerppola, 1996; Kerppola, 1997; Strauss and Maher, 1994). In addition, both natural and artificial constructs containing A-tracts have been used to understand the functional role of DNA bending, for example in the transcription initiation complex (Bracco *et al.*, 1989; Pazin *et al.*, 1997; Gartenberg and Crothers, 1991; Plaskon and Wartell, 1987; Perez-Martin *et al.*, 1994) as well as other biologically important processes (Goodman and Nash, 1989; Goodman *et al.*, 1992; Hagerman, 1990). A complete understanding of the process of recombination or transcription activation in these complexes will require a detailed description of the structure of the multicomponent complexes involved, including the path of DNA in each respective complex.

Essential to such a description will be an accurate value for the bend angle of an A-tract present in both the artificial and natural constructs. However, estimates in the literature range from 11° (Calladine *et al.*, 1988) to 42° (Chan *et al.*, 1993). Some of this variability derives from the use of indirect methods, for example electrophoretic analysis that requires significant assumptions about the behaviour of DNA fragments during gel electrophoresis. We have developed a method that uses topology to measure DNA bend angles from first principles, avoiding the need for assumptions about DNA migration in gel electrophoresis or requirements for comparison to DNA bend standards (Lutter *et al.*, 1996; Lutter *et al.*, 2000). In the current study we employ a modification of that method to measure the bend angle of an A6 tract. The original method measured the bend angle induced by catabolite gene activator protein (CAP) when it binds to its site in DNA (Lutter *et al.*, 1996). In that study, multiple plasmids were constructed that contain tandem repeats of the CAP binding site, with the repeat length progressively increased in 1-2 bp increments in successive plasmids. Binding CAP to each of these constructs bends the DNA of its site. This results in the generation of a series of superhelices, with the number of superhelical turns dependent on the acuteness of the bend angle. Relaxation of each plasmid with topoisomerase results in a linking number change that reflects the number and hand of these superhelices, and these changes can be quantified by electrophoresis in an agarose gel. From these data the bend angle can be calculated. The advantage of this method is that the bend angle calculation does not depend on use of external standards, and the DNA bend angle measured is that of the complex in solution at the relaxation (not electrophoresis) conditions and in a relaxed state.

The method measures the linking number difference of each respective plasmid in the bent (protein bound) and straight (unbound) state. For the A-tract measurement here, a modification needed to be developed to achieve the straight state of the DNA, since the natural state of the DNA under normal conditions is bent. This was accomplished by relaxing the plasmid containing the A-tract at high temperature with a hyperthermophile topoisomerase. The resultant analysis measures an A-tract bend angle at 14° C of 24° ± 2°.

2. RESULTS

A detailed description of the rotational variant method for bend angle determination is presented in (Lutter *et al.*, 1996; Lutter *et al.*, 2000). In brief, multiple plasmids are constructed that contain tandem repeats of the sequence to be tested. The repeat length is progressively increased in 1-2 bp increments in this series of rotational variant plasmids. Each plasmid is relaxed with topoisomerase with the sequence in its bent and straight state, respectively. The difference in linking number between these two states is determined by electrophoresis in an agarose gel, and the dependence of the linking number change on repeat length is used to calculate the bend angle for the site.

The fundamental repeat unit analysed is composed of 5 tracts of A6 with alternating spacers of CGGC and CGGGC. Since the screw in such a sequence is about 10.5 bp/turn, this results in an approximately planar overall bend composed of five helically phased A6 tracts (Koo *et al.*, 1990). This fragment was then cloned into the *Sma*I site of the vector pBluescript. This clone was used as a template in a PCR reaction along with various combinations of the primers shown to generate a series of fragments that increase in length in 1-2 bp increments. Each PCR product was then cloned and plasmids were generated containing 10 tandem repeats of each respective fragment (Radlinska *et al.*, 2001). This resulted in a series in which each plasmid contains 50 A6 tracts arranged in planar blocks of 5 per repeat, with the repeat length increasing by 1-2 bp through the plasmid series.

Each plasmid was then relaxed by *Drosophila* topoisomerase II at low temperature (14° C) and by a hyperthermophile topoisomerase V at high temperature (70° C and 85° C). At 14° C the A-tract is in its bent state, with the result that each plasmid will contain a specific form of superhelix depending on the plasmid's characteristic rotational setting of successive repeats. These various superhelical structures will be reflected in the linking number of the relaxed plasmid. Briefly, one turn of right-handed superhelix will contribute +1 to the linking number, a left-handed superhelix will contribute –1, and the zigzag structure will contribute 0 since it is planar (Lutter *et al.*, 1996; Lutter *et al.*, 2000). Plasmids relaxed at high temperature will not have these superhelices because the bends will be melted. This eliminates the contribution of any of the superhelices to the linking number of each respective plasmid construct. The difference in linking number between the bent and straight states of each respective plasmid is then used to calculate the bend angle (Lutter *et al.*, 1996; Lutter *et al.*, 2000).

The linking number differences for each plasmid relaxed at 14° C, 70° C and 85° C were determined by fractionation of the samples by electrophoresis in agarose gels (Lutter *et al.*, 1996; Lutter *et al.*, 2000). The unwinding of the duplex caused by the large temperature difference causes a substantial linking number difference between the samples relaxed at low and high temperatures. This required generation of a set of samples of intermediate linking numbers analysed on multiple gels to relate the linking numbers of the samples relaxed at low and high temperature (Keller, 1975; Drabik *et al.*, 1997). Analysis of the linking number differences between the 70° C and 85° C samples of each respective rotational variant showed this difference to be very near to constant, indicating the DNA at both temperatures is near to straight (see below). Therefore the 70° C and 85° C values were averaged, and this straight DNA average value was subtracted from the 14° C bent DNA value. This linking number difference between straight and bent DNA for each rotational variant is plotted as a function of repeat length in Figure 1. The resultant curve is very similar in shape to that obtained for the similar experiment analysing CAP-induced DNA bending (Lutter *et al.*, 1996). Thus there is a comparatively gradual decrease in the ΔL values from repeat length of 95 bp to 102 bp, followed by a rapid increase through 105 bp [similar to the 85-87 bp constructs in (Lutter *et al.*, 1996)], after which the decrease begins again. The rapid increase between repeat lengths 102 to 105 represents the transition through the "crossover point", where the tandemly repeated DNA bends change from a left-handed superhelix (102 bp repeat) to a right-handed superhelix (105 bp repeat). The 98 bp-99 bp repeats form an approximately planar, zigzag structure [see (Lutter *et al.*, 1996; Lutter *et al.*, 2000) for details of this description]. A preliminary estimate for the bend angle can be obtained from the difference between the ΔL values for the 102 and 105 bp repeats, which is about 4 linking number units. This "ΔΔL" value, reflecting the change in hand of the superhelix, represents twice the number of superhelical turns present in the structure (Lutter *et al.*, 1996; Lutter *et al.*, 2000). Thus there are about 2 superhelical turns, or about 720° of total bending, induced by the 50 A6 tracts, or about 14° per A6 tract. However, this preliminary value is necessarily an underestimate of the bend angle because it does not take into account the torsional flexibility of the DNA (Lutter *et al.*, 1996; Lutter *et al.*, 2000).

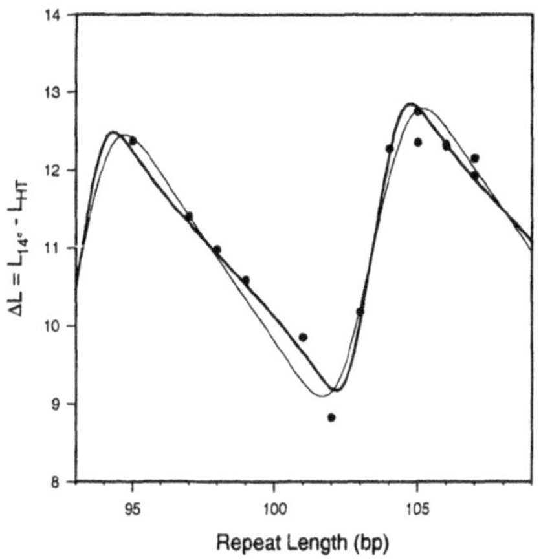

Figure 1. Linking number change versus repeat length. The difference in the linking number of the straight state (L_{HT}, the average of the linking numbers of the 70° and 85° high temperature relaxations) and the bent state ($L_{14°}$) is plotted against repeat length for each of the rotational variants. The solid line is the fit of the equation relating linking number to repeat length (Section 4). The parameters resulting from this fit are given in the text.

2.1 Fitting the function relating ΔL and repeat length to the data

To include the torsional flexibility as well as to use the entire set of data in determining the bend angle, the equation relating ΔL to repeat length was fitted to this data set (Section 4). This fit is shown as the continuous line in Figure 1. The parameters resulting from this fit were: bend angle b (±SD) = 23.9°±2.2°; the screw S_o (±SD) = 10.38±0.01 bp/turn (at 14°); the torsional stiffness [c.f. σ_T in (Lutter *et al.*, 1996)] value C = 105 (51-181) pN nm^2, with the 64% confidence region (corresponding to ±SD in the symmetric case) given in parentheses; and the root-mean-square of the deviations between the data and the fit, rms = 0.2266 turns.

The use of a hyperthermophilic topoisomerase enabled experiments at elevated temperatures (70° C and 85° C) which, when compared with experiments performed at 14° C, appeared identical within the experimental scatter (there was more scatter between replicate experiments at the elevated temperatures for reasons that are not clear.) Accordingly, we took the mean of the observations at high temperature and assigned a temperature of 77.5° C to the averaged values for the purpose of data analysis. This procedure led to the estimate of 24° for the bend at 14° C.

3. DISCUSSION

The initial application of the rotational variant method measured the bend and duplex twist angles induced when catabolite gene activator protein binds to its recognition site (Lutter *et al.*, 1996). We have employed the principles of this approach in the current work to measure the bend angle of an A6 tract. The method compares the topologies of plasmid DNA in which the site of interest is in a straight state versus that in which it is bent. Since the A-tract is always bent under typical solution conditions, a condition needed to be identified under which the A-tract is straight. This was achieved by performing the relaxation at elevated temperature, a condition that has been reported to melt the A-tract bend (Diekmann, 1987; Chan *et al.*, 1993; Diekmann and Wang, 1985). This modification in turn required the use of a hyperthermophile topoisomerase. The linking number difference at two different high temperatures (70° C and 85° C) was found to be very close to constant, indicating the A-tract is straight at both temperatures. An average of these linking number values was then used as the straight state. For each rotational variant construct, the linking number difference between this high temperature straight state and the 14° C bent state was measured and plotted versus repeat length, Figure 1.

The function relating linking number change to repeat length (Lutter *et al.*, 1996; Lutter *et al.*, 2000) was modified as described in Section 4 and fitted to this data set. The temperature coefficient obtained from this fit (–0.0117 ± .0003 degrees/bp•°C) agrees with prior determinations (Pulleyblank *et al.*, 1975; Depew and Wang, 1975; Shore and Baldwin, 1983). The value of 105 (range of 50-180) pN•nm^2 obtained for the torsional stiffness *C* also falls within the reported range for this value (Hagerman, 1988; Horowitz and Wang, 1984; Shore and Baldwin, 1983; Barkley and Zimm, 1979). The value obtained for the DNA screw of the repeat unit (10.38 ± 0.01 bp/turn) is within the range determined for various DNA sequences, including those containing A-tracts (Wang, 1979; Rhodes and Klug, 1980; Rhodes and Klug, 1981; Drak and Crothers, 1991).

An alternative method that has been used to measure the solution bend of the A-tract is ligase-catalysed circularization. We have used here the same poly-A-tract sequence that was employed by Koo et al. (Koo *et al.*, 1990) in their ligase-catalysed cyclization analysis. They reported a bend angle of 17°-21° in that study, which is significantly less than our value of 24°±2°. However, our measurement was performed at 14° C, while their study was performed at 22° C. We have recently analysed the temperature-dependence of the bend using our topological method, and the preliminary result is that the bend angle at 22° C is about 21°, a value consistent with that found by the cyclization study (Koo *et al.*, 1990).

4. METHODS

4.1 Data processing: Relating DNA geometry to linking number

The basic theory of this method has been presented previously (Lutter *et al.*, 1996; Lutter *et al.*, 2000). The experimental data require that the rigid body derivation be convolved with a distribution function for torsional fluctuations, which we did by numerical (Monte Carlo) methods. For the work reported here the convolution was performed analytically, speeding up the calculations and making it possible to consider more involved problems.

The starting point is the angle θ of Eq 8 in (Lutter *et al.*, 1996). It is more convenient to express this in terms of ϕ, the angle relating two bends ($\phi = 2\theta$). The angle ϕ can encompass a large number of duplex turns but it is only the part representing a fractional turn whose contribution is of interest, so we introduce ϕ^*, a positive angle less than one complete turn: $\phi^* = \phi - 2\pi \lfloor \phi / 2\pi \rfloor$, where $\lfloor x \rfloor$ ("floor(x)"') is the nearest integer to the left of x on the number line. In words, ϕ^* repetitively goes from 0 to 2 π as ϕ increases indefinitely. This has the advantage of removing the cumbersome "sign(x)"' function from Eq 8 of (Lutter *et al.*, 1996).

Analytic convolution of ϕ^* with a Gaussian distribution of fluctuations leads to a simple result from a Lebesque integral. The variable ϕ is replaced by ϕ_{av} where

$$\phi_{av} = \phi^* + (1/2)\left[erf\left((2\pi - \phi^*)/\varsigma\sqrt{2}\right) - erf\left(\phi^*/\varsigma\sqrt{2}\right)\right]$$

s is the standard deviation of the Gaussian distribution, and "erf( )" is the error function. This is a two-term truncation of an infinite series. It fails to be adequate when *s* becomes so large that there is an appreciable probability of a random fluctuation of one complete turn. In such an event the smearing is so severe that there is nothing to analyse. Further, it would take an unrealistic value of σ_T ($\approx 36^o$) to create such a situation.

As a result, Eq 8 of (Lutter *et al.*, 1996) can be replaced with

$$\Delta L = a\cos[\cos(B/2)\cos(\phi_{av}/2)]/\pi - \phi_{av}/\pi$$

where the convolution has been performed in the evaluation of ϕ_{av}.

4.2 Analysis

The data were analysed by non-linear least squares with a modified Marquardt-Levenberg algorithm. The analysis includes consideration of the correlation among the estimates of the parameters in evaluating the confidence region. The magnitude of the confidence region corresponds to a probability of 65% that this region contains the true value of the parameter; for a linear problem this would be plus or minus one standard deviation. Confidence regions for non-linear problems are asymmetric, but if they are not markedly so the results are given as plus or minus 1/2 the width of the region.

The parameters were expressed as *b*, the bend angle for one A-tract, S_o, the helical repeat or duplex screw (bp/turn) under reference conditions (14° C), and *C* the torsional stiffness of the DNA. The temperature difference experiment has one additional parameter, the unwinding of duplex DNA per kelvin, for a total of four adjustable parameters.

References

Barkley, M.D. and Zimm, B.H. (1979) Theory of twisting and bending of chain macromolecules; analysis of the fluorescence depolarization of DNA. *J.Chem.Phys.* **70,** 2991-3007.

Bracco, L., Kotlarz, D., Kolb,A., Diekmann, S., and Buc, H. (1989) Synthetic curved dna sequences can act as transcriptional activators in escherichia coli. *EMBO Journal* **8,** 4289-4296.

Calladine, C.R., Drew, H.R., and McCall, M.J. (1988) The intrinsic curvature of dna in solution. *Journal of Molecular Biology* **201,** 127-137.

Chan, S.S., Breslauer, K.J., Austin, R.H., and Hogan, M.E. (1993) Thermodynamics and premelting conformational changes of phased (da)5 tracts. *Biochemistry* **32,** 11776-11784.

Crothers, D.M. and Shakked, Z. (1999) DNA Bending by adenine-thymine tracts. In *Oxford handbook of nucleic acid structure* (Edited by Neidle, S.) Pp. 455-469. Oxford University Press, Oxford, New York.

Depew, D.E. and Wang, J.C. (1975) Conformational fluctuations of DNA helix. *Proc. Natl. Acad. Sci. USA* **72,** 4275-4279.

Diekmann, S. (1987) Temperature and salt dependence of the gel migration anomaly of curved dna fragments. *Nucleic Acids Research* **15,** 247-265.

Diekmann, S. and Wang, J.C. (1985) On the sequence determinants and flexibility of the kinetoplast DNA fragment with abnormal gel electrophoretic mobilities. *J. Mol. Biol.* **186,** 1-11.

Drabik, C.E., Nicita, C.A., and Lutter, L.C. (1997) Measurement of the linking number change in transcribing chromatin. *J. Mol. Biol.* **267,** 794-806.

Drak, J. and Crothers, D.M. (1991) Helical repeat and chirality effects on DNA gel electrophoretic mobility. *Proc. Natl. Acad. Sci. USA* **88,** 3074-3078.

Gartenberg, M.R., Ampe, C., Steitz, T.A., and Crothers, D.M. (1990) Molecular characterization of the GCN4-DNA complex. *Proc. Natl. Acad. Sci. USA* **87,** 6034-6038.

Gartenberg, M.R. and Crothers, D.M. (1991) Synthetic DNA bending sequences increase the rate of in vitro transcription initiation at the Escherichia coli lac promoter. *J.Mol.Biol.* **219,** 217-230.

Goodman, S.D. and Nash, H.A. (1989) Functional replacement of a protein-induced bend in a dna recombination site. *Nature* **341,** 251-254.

Goodman, S.D., Nicholson, S.C., and Nash, H.A. (1992) Deformation of dna during site-specific recombination of bacteriophage lambda: replacement of ihf protein by hu protein or sequence-directed bends. *Proceedings of the National Academy of Sciences of the United States of America* **89,** 11910-11914.

Hagerman, P.J. (1988) Flexibility of DNA. *Annual Review of Biophysics & Biophysical Chemistry* **17,** 265-286.

Hagerman, P.J. (1990) Sequence-directed curvature of DNA. *Annual Review of Biochemistry* **59,** 755-781.

Hagerman, P.J. (1992) Straightening out the bends in curved DNA. *Biochim. Biophys. Acta* **1131,** 125-132.

Harrington, R.E. (1993) Studies of dna bending and flexibility using gel electrophoresis. *Electrophoresis* **14,** 732-746.

Harvey, S.C., Dlakic, M., Griffith, J., Harrington, R., Park, K., Sprous, D., and Zacharias, W. (1995) What is the basis of sequence-directed curvature in DNAs containing A tracts?. *Journal of Biomolecular Structure & Dynamics* **13,** 301-307.

Horowitz, D.S. and Wang, J.C. (1984) Torsional rigidity of DNA and length dependence of the free energy of DNA supercoiling. *J. Mol. Biol.* **173,** 75-91.

Keller, W. (1975) Determination of the number of superhelical turns in simian virus 40 DNA by gel electrophoresis. *Proc. Natl. Acad. Sci. USA* **72,** 4876-4880.

Kerppola, T.K. (1996) Fos and Jun bend the AP-1 site: effects of probe geometry on the detection of protein-induced DNA bending. *Proceedings of the National Academy of Sciences of the United States of America* **93,** 10117-10122.

Kerppola, T.K. (1997) Comparison of DNA bending by Fos-Jun and phased A tracts by multifactorial phasing analysis. *Biochemistry* **36,** 10872-10884.

Kerppola, T.K. and Curran, T. (1991a) DNA bending by Fos and Jun: the flexible hinge model. *Science* **254,** 1210-1214.

Kerppola, T.K. and Curran, T. (1991b) Fos-Jun heterodimers and Jun homodimers bend DNA in opposite orientations: implications for transcription factor cooperativity. *Cell* **66,** 317-326.

Koo, H.S., Drak, J., Rice, J.A., and Crothers, D.M. (1990) Determination of the extent of DNA bending by an adenine-thymine tract. *Biochemistry* **29,** 4227-4234.

Lutter, L.C., Drabik, C.E., and Halvorson, H.R. (2000) Use of topology to measure protein-induced DNA bend and unwinding angles. In *Protein - DNA Interactions : A Practical Approach* (Edited by Travers, A. and Buckle, M.) Pp. 47-64. Oxford University Press, Oxford.

Lutter ,L.C., Halvorson, H.R., and Calladine, C.R. (1996) Topological measurement of protein-induced DNA bend angles. *J. Mol. Biol.* **261,** 620-633.

Pazin, M.J., Bhargava, P., Geiduschek, E.P., and Kadonaga, J.T. (1997) Nucleosome mobility and the maintenance of nucleosome positioning. *Science* **276,** 809-812.

Perez-Martin, J., Rojo, F., and de Lorenzo, V. (1994) Promoters responsive to DNA bending: a common theme in prokaryotic gene expression. *Microbiological Reviews* **58,** 268-290.

Plaskon, R.R. and Wartell, R.M. (1987) Sequence distributions associated with dna curvature are found upstream of strong e. coli promoters. *Nucleic Acids Research* **15,** 785-796.

Pulleyblank, D.E., Shure, M., Tang, D., Vinograd J., and Vosberg, H.P. (1975) Action of nicking-closing enzyme on supercoiled and nonsupercoiled closed circular DNA: formation of a Boltzmann distribution of topological isomers. *Proc. Natl. Acad. Sci. USA* **72,** 4280-4284.

Radlinska, M., Drabik, C.E., Tong, W.S., and Lutter, L.C. (2001) Generating Tandem Repeats by Cloning with Double Initiator Fragments. *Biotechniques* **31,** 340-347.

Rhodes, D. and Klug, A. (1980) Helical periodicity of DNA determined by enzyme digestion. *Nature* **286,** 573-578.

Rhodes, D. and Klug, A. (1981) Sequence-dependent helical periodicity of DNA. *Nature* **292,** 378-380.

Shore, D. and Baldwin, R.L. (1983) Energetics of DNA twisting II. Topoisomer analysis. *J. Mol. Biol.* **170,** 983-1007.

Sitlani, A. and Crothers, D.M. (1996) Fos and jun do not bend the ap-1 recognition site. *Proceedings of the National Academy of Sciences of the United States of America* **93,** 3248-3252.

Strauss, J.K. and Maher,L.J.3. (1994) Dna bending by asymmetric phosphate neutralization. *Science* **266,** 1829-1834.

Thompson, J.F. and Landy, A. (1988) Empirical estimation of protein-induced DNA bending angles: applications to lambda site-specific recombination complexes. *Nucleic Acids Res.* **16,** 9687-9705.

Wang, J.C. (1979) Helical repeat of DNA in solution. *Proc. Natl. Acad. Sci. USA* **76,** 200-203.

Zinkel, S.S. and Crothers, D.M. (1990) Comparative gel electrophoresis measurement of the DNA bend angle induced by the catabolite activator protein. *Biopolymers* **29,** 29-38.

STRUCTURAL INSIGHT INTO THE MECHANISM OF SUPERCOILING OF THE BACTERIAL FLAGELLAR FILAMENT

Keiichi Namba
Protonic NanoMachine Project, ERATO, JST
& Graduate School of Frontier Biosciences, Osaka University
3-4 Hikaridai, Seika, Kyoto 619-0237 Japan
keiichi@fbs.osaka-u.ac.jp

Abstract The bacterial flagellar filament is a helical propeller. Its tubular structure is made of 11 protofilaments of a single protein, flagellin, and yet, the filament switches between left- and right-handed supercoiled forms to allow bacteria to switch their swimming mode between "run" and "tumble". The mechanism of supercoiling and switching has been shown to involve an axial sliding switch between neighboring protofilaments coupled with a highly precise lengthwise switch of individual protofilament. The structure of the two types of straight filaments and a crystal structure of the protofilament are described in detail, which explains how flagellin can build such a dynamic structure.

Key words: Flagellin, Macromolecular complex, Supercoiling, Molecular Switch

1. INTRODUCTION

Bacteria swim by rotating helical flagellar filaments, which grow to 10 μm or longer, but are only about 20 nm in diameter. The rotary motor at the base of the filament drives the rotation of this helical propeller (Berg and Anderson, 1973; Silverman and Simon, 1974) at hundreds of revolutions per second (Kudo et al., 1990; Ryu et al, 2000). For chemotaxis and thermotaxis, the swimming pattern of bacteria such as *Salmonella* and *E. coli* alternates between "run" and "tumble"; a "run" lasts for a few seconds and a "tumble" for a fraction of second. During a "run", the motor rotates counterclockwise (as it is viewed from outside the cell), and several flagellar filaments having a left-handed helical shape form a bundle and propel the cell. A "tumble" is caused by quick reversal of the motor to the clockwise rotation (Larsen et al., 1974). The twisting force produced by quick reversal of the motor

H.R. Drew and S. Pellegrino (eds.),
New Approaches to Structural Mechanics, Shells and Biological Structures, 485–497.

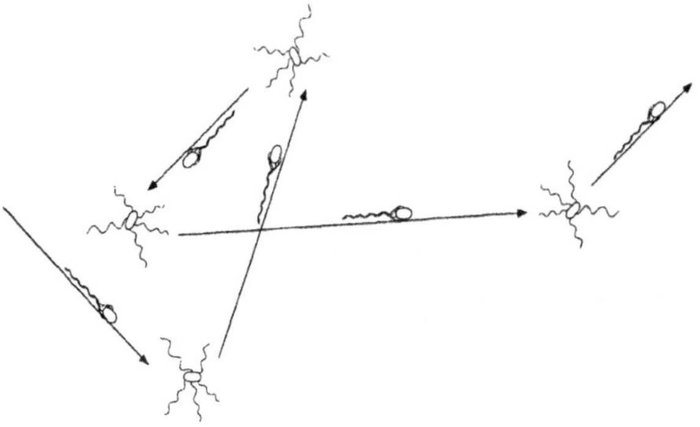

Figure 1. Swimming pattern of bacteria with "run" and "tumble".

transforms the left-handed helical form of the filament into right-handed ones (Macnab and Ornston, 1977; Turner et al., 2000), which causes the bundle to fall apart rapidly. The separated filaments generate forces in an uncoordinated way, which changes the orientation of the cell. Thus, the structure of the flagellar filament and its dynamic properties play an essential role in the tactic behaviour of bacteria, Figure 1.

The filament is a helical assembly of a single protein, flagellin, with approximately 11 subunits per two turns of 1-start helix. The filament structure can also be described as a tubular structure composed of 11 strands of protofilaments, which are nearly longitudinal helical arrays of subunits (O'Brien and Bennett, 1972), Figure 2. The helical forms of the filament are explained by supercoiling, the mechanism of which has been proposed assuming a switching of conformation or packing interaction of the subunits

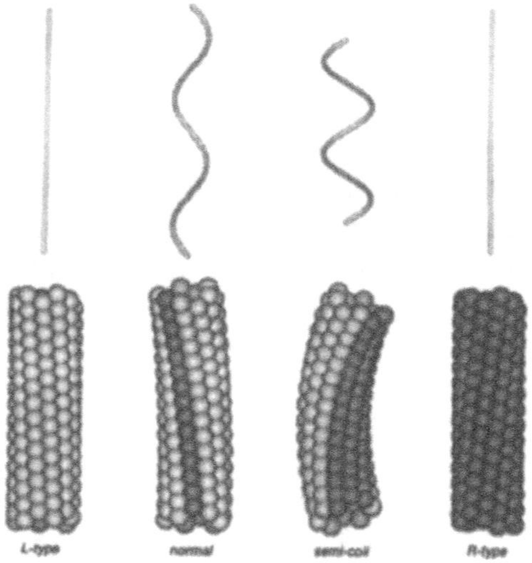

Figure 2. Model of the flagellar supercoiling. Different colours of protofilaments represent different states.

between two distinct states and their non- and quasi-equivalent intersubunit interactions (Asakura, 1970; Calladine, 1975, 1976, 1978). Asakura (1970) proposed a two-state protofilament model to explain how chemically identical molecules could build tubular structures that have different curvature and twist to be a switchable helical propeller. The model was based on the following assumptions: each protofilament is a cooperative switching unit that can take either one of the two different subunit repeat distances; each of neighbouring pair of protofilaments can take either one of the two lateral dispositions; and these two switches, one in the repeat distance and the other in the lateral disposition, are tightly coupled with each other. These two switches were required for producing the curvature and twist, respectively, of helical forms of the filament tube, and the coupling of these two switches was necessary to produce a limited number of well-defined helical forms. A few years later, Calladine (1975, 1976, 1978) proposed a more elaborate model in which the two switches were coupled in a nice and simple way, and this model was able to reproduce the observed helical forms quantitatively. The design of this model was elegant in the way the two switches were coupled, where a single action changed the axial and lateral interactions of subunits at once and the axial cooperativity along each protofilament was incorporated as well.

Apart from these elaborate designs of the filament model, the essence of the mechanism of supercoiling is depicted in Figure 2. All these models predict the presence of two straight forms, one with a left-handed twist of the protofilament bundle and the other with a right-handed twist, and the two distinct straight forms were actually identified under various solution conditions and in the filaments isolated from mutant strains of *Salmonella* (Kamiya et al., 1979). They were named L-type and R-type, respectively, according to their helical handedness. Two straight filaments isolated from *Salmonella typhimurium* mutant strains, SJW1660 and SJW1655 (Kamiya et al., 1980), were particularly suitable for structure analysis, because they were stable. When monomeric flagellins from the two strains were mixed in solution at a various ratio for copolymerization, the filament produced had well-defined helical forms depending on the mixing ratio. This indicated that the two states of subunit interactions in the two straight filaments represent the two that coexist in supercoiled filaments. We therefore focused on these two straight filament structures and carried out the structure analysis by electron cryomicroscopy, X-ray fiber diffraction and X-ray crystallography.

2. DOMAIN ORGANIZATION OF FLAGELLIN

By electron cryomicroscopy and helical image reconstruction the three-dimensional structure of the straight filaments were analyzed at around 10 Å resolution, and the overall shape of the flagellin subunit and its domain organization was revealed in both L- and R-type straight filaments (Mimori et al., 1995; Morgan et al., 1995; Mimori-Kiyosue et al., 1996). The density maps of the two filaments are compared in Figure 3a in solid surface representation. In the cross-section of the filament, domains of flagellin are tightly packed to form the core of the filament, which has an outer radius of about 60 Å and a central channel with a diameter of 30 Å. Two domains are identified in this core region, where the two domains form a coaxial double tubular structure that has axially continuous densities, Figure 3b. These two tubular structures were named the inner tube and outer tube, respectively. On the outer surface of the filament core, each subunit shows up as a projection composed of two domains. Thus, the flagellin subunit has four domains, and they were named D0, D1, D2 and D3 in the order of their radial positions from the filament axis. Obviously, interactions between domains D0 and D1 of neighboring subunits are responsible for the design and stability of the filament structure.

There are no significant differences between the L-type and R-type filaments either in the shape of the subunits or in their orientation within each protofilament, suggesting that the protofilament structure would go through only a subtle change in its conformation during the switching between the two states. The difference was so subtle that it was not possible

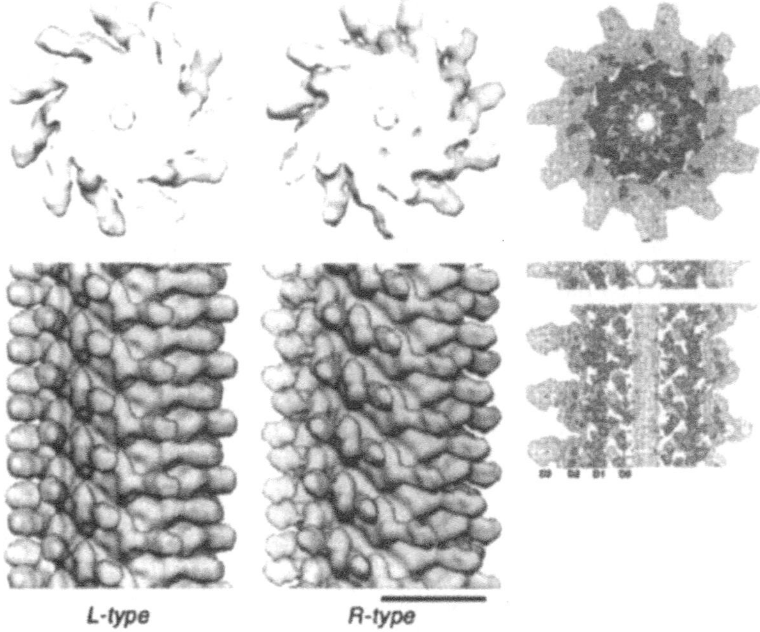

Figure 3. Comparison of the L-type and R-type filament, and contoured density map. Scale bar, 100 Å.

to see the structural changes involved at this resolution. Obviously, much higher resolution was necessary to actually see the changes.

We found that both terminal regions are unfolded in monomeric flagellin in solution (Vonderviszt et al., 1989). Of 494 amino acid residues of flagellin from *Salmonella typhimurium* SJW1103 strain, about 65 N-terminal and 45 C-terminal residues are disordered in the monomeric form (Vonderviszt et al., 1989), but these regions become folded upon polymerization into the filament structure (Aizawa et al., 1990). It was therefore relatively easy to prepare terminally truncated fragments of flagellin missing various lengths of either or both terminal regions by limited proteolysis with enzymes of various specificities. These fragments still polymerized in solutions of high ionic strength, but into straight filaments (Vonderviszt et al., 1991), and the structures analyzed by electron cryomicroscopy revealed that the unfolded terminal regions form domain D0 (Mimori-Kiyosue et al., 1997). When the fragments were polymerized onto the helical filaments from wild-type strain, however, they polymerized into a helical form called "coil" (Vonderviszt et al., 1991). This indicated that the interactions between domains D1 of neighboring subunits alone are responsible for the switching involved in the supercoiling. Roles of interactions between domains D0 were interpreted to simply stabilize the filament structure, because the more terminal residues were removed, the higher ionic strength solution was required to stabilize the filament.

3. TWO CYLIINDRICAL SURFACE LATTICES FORMED BY DOMAIN

To detect the structural differences between the L-type and R-type straight filaments in further detail, we analyzed the structures of the two filaments by X-ray fiber diffraction (Yamashita et al., 1998). The X-ray fiber diffraction method has a potential to resolve the filament structure with helical symmetries in atomic detail provided that highly well oriented liquid crystalline sols of the filament can be prepared. We devised a method to prepare such specimen by stepwise application of slow centrifugation and magnetic orientation, and successfully obtained oriented sols with disorientation angle as small as 0.6 degree (Yamashita et al., 1998). We measured layer line positions in their X-ray diffraction patterns precisely by using a two-dimensional profile fitting procedure and deduced the cell dimension of the cylindrical surface lattice at a radius of 45 Å, which corresponds to the position of the outer-tube domain, D1. The repeat distance along the L-type and R-type protofilament are 52.7 Å and 51.9 Å, respectively, Figure 4. And, this difference is coupled with a mutual sliding shift between neighboring protofilaments by 2.6 Å, as measured for the two neighboring subunits along the left-handed 5-start helix, which is the major lateral contact. The difference in the subunit repeat distance along the protofilament is responsible for the curvature formation, and the mutual axial shift of protofilaments at their lateral contact is for producing various twist angle of the bundle of 11 protofilaments.

Following the mathematical procedure worked out by Calladine (1996), we used these lattice parameters to simulate the shape of supercoiled filaments by mixing the L-type and R-type lattices (Yamashita et al., 1998) at every possible number ratio in building the 11-stranded protofilament tube. This experiment was to confirm that even such a subtle change as 0.8 Å is used for the actual design of this nano-scale actuator. Comparison with the shape of observed helical flagellar filaments (Kamiya and Asakura, 1976, 1977) showed a very nice fit between calculated and observed curvature and twist, indicating that these structural parameters and the model of supercoiling are both physically sound, Figure 5. This also indicated that the level of precision at which flagellin works as a lengthwise mechanical switch is on the order of a tenth of angstrom.

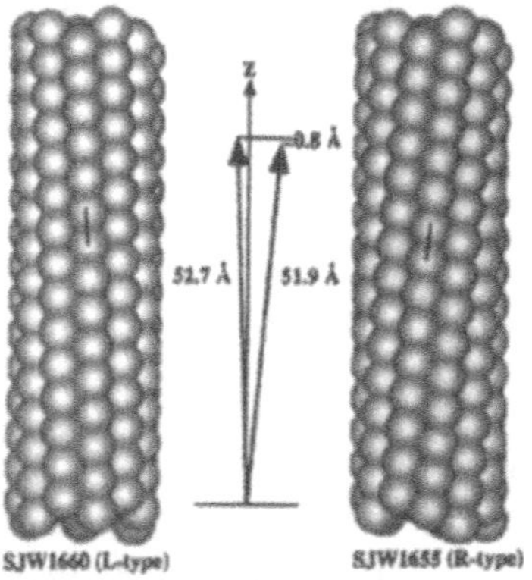

Figure 4. Comparison of the protofilament repeat distance between L-type and R-type.

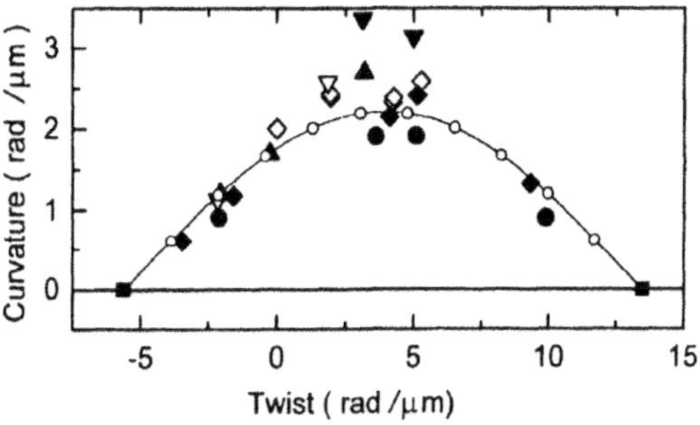

Figure 5. Twist-Curvature plot for simulated and observed supercoils. Open circles are those calculated and all the other symbols are for those observed, and solid squares at zero curvature represent L-type (negative twist) and R-type (positive twist).

4. ATOMIC STRUCTURE Of THE PROTOFILAMENT

We then tried to solve the atomic structure of flagellin by X-ray crystallography, hoping to find clues to the conformational changes responsible for these relatively small changes of lattice parameters. Crystallization of flagellin, however, turned out to be impossible, because under any crystallization conditions, which reduce the protein solubility, flagellin polymerized into filaments. It did not take long to realize that flagellin is a protein molecule designed to polymerize. But, by truncating unfolded terminal regions of flagellin, we were able to suppress its strong polymerization ability. We prepared a fragment F41, which was named after its apparent molecular mass of 41 kDa, by removing 52 N-terminal and 44 C-terminal residues and found a solution condition in which crystals grew to a size sufficient for crystallographic data collection.

Another difficulty with this F41 crystal was its thickness; it was only about 5 μm thick. Frozen crystals prepared for cryocrystallographic data collection did not diffract well because these extremely thin crystals were easily bent during the freezing process. However, annealing of the bent crystals made them flatten nicely, and these flattened crystals diffracted beyond 2.0 Å resolution (Samatey et al., 2000). The third generation synchrotron radiation facilities, such as SPring-8 or ESRF, which started their operation for diffraction data collection just about the time, was essential for collecting high-resolution data from such thin crystals. After collecting many data sets at SPring-8 and ESRF, which we used to refine the methods of heavy atom derivative preparation and data collection, we finally obtained an electron density map at 2.0 Å resolution from a set of multiple wavelength anomalous diffraction data collected at SPring-8 and built an atomic model of F41 (Samatey et al., 2001).

The folding of the Cα backbone of F41 is shown in Figure 6. F41 consists of three domains, in which a vertically elongated domain has both termini and is made of three α-helices and a β-hairpin, and the other two domains are made mostly of β-strands and β-hairpins. One of the four domains of flagellin, domain D0, is completely missing in this model because the truncated terminal regions construct this domain. The most interesting finding in this crystal structure was, as shown in the crystal packing arrangement, Figure 7a, that its a-axis had a repeat distance of 51.9 Å, which is exactly the same as that of the R-type protofilament. The a-axis

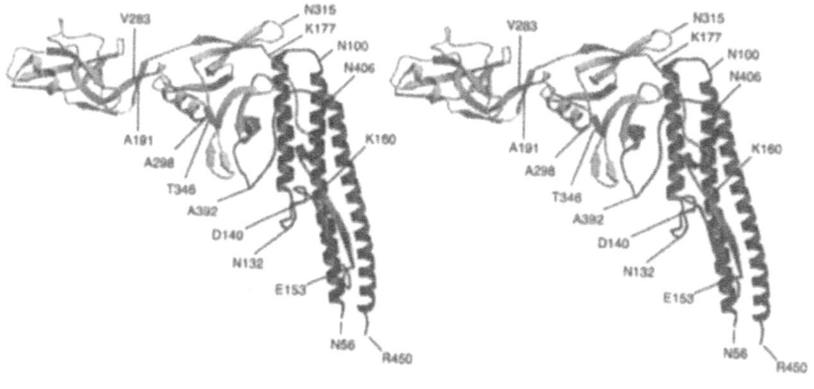

Figure 6. Cα backbone model of F41 in stereo view.

array of F41 molecules looked exactly like the protofilament revealed in the filament density map deduced by electron cryomicroscopy, except that it was perfectly straight, not even gently twisted as in the filament. We docked this a-axis array of F41 molecules into the filament density map, and the fit was nearly perfect at least for a few consecutive subunits because the twist of the protofilament is relatively small even in the R-type filament, Figure 7b. This proved that this a-axis array was the protofilament, and domains D1, D2 and D3 were assigned to the atomic model with no ambiguity. In the crystal, the protofilament is straight and packed antiparallel to form a layer of two-dimensional sheet, and many layers of these sheets are stacked to form the three-dimensional crystal. Domain D1 is the only one in axial contact with neighbors along the protofilament. This meant that we had a unique opportunity to study the axial interactions of subunits that are responsible for building the protofilament structure and its switching between the two states having different repeat distances.

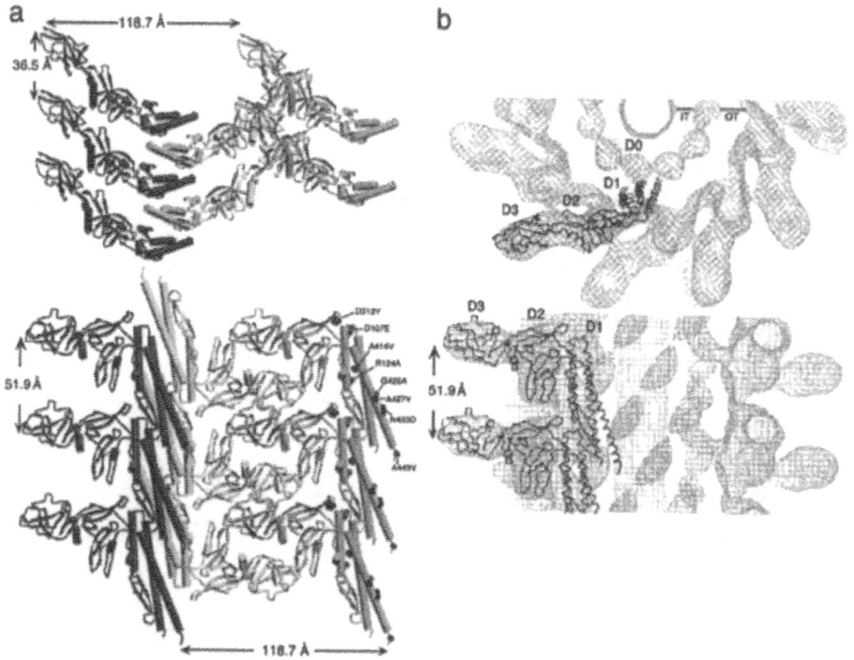

Figure 7. Crystal packing of F41 and docking of a protofilament into a filament density map. In (a), upper panel is viewed along the a-axis; lower panel, along the b-axis.

5. SIMULATED EXTENSION OF THE PROTOFILAMENT MODEL

Since the protofilament is an independent mechanical switch and what we identified in the crystal structure was the R-type protofilament, a model of the protofilament with a shorter repeat distance, an immediate anticipation was that we should be able to see how the model goes through a structural transition from this R-type to the L-type state with a longer repeat distance by forced extension. We therefore took three consecutive F41 molecules from the a-axis array in the crystal and treated it as a model of the protofilament, Figure 8a. We fixed the Cα atoms of an F41 subunit at one end (the upper in Figure 8a), displaced a subunit at the other end (the lower in Figure 8a) to extend the protofilament, then fixed the Cα atoms of the lower subunit as well as those of the upper, and carried out energy minimization of the whole model to see what would occur in the subunit in between. The distance of displacement in each step was 0.1 Å, and we repeated this process up to 60 steps, which corresponds to an extension up to 6.0 Å.

By extension up to 4.5 Å, every portion of the middle subunit was gradually stretched as typically displayed in the elongation of the α-helical pitch in domain D1, indicating that the protein molecule is relatively flexible and elastic as well. Then, suddenly after this point, by further extension from 4.5 Å to 4.7 Å, the β-hairpin of domain D1 showed an abrupt transition in its conformation in a direction by which the accumulated strain was released, Figure 8b. Small movements of the terminal α-helices accompanied this

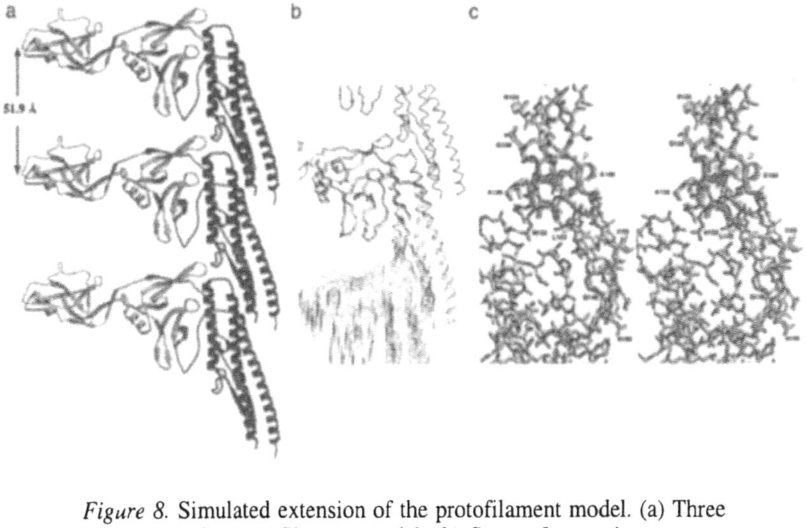

Figure 8. Simulated extension of the protofilament model. (a) Three subunits as the protofilament model. (b) Steps of extension at every 0.5 Å superimposed. (c) Stereo view of the β-hairpin in two conformations at extension of 4.5 Å and 4.7 Å superimposed.

conformational transition. Further extension caused further gradual distortion of the middle subunit as observed before the transition. The two conformations before and after the transition are superimposed for direct comparison in Figure 8c. The characteristic nature of this conformational switching is a small but significant movement of the β-hairpin that slightly pushes out and down the lower subunit at the axial interface, making the axial subunit dispositions with a slightly longer repeat distance stable. This result strongly suggested that the conformational switching of the β-hairpin is responsible for the two distinct states of flagellin packing with two distinct repeat distances at an accuracy of sub-Å.

6. COUPLING MECHANISM OF TWIST AND CURVATURE

We thus identified the structural motif responsible for the lengthwise mechanical switch of the flagellar protofilament. The reason that the β-hairpin in domain D1 can switch its conformation by mechanical strains appears to be in the structural design of this domain, which is comprised of three α-helices paired with a β-hairpin. Because of this unique design, the packing interactions of side chains in the hydrophobic core have a local irregularity near the base of the β-hairpin, thereby the hydrophobic core has a small gap that accommodates a water molecule. Probably because of this irregularity in the side-chain packing interactions, domain D1 of flagellin cannot behave as a single rigid domain as most protein domains do.

We still need to find the structural mechanism that tightly couples this lengthwise switch with the axial shift switch at the lateral interface of neighboring protofilaments. This coupling mechanism is essential for driving the sequential process of the polymorphic transitions that occur during the bacterial swimming with "run" and "tumble". When the motor reverses it rotation quickly, the long helical propeller cannot follow such abrupt change of rotation, and therefore a twisting force is accumulated at the base of the filament. This twisting force is then converted into a shear force between neighboring protofilaments, which drives the axial shift switch at lateral interfaces of a few consecutive neighbors. Here, the coupling mechanism must drive the lengthwise switch of the protofilament on one side of each of these lateral interfaces. We should be able to see how this coupling works in the high-resolution structures of the L-type and R-type filaments, for which the structure analyses by electron cryomicroscopy and X-ray fiber diffraction are underway. Also, by building the atomic models of the two filaments based on currently available structural data and the F41 model, we might be able to carry out a molecular dynamics simulation to see what structural changes would occur by twisting the filament tube.

In any case, we will have to visualize the atomic detail of what goes on during the switching. A simple and elegant mechanism of the twist-curvature coupling proposed by Calladine (1975, 1976, 1978), with just a single action of subunit movement at a diagonal corner of intersubunit interface, was apparently not used by nature. This mechanism required so called contra-rotation in the cylindrical plane, which appears to be not possible because of the tight packing of axially elongated domain D1. However, it is worth noting that a local conformational switching within a domain was pointed out as a possibility in one of his models (1978), and it is actually used for the structural design of switching mechanism. A model we proposed previously (Yamashita et al., 1998), based on the precise measurement of the structural parameters, careful analysis of the structures at around 10 Å resolution, and a reasonable assumption that protein domains mostly behave as a rigid entity, turned out to be wrong, too. We now know that domain D1 is not a rigid unit. Without actually looking at the structure at atomic detail, it was not possible to predict the presence of a structural motif having a local and small conformational switching ability that realizes the lengthwise mechanical switch within the domain. It seems, in a way, that nature takes any structural designs available and often does not spend extra time to consider what is the most elegant and simple way of achieving the goal. But, it is also possible that the current structural designs of flagellin and the flagellar filament are the best possible ones that realize the rigidity, stability, flexibility, and dynamic switching ability all at once for its function as a switchable propeller.

Acknowledgements

I would like to thank Chris Calladine, Ritsu Kamiya, Sho Asakura and Fumio Oosawa for their continuous interest, support, encouragement and stimulating discussion over the years.

References

Aizawa, S.-I., Vonderviszt, F., Ishima, R. and Akasaka, K. (1990) Termini of Salmonella flagellin are disordered and become organized upon polymerization into flagellar filament. *J. Mol. Biol.* **211**, 673-677.

Asakura, S. (1970) Polymerization of flagellin and polymorphism of flagella. *Advan. Biophys.* **1**, 99-155.

Berg, H. C. and Anderson, R. A. (1973) Bacteria swim by rotating their flagellar filaments. *Nature* **245**, 380-382.

Calladine, C. R. (1975) Construction of bacterial flagella. *Nature* **225**, 121-124.

Calladine, C. R. (1976) Design requirements for the construction of bacterial flagella. *J. Theoret. Biol.* **57**, 469-489.

Calladine, C. R. (1978) Change of waveform in bacterial flagella: The role of mechanics at the molecular level. *J. Mol. Biol.* **118**, 457-479.

Kamiya, R. and Asakura, S. (1976). Helical transformations of Salmonella flagella in vitro. *J. Mol. Biol.* **106**, 167-186

Kamiya, R. and Asakura, S. (1977) Flagellar transformations at alkaline pH. *J. Mol. Biol.* **108**, 513-518.

Kamiya, R., Asakura, S., Wakabayashi, K. and Namba, K. (1979) Transition of bacterial flagella from helical to straight forms with different subunit arrangements. *J. Mol. Biol.* **131**, 725-742.

Kamiya, R., Asakura, S. & Yamaguchi, S. (1980) Formation of helical filaments by copolymerization of two types of ' straight' flagellins. *Nature* **286**, 628-630.

Kudo, S., Magariyama, Y. and Aizawa, S.-I. (1990). Abrupt changes in flagellar rotation observed by laser dark-field microscopy. *Nature* **346**, 677-680

Larsen, S. H., Reader, R. W., Kort, E. N., Tso, W. W. and Adler, J. (1974). Change in direction of flagellar rotation is the basis of the chemotactic response in Escherichia coli. *Nature* **249**, 74-77

Macnab, R. M. and Ornston, M. K. (1977) Normal-to-curly flagellar transitions and their role in bacterial tumbling. Stabilization of an alternative quaternary structure by mechanical force. *J. Mol. Biol.* **112**, 1-30.

Mimori, Y., Yamashita, I., Murata, K., Fujiyoshi, Y., Yonekura, K., Toyoshima, C. and Namba, K. (1995) The structure of the R-type straight flagellar filament of Salmonella at 9 Å resolution by electron cryomicroscopy. *J. Mol. Biol.* **249**, 69-87.

Mimori-Kiyosue, Y., Vonderviszt, F., Yamashita, I., Fujiyoshi, Y. and Namba, K. (1996) Direct interaction of flagellin termini essential for polymorphic ability of flagellar filament. *Proc. Nat'l. Acad. Sci. USA* **93**, 15108-15113.

Mimori-Kiyosue, Y., Vonderviszt, F. and Namba, K. (1997) Locations of terminal segments of flagellin in the filament structure and their roles in polymerization and polymorphism. *J. Mol. Biol.* **270**, 222-237.

Morgan, D. G., Owen, C., Melanson, L. A. and DeRosier, D. J. (1995) Structure of bacterial flagellar filaments at 11 Å resolution: Packing of the α-helices. *J. Mol. Biol.* **249**, 88-110.

O'Brien, E. J. and Bennett, P. M. (1972) Structure of straight flagella from a mutant Salmonella. *J. Mol. Biol.* **70**, 133-152.

Ryu, W. S., Berry. R. M. and Berg, H. C. (2000) Torque-generating units of the flagellar motor of Escherichia coli have a high duty ratio. *Nature* **403**, 444-447.

Samatey, F.A., Imada, K., Vonderviszt, F., Shirakihara, Y. and Namba, K. (2000) Crystallization of the F41 Fragment of Flagellin and Data Collection from Extremely Thin Crystals. *J. Struct. Biol.* **132**, 106-111.

Samatey, F. A., Imada, K., Nagashima, S., Kumasaka, T., Yamamoto, M., Vonderviszt, F., and Namba, K. (2001) Structure of the bacterial flagellar protofilament and implication for the switch with sub-Å precision. *Nature* **410**, 331-337.

Silverman, M. and Simon, M. (1974) Flagellar rotation and the mechanism of bacterial motility. *Nature* **249**, 73-74.

Turner, L., Ryu, W. S. and Berg, H. C. (2000) Real-time imaging of fluorescent flagellar filaments. *J. Bacteriol.* **182**, 2793-2801.

Vonderviszt, F., Kanto, S., Aizawa, S.-I. and Namba, K. (1989) Terminal region of flagellin are disordered in solution. *J. Mol. Biol.* **209**, 127-133.

Vonderviszt, F., Aizawa, S.-I. and Namba, K. (1991) Role of the disordered terminal regions of flagellin in filament formation and stability. J. Mol. Biol. **221**, 1461-1474.

Yamashita, I., Hasegawa, K., Suzuki, H., Vonderviszt, F., Mimori-Kiyosue, Y. & Namba, K. (1998) Structure and switching of bacterial flagellar filament studied by X-ray fiber diffraction. *Nature Struct. Biol.* **5**, 125-132.

Yamashita, I., Suzuki, H. and Namba, K. (1998) Multiple-Step Method for Making Exceptionally Well-Oriented Liquid-Crystalline Sols of Macromolecular Assemblies. *J. Mol. Biol.* **278**, 609-615.

COILS AND SUPERCOILS IN PROTEINS

Murray Stewart and A.D. McLachlan
MRC Laboratory of Molecular Biology,
Hills Road, Cambridge, CB2 2QH, U.K.
ms@mrc-lmb.cam.ac.uk, admcl@mrc-lmb.cam.ac.uk

Abstract Coiled coils of α-helices are an important structural motif in protein molecules. They act as structural components in fibrous proteins, as measuring rods, as elastic elements and as control switches. The most important type of control switch is the dimerization switch provided by two helices that can pair up or separate under prescribed conditions, as in DNA gene transcription factors.

Keywords: Coiled coils, Protein structure, Dimerization switch

1. PROTEINS AND THEIR FUNCTIONS

Protein molecules are the working parts of all living organisms, and they have evolved the power to carry out an astonishing variety of tasks in the cell. Some proteins are simple passive structural components, in an engineering sense, which merely act as supporting frameworks, struts or separating surfaces. Others, the enzymes, with chemically active catalytic clusters of side chains, are used to carry out chemical reactions.

In the simplest types of enzyme a single substrate is recognised by a lock-and-key mechanism and then converted into a unique product. More elaborate enzyme systems can carry out a whole sequence of co-ordinated reactions in succession, to act as a miniature "production line". These systems are often built of many individual protein subunits and sometimes include DNA or RNA molecules. The most elaborate known production line is the ribosome, which synthesises new polypeptide chains under the genetic control of the messenger RNA.

Another class of proteins convert chemical energy into mechanical energy and produce controlled movements of protein domains. Many so-called "motor proteins" use energy-rich adenosine triphosphate (ATP) molecules as a fuel. When ATP is converted into adenosine diphosphate the local small side-chain movements of $2 - 5$Å at the active site cause

H.R. Drew and S. Pellegrino (eds.),
New Approaches to Structural Mechanics, Shells and Biological Structures, 499–511.

whole sections of protein to move as rigid bodies, in rolling or hinge bending motions, which finally lead to larger levered movements of 50 – 100Å. Here we use the Angstrom unit of length : $1Å = 10^{-10}$m. Examples of motor proteins are myosin in muscle, kinesin in microtubules and the rotating flagellar motor in bacteria (Samatey et al., 2001), whose molecular mechanism has been analysed thoroughly by Calladine (Calladine, 1978). Cell membranes contain other kinds of mechanically active protein, such as pumps and transporters, which use energy to move other molecules about. One of the simplest active membrane proteins is the proton pump bacteriorhodopsin, where the energy of sunlight, trapped by the conjugated double bonds of a buried retinal molecule, drives a hydrogen ion through a narrow channel between four α-helices.

This division of the protein world into inert structural components, chemical reactors, and mechanical engines is seriously incomplete, because some proteins have logical genetic control functions which are vital for the intelligent switching on and off of genes. These transcription factor proteins bind specifically to DNA or RNA in the promoter regions of genes and control the synthesis of the particular gene product. Transcription factors normally contain a recognition site, such as a short α-helix, which is able to recognise a specific piece of DNA or RNA. When this site binds to the promoter a new molecular complex is formed, which allows the transcription machinery to come into operation.

We shall see that helical coiled-coil structural elements play an essential part in proteins from each of these general classes. This is because these elements can combine the well-defined architectural features of a regular hydrogen-bonded framework with more subtle chemical and biological functions which are provided by networks of charged side chains in contact.

2. PROTEIN FAMILIES

A complete plant or animal may contain many tens of thousands of different kinds of protein molecules, which are each folded into their own characteristic native structure. But all the members of the terrestrial protein universe are evolutionary variants on a much smaller number of basic folding patterns. These patterns all use the same 20 amino acids, with a few additional attached groups, such as ATP. The basic patterns have evolved from a rather small number of primeval proteins. The minimum number of proteins thought to be necessary for the life of the simplest self-sufficient bacteria is only 500-1000. Genome analysis and X-ray diffraction studies show that all the known proteins are modelled on only a few thousand folding patterns, which can be grouped together into families and superfamilies with characteristic types of architecture.

The majority of proteins are globular molecules, compact rounded structures with polar or charged amino-acid side chains covering the polar outer surface and with a hydrophobic non-polar core. Globular membrane proteins may also have external hydrophobic groups submerged in the lipid bilayer. The simplest type of globule is a single domain, which is formed from one coherent backbone of hydrogen-bonded main-chain atoms. The hydrogen bonds assemble themselves into α-helices and parallel or anti-parallel β-sheets, linked by irregular connecting regions. Thus each type of domain fold has a characteristic secondary structure framework of hydrogen bonds.

Large globular proteins have often evolved by joining several domains end-to-end into a single long polypeptide chain. The various domains sometimes pack closely together, but they are often loosely linked through flexible connections, like a string of beads. A very common way to generate a multi-domain protein is by gene duplication. Here the DNA sequence of one protein unit is copied into several serial repeats to form a new gene which codes for an elongated protein with two, three, or more copies of the same unit chained together. Mutations occur in later generations, so that each repeat unit evolves its own distinctive sequence and biological function.

3. FIBROUS PROTEINS

Not all proteins are globular. There are a number of small specialised metalloprotein cage structures and a variety of small peptides or proteins without much intrinsic secondary structure. But the most important remaining structural group is the class of fibrous proteins. These are long thin molecules which are generally built on a simple repetitive structural pattern: a few parallel α-helices or a repetitive β-sheet conformation. Historically the fibrous proteins from hair, silk and collagen were the first biological structures to be analysed by X-ray analysis, in research which led directly to the discovery of the α-helix in 1950, its fibrous variant - the α-helical coiled-coil - in 1953 and the two types of β-sheet (Fraser and MacRae, 1973).

4. HELICAL COILED COILS

Imagine that two identical long right-handed α-helices, with the same specified amino-acid sequence (of left-handed amino acids) are placed vertically side by side in parallel contact, so that they stick together. The result is a two-stranded rope or coiled coil. The wide face of the rope is not flat, like the surface of a ladder, but becomes twisted into a supercoil with a slight left-handed overall twist. Normally the two component helices in a supercoil lie side by side in register, so that the combined molecule is symmetrical. There is a two-fold dyad axis along the length of the coil, running down the centre line.

Coiled-coils occur widely in nature, not only as free-standing α-helices, as for example in the α-keratin of animal hair and horn or in the muscle control protein tropomyosin, but also as long rod-like linkers joined to globular protein domains. The muscle motor protein myosin, for example, consists of a mechanically active globular head region at the amino-terminal end, joined to a long coiled-coil rodlike tail at the carboxy-terminal end. The myosin tail has two functions: while the N-terminal part acts as a flexible linker between the moveable head and the standing part of the thick filament, the C-terminal part is incorporated into the core of the standing part and helps to anchor the heads in place on a regular periodic array (Squire, 1981).

A much more subtle function for coiled-coil segments, either alone, or when they are part of a large protein with other globular regions, is as a dimerization switch. These switches are used to control the assembly of large protein complexes from their parts. Suppose that there is a need to direct the assembly of two globular proteins into a complex but the globules do not stick together well of their own accord. If each globule is linked to a piece of α-helix and the helices then pair up naturally to form a coiled coil, the dimerization can be achieved quite easily. The amino-acid sequences of the paired coil region helices may then evolve their own special control mechanisms, using buried charged groups and other devices to respond to pH or other external factors. Two interesting examples of dimerization switches are the "leucine zipper" regions found in the bZIP class of transcription factors (O'Shea et al., 1991) and the coiled coil of cortexillin, a protein which controls the assembly of actin (Burkhard et al., 2000). High-resolution X-ray studies of these coiled-coil segments reveal the presence of surprisingly subtle and interesting assembly controls, which could never have been guessed at from simple model-building experiments. The coiled coil motif displays considerably versatility in its interactions with other molecules, using the same amino acid side chains both as functional groups and as structural elements of the framework.

Nature is rarely simple: coiled coils can have three, four or even five strands as well as two. There are also coils with antiparallel α-helices and coils made of non-identical helices with different sequences. One of the most remarkable coiled coil architectures yet found is a 12-stranded hollow α-helical barrel or tube, built from antiparallel sets of helices

(Calladine et al., 2001). This is part of a membrane transport system found in the TolC membrane protein of *Escherichia coli* and forms a pore that is designed to carry selected molecules out of the cell.

5. STRUCTURAL PROPERTIES OF THE COILED COIL

A single straight standard α-helix has a unit step height h = 1.5Å along the axis and a unit twist angle $t = 100^0$. Thus there are 3.6 amino acid residues per turn and the helix pitch is P = 5.4Å. The X-ray diffraction pattern of a perfectly oriented vertical α-helical fibre bundle therefore has characteristic layer-line spacings of $(1/5.4)\text{Å}^{-1}$ and meridional spacings of $(1/1.5)\text{Å}^{-1}$. Many α-helical proteins, however, lack a $(1/5.4)\text{Å}^{-1}$ reflection, but have a strong meridional reflection at $(1/5.1)\text{Å}^{-1}$ instead. Crick (1953) showed that this type of diffraction pattern was characteristic of a two-stranded coiled structure in which two α-helices lie side by side in close contact, twisted and bent so that they wrap continuously round another, with their centres following a long-pitched helical path. The centres are about $10 - 11$Å apart and the axes are each tilted at a small angle to the vertical (about 10^0). Crick saw that the molecular surface of each helix is like a smooth cylinder of peptide bonds with projecting "knobs" formed by the amino-acid side chains. The patterns of knobs on the two paired helices mesh together along the seam of the contact region, like cog wheels, so that the knobs on one molecule fit into the holes between the opposed knobs on the other. The meshing pattern repeats itself every seven residues, so that the contacts lie along a gently sloping line along each α-helix surface. The knobs which mesh into the seam lie at alternate intervals of $3 - 4 - 3 - 4\ldots$ residues along the polypeptide chain and form a zigzag raised edge of projections. Crick predicted that the coiled coil would be held together by hydrophobic interactions along the seam and that the amino acid sequence would have a quasi-periodic character, based on a heptad pattern **(a,b,c,d,e,f,g)**, in which the side chains at positions **(a)** and **(d)** are normally hydrophobic.

To picture the local pattern of contacts more simply, imagine the entire supercoil to be unwound slightly until the curved helix axes are converted back into straight lines. The transformation converts the strands into vertical ropes with 3.5 residues per turn, in contact along a vertical seam which is now flattened. We deduce that the number of residues per turn, N_{Coil}, in the (left-handed) supercoil is given by

$$\frac{1}{N_{\text{Coil}}} = \frac{1}{N_{\text{Contact}}} - \frac{1}{N_{\text{Alpha}}} \tag{1}$$

where $N_{\text{Contact}} = 3.5$ and N_{Alpha} is the actual value for the real non-standard α-helix, being close to 3.6. Thus a first estimate is that $N_{\text{Coil}} = 126$, but the result is highly sensitive to small variations in N_{Alpha}. The unit step height of the tilted α-helices in a coiled coil is close to 1.485Å, so that the ideal pitch is $P_{\text{Coil}} = 1.485 N_{\text{Coil}} = 187$Å. In practice the pitch of real coiled coils is rather shorter, being closer to 140Å.

6. TROPOMYOSIN AND ACTIN

Tropomyosin was one of the earliest coiled-coil proteins to be discovered and is a rodlike protein about 420Å long and 20Å in diameter that attaches itself to the actin filaments in muscle and controls the contractile reaction between actin and myosin (Squire, 1981). Tropomyosin lies in a long vertical groove between the subunits of the actin helix and its position is regulated by another protein complex called troponin which responds to calcium ions. When a pulse of calcium ions stimulates a relaxed muscle cell to contract, the troponin complex changes shape and moves the tropomyosin rods across the actin helix surface, thus making way for the myosin heads to bind onto individual actin molecules. Each tropomyosin is long enough to interact with seven successive actins spaced 55Å apart.

The 284-residue amino-acid sequence of tropomyosin is a perfect example of a coiled coil structure (Stone et al., 1976) and confirms Crick's predictions in a remarkable way. There are 40 complete and uninterrupted heptads in which positions **(a,d)** almost always contain non-polar residues (Leu,Val,Ala,Ile). A typical heptad unit in the molecule has the sequence

Leu-Lys-Val-Leu-Glu-Glu-Lys.

The sequence contains many positively and negatively charged residues in positions **(e,g)** which can form pairs of salt bridges between the two strands of the coiled coil, linking an **(e)** position on one α-helix to a **(g)** position on the other. The crystal structure of tropomyosin (Whitby and Phillips, 2000) shows a clear outline of the twisted helical strands, in which the supercoil makes exactly three turns along the molecular length, with a pitch of 140Å. Successive lengths of molecule join together with a short end-to-end overlap. Thus the mean value of N_{Coil} is close to $280/3 = 93.3$ and the number of residues per turn deduced for the constituent α-helices is $N_{\text{Alpha}} = 3.636$. The rods lie in layers within the crystal and follow a wavy curved path.

The repeated heptad pattern in tropomyosin has probably evolved by gene duplication, being constructed by stringing together seven variations on a primitive 42-residue motif. Tropomyosin also has a strongly periodic pattern of negatively charged acidic residues along its whole length, comprising 14 bands spaced 19.7 residues apart. These charges are thought to be parts of the recognition sites for binding to seven consecutive actin molecules, with two bands allocated to each actin, accounting for a length of 39.4 residues or 58.6Å (Stewart and McLachlan, 1975). The actin helix has approximately 13 subunits of actin in every six turns, and so the long groove makes approximately 1/13 of a backward turn for each of the seven actins which flank one tropomyosin molecule, or 7/13 backward turns in all. Thus the groove makes about half a turn along an actin axial length of 385Å. If we imagine a slight unwinding of the actin helix and its bound tropomyosin by half a turn to make the groove run straight up a vertical edge of the actin we can see that the supercoil twists of the tropomyosin molecule make a relative rotation of seven half-turns against the actin, or one quarter turn across each band of negative charge.

A simple model for the regulation of muscle by tropomyosin proposes that tropomyosin has two alternative positions named "on" or "off" on the actin helix and makes approximately equivalent contacts with each of seven actins. The "on" and "off" positions are related by a rolling motion through one quarter turn. The precise stereochemistry of this regulatory transition is still uncertain, but the model suggests that the tropomyosin model must need to bend and unbend easily as it moves across the undulating surface of actin. Three-dimensional images reconstructed by electron microscopy now suggest a more complicated picture, with the possibility of three different binding positions (Narita et al., 2001),(McKillop and Geeves, 1993).

7. SUPERCOIL WRIGGLING

The recent high-resolution 2.0Å crystal structure of a short fragment of tropomyosin (residues 1-81 only) throws an interesting light on the flexibility of coiled coils in general and of tropomyosin in particular (Brown et al., 2001). The detailed "knobs-into-holes" interlocking of the helices in this fragment depends rather critically on the sizes of the non-polar amino-acid side chains which occupy positions **(a)** and **(d)**. Leucine side chains generally give a tightly packed seam when the helices are in precise register axially (unstaggered) and spaced about 10Å wide apart. The much smaller side chains of alanine, especially in position **(d)**, are however too small to fill completely the holes that would be filled by leucine. Instead they prefer to sit in a narrower coiled coil only 8Å wide, in which the two helices are slightly staggered, or sheared unsymmetrically across the axis, being out of register by about 1.2Å. The result is that the diameter and pitch of the tropomyosin supercoil both vary along its length. The path of the filament through the crystal turns sharply in discrete regions which have alanines clustered at successive **(d)** positions. The whole tropomyosin molecule contains seven localised alanine clusters, and these probably act as prescribed hinge points that allow the filament to bend flexibly when it follows the surface profile of the actin filament. The local deformation is a wriggling motion in which the broad flat face of the supercoil moves from side to side in its own plane rather than up and down.

8. MULTIPLE PERIODIC PATTERNS IN THE MYOSIN ROD

There are many forms of myosin, but we take one example, from the striated body-wall muscle of the nematode worm, coded by the unc-54 gene. This is a long protein chain of 1966 residues which has a globular head and a long rodlike tail. The rod portion of the molecule is a very long two-stranded coiled coil of 156 heptads which is remarkable for the many overlapping periodic patterns in its sequence (McLachlan and Karn, 1983),(McLachlan and Karn, 1982). The fundamental motif is a repeat unit of 28 residues (four heptads) with an axial length of 41.6Å, which has a characteristic pattern of positive and negative electrostatic charges. This pattern has the property that two rods lying side by side are strongly attracted together if they are staggered by a half-period, that is by a half-integral multiple of 28 residues $(14, 42, 70, \dots)$. These half-integral staggers give the fundamental assembly rule for building the thick myosin filament out of a large number of parallel rods. However not all of the possible staggers are used in the actual filament. In fact the rod sequence seems to be built from longer segments of 196 residues, which consist of 7 consecutive 28-residue zones, each slightly different from the other. These long segments specify longer half-integral staggers of 98 and 294 residues which account for the observation that the myosin heads in the thick filaments of muscle are regularly spaced along the shaft at intervals of 143Å and 430Å. The pitch of the coiled coil of the myosin rod determined from crystals (Quinlan and Stewart, 1987) is close to 98 residues, or 143Å , which is close to the pitch found in tropomyosin. This pitch would correspond to $N_{\text{Alpha}} = 3.630$ for the component α-helix. Because the supercoil pitch of 98 residues matches the axial stagger it is possible to assemble model myosin rod arrays in which rows of staggered heads all point in equivalent directions. The mode of packing of the rods in the filament shaft is still unknown in detail, but the closest packing of parallel coiled coils is normally achieved when the broad face of each coil at any axial level is rotated by 90^0 relative to its neighbours.

An evolutionary cousin of myosin is the rodlike protein paramyosin which has no head, but forms the internal core of some thick filaments, especially in shellfish. Its sequence (Kagawa et al., 1989) is similar to the myosin rod and paramyosin forms assemblies with staggers of 98 and 490 residues (145Å and 725Å). As yet there are no high-resolution structures which can show the atomic details of the molecular packing. The analysis of the myosin amino-acid sequences does however illustrate how a set of overlapping periodic modulations imposed on the simple repeated framework of a coiled-coil backbone can act as a vernier recognition device and specify a set of much longer repeat distances.

9. A DIMERIZATION SWITCH: CORTEXILLIN

Cortexillin I is a dimer-monomer control protein from *Dictyostelium discoideum* which is able to organise actin filaments into cross-linked bundles. Cortexillin has a typical actin-binding head globule similar to spectrin, joined to a long coiled-coil tail region of 18 continuous heptads. The coil has a sequence similar to the tail of muscle myosin II. The X-ray structure of the tail portion at 2.7Å resolution (Burkhard et al., 2000) provides a wealth of information about the role of salt bridges in the assembly of coiled coils, but the most interesting discovery is that there is a special dimer-trigger site of 14 residues close to the centre of the tail which is essential to initiate the formation of the dimer. The trigger site is a bulged section of coiled coil in which the two strands of helix are prised apart asymmetrically by an arginine residue which jams into the seam. The bulge is stabilised by a very complex pattern of salt bridges and these structural features suggest that the dimerization is able to be regulated by external factors, such as pH.

10. LEUCINE ZIPPER TRANSCRIPTION FACTORS

The transcriptional activator protein GCN4 is one of the bZIP class of transcription factors, which are all active as dimers. That is, they only bind to DNA when they are in a dimeric form. Dimerization depends on a coiled-coil sequence in the last 33 residues of the protein, where there is a characteristic heptad repeat of four consecutive leucine residues, the so-called "zipper". The crystal structure of the GCN4 leucine zipper fragment at 1.8Å resolution (O'Shea et al., 1991) has provided the first precise atomic picture of a small coiled-coil protein. The parallel helices are 9.3Å apart and 45Å long and they make one quarter of a turn of a supercoil with a pitch of about 180Å. The helices maintain contact along this length by bending slightly inwards and they cross over at an angle of 18^0. The hydrophobic core of the zipper dimer contains tightly packed leucines and valines, each in its lowest energy side chain rotamer, and every "knob" is flanked by four other side chains which mark out the boundaries of the "holes". The shapes of the holes at the (**a**) and (**d**) positions are systematically different in detail. A surprise feature of the structure is that in positions 16(**a**) midway along the lengths of the two helices there are a pair of juxtaposed Asn side chains which are hydrogen bonded to one another in an asymmetrical manner and distort the core. The outer molecular surface contains the typical expected (**e**)-(**g**) salt bridges between helices. The long side chains of these charged groups help to cover the hydrophobic core. One potential salt bridge is blocked by the side chain of Asn 16(**a**). The GCN4 structure confirms Crick's knobs-into-holes scheme in a remarkable way. It shows how even a short coiled-coil segment is able to form a highly stable dimer by using a combination of non-polar contacts and salt bridges. Engineered mutants of GCN4 (Harbury et al., 1993) have been used to test the effect of changes in the hydrophobic core of the zipper. Simple alterations can convert the molecule into a three-stranded parallel coil or an antiparallel four-helix bundle.

Other transcription factors with sequences related to GCN4 are Fos and Jun. These become active only in the form of a Fos-Jun heterodimer, whereas Fos-Fos and Jun-Jun dimers are unstable. The crystal structure of the Fos-Jun complex (Glover and Harrison, 1995) shows how this transcription factor recognises DNA. The bZIP region of the mixed dimer is a Y-shaped molecule built from two continuous bent α-helices, and shaped like a tuning fork. The stem of the Y is the zipper coiled coil, while each of the two separated helical arms binds to opposite sides of a short double-stranded piece of DNA (the AP-1 recognition site, just seven bases long). The coiled-coil part of Fos-Jun is very similar to GCN4 and is held together by eight salt bridges (four between (**g**)-(**e**) and four others between (**g**)-(**a**)). There is only one copy of Asn 16(**a**), sited on the Jun side of the coil, while Fos has a Lys 16(**a**) instead. Both these residues take part in (**g**)-(**a**) salt bridges. Analogous models of Fos-Fos and Jun-Jun dimers suggest that these other pairings are less stable because they cannot form as many of the essential salt bridges between helices.

11. VARIATIONS ON THE REGULAR COILED COIL STRUCTURE

A thorough understanding of coiled coil proteins requires a searching analysis of the stereochemistry of their non-polar cores and their surface salt bridges (Burkhard et al., 2001). Computer programs have been designed that detect potential coiled coil regions in protein sequences by looking for consecutive heptad motifs (Lupas, 1996), (Woolfson and Alber, 1995). These methods use empirical profile tables of the amino-acid preferences at each heptad position to distinguish between dimeric and trimeric or between parallel and antiparallel strands. The observed preferences can often be accounted for simply in terms of packing geometry.

Not all coiled coil structures are regular straight rods. Calladine has shown how the supercoiled filament may bend as an elastic element (Stewart et al., 1987), particularly across its narrow face, and it may wriggle by shearing loose Ala-Ala contacts across the broad face, as we noted above in tropomyosin (Stewart, 2001). But there can also be irregularities in the sequence, with breaks or discontinuities in the pattern of heptads. Some regular parallel coils are based on an 11-residue motif and the twist is then in the reverse (right-handed) direction. The seam of contact has $N_{\mathrm{Contact}} = 11/3 = 3.67$ residues per turn and if we use Equation (1) with a realistic value of $N_{\mathrm{Alpha}} = 3.63$ the estimated supercoil pitch is very long at $N_{\mathrm{Coil}} \approx 360$. In heptad-based coiled coils three main types of break are observed, which shift the azimuthal position of the hydrophobic seam across the surface of the α-helix and thereby advance or retard the normal twisting of the supercoil (Brown et al., 1996). These breaks are classified according to their perturbed separations between the consecutive non-polar core residues:

$$3-4:3-4:(4):3-4:3-4 \qquad \text{A "stutter".}$$
$$3-4:3-4:(3):3-4:3-4 \qquad \text{A "stammer".}$$

A stutter inserts 4 residues or deletes 3. Conversely a stammer inserts 3 residues or deletes 4. The third kind of break, first observed in the myosin rod, is the insertion of a single residue, and behaves like two stutters in close succession. This is called a "skip residue" and one example of an irregular skip pattern is:

$$3-4:(4):3-4:3-4:(4):3-4 \qquad \text{A "skip".}$$

A stutter retards the supercoil twist over a short local transition region and a stammer advances it. Because a $[3-4:(4):]$ unit is very like a piece of the regular 11-residue motif it is likely that the broad face of the supercoil in a stutter region is almost free from twist over a range of about 11 residues, whereas a regular left-handed heptad coil with a pitch of 98 residues would twist by about 40^0 over this length. Therefore a skip region probably results in a loss of 80^0 or almost a quarter turn of twist, and the flattened supercoil would also be easy to bend in the vicinity of the skip. However these arguments are only valid if the constituent alpha helices retain a regular hydrogen-bonded backbone and do not contain tightened fragments of 3_{10} helix.

12. CONCLUSION

Coiled coil structures are basically simple and easy to understand. They are able to combine useful engineering properties of accurately controlled length, twist and bendability with more subtle biological functions that are mediated by networks of salt bridges. These networks are easy to recognise in X-ray structures but quite hard to predict from rational model-building. Nevertheless helical coiled coils and helix bundles are attractive workbench platforms for the design and testing of engineered proteins (Harbury et al., 1993). Even such a simple paired helical construct as the Fos-Jun leucine zipper is able to recognise double-stranded DNA and thus act as a control over gene transcription.

Acknowledgments

We thank Chris Calladine for stimulating and enlightening collaboration over many years and wish him well in his retirement.

References

Brown, J., Cohen, C., and Parry, D. (1996). Heptad breaks in α-helical coiled coils: Stutters and stammers. *Proteins-Structure,Function, and Genetics*, 26:134–145.

Brown, J., Kim, K., Jun, G., Greenfield, N., Dominguez, R., Volkmann, N., Hitchcock-DeGregori, S., and Cohen, C. (2001). Deciphering the design of the tropomyosin molecule. *Proc. Natl. Acad. Sci. USA*, 98:8496–8501.

Burkhard, P., Kammerer, R., Steinmetz, M., Bourenkov, G., and Aebi, U. (2000). The coiled-coil trigger site of the rod domain of cortexillin I unveils a distinct network of interhelical and intrahelical salt bridges. *Structure*, 8:223–230.

Burkhard, P., Stetefeld, J., and Strelkov, S. (2001). Coiled coils: a highly versatile protein folding motif. *Trends in Cell Biol.*, 11:82–88.

Calladine, C. (1978). Change of waveform in bacterial flagella: The role of mechanics at the molecular level. *J. Mol. Biol.*, 118:457–479.

Calladine, C., Sharff, A., and Luisi, B. (2001). How to untwist an α-helix: Structural principles of an α-helical barrel. *J. Mol. Biol.*, 305:603–618.

Crick, F. (1953). The packing of α-helices: simple coiled coils. *Acta Cryst.*, 6:689–697.

Fraser, R. and MacRae, T. (1973). *Conformation in Fibrous Proteins and Related Synthetic Polypeptides*. Academic Press, New York.

Glover, J. and Harrison, S. (1995). Crystal structure of the heterodimeric bZIP transcription factor c-Fos-c-Jun bound to DNA. *Nature*, 373:257–261.

Harbury, P., Zhang, T., Kim, P., and Alber, T. (1993). A switch between two-, three-, and four-stranded coiled coils in GCN4 leucine zipper mutants. *Science*, 262:1401–1407.

Kagawa, H., Gengyo, K., McLachlan, A., Brenner, S., and Karn, J. (1989). The paramyosin gene (unc-15) of *caenorhabditis elegans* :Molecular cloning nucleotide sequence and models for thick filament assembly. *J. Mol. Biol.*, 207:311–333.

Lupas, A. (1996). Coiled coils: new structures and new functions. *Trends in Biological Science*, 21:375–382.

McKillop, D. and Geeves, M. (1993). Regulation of the interaction between actin and myosin subfragment 1: evidence for three states of the thin filament. *Biophys. J.*, 65:693–701.

McLachlan, A. and Karn, J. (1982). Periodic charge distributions in the myosin rod amino acid sequence match cross-bridge spacings in muscle. *Nature*, 299:226–231.

McLachlan, A. and Karn, J. (1983). Periodic features in the amino acid sequence of nematode myosin rod. *J. Mol. Biol.*, 164:605–626.

Narita, A., Yasunaga, T., Ishikawa, T., Mayanagi, K., and Wakabayashi, T. (2001). Ca^{2+}-induced switching of troponin and tropomyosin on actin filaments as revealed by electron cryomicroscopy. *J. Mol. Biol.*, 308:241–261.

O'Shea, E., Klemm, J., Kim, P., and Alber, T. (1991). X-ray structure of the GCN4 leucine zipper, a two-stranded, parallel coiled coil. *Science*, 254:539–544.

Quinlan, R. and Stewart, M. (1987). Crystalline tubes of myosin subfragment-2 showing the coiled-coil and molecular interaction geometry. *J. Cell Biol.*, 105:403–415.

Samatey, F., Imada, K., Nagashima, S., Vonderviszt, F., Kumasaka, T., Yamamoto, K., and Namba, K. (2001). Structure of the bacterial flagellar protofilament and implications for a switch for supercoiling. *Nature*, 410:331–337.

Squire, J. (1981). *The Structural Basis of Muscular Contraction*. Academic Press, New York.

Stewart, M. (2001). Structural basis for bending tropomyosin around actin in muscle thin filaments. *Proc. Natl. Acad. Sci. USA*, 98:8165–8166.

Stewart, M. and McLachlan, A. (1975). Fourteen actin-binding sites on tropomyosin? *Nature*, 257:331–333.

Stewart, M., McLachlan, A., and Calladine, C. (1987). A model to account for the elastic element in muscle crossbridges in terms of a bending myosin rod. *Proc. Roy. Soc. London*, B229:381–413.

Stone, D., Sodek, J., Johnson, P., and Smillie, L. (1976). *Tropomyosin: correlation of amino-acid sequence and structure*, volume 31 of *Proteins of Contractile Systems*, pages 125–136. North-Holland, Amsterdam.

Whitby, F. and Phillips, G. (2000). Crystal structure of tropomyosin at 7 Angstroms resolution. *Proteins-Structure Function and Genetics*, 38:49–59.

Woolfson, D. and Alber, T. (1995). Predicting oligomerisation states of coiled coils. *Protein Science*, 4:1596–1607.

SUPERCOILING OF DNA MOLECULES

J. M. T. Thompson
Centre for Nonlinear Dynamics, Civil Engineering Building,
University College London, Gower Street, London, WC1E 6BT
jmtt@ucl.ac.uk

Abstract The mechanics of ply formation in DNA supercoils is examined in the light of recent progress. Derived constitutive relations are confirmed by laboratory tests on mechanical models. Derivation of writhe and link from a DNA micrograph is demonstrated.

Key words: DNA molecule, Super-coiling, Plasmids, Mechanics of a ply

1. INTRODUCTION

In order to transcribe the genetic code, a DNA molecule must screw through an RNA polymerase, and its rapid rotation can induce large twisting stresses. If these induce excessive writhing the DNA is unable to function. So important is this, that the body has a special enzyme, the *topoisomerase*, which can cut the molecule, untwist it to alleviate the stress, and re-join it.

On the scale of this writhing, the internal double-helix of the DNA is irrelevant, and a long strand behaves as if it were a slender elastic rod. Modelling at this level is being pursued by many biologists and mathematicians [5, 6, 11-14, 16]. One significant topic is super-coiling, in which the fibre representing the DNA adopts a configuration akin to those of Figs 1 & 2. The interwound regions of these are said to form a *ply*. The simplest way to observe a ply physically is to twist a long rubber rod. If, after imposing the twist, the ends are brought together, the rod will buckle locally and jump into the familiar ply-plus-loop form of Fig. 1.

An electron micrograph of DNA from *E. coli* reproduced by Calladine & Drew [2], shows a supercoiled and interwound form. In this picture the molecule forms a closed loop, called a *plasmid*.

H.R. Drew and S. Pellegrino (eds.),
New Approaches to Structural Mechanics, Shells and Biological Structures, 513–524.

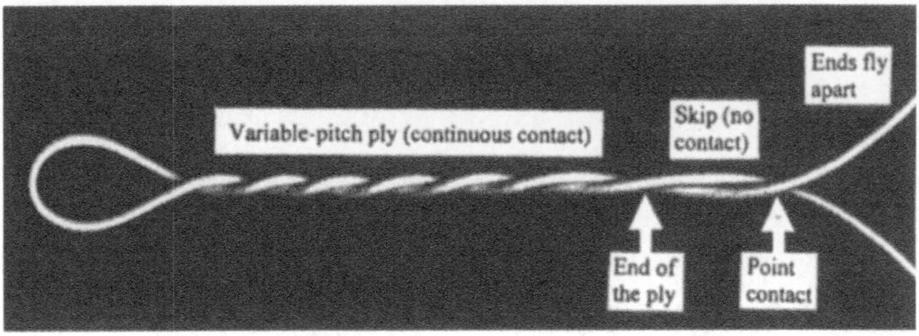

Figure 1. A rubber rod forms a left-handed variable balanced ply with a free end loop. The helical angle varies along the ply, and at the right we see the skip-fly phenomenon.

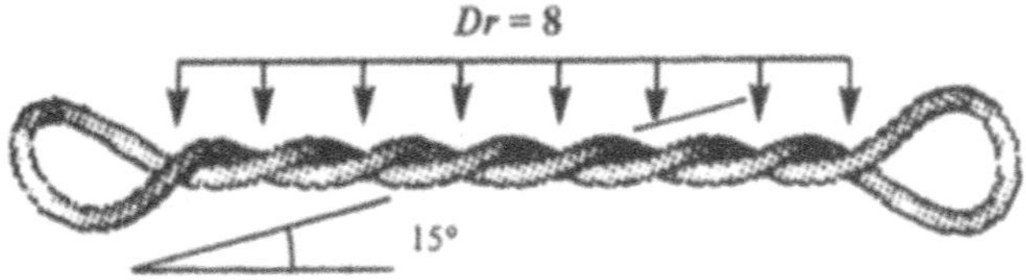

Figure 2. A variable balanced ply due to Coleman & Swigon [5] with Lk =10, Wr = 7.548.

A basic configuration of a twisted plasmid, derived analytically from rod theory by Coleman & Swigon [5], is shown in Fig 2. The straight central region forms a left-handed variable ply, closed by two free end loops.

2. MODELLING AND KINEMATICS

2.1 Modelling a Ply

A long length of DNA can be modelled as an elastic rod, with circular cross-section of radius r. All we need to know about its mechanical properties is the ratio, γ, of its torsional stiffness, C, to its bending stiffness, B. For DNA, this is thought to lie in the interval $0.7 < \gamma < 1.5$. A possible nonlinear coupling between the bending and twisting of DNA is discussed by Calladine [1].

Now a solid circular rod has $\gamma = C/B = 1/(1+\nu)$, where ν is Poisson's ratio (for metal $\nu = 1/3$, $\gamma = 3/4$; for rubber $\nu = 1/2$, $\gamma = 2/3$). So as γ varies from 0.7 - 1.5, the *equivalent* Poisson's ratio varies from +0.43 to -1/3. Such a negative ν is not observed for any normal material.

Recent work on ply mechanics should be noted. First, Fraser and co-workers [7, 13] derived the equation of a *uniform balanced ply*. This has two rods winding around themselves, and touching on a straight central line, the *ply axis*, with the centre-line of each rod forming a helix of radius r and helical angle θ (the helical angle in Fig. 2 is 15°). This solution is only observed if the correct boundary conditions are applied at the ends. Second, Coleman & Swigon [5] derived the equation of a *variable balanced ply*, which allows θ to vary along the ply, with the two rods lying as if wound on a cylinder of radius r. The extra flexibility of the variable θ allows the rods to separate at the ends without the unphysical point moment that had to be introduced by Fraser *et al.* Third, Thompson *et al* [16] relaxed the condition of balance, and presented the equations of a *variable loaded ply* carrying an end wrench, comprising a tension, G and a twisting moment, N. This solution also arises as a special case of a rod constrained to lie on a cylinder as formulated by van der Heijden [17].

In this brief article, some of the results of the above contributions, especially [5] and [16], are explained and illustrated.

2.2 Topology of a Ply

We consider an inextensional rod of circular cross-section, with length L and radius r. To discuss torsion, we imagine parallel stripes drawn on the surface of the straight and unstressed rod parallel to its centre-line. While it remains straight, let the rod be twisted uniformly by applying a rotation about the centre-line, ϕ, (in radians) to one end. The kinematic twist rate is then $\tau = \phi / L$, positive if the stripes form a *right-handed helix*. Supposing the cylindrical surface of the twisted rod to be unrolled, the planar *stripe angle*, ψ, between the stripes and the centre-line is a convenient measure of the twist, given by $\tan \psi = \tau r$. We take this as our definition of ψ, but note that it will not be the unrolled angle on a *bent* rod due to significant outer-fibre strains. The total twist, Tw (in turns), is the integral of $\tau / 2\pi$ over L, which for constant τ gives $Tw = \tau L / 2\pi$.

We now imagine gluing the two ends of our rod together to form a closed loop. If we just bend the rod in a plane and glue the ends together without inserting twist, the centre-line and stripes will be circles. Suppose, however, that having bent it in a plane and brought the ends together we insert a number of turns of twist before gluing. We define this number as the link, Lk (positive if it induces positive τ), and as long as the glued ring remains planar, we have $Lk = Tw$.

Strictly, the link of a plasmid will vary by integer increments because each sugar-phosphate chain must join to itself, but in the context of rod theory it is convenient to suppose that Lk can vary continuously.

If we now get hold of the glued ring and distort it out of the plane, the link will, by definition, not change. The total twist does however change, and the two are related by the important result [8],

$$Lk = Tw + Wr$$

Writhe, Wr, is a property of the shape of the rod's centre-line and can be calculated as the number of (signed) crossings in a view, averaged over all views. The number of crossings in a single view is called the directional writhe, Dr : in the side view of Fig. 2 a count of the crossings gives $Dr = 8$.

2.3 A Laboratory Ply

We must consider how we could make a ply, either conceptually or in a laboratory experiment. To make a *right-handed* ply, we take two rods of length L and place them side by side with their left-hand ends fixed, as illustrated (later) in Fig. 6. While straight, each rod is given a *left-handed* twist by turning the right-hand ends through a positive angle A, giving the initial twist rate $\tau_0 = -A/L$. We refer to this straight twisted configuration as the pre-ply (or α-state) and write the stripe angle as $\psi_\alpha = \arctan(\tau_0 r)$. To make a fairly uniform ply, angle A should correspond to at least 5 turns. The rods are now joined together at both ends, using a paper clip or a more careful moulding. Thinking of these two straight, planar rods and their mouldings as one closed entity, we have $Wr = 0$ and $Lk = \tau_0 2L/2\pi$. Note that this straight pre-ply already forms an example of our wrench-loaded ply, because the resultant constraint is simply a twisting moment, $N = 2C\tau_0$, with zero tension.

If we now let go of the rods, still moulded together at their ends, they might jump into a variety of spatial forms, one of which will be a *right-handed* balanced ply, with approximately constant θ. A controlled way to get this ply would be to hold the two mouldings, and just let them rotate slowly about the ply axis until the ply reaches equilibrium under zero wrench. We can finally load this ply by applying any wrench, (G, N), to the mouldings, the link Lk remaining unchanged throughout. Once the rods are released, the moulding will not provide the correct end conditions for a *uniform* ply, but significant variation of θ is usually localised near the ends as we shall see.

For constant θ we can easily write down the writhe of the two helical rods. To do this we use the result of Fuller [9] that for a single rod $1 + Wr =$ area$/2\pi$ (mod 2) where the area is that enclosed, cumulatively, by the orbit of the unit tangent vector on the unit sphere.

Application of this result gives for our two-strand ply,

$$Wr = K(1 - \cos\theta) - K(1 + \cos\theta) = -L\sin 2\theta/2\pi r$$

Here K is the number of helical waves in one rod given by $K = L\sin\theta/2\pi r$. Using the conservation of link, we can also write the kinematic result

$$Lk\,(\pi r/L) = \tau_0 r = \tan\psi_\alpha = \tau r - \tfrac{1}{2}\sin 2\theta = \tan\psi - \tfrac{1}{2}\sin 2\theta$$

which relates the initial twist rate, τ_0, to the final current twist rate, τ.

3. ANALYSIS OF A UNIFORM LOADED PLY

3.1 Energy Formulation

We give here an energy analysis of the *uniform loaded ply*, comprising our two rods in continuous contact along a straight *ply axis*. Their centre-lines wind around this axis as a pair of identical helices of radius r. Moulding at the ends are loaded by tension G and twisting moment, N, the latter tending to tighten the mutual winding of the rods. We ignore friction and gravity. Each rod in the ply has length L, and in the balanced state with $G = N = 0$ the helical angle is denoted by β. We want to find the elongation and end rotation of the ply as functions of G and N, these being the *corresponding displacements* through which G and N do work.

A right-handed uniform ply has one degree of freedom which we take to be the helical angle of the ply, θ. This angle will vary during the loading process, but we assume the end conditions are such that the ply always remains (at least approximately) uniform, with θ constant along its length. We define g as the variable length of the ply, $L \cos \theta$, divided by the radius, r: and n as the total rotation (in radians) of the whole ply, measured from the datum in which the 2 rods lie straight side by side. In terms of the aspect ratio $a \equiv L/r$, we then have $g(\theta) = a \cos\theta$, $n(\theta) = a \sin \theta$, with the constraint that $g^2 + n^2 = a^2$.

The curvature of a helical rod is $\kappa = \sin^2 \theta / r$ and as in §2.3 the writhe of the two rods is $Wr = -\tfrac{1}{2} (a/\pi) \sin 2\theta$. The twist rate, using $Lk = Tw + Wr$, is given by $\tau r = \tau_0 r + \tfrac{1}{2} \sin 2\theta$. The bending energy of the two rods, U_B, the torsional energy, U_C, and the energy of the end loads, U_L, are given by

$$U_B = \tfrac{1}{2} B 2 L \kappa^2 = (BL/r^2) \sin^4 \theta$$

$$U_C = \tfrac{1}{2} C 2 L \tau^2 = (CL/r^2) (\tau_0 r + \tfrac{1}{2} \sin 2\theta)^2$$

$$U_L = -Ggr - Nn = -GL \cos \theta - Na \sin \theta$$

Adding these contributions, the total potential energy is $V(\theta) = U_B + U_C + U_L$ and for equilibrium we can set $\partial V / \partial \theta = 0$, giving

$$4 \sin^3\theta \cos\theta + 2\gamma (\tau_0 r + \tfrac{1}{2} \sin 2\theta) \cos 2\theta + (Gr^2/B) \sin \theta - (Nr/B) \cos \theta = 0$$

This is the governing equation for a loaded uniform ply. Note that the bracketed term after γ is simply τr. For given γ and $\tau_0 r$, together with the load parameters (Gr^2/B) and (Nr/B), this equation can be solved for the ply angle θ. Limits on the range of applicability of this analysis are that the contact force should remain compressive, and that θ should be less than 45°, at which the helices lock together [16].

3.2 Experimental Verification

The theoretical variation of n with (Gr^2/B) in the absence of any applied moment ($N = 0$) is compared with experimental results due to my grandson Ben Thompson in Fig. 3. In the experiment, a length of ply was made from two silicone rubber rods, each of length $L = 21$ cm, lubricated with oil to minimise frictional forces. The diameter of the rods was 3.515 mm, and Young's Modulus of the rubber was estimated to be 1.97×10^6 N/m^2. The maximum load applied was 963 g, corresponding to $Gr^2/B = 1.977$.

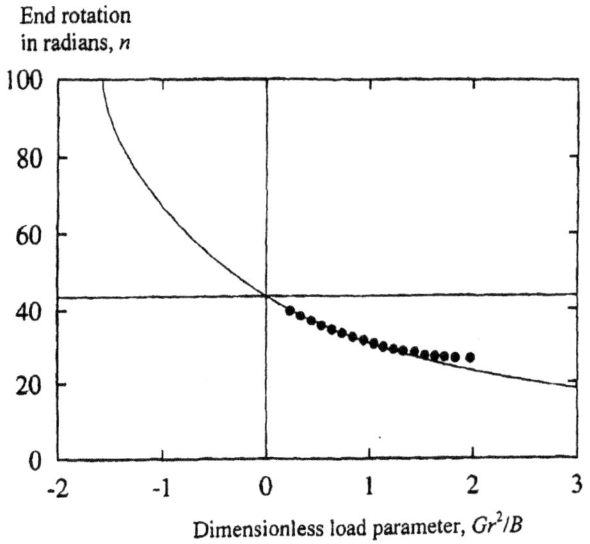

Figure 3. Experimental and theoretical results for a uniform ply under tension. Experimental results due to Ben Thompson are shown as solid circles. Theoretical result is the curve.

The graph shows a plot of the end rotation, n, against the dimensionless tension parameter. The experimental points are denoted by solid circles, and agree well with the theoretical curve. The slight disagreement at high load is probably due to factors not considered in the theory, notably the axial extensibility of the rubber rod, and the Poisson contraction of its radius.

3.3 The Uniform Balanced Ply

Setting $G = N = 0$, the solution for the *uniform balanced ply* (the β-state) can be written as

$$\tau r \gamma = \gamma \tan \psi_\beta = -\sin^2\beta \tan 2\beta$$

or in terms of the initial twist rate as

$$\tan \psi_\alpha = \tau_0 r = -\tfrac{1}{2} \tan 2\beta \,(1 + 2\nu \sin^2\beta)$$

where ψ_α is the initial stripe angle, and β is the balanced solution for θ. For convenience in the second equation we have replaced the stiffness ratio, γ, by Poisson's ratio using $\gamma = 1/(1+\nu)$. If we plot the given $\tau_0 r$ against the ply angle β we see that $\tau_0 r$ goes to infinity as β approaches 45°. For small angles we find $\psi_\alpha \approx -\beta$ and the stripe angle in the β-state, $\psi_\beta \approx 0$. To this *first order*, as we go from the α-state to the β-state we have:

The stripe angle vanishes and is converted into a ply angle of opposite sign.

4. ANALYSIS OF A VARIABLE LOADED PLY

4.1 Energy Formulation

It is useful to see how the energy analysis of the uniform ply can be generalised to embrace the variable ply, and we give in this section a brief summary of the theory following van der Heijden *et al* [19]. The total potential energy of the generalised ply is

$$V = 2\int_0^L \left(\frac{1}{2}B\kappa^2 + \frac{1}{2}C\tau^2\right) ds - Ggr - Nn\,.$$

With θ measuring the deviation of the tangent vector from the ply axis, and ϕ representing the internal twist, the total curvature is given by

$$\kappa^2 = \theta'^2 + \frac{1}{r^2}\sin^4\theta$$

the twist is

$$\tau = \phi' + \frac{1}{r}\sin\theta\cos\theta$$

and the potential energy can be written as

$$V = \int_0^L \Psi(\theta,\theta',\phi,\phi')\,ds$$

Here Ψ is the known integrand of V, and the Euler-Lagrange equations are

$$\frac{d}{ds}\frac{\partial\Psi}{\partial\theta'} = \frac{\partial\Psi}{\partial\theta} \qquad \frac{d}{ds}\frac{\partial\Psi}{\partial\phi'} = \frac{\partial\Psi}{\partial\phi}$$

Notice that because Ψ is independent of ϕ the latter is an *ignorable* variable and the second equation proves immediately that τ = constant. The first equation gives our result for the loaded variable ply,

$$\theta'' r^2 = 2\sin^3\theta\cos\theta + \tau r\gamma\cos 2\theta + ½(Gr^2/B)\sin\theta - ½(Nr/B)\cos\theta$$

This is a differential equation for the variation of $\theta(s)$ along the ply, with a prime denoting differentiation with respect to arc length along a rod, s.

4.2 The Phase Portrait

With no end loads ($G = N = 0$) the above equation agrees with the result of Coleman & Swigon [5] for the variable balanced ply. To examine the solutions, we start by using the static-dynamic analogy, and visualise solutions of the differential equation in the phase space obtained by replacing arc length s by time t. Note that the current twist rate, τ, is constant along a rod, having been shown to be not a function of s. We do not need its magnitude here, but we could find it in terms of the initial twist rate using a suitable generalisation of the kinematic results of §2.3 to conditions of variable θ.

The phase portrait, equivalent to that of an undamped nonlinear oscillator, is shown in Fig. 4. The saddle point (solid circle) of this oscillator with $\theta = \beta$, found by setting $\theta'' = 0$, is of course just the *uniform balanced ply*. This uniform ply will only exist if the correct end conditions are applied. In general this will not be the case, as for example when the ply is closed by end loops.

However, a long ply will often have $\theta(s) \approx \beta$ over a long central section. In phase space, this will correspond to a solution close to the saddle, S, with small θ'. The divergence before and after the slow transit near S gives a *boundary layer* in which $\theta(s)$ adjusts relatively quickly to accommodate the conditions at the end of the ply.

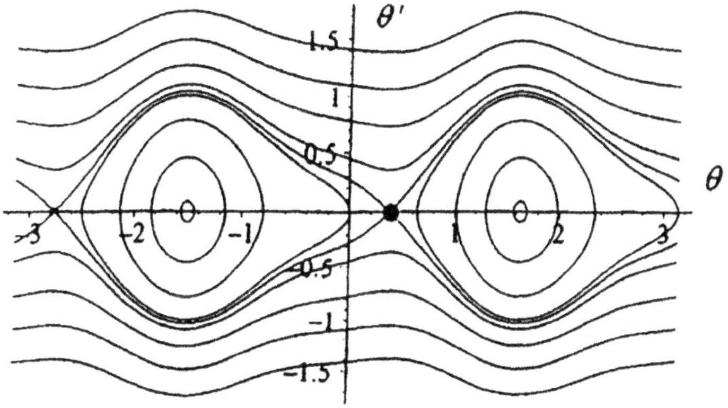

Figure 4. Phase portrait of the variable ply. The saddle denoted by a solid circle is the uniform ply with constant θ (here in radians).

4.3 Plasmid Solutions

Some precise numerical solutions of Coleman & Swigon [5] for the writhing of a DNA plasmid are shown in Fig. 5. The ends of the rod are imagined to be glued together after some link, *Lk*, has been inserted (Coleman & Swigon call this the excess link because they use a datum that includes the twisting of the double helix itself). For a given controlled input of *Lk* they calculate the spatial equilibrium configuration of the plasmid, taking full account of all self-contacts. As their measure of deformation they adopt the writhe, *Wr*. The response under slowly varied *Lk* is shown in the left-hand picture of Fig. 5. The rod is stable in its planar circular state up to the subcritical bifurcation at A^0, from which a dynamic jump would carry the ring to a state of self-contact as illustrated by the double arrow. The bifurcation at A^0 is at $Lk = (B/C)\sqrt{3}$, a result due to Zajac [20].

Ignoring for the moment the physical jumping behaviour, we can focus on the post-buckling path that emerges from A^0, noting that the number of self-contact points is shown in square brackets. A solid (or broken) line denotes a stable (or unstable) path under controlled *Lk*. Between A^0 and A^1 we have an unstable falling path with no self-contact.

Between A^1 and A^2 we have a path with one self-contact at a point; between A^2 and A^3 a path with self-contact at two points; and between A^3 and A^4 a path with self-contact at three points. After A^4 a continuous line of self-contact, namely a ply, is observed, together with self-contact at two points. Physical jumps under slowly varying *Lk* are indicated, and sample computed shapes are displayed.

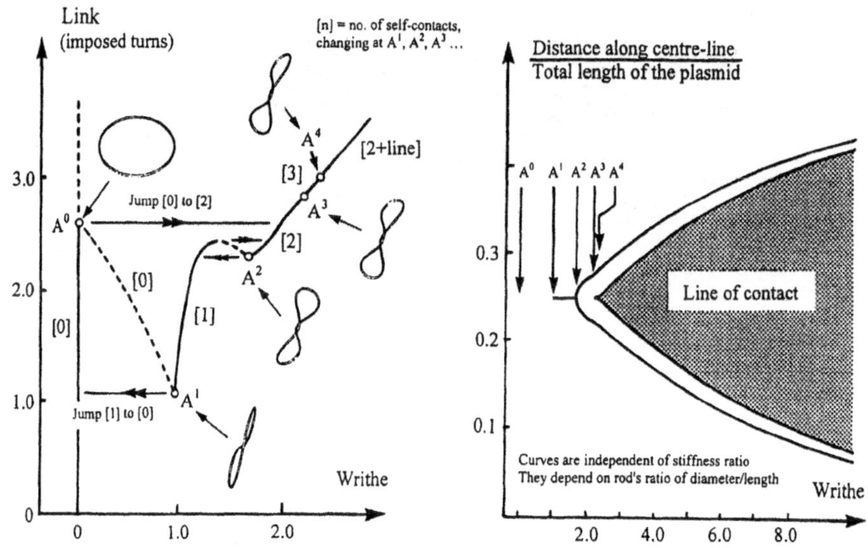

Figure 5. Super-coiled plasmids analysed by Coleman & Swigon [5], corresponding to DNA of diameter 20 Å with 718 base pairs ($\gamma = 2/3$).

The right-hand diagram shows details of the self-contacts as we progress along path $A^0\ A^1\ A^2\ A^3\ A^4$ .... We see that a single central point-contact splits into two at A^2, followed by a new central point-contact at A^3. At A^4 this central point-contact starts to spread to give us finally a central symmetric ply, which *skips* off at each end, only to re-contact at a point shortly after, before the two rods finally *fly* apart. This skip-fly phenomenon at the end of a variable ply is illustrated in Fig. 1. A sample shape computed by Coleman & Swigon at higher *Lk* is shown in Fig. 2.

Somewhat analogous to the writhing of a closed plasmid is the response of a single stretched and twisted rod, analysed extensively using the static-dynamic analogy in which structural localisation corresponds to a homoclinic orbit [3, 4, 10, 15, 18].

4.4 Clamped Solutions

In continuation of the present work, van der Heijden *et al* [19] have computed variable ply solutions with clamped ends, as shown in Fig. 6.

It is clear that the effect of the clamps is localised into a *boundary layer*, with most of the ply in a fairly uniform state of roughly constant helical angle.

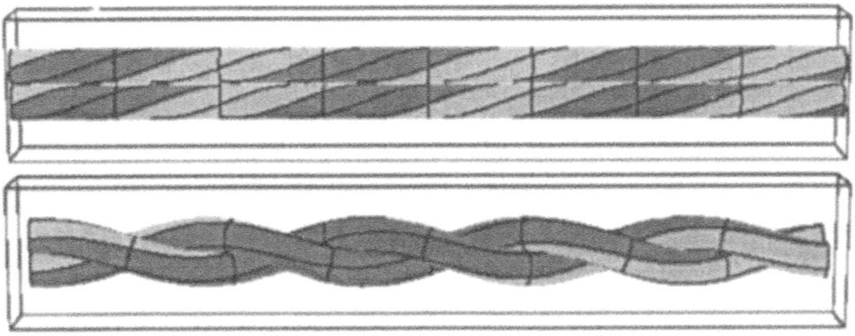

Figure 6 A balanced variable ply satisfying the clamped boundary conditions of a moulding.

5. APPLICATION TO A PLASMID MICROGRAPH

To alleviate twist, a DNA plasmid will writhe in space to form a supercoil, a common form of which is the interwound configuration shown in Fig. 2. To illustrate how the new results can be used to assess the link of a plasmid based on its micrograph image, we will suppose that Fig. 2 (showing a form computed by Coleman & Swigon) is such a micrograph. Can we deduce from the image their values of $Lk = 10$, $Wr = 7.548$?

Observing that most of the rod is in the ply, it is tempting to use just the ply for our estimate, ignoring the length of rod in the end loops. Moreover the angle, θ, varies only minutely along the ply, so we assume it to be constant, with the rods in the form of a helix. The result in §2.3 for the writhe in two interwound helices then gives Wr = $Dr^* \cos\theta = \pm 2K \cos\theta$. Here $Dr^*=2K$ is the particular directional writhe corresponding to the number of crossings in a *side view of a (long) ply*, equal to 8 in Fig. 2. Now from this picture we measure the balanced ply angle as $\beta \approx -15°$, so $\cos\beta = 0.965$ and we find $Wr \approx 7.73$, comparing well with the value used in [5].

We next use the new mechanics results to augment the geometrical analysis. The first equation of §3.3 gives us the total twist in our plasmid of length S as $Tw\,(2\pi r\,\gamma/S) = \tau r\gamma = -\sin^2\beta \tan 2\beta$. Setting the stiffness ratio as $\gamma = 2/3$, and radius to length ratio as $r/S = 4.1 \times 10^{-3}$, as in [5], we have finally $Tw = +\,2.25$, giving our estimate for the link as $Lk = 9.98$, in excellent agreement with the value of 10 used by Coleman and Swigon.

References

1. Calladine, C. R. 1980 Toroidal elastic supercoiling of DNA, *Biopolymers*, **19**, 1705-1713.
2. Calladine, C. R. & Drew, H. R. 1997 *Understanding DNA: the molecule and how it works*, Second Edition, Academic Press, San Diego.

3. Champneys, A. R. & Thompson, J. M. T. 1996 A multiplicity of localized buckling modes for twisted rod equations, *Proc. R. Soc. Lond.*, A **452**, 2467-2491.
4. Champneys, A. R., van der Heijden, G. H. M. & Thompson, J. M. T. 1997 Spatially complex localization after one-twist-per-wave equilibria in twisted circular rods with initial curvature, *Phil. Trans. R. Soc. Lond.*, A **355**, 2151-2174.
5. Coleman, B. D. & Swigon, D. 2000 Theory of supercoiled elastic rings with self-contact and its application to DNA plasmids, *J. Elasticity* **60**, 173-221.
6. Coleman, B. D., Swigon, D. & Tobias, I. 2000 Elastic stability of DNA configurations. II Supercoiled plasmids with self-contact, *Phys. Rev. E* **61**, 759-770.
7. Fraser, W. B. & Stump, D. M. 1998 The equilibrium of the convergence point in two-strand yard plying, *Int. J. Solids Structures* **35** 285-298.
8. Fuller, F. B. 1971 The writhing number of a space curve, *Proc. Natl. Acad. Sci. USA.* **68** 815-819.
9. Fuller, F. B. 1978 Decomposition of the linking of a closed ribbon: a problem from molecular biology, *Proc. Natl. Acad. Sci. USA.* **75** 3557-3561.
10. Goriely, A. & Tabor, M. 1998 Nonlinear dynamics of filaments. IV Spontaneous looping of twisted elastic rods, *Proc. R. Soc. Lond.* A **454** 3183-3202.
11. Manning, R. S., Maddocks J. H. & Kahn, J. D. 1996 A continuum rod model of sequence-dependent DNA structure, *J. Chem. Phys.* **105** 5626-5646.
12. Stump, D. M. & Fraser, W. B. 2000 Multiple solutions for writhed rods: implications for DNA supercoiling, *Proc. R. Soc. Lond.* A **456** 455-467.
13. Stump, D. M., Fraser, W. B. & Gates, K. E. 1998 The writhing of circular cross-section rods: undersea cables to DNA supercoils, *Proc. R. Soc. Lond.* A **454** 2123-2156.
14. Swigon, D. 1999 *Configurations with self-contact in the theory of the elastic rod model for DNA*, PhD Dissertation, Rutgers, State University of New Jersey, USA. January.
15. Thompson, J. M. T. & Champneys, A. R. 1996 From helix to localized writhing in the torsional post-buckling of elastic rods, *Proc. R. Soc. Lond..* A **452**, 117-138.
16. Thompson, J. M. T., van der Heijden, G. H. M. & Neukirch, S. 2002 Supercoiling of DNA plasmids: mechanics of the generalized ply, *Proc. R. Soc. Lond..* A **458** 959-985.
17. van der Heijden, G. H. M. 2001 The static deformation of a twisted elastic rod constrained to lie on a cylinder, *Proc. R. Soc. Lond.* A **457**, 695-715.
18. van der Heijden, G. H. M. & Thompson, J. M. T. 2000 Helical and localised buckling in twisted rods: a unified analysis of the symmetric case, *Nonlinear Dynamics* **21**, 71-99.
19. van der Heijden, G. H. M., Thompson, J. M. T. & Neukirch, S. 2001 A variational approach to loaded ply structures, *J. Vibration & Control*, in press.
20. Zajac, E. E. 1962 Stability of two planar loop elasticas, *J. Appl. Mech.* **29** 136-142.

AUTHOR INDEX

Batch number: 10809436

Printed by Printforce, the Netherlands